T0192618

# Hands-On Accelerator Physics Using MATLAB®

# Hands-On Accelerator Physics Using MATLAB®

Volker Ziemann

CRC Press
Taylor & Francis Group
Boca Raton  London  New York

CRC Press is an imprint of the
Taylor & Francis Group, an **informa** business

Front cover image reproduced with permission from Marek Jacewicz.

CRC Press
Taylor & Francis Group
6000 Broken Sound Parkway NW, Suite 300
Boca Raton, FL 33487-2742

First issued in paperback 2021

Version Date: 20190410

ISBN 13: 978-0-367-77989-4 (pbk)
ISBN 13: 978-1-138-58994-0 (hbk)

**Visit the Taylor & Francis Web site at**
http://www.taylorandfrancis.com

**and the CRC Press Web site at**
http://www.crcpress.com

# Contents

Preface                                                                   xi

Acknowledgments                                                           xiii

CHAPTER  1 ▪ Introduction and History                                      1

CHAPTER  2 ▪ Reference System                                             13

2.1    THE REFERENCE TRAJECTORY                                           13
2.2    COORDINATE TRANSFORMATIONS                                         16
2.3    PARTICLES AND THEIR DESCRIPTION                                    18
2.4    PARTICLE ENSEMBLES, BUNCHES                                        19

CHAPTER  3 ▪ Transverse Beam Optics                                       27

3.1    MAGNETS AND MATRICES                                               28
       3.1.1    Thin quadrupoles                                          29
       3.1.2    Thick quadrupoles                                         31
       3.1.3    Sector dipole                                             32
       3.1.4    Combined function dipole                                  34
       3.1.5    Rectangular dipole                                        35
       3.1.6    Coordinate rotation                                       36
       3.1.7    Solenoid                                                  37
       3.1.8    Non-linear elements                                       37
3.2    PROPAGATING PARTICLES AND BEAMS                                    38
3.3    TWO-DIMENSIONAL                                                    40
       3.3.1    Beam optics in MATLAB                                     40
       3.3.2    Poincarè section and tune                                 42
       3.3.3    FODO cell and beta functions                             44
       3.3.4    A complementary look at beta functions                   47
       3.3.5    Beam size and emittance                                   49
3.4    CHROMATICITY AND DISPERSION                                        52
       3.4.1    Chromaticity                                              52
       3.4.2    Dispersion                                                54
       3.4.3    Emittance generation                                      58

|   | 3.4.4 | Momentum compaction factor | 60 |

| 3.5 | FOUR-DIMENSIONAL AND COUPLING | 60 |
| 3.6 | MATCHING | 65 |
|   | 3.6.1 | Matching the phase advance | 65 |
|   | 3.6.2 | Match beta functions to a waist | 66 |
|   | 3.6.3 | Point-to-point focusing | 68 |
| 3.7 | BEAM-OPTICAL SYSTEMS | 70 |
|   | 3.7.1 | Telescopes | 70 |
|   | 3.7.2 | Triplets | 71 |
|   | 3.7.3 | Doublets | 74 |
|   | 3.7.4 | Achromats | 75 |
|   | 3.7.5 | Multi-bend achromats | 77 |
|   | 3.7.6 | TME cell | 78 |
|   | 3.7.7 | Dispersion suppressor | 79 |
|   | 3.7.8 | Interaction region | 80 |
|   | 3.7.9 | Bunch compressors | 81 |

CHAPTER 4 ■ Magnets     85

| 4.1 | MAXWELL'S EQUATIONS AND BOUNDARY CONDITIONS | 85 |
| 4.2 | 2D-GEOMETRIES AND MULTIPOLES | 87 |
| 4.3 | IRON-DOMINATED MAGNETS | 89 |
|   | 4.3.1 | Simple analytical methods | 89 |
|   | 4.3.2 | Using the MATLAB PDE toolbox | 91 |
|   | 4.3.3 | Quadrupoles | 98 |
|   | 4.3.4 | Technological aspects | 99 |
| 4.4 | SUPER-CONDUCTING MAGNETS | 101 |
|   | 4.4.1 | Simple analytical methods | 102 |
|   | 4.4.2 | PDE toolbox | 103 |
| 4.5 | PERMANENT MAGNETS | 106 |
|   | 4.5.1 | Multipoles | 108 |
|   | 4.5.2 | Segmented multipoles | 110 |
|   | 4.5.3 | Undulators and wigglers | 112 |
| 4.6 | MAGNET MEASUREMENTS | 114 |
|   | 4.6.1 | Hall probe | 114 |
|   | 4.6.2 | Rotating coil | 115 |
|   | 4.6.3 | Undulator measurements | 116 |

CHAPTER 5 ■ Longitudinal Dynamics and Acceleration     119

| 5.1 | PILL-BOX CAVITY | 120 |
| 5.2 | TRANSIT-TIME FACTOR | 124 |
| 5.3 | PHASE STABILITY AND SYNCHROTRON OSCILLATIONS | 124 |

5.4   LARGE-AMPLITUDE OSCILLATIONS                                    127

5.5   RF GYMNASTICS                                                   133

5.6   ACCELERATION                                                    134

5.7   A SIMPLE WORKED EXAMPLE                                         137

CHAPTER   6 ▪ Radio-Frequency Systems                                143

6.1   POWER GENERATION AND CONTROL                                   143

6.2   POWER TRANSPORT: WAVEGUIDES AND TRANSMISSION LINES             145

6.3   COUPLERS AND ANTENNAS                                          153

6.4   POWER TO THE BEAM: RESONATORS AND CAVITIES                     156

   6.4.1   Losses and quality factor $Q_0$ of a pill-box cavity      156

   6.4.2   General cavity geometry with the PDE toolbox              159

   6.4.3   Disk-loaded waveguides                                    160

6.5   TECHNOLOGICAL ASPECTS                                          163

   6.5.1   Normal-conducting                                         163

   6.5.2   Super-conducting                                          164

6.6   INTERACTION WITH THE BEAM                                      165

   6.6.1   Beam loading                                              165

   6.6.2   Steady-state operation                                    166

   6.6.3   Pulsed operation and transient beam loading               167

   6.6.4   Low-level RF system                                       170

CHAPTER   7 ▪ Instrumentation and Diagnostics                        173

7.1   ZEROTH MOMENT: CURRENT                                         173

7.2   FIRST MOMENT: BEAM POSITION AND ARRIVAL TIME                   175

7.3   SECOND MOMENT: BEAM SIZE                                       180

7.4   EMITTANCE AND BETA FUNCTIONS                                   183

7.5   SPECIALTY DIAGNOSTICS                                          185

   7.5.1   Turn-by-turn position monitor data analysis               186

   7.5.2   Beam-beam diagnostics                                     188

   7.5.3   Schottky diagnostics                                      189

CHAPTER   8 ▪ Imperfections and Their Correction                     193

8.1   SOURCES OF IMPERFECTIONS                                       194

   8.1.1   Misalignment and feed down                                194

   8.1.2   Tilted components                                         196

   8.1.3   Rolled elements and solenoids                             197

   8.1.4   Chromatic effects                                         197

   8.1.5   Consequences                                              197

8.2   IMPERFECTIONS IN BEAM LINES                                    198

   8.2.1   Dipole kicks and orbit errors                             198

| | | |
|---|---|---|
| 8.2.2 | Quadrupolar errors and beam size | 198 |
| 8.2.3 | Skew-quadrupolar perturbations | 200 |
| 8.2.4 | Filamentation | 201 |
| 8.3 | IMPERFECTIONS IN A RING | 203 |
| 8.3.1 | Misalignment and dipole kicks | 203 |
| 8.3.2 | Gradient imperfections | 204 |
| 8.3.3 | Skew-gradient imperfections | 205 |
| 8.4 | CORRECTION IN BEAM LINES | 207 |
| 8.4.1 | Trajectory knobs and bumps | 208 |
| 8.4.2 | Orbit correction | 209 |
| 8.4.3 | Beta matching | 212 |
| 8.4.4 | Dispersion and chromaticity | 213 |
| 8.5 | CORRECTION IN RINGS | 213 |
| 8.5.1 | Orbit correction | 214 |
| 8.5.2 | Dispersion-free steering | 215 |
| 8.5.3 | Tune correction | 215 |
| 8.5.4 | Chromaticity correction | 216 |
| 8.5.5 | Coupling correction | 217 |
| 8.5.6 | Orbit response-matrix based methods | 217 |
| 8.5.7 | Feedback systems | 218 |

CHAPTER 9 ▪ Targets and Luminosity — 221

| | | |
|---|---|---|
| 9.1 | EVENT RATE AND LUMINOSITY | 221 |
| 9.2 | ENERGY LOSS AND STRAGGLING | 222 |
| 9.3 | TRANSVERSE SCATTERING, EMITTANCE GROWTH, AND LIFE-TIME | 226 |
| 9.4 | COLLIDING BEAMS | 228 |
| 9.5 | BEAM-BEAM LUMINOSITY | 229 |
| 9.6 | INCOHERENT BEAM-BEAM TUNE SHIFT | 233 |
| 9.7 | COHERENT BEAM-BEAM INTERACTIONS | 235 |
| 9.8 | LINEAR COLLIDERS | 237 |

CHAPTER 10 ▪ Synchrotron Radiation and Free-Electron Lasers — 241

| | | |
|---|---|---|
| 10.1 | EFFECT ON THE BEAM | 242 |
| 10.1.1 | Longitudinally | 242 |
| 10.1.2 | Vertically | 244 |
| 10.1.3 | Horizontally | 244 |
| 10.1.4 | Quantum lifetime | 246 |
| 10.2 | CHARACTERISTICS OF THE EMITTED RADIATION | 247 |
| 10.2.1 | Dipole magnets | 248 |
| 10.2.2 | Undulators and wigglers | 249 |

| | | |
|---|---|---|
| 10.3 | SMALL-GAIN FREE-ELECTRON LASER | 251 |
| | 10.3.1 Amplifier and oscillator | 251 |
| 10.4 | SELF-AMPLIFIED SPONTANEOUS EMISSION | 254 |
| 10.5 | ACCELERATOR CHALLENGES | 257 |

CHAPTER 11 ■ Non-linear Dynamics | | 259

| | | |
|---|---|---|
| 11.1 | A ONE-DIMENSIONAL TOY MODEL | 259 |
| 11.2 | TRACKING AND DYNAMIC APERTURE | 261 |
| 11.3 | HAMILTONIANS AND LIE-MAPS | 263 |
| | 11.3.1 Moving Hamiltonians | 265 |
| | 11.3.2 Concatenating Hamiltonians | 267 |
| 11.4 | IMPLEMENTATION IN MATLAB | 267 |
| 11.5 | TWO-DIMENSIONAL MODEL | 271 |
| 11.6 | KNOBS AND RESONANCE-DRIVING TERMS | 272 |
| 11.7 | NON-RESONANT NORMAL FORMS | 275 |

CHAPTER 12 ■ Collective Effects | | 279

| | | |
|---|---|---|
| 12.1 | SPACE CHARGE | 279 |
| 12.2 | INTRABEAM SCATTERING AND TOUSCHEK-EFFECT | 282 |
| 12.3 | WAKE FIELDS, IMPEDANCES, AND LOSS FACTORS | 283 |
| 12.4 | COASTING-BEAM INSTABILITY | 286 |
| 12.5 | SINGLE-BUNCH INSTABILITIES | 288 |
| 12.6 | MULTI-BUNCH INSTABILITIES | 291 |

CHAPTER 13 ■ Accelerator Subsystems | | 295

| | | |
|---|---|---|
| 13.1 | CONTROL SYSTEM | 295 |
| | 13.1.1 Sensors, actuators, and interfaces | 295 |
| | 13.1.2 System architecture | 296 |
| | 13.1.3 Timing system | 297 |
| | 13.1.4 An example: EPICS | 297 |
| 13.2 | PARTICLE SOURCES | 299 |
| | 13.2.1 Electrons | 299 |
| | 13.2.2 Protons and other ions | 301 |
| | 13.2.3 Highly charged ions | 302 |
| | 13.2.4 Negatively charged ions | 302 |
| | 13.2.5 Radio-frequency quadrupole | 303 |
| 13.3 | INJECTION AND EXTRACTION | 304 |
| 13.4 | BEAM COOLING | 306 |
| 13.5 | VACUUM | 307 |
| | 13.5.1 Vacuum basics | 307 |
| | 13.5.2 Pumps and gauges | 308 |

|  | 13.5.3 Vacuum calculations | 310 |
| 13.6 | CRYOGENICS | 312 |
| 13.7 | RADIATION PROTECTION AND SAFETY | 314 |
|  | 13.7.1 Units | 315 |
|  | 13.7.2 Range of radiation in matter | 315 |
|  | 13.7.3 Dose measurements | 316 |
|  | 13.7.4 Personnel and machine protection | 316 |
| 13.8 | CONVENTIONAL FACILITIES | 317 |
|  | 13.8.1 Electricity | 317 |
|  | 13.8.2 Water and cooling | 317 |
|  | 13.8.3 Buildings and shielding | 318 |

CHAPTER 14 ■ Examples of Accelerators — 319

| 14.1 | CERN AND THE LARGE HADRON COLLIDER | 319 |
| 14.2 | EUROPEAN SPALLATION SOURCE | 321 |
| 14.3 | SLAC AND THE LINAC COHERENT LIGHT SOURCE | 322 |
| 14.4 | MAX-IV | 323 |
| 14.5 | TANDEM ACCELERATOR IN UPPSALA | 324 |
| 14.6 | ACCELERATORS FOR MEDICAL APPLICATIONS | 325 |
| 14.7 | INDUSTRIAL ACCELERATORS | 327 |

APPENDIX A ■ The Student Labs — 331

| A.1 | BEAM PROFILE OF LASER POINTER | 331 |
| A.2 | EMITTANCE MEASUREMENT WITH A LASER POINTER | 334 |
| A.3 | HALBACH MULTIPOLES AND UNDULATORS | 335 |
| A.4 | MAGNET MEASUREMENTS | 339 |
| A.5 | COOKIE-JAR CAVITY ON A NETWORK ANALYZER | 341 |

APPENDIX B ■ Appendices Available Online — 345

| B.1 | LINEAR ALGEBRA | 345 |
| B.2 | MATLAB PRIMER | 345 |
| B.3 | OPENSCAD PRIMER | 345 |
| B.4 | LIGHT OPTICS, RAYS, AND GAUSSIAN | 345 |
| B.5 | MATLAB FUNCTIONS | 345 |

Bibliography — 347

Index — 355

# Preface

A little over 10 years ago I moved from our lab that operated the CELSIUS storage ring for nuclear physics experiments to the high-energy physics department at Uppsala University. There I followed a request to develop a course on Accelerator Physics. The main purpose at the time was to teach graduate students in nuclear and high-energy physics how the accelerators, on which they perform the experiments for their theses, work. Of course, I was also interested to attract students that are interested in my own field of research—accelerator physics. When designing the course I had two types of students in mind, particle-physics experimentalists and accelerator physicists, and what they should know after completing my course.

As the archetypical high-energy or nuclear physics student, I visualized a detector-liaison physicist, similar to the ones that populated the accelerator control-room of the SLAC Linear Collider, where I used to work as a Post-Doc. At that time, in the early nineties, the SLD detector "consumed" the beams and the detector collaboration deployed a physicist in our control-room, where we, the accelerator crew, prepared the beams. The idea was to tell us when our beams were not good enough for the detector or when we could experiment with the beams ourselves, while they changed data collection tapes. These detector liaisons were extremely beneficial for smooth operation and our interaction worked better as we learned each other's systems. In short, I want my students from high-energy physics to have a sound understanding of the system, we accelerator physicists operate, in order to efficiently interface the operation of detector and accelerator—a sort of survival guide for detector liaisons. Over the years giving the course a variant of the liaisons, colleagues from neighboring departments, participated and I realized the course gives these experts in their respective fields a wider perspective of accelerator physics and technology.

As to my prospective accelerator physics students, I visualized them marooned on an island with a computer that is well-equipped with software to design accelerators and their subsystems. Is there a better way to beat the boredom on a deserted island than to design prototype accelerators? The software I selected at the time was the MADX code for beam optics calculations, the student version of the Maxwell code from ANSYS for magnet calculations, and my own VAKTRAK for vacuum calculations. I used these codes on very simple problems to teach how to get started designing beam optics, magnets, or vacuum systems. This hands-on approach to really understand how to work with real-world, though simple, problems was appreciated by the students, both with detector- and with accelerator-interest, because they learned things useful beyond the direct use in accelerators. This worked quite well for a number of years.

But over time, some of the software was not supported anymore and the installation of large software packages took too much time to make it work for a number of students with computers running all sorts of operating systems. Moreover, for some short examples, learning an arcane input language distracts too much from the underlying physics. In the past few years I therefore moved to using MATLAB® [1] and Octave [2] for a number of topics. This had the additional benefit that the students could play with the software more easily, add features and inspect what goes on inside the code. Moreover, only a single software package, available on many operating systems, has to be installed and maintained.

Octave is open source and freely available, and most technical universities have site licenses for MATLAB.

This book grew out of the move towards MATLAB and Octave and teaches the core topics of accelerator design, but not limited to beam optics, magnets, radio-frequency and vacuum systems with the help of small or moderately complex code examples—each discussed in detail. All MATLAB functions referred to in this book, to prepare figures or otherwise, as well as an instructor's manual with complete solutions to all exercises, are available on the book's web page at https://www.crcpress.com/9781138589940. Most examples work equally well in MATLAB or in Octave. Only the examples to numerically calculate electro-magnetic fields in Chapters 4, 6, and 7 require solving partial differential equations and are based on MATLAB's PDE toolbox. Furthermore, the interface to the functions of the PDE toolbox (the workflow in MATLAB's parlance) has changed. Therefore, two versions are provided: one for R2018b, which is covered in the text; and one for R2015b, which is explained in the online material.

The best way of learning about accelerators is to spend time in the control-room, observe what goes on, and eventually turn knobs oneself. But that is unfortunately not always possible; either there is no accelerator close by, or the one that is, is tied up running for users, either high-energy physicists, users of synchrotron radiation, or for treating patients. A way out of this dilemma is to use equivalent systems that show key features of the corresponding accelerator systems. I follow this approach in the lab sessions described in the appendix. There I use a laser-pointer to illustrate beam size and emittance measurements, design and measure magnets with small permanent magnets, and analyze a radio-frequency resonator made of a closed metallic cylinder—a cookie-jar with antennas.

For product information on MATLAB, please contact:
    The Mathworks, Inc.
    3 Apple Hill Drive
    Natick, MA 01760-2098 USA
    Tel: 508-647-7000
    Fax: 508-607-7001
    Email: info@mathworks.com
    Web: www.mathworks.com

# Acknowledgments

Writing a book on a technical subject is impossible to do alone. Foremost, I want to thank all my colleagues for their generosity to share their knowledge with me throughout my career. This book would not be possible without them.

I am particularly grateful to my colleagues who discussed the manuscript during various stages with me and helped me immensely by proof-reading various parts of the manuscript. I am particularly indebted to R. Ruber, B. Holzer, J. Ögren, D. Primetzhofer, M. Aiba, B. Riemann, H. Li, R. Santiago-Kern, K. Gajewski, M. Jobs, and G. Kipp. They did a heroic effort to make my writing comprehensible. Any deficiencies and remaining errors are entirely my responsibility. The help of H. Li with measuring the cookie-jar in Appendix A.5, as well as J. Erikssons help with 3D-printing the frames for the magnets, was invaluable. M. Jacewicz's permission to use one of his photos for the book cover is highly appreciated.

I am indebted to H. Schmickler and W. Herr who gave me the chance to use the hands-on beam optics software during the tutorial sessions of the CERN Accelerator School in Constanta, Romania.

I really need to thank my family for their patience to cope with my single-mindedness during the rather intense period of writing.

# CHAPTER 1

# Introduction and History

Charged particle accelerators drive a large sector of modern research in sub-atomic physics, in material, and in life sciences. Moreover, they play an ever increasing role in medical applications, both for therapy and for diagnostic purposes. Most of these accelerators are large installations, but also smaller accelerators are used to modify material properties or to sterilize medical waste. Even old-fashioned thick-screen television sets accelerate and guide charged particles, electrons, in an evacuated vessel and guide them to a screen, where they produce light that we experience as TV images. In this book we answer the question why and how accelerators acquired such a prominent role in modern society and which technologies are used to satisfy this role. But before discussing the scientific and technical aspects, we briefly recall the history of charged particle accelerators.

The first technical device, which accelerated charged particles, electrons, is the *cathode-ray tube,* invented by Braun. Already in 1897, he used electric and magnetic deflecting fields to make electrical oscillations visible by directing accelerated electrons, moving in a vacuum tube, onto a luminescent screen. Figure 1.1 shows the schematics of such a tube. The electrons are created in a thermionic cathode, shown on the left. It is basically a heated wire from which electrons are extracted by the electric field caused by the voltage between the anode and the cathode. A second voltage, which can be time-varying, is applied to deflection plates and directs the beam to the luminescent screen, where the electrons are detected. The entire setup is embedded in a vacuum tube to avoid collisions of the electrons with the residual gas.

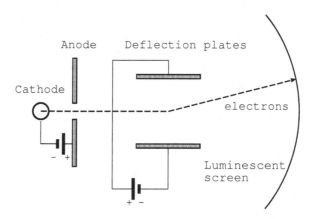

Figure 1.1  Schematic of a cathode ray tube.

As pointed out before, we already have the essential ingredients of an accelerator present: particle creation, acceleration, guide field, detection or diagnostic, and vacuum. Note that in a TV tube the deflection plates are replaced by magnetic coils and the deflection is done both horizontally and vertically. Cathode ray tubes were used in TV sets and oscilloscopes into the early years of the third millennium, but are largely superseded by systems based on flat-screen displays.

The energy of the electrons in cathode ray tubes is rather modest after being accelerated through voltages on the order of a few tens of kV, but is sufficient to produce images. The development of accelerators did not advance significantly until the late 1920s when several new technologies, which dramatically increased the kinetic energy of charged particles, appeared. Of course this development was stimulated, like so much else, by the development of quantum- and nuclear-physics during the first three decades of the previous century. In 1911, Rutherford used charged particles, emitted by a radioactive source, to probe atoms located in thin foils. He detected very large deflection angles and inferred that atoms are made up of positively charged nuclei and negatively charged electrons, an observation that led Bohr to publish his model of the atom two years later. Using high-velocity charged particles to probe the substructure of the target material thus proved highly useful and inspirational.

In 1913 father and son Bragg used X-rays in order to probe the structure of crystals because the X-rays have a wavelength comparable to the distance between atoms in the crystal and cause diffraction patterns on photographic plates. Earlier, Einstein and Planck had postulated that photons, such as X-rays, can behave like waves or like particles, depending on the experiment performed. By exploiting this analogy, de'Broglie suggested in 1924 that even electrons, or actually any matter, can behave like either wave or particle with wavelength $\lambda$, given by the momentum $p$ of the particle

$$\lambda = \frac{h}{p} \,, \tag{1.1}$$

where $h$ is Planck's constant. This suggestion was picked up by Schrödinger, who used it to develop wave mechanics and the equation that bears his name. The inverse proportionality of the wavelength on the momentum indicates that high momenta are required to generate diffraction patterns from small structures, such as atomic nuclei. The bottom-line is that nuclear probes with large momenta are needed. And this resulted in the development of a number of different accelerators, all of which were invented in the late 1920s.

In 1928 Widerøe demonstrated a method to accelerate electrons using voltages that reverse their polarity at very high (radio-)frequencies. Figure 1.2 illustrates the idea. So-called drift-tubes are connected to a voltage generator that rapidly reverses their polarity, such that an alternating electric field $E$ exists between adjacent drift-tubes. In the figure,

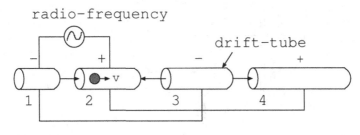

Figure 1.2  Schematic of a drift-tube linear accelerator.

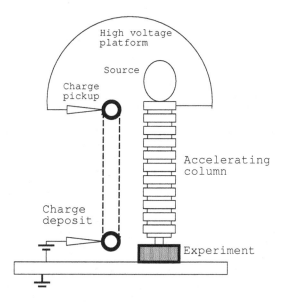

Figure 1.3 Schematics of a Van-de-Graaff accelerator.

the negatively charged electron (grey) was just accelerated by the electric field in the gap between the first two drift-tubes. The field between drift-tubes two and three at this instance points in the wrong direction and would decelerate the electron. But the length of the drift-tube is calculated such that the electron appears at the exit of tube two at the instance, the field has reversed its polarity and points towards the right, thus accelerating the electron once more. This mechanism repeats in every drift-tube with tubes of increasing length in order to account for the increased velocity of the electron, which "hides" inside the tubes, while the polarity is reversed. Since electrons are very light, the velocity increases rapidly and the drift-tubes become very long for the radio-frequencies that could be generated in the late 1920s. This made this method of acceleration impractical for electrons and it was not used in experiments. On the other hand, even today it is used to accelerate protons or heavier ions. An important modification of Wideröe's principle of using drift-tubes is due to Alvarez. In the late 1940s, he embedded the drift-tubes in a large tank and used high-frequency radio waves to excite standing waves in the tank. The geometry was adjusted in such a way that the longitudinal electric field components in the gap between adjacent drift tubes have opposite signs. In this way particles "hide" in the drift-tubes, while the standing wave reverses polarity before they are accelerated again in the following gap. This method is used in early stages of linear proton accelerators, such as the European Spallation Source.

From 1929 on, *Van-de-Graaff* developed the electrostatic accelerator, shown in Figure 1.3, which is based on charge being deposited on a belt, made of isolating elastic material, such as silk or rubber, by a corona-discharge. The belt is mechanically moved by pulleys and transports charges to the upper platform where they are deposited at a higher potential. Ions, created in a source at the high potential, are accelerated towards ground potential in the accelerating column, where a resistor cascade ensures a linearly changing potential and a constant accelerating field. Back at ground potential, the accelerated ions impinge on a target that is part of an experimental station. A modern version of this device is called a Pelletron and is, for instance, used for experiments to determine the surface properties of materials, but also in high-energy electron coolers. A *tandem accelerator*

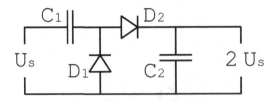

Figure 1.4 The Greinacher voltage multiplier that is the base of a Cockroft-Walton accelerator.

uses the high voltage generated by a Van-de-Graaff accelerator twice. Once to accelerate negatively charged ions, strip one or several electrons from the ion, and to use the same acceleration potential a second time to accelerate the, now positively charged ions, towards the experiment, conveniently located at ground potential.

A second method to generate constant high voltages was employed by *Cockroft and Walton*. They used a cascaded network of capacitors and diodes, invented by Greinacher in the 1920s, to reach very high accelerating voltages. Figure 1.4 explains the basic operating principle of the voltage multiplier. The negative voltage of the AC voltage $U_s$ that is applied at the left-hand side charges capacitor $C_1$ to $U_s$ and the positive voltage half an oscillation period later charges $C_2$ to $2U_s$. Cascading these units allows us to reach voltages into the MV range. In 1932 Cockroft and Walton used their apparatus to impinge accelerated protons on to a Lithium target and induce the first artificial nuclear reaction: p + Li → 2 He. For this discovery they received the Nobel prize in Physics in 1951.

Sustaining high multi-MV constant voltages to accelerate particles, as used in both Van-de-Graaff and Cockroft-Walton accelerators, is technologically demanding and limited by the available insulation. In 1929 Lawrence had the ingenious idea to use alternating radio-frequency voltages to accelerate protons in a wound-up drift-tube linac. Figure 1.5 illustrates the operating principle of the device, called the *cyclotron*. Charged particles are forced on a circular orbit by a static magnetic field and are accelerated by repeatedly changing the polarity of two accelerating "dees," which have the same role as the drift-tubes. The particles "hide" in the "dees," while the polarity is reversed, such that they are accelerated once again in the second gap. The increase of the radius of the particles must exactly compensate their increased speed. For non-relativistic particles at radius $\rho$, the speed $v$, and revolution time $T$ are related by $v = 2\pi\rho/T = 2\pi f\rho$, where we introduce the revolution frequency $f = 1/T$. Moreover, on the circular path, the centrifugal and centripetal forces must balance

$$\frac{mv^2}{\rho} = evB \tag{1.2}$$

and inserting the relation between speed $v$ and frequency $f$ in order to eliminate the radius $\rho$, we find that choosing the frequency to be

$$f = \frac{eB}{2\pi m} \tag{1.3}$$

guarantees synchronicity, independent of the radius and the speed of the particle. For obvious reasons $f$ is called the cyclotron frequency. Since the magnetic field is constant, the radius of the orbit increases as is indicated by the spiral drawn in Figure 1.5. Synchronism is guaranteed as long as the particles move at non-relativistic speeds up to energies of about 100 MeV for protons.

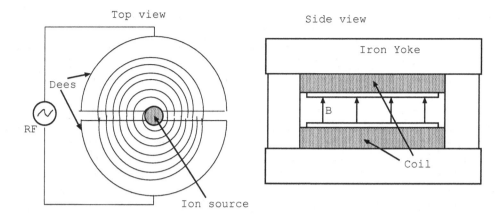

Figure 1.5 Schematics of a cyclotron.

The synchronous condition is violated, if the mass of the proton changes due to relativistic effects as $m = \gamma m_0$, where $m_0$ is the particle rest mass, $\gamma = 1/\sqrt{1 - v^2/c^2}$, and $c$ is the speed of light. At higher speed or energy the radio-frequency needs to be reduced in order to maintain synchronicity. This is actually done in synchro-cyclotrons that modulate the frequency in such a way as to compensate the relativistic increase in the mass. That the particles follow the changing radio-frequency in a stable fashion is a consequence of *phase focusing*, discovered in the mid-1940s by Vecksler and McMillan; particles with speed too low arrive a little late and are accelerated a bit more and consequently are pushed closer to their desired arrival time. We will discuss *phase stability* in detail in Chapter 5. A consequence of modulating the frequency is a pulsed emission of accelerated particles from the cyclotron, typically a few hundred times per second. The frequency is suitable to capture particles in the center but then it decreases to guide the captured particles to higher energies. While one particle batch travels outwards in the cyclotron to higher energies, no other particles can be captured in the center, because the frequency is non-synchronous. Synchro-cyclotrons with specially shaped magnetic fields in addition to the modulating generators allow us to reach moderately relativistic energies up to about 600 MeV in the large cyclotrons at PSI near Zürich, Switzerland and TRIUMF in Vancouver, Canada.

The magnetic field in a cyclotron is constant and the accelerated particles circulate with increasing radii, which requires huge magnets. In 1942 Oliphant suggested to maintain a fixed radius by synchronously increasing the magnetic field and the frequency of the RF system used to accelerate the particles. Phase-stability or phase-focusing, as it is also sometimes called, guaranteed that such an accelerator operates stably and particles with a slightly "wrong" arrival time or energy experience a force that restores them towards their design values. Adjusting magnets and RF synchronously led to calling these accelerators *synchrotrons*. Early machines were built for electrons and later, in the 1950s, for protons. The early synchrotrons such as the Cosmotron in Brookhaven and the Bevatron in Berkeley, used specially shaped magnet-poles to provide focusing towards the particle's design trajectory, which is called weak-focusing (more on that in later chapters). In 1952, however, Courant and his colleagues found out that splitting the magnets into functional blocks, dipole magnets to deflect, and quadrupole magnets to focus, allowed a more compact construction of the accelerator. Since Maxwell's equations force the quadrupoles to focus in one plane, say the horizontal, and defocus in the other, vertical plane, a sequence of focusing and defocusing magnets is needed, which led to the name *alternating gradient focusing*. Moreover, the focusing magnets could use much stronger fields, which led to calling this method *strong focusing*. The first large strong-focusing synchrotrons were the AGS in Brookhaven and the PS at CERN in Geneva, the latter designed to accelerate protons to 25 GeV, an energy that corresponds to a little under 30 times the rest-energy $m_p c^2$ of the proton.

At this point, we need to look at the scientific output of the high-energy accelerators, the developments of particle physics into the 1960s, and how that determined the further development of accelerators. Life as a particle physicist was easy in the 1920s; only *photons,* the carriers of light quanta, *protons* and *electrons* had to be considered. But in 1928 Dirac predicted that the electron has a sibling, the *positron,* which can be considered as its mirror image. Shortly afterwards, in 1932, Anderson discovered the positron and in the same year Chadwick and Urey discovered of the *neutron.* By analyzing radioactive decay and in order to explain the decay of the neutron, Pauli postulated the existence of yet another particle, the *neutrino.* Further subatomic particles joined the zoo with the discovery of the *muon* in 1936. Earlier, Yukawa had predicted the existence of elementary particles to account for the stability of atomic nuclei, so-called *mesons.* The number of elementary particles proliferated dramatically after accelerators started producing them in abundance and detection methods for new particles improved. In the 1950s and 1960s, a veritable *particle-zoo* emerged with a large number of different mesons and heavier particles, the *baryons.* The abundant proliferation of new particles initially had no discernable structure, but order was partially restored, when Gell-Mann and Zweig identified underlying symmetries and proposed that mesons and baryons are composed of more fundamental entities, the *quarks.* This discovery not only explained the properties of the many recently discovered particles, but also predicted new ones, which were promptly found. At this point the question arises how the quarks interact with each other.

And this forces us to briefly discuss the *fundamental forces* that govern all interactions. First known was the gravitational force, put on a sound theoretical foundation, first by Newton in 1687, and later by Einstein with the general theory of relativity in 1915. The second type of fundamental forces known were electric and magnetic forces. In 1865, Maxwell published a theory that explains these forces in a unified way as a single underlying force, the *electro-magnetic interaction.* Later these forces were associated with the exchange of particles, the photons. The fundamental force responsible for radioactive decay is called the *weak interaction* and we now know that it is mediated by $Z$ and $W$–bosons. The last fundamental force is called the *strong interaction* and is responsible for the forces inside the nucleus and between quarks. Since the 1970s we know that the strong interaction is mediated by carriers called *gluons.* Here we already see another ordering scheme. There are particles that constitute matter and there are force-carriers, the interaction bosons such as photons, Zs, or gluons. The theory that was developed during the 1960s and 1970s that places the electro-magnetic, weak and strong interaction in a coherent framework is called the *standard model.* This model was extremely successful in explaining and predicting subatomic phenomena, culminating in the discovery of the Higgs-boson in 2012.

The masses of the predicted particles were sometimes unknown and often beyond the reach of existing accelerators. This triggered the construction of bigger accelerators to extend the range towards *higher energies* in order to find the particles and probe their properties. The accelerators dramatically grew in size because the field in conventional iron-dominated electro-magnets is limited to approximately 2 T. This high-energy frontier was explored with large synchrotrons in Serpukhov, Russia, at Fermilab near Chicago, and at CERN with the

Super-Proton Synchrotron (SPS). These accelerators reached proton energies of several 100s of GeV and had circumferences measured in km.

A second feature of the predicted particles is the very small probability to produce "wanted" new particles by smashing a proton into a target. This requires us to increase the rate of collisions of beam particles with the target. Moreover, when smashing particles into a fixed target, momentum conservation dictates that some of the available energy is converted to the kinetic energy of the reaction products. On the other hand, when colliding particles head-on, with momentum of equal magnitude, but opposite direction, the total momentum is zero in both laboratory and rest frame, such that all energy is available for the reaction products. Additionally, by circulating two beams in opposite directions and repeatedly colliding them head-on increases the collision rate, and thus the ability to observe rare reactions. This reasoning led to the concept of *colliding beam storage rings,* which was first investigated with electrons already in the early 1960s with the small electron-positron collider ADA in Frascati near Rome. In the following years the concept was tested at the VEPP colliders in Novosibirsk. For protons, the first collider was the Intersecting Storage Ring (ISR) at CERN, where the ideas were tested further and the technology mastered throughout the 1970s.

Note that the ISR collided protons on protons due to the unavailability of a high-quality source for anti-protons at the time. But the situation changed with the invention of *stochastic cooling* by van der Meer, which permitted us to greatly improve the beam quality of the anti-protons that were generated by smashing high-energy protons into a target. Once the production of high-quality anti-protons was under control, the large synchrotrons at Fermilab and CERN were replaced or converted to colliders for protons and anti-protons, leading to the discovery of the mediators of the weak-force, the Z- and W-bosons at CERN in 1983 and the top-quark in 1995 in the Tevatron at Fermilab. More recently, in operation since the year 2000, the Relativistic Heavy-Ion Collider (RHIC) in Brookhaven collides, well, heavy ions, to probe how matter behaved immediately following the big-bang. At CERN, the Large Hadron Collider (LHC), in full operation since 2010, was instrumental in the discovery of the Higgs-boson, which explains the masses of elementary particles. The FAIR-facility, based on a large synchrotron, dedicated to many aspects of nuclear physics, is under construction at GSI near Frankfurt.

Most of the accelerators discussed so far accelerate protons and sometimes heavier ions in order to induce sub-nuclear reactions and discover new particles. The large mass of the proton helps to reach high energies used to stimulate nuclear reactions in targets, either a fixed target or the counter-propagating beam in a collider. The high energy density makes them the perfect tool to discover new particles or excited states of known particles. Thus proton accelerators can be considered *discovery machines.* They allow us to create energy densities, not accessible in any other way. But carefully probing a target with protons is difficult, because protons are composite particles made of quarks and gluons. Electrons, on the other hand are, as far as we know today, point-like and are much better probes of the internal structure of the targets, such as nuclei. Therefore, high-energy electron accelerators can be considered as *precision probes.* A hybrid-machine that uses electrons to probe protons is the Hadron Electron Ring Accelerator (HERA) that operated at DESY in the 1990s. But now let us turn to plain electron or electron-positron accelerators.

After the first electron accelerator, the cathode ray tube and initial experiments by Wideröe in the late 1920s, Kerst developed the *betatron* in the 1940s. This accelerator uses the induction voltage generated by a time-varying magnetic field. In this way the betatron resembles a transformer, where the magnet coil is the primary winding and the accelerated electron beam the secondary winding. Betatrons can accelerate electrons up to about 200 MeV but are not much used today, except in some cases to generate X-rays by impinging the accelerated electrons on a target made of, for example, tungsten.

As we saw in the previous paragraphs, electron accelerators were developed in parallel to the high-energy proton machines. But the small mass of the electrons makes reaching high energies more difficult compared to protons. Moreover, it was already known in the late 1940s that forcing electrons on a circular orbit causes them to emit synchrotron radiation. The emitted energy is inversely proportional to the bending radius $\rho$ and to the fourth power of the mass of the radiating particle. Electrons therefore require either large rings or linear accelerators. But in linear accelerators, the acceleration structures can only be used once and many of them are required in order to reach high energies. To power the structures, efficient sources of radio-frequency power are needed. This technology became available after the Second World War through developments during the war, both in vacuum and in radio-frequency technology for radar. In particular, power-amplifiers for RF signals, called *klystrons,* became available to generate radio-frequencies in the multi-MHz and even GHz rage at MW power levels. This development triggered the construction of the 3 km linear accelerator at the Stanford Linear Accelerator Center (SLAC) from 1962 onwards. In the SLAC linac, 240 klystrons were used to accelerate electrons to a maximum energy of 20 GeV.

The SLAC linac proved to be a precision probe for nuclear matter. It smashed point-like electrons into target materials to probe the sub-structure of atomic nuclei. Their constituents were originally called *partons,* but could later be identified as quarks and gluons, thus proving their existence. In the 1980s the acceleration system was upgraded to increase the maximum beam energy to 50 GeV. This energy is suitable to probe the $Z$–bosons, earlier found in the SPS, and the SLAC linac was converted to a linear collider. Both electrons and positrons were first accelerated in the linac and then guided through two arcs to collide head-on with micron-sized beams. In the collisions, the details of the Z-boson were investigated. In parallel, also in the 1980s, a large synchrotron with a circumference of 27 km to collide electrons and positrons, named LEP, was built at CERN. It was also used for precision studies of Z and W-bosons. The copious amount of synchrotron radiation emitted by LEP, which is proportional to $E^4/\rho$ where $E$ is the beam energy and $\rho$ the radius of curvature limited the energy of the particles to 100 GeV, in which case the beam lost about 3 % of its energy *per turn.* An electron collider at higher energies therefore needs to be straight—a linear collider, or have a much larger circumference than LEP. There are presently three candidates for accelerators to do precision measurements of the physics discovered at the LHC. One is the International Linear Collider (ILC), a 30 km long linear accelerator that uses super-conducting radio-frequency accelerating structures. The second candidate is the Compact Linear Collider (CLIC). It uses normal-conducting structures, is 50 km long, but promises to reach much higher energies than ILC. The third candidate is the Future Circular Collider (FCC), a ring with a circumference of around 100 km that might first house an $e^+e^-$–collider, later to be replaced by a proton collider, to repeat the trick to replace LEP with LHC in the same tunnel.

Of course the large electron ring for the FCC is not the first of its kind; several electron synchrotrons were constructed at DESY in Hamburg, among them DESY, DORIS, and PETRA, to explore the physics of quarks. In Japan, the KEK laboratory was established in 1971 and operated, among other machines, the TRISTAN synchrotron and the KEKB B-factory. Even at SLAC the SPEAR and PEP colliders were in operation and some of them live on in different incarnations as synchrotron light sources or as *factories.* Nowadays, we call colliders with very high beam currents "factories," because they allow us to probe extremely rare events, involving charm-quarks at Daphne in Frascati, or bottom(B)-quarks at KEKB, mentioned above, and PEP-II at SLAC.

A second use of electron rings emerged from using the emitted synchrotron radiation in material and life-science experiments. Electron rings, dedicated to this purpose, are called *synchrotron light sources* and are in recent times custom-designed and built, but older machines are often refurbished electron-positron colliders. As a prototypical example let us consider SPEAR at SLAC, which underwent several stages. It was initially constructed as a collider, where the $J/\psi$ particle was discovered in 1973. The spectrum of the radiation emanating from the dipole bending magnets extends into the X-ray regime and was used for material science and medical applications since the late 1970s. Thus SPEAR served as both a high-energy collider and as a first-generation light source. Soon it was realized that the radiation can be dramatically enhanced by placing undulator and wiggler magnets in the ring. They do not change the overall geometry of the ring. Within the wigglers, alternating magnetic dipole fields cause the electrons to wiggle back and forth, which leads to an increased emission of synchrotron radiation. Adding these specialty magnets transformed SPEAR into a second-generation light source. In the 1990s SPEAR was completely rebuilt with a dedicated magnet sequence to optimize the generation of synchrotron radiation, which turned it into a third-generation light source. The expertise in using synchrotron radiation later led to the conversion of part of the SLAC linac into the Linac Coherent Light Source (LCLS), the first free-electron laser producing X-rays, which is often considered a fourth-generation light source.

World-wide, there are numerous third-generation light sources, specifically built to serve a huge user base. Examples are ALS in Berkeley, BESSY-II in Berlin, the Shanghai Light Source, the Swiss Light Source near Zürich, Diamond near Oxford, NSLS in Brookhaven, TPS in Taipeh, PLS in South-Korea, the APS in Argonne, ESRF in Grenoble, and the MAX IV laboratory in Sweden. Not only are the ring-based light sources proliferating, also more and more Free-electron lasers are appearing. Examples are FLASH and the European XFEL in Hamburg, SACLA in Japan and at PLS in Korea.

Whereas synchrotron light sources produce radiation to probe materials, high-energy electron accelerators probe the sub-nuclear world on the level of quarks and below. For example, microtrons produce a continuous stream of electrons with moderate energies, up to the GeV range, for nuclear physics. A close relative to microtrons and reaching energies of soon up to 12 GeV, is the CEBAF accelerator at Jefferson Laboratory in the US. *Energy recovery linacs* (ERL) are close relatives to microtrons. They use the same structures, which accelerate the beam, to later decelerate it, and thereby recover the energy that is carried by the beam.

In life and material sciences, the photons, produced so copiously in synchrotron-based light sources, are used to probe the distribution of electrons in matter. A complementary view is provided by neutrons that mostly scatter from light atoms such as hydrogen. Moreover, being electrically neutral, and, at the same time, carrying a magnetic moment, neutrons are the perfect probes for magnetic properties of materials. Historically, many experiments with neutrons were performed on nuclear reactors, but they only provide a comparatively low neutron flux and the neutrons arrive continuously, making time-of-flight experiments difficult or wasteful by chopping the neutron beam. A complementary approach is pursued by dedicated accelerator-based neutron sources, such as ISIS near Oxford, the cyclotron at PSI near Zürich, the SNS near Oak Ridge, and soon the European Spallation Source in Sweden. These machines accelerate protons and direct them onto targets, where they cause nuclear reaction cascades, resulting in a large number of neutrons. These are moderated to low energies in a large block of hydrogen-rich material, and directed to experimental stations. There, the arrival time of the neutrons is related to their energy and a further selection is achieved by monochromators, before directing the neutrons on a sample to probe its properties.

Large numbers of accelerators are used for medical purposes. Small linear electron-accelerators with energies of around 10 MeV operate in many hospitals, either to produce high-energy photons or to directly irradiate tumors. Since in this case the depth-profile of the deposited dose decays exponentially, nowadays even protons or heavier ions such as carbon are used, because they deposit most of the dose at a certain depth, as we shall see in Chapter 9. Protons with energies around 200 MeV deposit most of their dose at a depth of about 28 cm, which allows us to irradiate tumors in any part of the human body. Since the 1990s, dedicated cancer clinics using protons are in operation. Small proton accelerators, often cyclotrons, are in use in order to produce radioactive isotopes for positron-emission tomography or to produce medical tracers.

Apart from the larger groups of accelerators for sub-atomic research, synchrotron radiation or medical applications, a wide variety of other specialty machines exist to sterilize food, implant ions, and there is even one in the Louvre in Paris for analyzing works of art. Van-de-Graaff Tandem accelerators are used to probe surfaces by detecting scattered products from the surface.

Despite the huge difference in beam energies and sizes of accelerators, the basic physical concepts that guide their design and operation are the same. In the following pages we discuss these concepts in detail and illustrate the methods with MATLAB and Octave code. We also discuss the technological choices to achieve a certain performance goals, such as why to use either normal- or super-conducting technology for magnets or for accelerating structures. We also address limits for different types of accelerators arising from achievable fields in magnets that typically limit high-energy proton accelerators or radio-frequency technology that may limit high-energy electron accelerators. But now we turn to a brief outline of things to come.

In the next chapter we first discuss how to formulate design criteria for an accelerator and define its geometry, which we illustrate with code to determine where to put magnets an other accelerator-components on the floor. We even produce input files for 3D modeling programs that allow us to visualize the layout of the accelerator. Dipole magnets and the distance between them define the reference trajectory and we continue to introduce the commonly used coordinate system relative to the reference trajectory. We use it to describe the positions of individual beam particles and even ensembles of very many particles, the entire beam.

In Chapter 3, we introduce the widely used transfer matrix formulation for beam optics and illustrate this with MATLAB code, first in one, the horizontal plane, and then in both horizontal and vertical planes. We discuss concepts such as the beam matrix, emittance and beta function, but also energy-dependent effects such as dispersion and chromaticity. Finally we apply the developed formalism to design simple beam optical systems, also illustrated with MATLAB. In the subsequent chapter we discuss methods to design the magnets needed to build the accelerator, designed in the previous section and also illustrated with MATLAB to solve Poisson's equation in simple geometries. We also discuss technological aspects pertaining to iron-dominated, super-conducting, and permanent magnets.

After having the magnets to guide and focus the beam, we need to discuss how to accelerate it in Chapter 5. Here we discuss the concept of phase-stability, alluded to above, and how to accelerate in linear and circular accelerators. After the beam physics we turn to radio-frequency technology in the next chapter, where we discuss power generation, transport in waveguides and coax lines, couplers and antennas and the accelerating structures, sometimes also called cavities, that transfer the power to the beam. For the analysis of the modes in waveguides and cavities we use the MATLAB-PDE toolbox. Now that we have the basics covered, we know how to accelerate and guide a perfect beam. But neither the world nor accelerators are perfect and we devote Chapter 7 to diagnostic methods to find out what is wrong and Chapter 8 to imperfections and to correction methods to fix the imperfections.

At this point we can design, build, control, and operate an accelerator and it is time to use it for experiments. In Chapter 9, we discuss beam-physics issues arising from colliding beams with targets or counter-propagating beams. In the following chapter we talk about the generation of synchrotron radiation, both about the amount and properties of the radiation and how this affects the beam. We also cover the basic theory and technological aspects of Free-electron lasers. Chapter 11 introduces non-linear dynamics including Hamiltonians, Lie methods, and normal forms, all illustrated with MATLAB code. In Chapter 12, we cover intensity-dependent limitations of accelerators and in Chapter 13 we consider many of the technological subsystems that are needed to operate an accelerator. Here we treat, among other topics, the control system, the particle sources, vacuum and cryogenics. In Chapter 14 we discuss a number of accelerators, such as the LHC, LCLS, ESS, as well as medical and industrial accelerators. There we point out how the topics covered in the earlier chapters are used to operate a real machine.

Several student labs are discussed in the appendix. Here we measure the beam profile of a beam from a laser pointer and determine the emittance, or $M^2$ in laser parlance, of the same laser beam. In the next labs, we calculate, build, and measure magnets, based on small permanent magnet cubes inserted in 3D-printed frames. In the last lab, we use a network analyzer to characterize a simple pill-box cavity made from a cookie-jar.

Several topics are relegated to online appendices, available from this book's web page at https://www.crcpress.com/9781138589940, which also contains the MATLAB source code that appears in this book. The appendices comprise an overview over methods from linear algebra, essential for our discussion of beam physics, and a short MATLAB tutorial. A further online appendix contains detailed descriptions of all MATLAB code used in this book, both in the simulations and to generate the figures. A short description of the relation between light optics and the optics of charged particles follows and a further appendix contains a very brief tutorial of OpenSCAD, the software used to design 3D-models of beam lines and the 3D-printed frames for the magnets.

After all the historical background and outline of things to come, let's get down to business...

# CHAPTER 2

# Reference System

## 2.1  THE REFERENCE TRAJECTORY

We start out by discussing how to place our accelerator into the world and that depends of course on the type of accelerator and the space available. Imaging the task to place the LHC in the 27 km tunnel originally occupied by LEP; it just had to fit. Or whether an ion-implanter fits into the basement of the lab; or whether the large parking lot in front of the office is big enough for a small synchrotron light source. In any case, we need to find out how to match the accelerator to the available space and where on the floor to put all those magnets and accelerating structures. This is the first problem we address in this chapter.

For this purpose we employ a somewhat formal language to describe the accelerator with its components and their relative positions. We give each element a code number, a repeat count, a length, and an additional parameter that contains other descriptive numbers, such as deflection angle or focusing strength. A line in our input file thus looks like

```
1   10   0.2   0
```

where the code is 1, indicating an empty piece of beam pipe that constitutes 10 segments that are 0.2 m long each. The fourth number, not used here, is set to zero. The reason we chose this format is that it can be easily represented as a $n \times 4$ matrix in MATLAB or Octave and concatenating is made trivial by using powerful built-in functions to manipulate matrices. Moreover, reading from and writing to external files can be handled by built-in functions, such as `dlmread` and `dlmwrite`, without having to spend many pages on the discussion of elaborate, though probably much more convenient, input parsers. On the other hand, the purpose of this book is to show the inner workings of beam dynamics codes. This helps us to understand, and to appreciate the power of other "big" codes, such as MADX [3] or TRANSPORT [4, 5], the grand-daddies of most beam optics codes, and what goes on under their hood.

We extend the number of codes to comprise several different elements in an accelerator, but here we confine ourselves to those that predominantly determine its geometry, and those elements are

- Empty beam pipe with code number 1, the same element we met in the previous paragraph.
- Thin-lens focusing element with code 2.
- Finite-length Quadrupole or focusing magnet with code 5.
- Dipole or bending magnet with code 4, that changes the direction of the reference trajectory. One line in the input file may look like "4 1 0.7 15," which describes a dipole with one segment of length 0.7 m, and deflection angle of 15 degrees.

- Coordinate rotation with code 20.

Additional elements can be easily accounted for by introducing additional codes. In the context of this chapter the reference beam goes straight through empty beam pipes, quadrupoles, solenoids, and acceleration structures, and the only property we care about is their length. In the dipole magnets on the other hand, the reference trajectory changes direction.

Here we always assume that the dipole magnets deflect in the horizontal plane, any other angle can be handled by sandwiching a dipole between coordinate rotations. The bending radius for a dipole with deflection angle $\phi$ and length $L$ is given by $\rho = L/\phi$. For a beam with momentum $p$ the magnetic flux density $B$ in the dipole is then given by $B = (p/e)\phi/L$. Note that we can rewrite the definition of the bending radius as $B\rho = p/e$ and immediately see that we obtain a convenient translation of physics units of momentum in eV/c to engineering units in Tm. This is often used in beam physics, where the momentum is given in units of Tm and voiced as "Bee-rho." Remembering the conversion that $1\,\mathrm{Tm}$ corresponds to a momentum $300\,\mathrm{MeV/c}$ often proves useful in estimating the required field and length of a magnet.

After these preliminary considerations we are ready to describe our first part of an accelerator, a short section with one dipole and two quadrupoles, often called a FODO cell for reasons that will become clear in the next chapter. The description of the beam-line is the following

```
fodo=[ 1  1  2.5  0 ;
       5  1  1.0  0 ;
       1  1  1.5  0 ;
       4  1  2.0  60;
       1  1  1.5  0 ;
       5  1  1.0  0 ;
       1  1  2.5  0 ]
```

where we simply define an array, called fodo, to describe the sequence of elements, first a single drift section (code 1) followed by a long quadrupole (code 5) and then another drift. Next comes a dipole magnet with a deflection angle of 60 degrees. Here we do not need to specify the strength of the quadrupoles, they are considered drifts in the context of finding the reference trajectory. Concatenating beam-line sections is as easy as writing

```
ring = [fodo; fodo; fodo; fodo; fodo; fodo];
```

or, more compactly, as ring=repmat(fodo,6,1); by using the built-in function repmat. In this way we define separate sections of a beam line and later combine them to a larger, composite beam line. Here six cells with 60-degree bending magnets make up an entire ring.

Having the description of the beam-line, we now turn to calculating where on the floor we have to put the magnets. This is most easily done by following a Frenet-Serret tripod on its journey along our beam line. This right-handed tripod rides on the reference trajectory and is determined by a vector $\vec{V}$ pointing to the position of its origin and a matrix $W$ that describes its orientation in terms of three angles with respect to the orientation at the coordinate origin.

All we have to do is to step through the beam line, calculate the vector $d\vec{V}$ that points from the entrance of that element to its exit, and calculate the matrix $dW$ that encodes the change in orientation. Using these changes we update $\vec{V}$ and $W$ after each element and record the starting and end positions to tell us where the elements are and what their orientation is. $d\vec{V}$ and $dW$ for the straight elements are given by

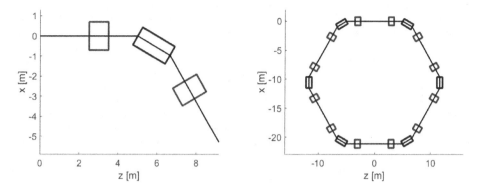

Figure 2.1  Geometry of a single FODO cell (left) with a 60-degree bending magnet and a ring that consists of six such cells (right).

```
dv=[0;0;beamline(line,3)]; dw=eye(3);
```

and in sector dipoles, we use

```
phi=beamline(line,4)*pi/180;   % convert to radians
if abs(phi)>1e-7
  rho=beamline(line,3)/phi;
  dv=[rho*(cos(phi)-1);0.0;rho*sin(phi)]; dw=wmake(0,-phi,0);
end
```

where the function wmake() creates the rotation matrix around the respective axis, here the second, vertical axis. Once the changes in position $d\vec{V}$ and orientation $dW$ are available, we update $\vec{V}$ and $W$ with the following function

```
% wprop.m, updates the new vector vv and matrix ww
function [vnew,wnew]=wprop(vv,ww,dv,dw)
vnew=vv+ww*dv;
wnew=ww*dw;
```

These functions, along with routines to draw the magnets and the beam pipe, are coded in the function layout.m. Inside the code, after some initializing tasks, a loop steps through the beam line, determines dv and dw, depending on the type of element, and uses wprop to update v and w. Then, again depending on the type of element, boxes and lines, representing the magnets, are drawn on the image. Figure 2.1 shows the 2D-rendering of a single FODO cell on the left-hand side and a complete ring on the right-hand side. The layout script, along with supporting functions, is described in detail in Appendix B.5 that are available online at this book's web page.

Running the layout script with a single fodo cell beamline=fodo; results in the MAT-LAB plot shown on the left in Figure 2.1 with the dipole in the center that deflects the trajectory and quadrupoles on either side of the dipole. Running the same program again, but this time with six fodo cells beamline=repmat(fodo,6,1); results in the plot shown on the right-hand side in Figure 2.1.

Simultaneously with the two-dimensional images, shown in Figure 2.1, the layout() function produces an input file for the 3D modeling program OpenSCAD [6]. In OpenSCAD

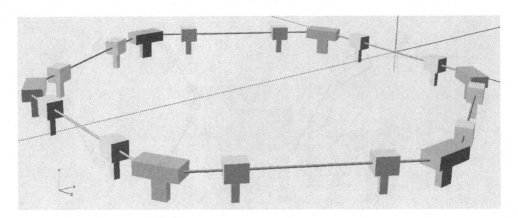

Figure 2.2 3D-view of the ring shown on the right-hand side in Figure 2.1.

simple geometric objects, such as cubes and cylinders, are placed in a "3D world" using the same vector $\vec{V}$ and matrix $W$, we used before to generate the plots in Figure 2.1. A brief introduction into using OpenSCAD can be found in the online Appendix B.3, available at this book's web page. The functionality to generate the 3D models is already built into `layout`, which generates an output file `layout.scad` that can be directly loaded into OpenSCAD with the command

```
openscad layout.scad
```

Figure 2.2 shows the 3D-model of the ring with six fodo-cells exported from OpenSCAD. We see the dipole magnets and quadrupoles, all with their small pedestal, and the beam pipe in the sections in between magnets. This simple example illustrates how to determine the reference trajectory and where to put all the magnets and other elements. The curious reader is encouraged to add more, and nicer looking, elements as well as additional features to the MATLAB script or port it to other programing languages. As a side-note, we point out that OpenSCAD can export the model in formats that are compatible with "slicer" programs to prepare the model for 3D-printing.

After having defined the reference trajectory, we know where the accelerated particles should move in an ideal world, namely on the reference trajectory. But since the real world is not ideal, they move around it and we address how to describe individual particles first, and then ensembles of particles, moving in the vicinity of the reference trajectory.

## 2.2  COORDINATE TRANSFORMATIONS

Since the beam particles move in the vicinity of the reference trajectory $\vec{r}_0$, it is convenient to describe their motion in a coordinate system that "rides on the reference particle." A natural choice for this coordinate system is the *Frenet-Serret* tripod, which is based on three normalized unit vectors: the first is the tangent vector to the reference trajectory $\vec{t} = d\vec{r}_0(s)/ds$. Here we assume that $\vec{r}_0(s)$ is parametrized by the arc length $s$, which guarantees that $\vec{t}$ has unit length. The second unit vector $\vec{n}$ is called the normal vector and, for planar trajectories without torsion, it is proportional to the rate of change of the tangent $d\vec{t}/ds$. The factor that makes $\vec{n}$ a unit vector is the *curvature* $\kappa(s)$ or, equivalently, the inverse of the bending radius $\kappa(s) = 1/\rho(s)$, such that we obtain $\vec{n} = -\kappa d\vec{t}/ds$. The minus sign causes $\vec{n}$ to point away from the center of the deflection and ensures that the

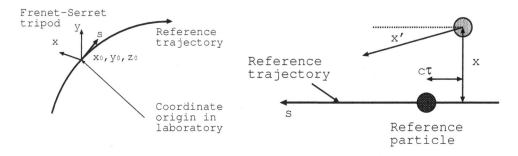

**Figure 2.3** Reference trajectory and the co-moving tripod (left) that is used to define the deviations (right) of particles with respect to the reference particle.

third unit vector $\vec{b}$, defined by $\vec{b} = \vec{t} \times \vec{n}$, points upwards. This construction causes the three base vectors $\vec{t}, \vec{n}$ and $\vec{b}$ to form a right-handed coordinate system. In a circular accelerator, for example, $\vec{t}$ points along the direction of propagation of the reference particle, $\vec{n}$ points towards the outside of the ring, and $\vec{b}$ points upwards. By convention, in the accelerator literature they are often denoted by $\vec{s}, \vec{x}$, and $\vec{y}$, respectively. The left-hand side in Figure 2.3 illustrates their orientation.

The geometry of components, such as magnets or radio-frequency devices, that affect the motion of beam particles, is normally specified in the laboratory system and also the forces that act on the beam are given in that frame. But now we transform these forces, which determine the equations of motion, to the reference system that "rides on the reference particle." Since the particles move with relativistic velocities $\vec{v}$ we choose to start our discussion with the relativistic invariant Lagrangian $L(\vec{r}, \vec{v})$ of a point charge with charge $e$, mass $m$, and coupled to electro-magnetic fields specified by their potentials $\Phi(\vec{r}, t)$ and $\vec{A}(\vec{r}, t)$ [7, 8], which is given by $L = -\gamma mc^2 - e\Phi + e\vec{v}\vec{A}$ with $\gamma = 1/\sqrt{1 - v^2/c^2}$. Hamilton's principle and the ensuing Euler-Lagrange equations then lead to the well-known Lorentz force $d(\gamma m\vec{v})/dt = e\vec{E} + e\vec{v} \times \vec{B}$, where the fields $\vec{E}$ and $\vec{B}$ are related to the potentials $\Phi$ and $\vec{A}$ by $\vec{E} = -\vec{\nabla}\Phi - \partial\vec{A}/\partial t$ and $\vec{B} = \vec{\nabla} \times \vec{A}$.

Instead of using the Lagrange function $L(\vec{r}, \vec{v})$, which depends on the positions $\vec{r}$ and velocities $\vec{v}$ of a particle, it is often more convenient to use the Hamiltonian $H(\vec{P}, \vec{r})$, which depends on positions $\vec{r}$ and the canonical momenta $\vec{P}$, defined by $\vec{P} = \vec{\nabla}_{\vec{v}}L$. As an illustration of the canonical momentum, we consider one dimension, where it is given by $P = \partial L/\partial v$. The Hamiltonian $H$ can be derived from the Lagrangian $L$ by a Legendre-transformation $H = \vec{P} \cdot \vec{v} - L$ and the equations of motion are transformed to a set of first-order differential equations—Hamilton's equations [8].

The Hamiltonian in the previous paragraph still depends on the coordinates $\vec{r}$ in the laboratory system, but we can transform it to the co-moving system $x, y, s$ with the help of a *canonical transformation* [8] with the generating function of type 3 [9] $F_3(\vec{P}, x, y, s) = -\vec{P} \cdot \left(\vec{r}_0(s) + x\vec{n}(s) + y\vec{b}(s)\right)$ that depends on the momentum $\vec{P}$ in the laboratory frame and the positions $x, y$, and $s$ in the co-moving frame. The momenta in the co-moving frame $\vec{p}$ are then given by the derivatives of the generating function $F_3$ with respect to $x, y$, and $s$, for example $p_s = -\partial F_3/\partial s = (1 + \kappa x)\vec{P} \cdot \vec{t}$, where we used $d\vec{n}(s)/ds = \kappa\vec{t}$ and $d\vec{b}(s)/ds = 0$, which is valid for planar reference trajectories. Note that $p_s$ is proportional to the projection of the momentum in the lab frame $\vec{P}$ onto the tangent $\vec{t}$ at $s$. The factor $1 + \kappa x = 1 + x/\rho$ accounts for the longer path $s$ at larger horizontal position $x$ in bending magnets, which have $\rho \neq 0$. Transforming the Hamiltonian $H$ to the new variables with $K = H + \partial F_3/\partial t$, expressing it in terms of the new variables, we obtain

$$K = e\Phi + c\sqrt{m^2c^2 + (p_x - eA_x)^2 + (p_y - eA_y)^2 + \left(\frac{p_s - eA_s}{1 + x/\rho}\right)^2} , \qquad (2.1)$$

where the potentials $\Phi$ and $\vec{A}$ depend on the positions $\vec{r} = \vec{r}_0(s) + x\vec{n}(s) + y\vec{b}(s)$. Hamilton's equations in positions $x, y, s$ and momenta $p_x, p_y, p_s$ then yield the equations of motion in the variables of the co-moving frame: $\dot{x} = \partial K/\partial p_x$, $\dot{p}_x = -\partial K/\partial x$, and similarly for $y$ and $s$.

These equations of motion still depend on time as the independent variable, but using the position on the trajectory $s$ as the independent variable makes interpreting the location of particles much more accessible. We do not want to know when they are somewhere, but rather want to know their transverse position at a specific place, say, an experimental station. Following [9], we solve Equation 2.1 for $p_s$ and can interpret $G = -p_s$ as a new Hamiltonian, resulting in

$$G = -\left(1 + \frac{x}{\rho}\right)\sqrt{\frac{(K - e\Phi)^2}{c^2} - m^2c^2 - (p_x - eA_x)^2 - (p_y - eA_y)^2} - eA_s , \qquad (2.2)$$

which depends on the variables $x, p_x, y, p_y, t$ and $-K$. Recognizing that $K - e\Phi$ equals the total energy $E = \sqrt{m^2c^4 + p^2c^2}$ of a particle with momentum $p$, we simplify Equation 2.2 to $G = -(1 + x/\rho)\sqrt{p^2 - (p_x - eA_x)^2 - (p_y - eA_y)^2} - eA_s$. In most accelerators the total momentum of particles are close to the reference momentum $p_0$ and we can write $p = (1 + \delta)p_0$ with $\delta = (p - p_0)/p_0$. Replacing $p$ then yields

$$G = -\left(1 + \frac{x}{\rho}\right)\sqrt{(1 + \delta)^2 p_0^2 - (p_x - eA_x)^2 - (p_y - eA_y)^2} - eA_s . \qquad (2.3)$$

Note that $\delta$ is a dynamical variable in this context, because it directly depends on $-K$ via $p$. Moreover, we observe that the kinematic state of a particle, which requires three spatial coordinates and three corresponding momenta or velocities, is described by the six variables $x, p_x, y, p_y, t$ and $\delta$. They form the base for our further discussions. In passing, we note that the longitudinal component of the vector potential $A_s$ describes the transverse fields of the magnets we discuss in Chapter 4. Moreover, $A_s$ will play a central role when treating non-linear motion in Chapter 11.

But now, we move on and discuss the description of particles and beams using the variables introduced in this section.

## 2.3 PARTICLES AND THEIR DESCRIPTION

Equation 2.3 is the base for a consistent description of the equations of motion that determines the propagation of particles in accelerators and we can use its independent variables to describe the state of particles. Since the transverse momenta $p_x$ and $p_y$ in most accelerators are much smaller than the total momentum $p \approx p_0$, we use $x' = p_x/p_0$ and $y' = p_y/p_0$ instead. These variables allow a systematic expansion of the Hamiltonian $G$ in Equation 2.3 and can be interpreted as angles with respect to the reference trajectory, which eases visualizing the motion of the particles, illustrated on the right-hand side in Figure 2.3. Finally, we shift the origin of the time variable $t$ to coincide with that of the reference particle $\hat{t}$ and use $\tau = t - \hat{t}$ as variable.

Thus we arrive at the following three position-like coordinates and three parameters related to momenta that are used to characterize particles in accelerators. Here we summarize the parameters for convenience.

$x$ , the horizontal distance to the reference particle;

$x'$ , the horizontal angle with respect to the trajectory of the reference particle;

$y$ , the vertical distance to the reference particle;

$y'$ , the vertical angle with respect to the trajectory of the reference particle;

$\tau$ , the arrival time with respect to the reference particle;

$\delta$ , the relative momentum difference with respect to the reference particle.

The parameters can be visualized as the differences of the particles' coordinates to those of a tripod that rides on the reference particle.

It should be noted that some programs, such as MADX [3], use slightly different variables, which, however, in the ultra-relativistic limit, agree with those mentioned here. An advantage of the variables from the table is that they describe geometric concepts like distances and angles, as shown on the right-hand side in Figure 2.3. The arrival time is relevant if time-varying electro-magnetic fields, for example, generated in accelerating structures, affect the particle. The relative momentum difference $\delta$ is convenient to use, because it describes the relative deviation from the design deflection angles of dipole magnets that is due to the deviation from the design momentum of the particles.

The longitudinal momentum is almost always much larger than the transverse momenta, which causes the angles $x'$ and $y'$ to be very small, and typical values are on the order of mrad. Many approximations, used when dealing with accelerators, therefore use the *paraxial approximation* and expand variables in power-series in the kinematic variables $x, x', y, y', \tau, \delta$. As a matter of fact, the typical magnitude of these parameters is on the scale of mm for $x, y$ and $c\tau$ and $10^{-3}$ for the angles $x', y'$ and the momentum deviation $\delta$. Of course, under special circumstances, also very different values can occur.

All possible values of the kinematic variables are commonly denoted as the six-dimensional *phase space* and a particular set of values are used to describe the *state* of a single particle. Sometimes it is not necessary to consider the full 6D phase space and it is sufficient to only cover the *horizontal phase space,* comprising the subspace of $x$ and $x'$ or the *vertical phase space* comprising $y$ and $y'$, respectively. The subspace spanned by $\tau$ and $\delta$ is denoted *longitudinal phase space*. In later chapters, we will often restrict ourselves to these subspaces, which is possible because the dynamics of the different subspaces is often independent (also called un-coupled). This also allows us to focus on the essential dynamics at hand without cluttering the notation. Occasionally, we will use a hybrid phase space comprising $x, x'$ and $\delta$ in order to account for momentum-dependent, also called *chromatic,* effects that mostly appear in the horizontal plane, because the deflection angle of dipole magnets depends on the momentum of the particles.

So far, we have considered ways to describe the state of a single particle in an accelerator, but in real accelerators large numbers of particles, often in the range $10^6$ to $10^{11}$ or more, propagate and we need efficient methods to describe these large ensembles of particles.

## 2.4 PARTICLE ENSEMBLES, BUNCHES

The simplest way to illustrate the distribution of a large number of particles is in the form of a *histogram*. For the time being, we only consider a single variable, say the horizontal position $x$, and we plot the number of particles having positions between $x$ and $x + \Delta x$ with $\Delta x = 0.02 \, \text{mm}$ which yields the histogram, shown on the left-hand side in Figure 2.4.

The values used in the histogram were chosen arbitrarily, but they show general features of many *distributions* anyway. First, the number of particles $N$ is finite, $N = 800$ in this case. This implies that the distribution can be normalized. Second, most particles have positions

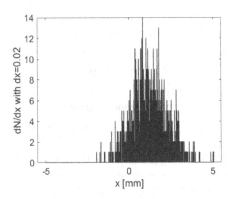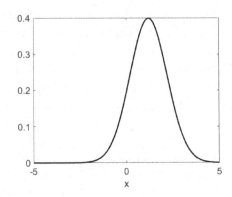

Figure 2.4 A histogram of the horizontal position of 1000 particles (left), where the vertical axis is given by the number of particles having position values between $x$ and $x + \Delta x$ with $\Delta x = 0.02$ mm. A normalized Gaussian distribution is shown on the right.

$x$ clustered around an average value, often denoted by $\langle x \rangle$. Finally, the distribution has a width in values, that we denote by $\sigma$. It is illustrative to visualize such a distribution as a collection of repeated measurements of the same quantity. In such cases we calculate the average value to represent the most probable value and the root-mean-square deviation as the spread in measured values or the uncertainty of the measurement.

Let us consider a discrete distribution in a histogram that is given as a table of discrete $x$−values, say $x_i$, where the index $i$ labels the bins, populations (the height) of the bins $b_i$, and a bin width $\Delta x$. We calculate the normalization, the average and the width in an almost self-evident way. First, the normalization $N$ is given by the sum over all bin-populations

$$N = \sum_i b_i \, , \tag{2.4}$$

where the sum extends over all bins. For the average $X = \langle x \rangle$, we have to weigh the position represented in bin $i$ at $x_i$ by its population

$$X = \langle x \rangle = \frac{1}{N} \sum_i x_i b_i \tag{2.5}$$

and normalize by using the normalization constant $N$. Another way to visualize this is that in each bin at position $x_i$ the fraction $b_i/N$ of all particles reside and we weigh the $x_i$−value with that fraction. The spread or width $\sigma$ of the distribution we calculate as the average of the squared distance $(x_i - X)^2$ from the average

$$\sigma^2 = \frac{1}{N} \sum_i (x_i - X)^2 b_i \, . \tag{2.6}$$

This quantity $\sigma$ is often called the root-mean-square, or RMS, of a distribution of values.

In MATLAB or Octave, built-in commands make it almost trivial to calculate these values. We assume that we are given two column vectors $x$ and $b$ that contain values $x_i$ and $b_i$ used in the previous paragraph. The vectors must have equal size and running the command `plot(x,b)` would produce a plot similar to the one shown in Figure 2.4. In that case `N=sum(b)` corresponds to Equation 2.4, `Xavg=sum(x.*b)/N` to Equation 2.5, and `sigma2=sum(b.*(x-Xavg).^2)/N` calculates $\sigma^2$. Note that we need to use the element-wise operations prepended with a dot. See Appendix B.2 for a discussion of basic MATLAB commands.

In the previous paragraphs we calculated powers $x^m$ of the variable $x$, weighted with the bin-populations. These quantities, calculated without subtracting the average value, are called *moments* of the distribution and are defined by

$$\langle x^m \rangle = \frac{1}{N} \sum_i x_i^m b_i \ .$$
(2.7)

They will prove to be very useful in later chapters. This equation serves to define the angle brackets to perform the average of the included quantity over the bin-population. Note that the average value is the same as the first moment and that $\sigma^2$ can be called the *central second moment* because the average is subtracted. It can also be expressed through the moments by

$$\sigma^2 = \langle (x - X)^2 \rangle = \langle x^2 - 2xX + X^2 \rangle = \langle x^2 \rangle - X^2$$
(2.8)

which is the second moment minus the first moment squared.

The discrete distributions normally shown in a histogram have continuous pendents, the continuous probability distribution functions, where "probability" essentially means that the distribution is positive semi-definite—all values are larger than or equal to zero—and it is normalized. Both requirements are obviously fulfilled for distributions that describe physical quantities. A general one-dimensional distribution function $\psi(x)$ that depends on the variable $x$ has a simple interpretation in the sense that the number of particles $N$ (or another quantity that the distribution function describes) in the interval between $x$ and $x + \Delta x$ is given by

$$N_x = \psi(x)\Delta x$$
(2.9)

and we immediately see that the distribution function $\psi(x)$ has physical units of the inverse of its independent variable, here $x$.

The prototypical probability distribution function, which is only characterized by its average value $X$ and its width $\sigma$, is the Gaussian distribution with its characteristic bell-shaped curve shown on the right-hand side in Figure 2.4. Gaussians appear in many contexts, because they are the limiting distributions of many, though not all, random processes, which is a consequence of the *central limit theorem*. We explore this further in the exercises, but here only point out that they are frequently used to describe the beams in accelerators. Their functional form is given by

$$G(x; X, \sigma) = \frac{1}{\sqrt{2\pi}\sigma} e^{-(x-X)^2/2\sigma^2} \ ,$$
(2.10)

which is represented as a MATLAB inline function with arguments specified via the `@()`–construction in the following lines of code

```
x=-5:0.01:5;
G=@(x,X,sigma)exp(-((x-X).^2)/(2*sigma^2))/(sqrt(2*pi)*sigma);
plot(x,G(x,1.2,1),'k');
```

which produces the right-hand plot in Figure 2.4. The peak value lies around 0.4 and the width of the Gaussian in Figure 2.4 appears to be around unity, but we can also prove that this is the case by first verifying that the distribution in Equation 2.10 is normalized, has average $X$, and RMS $\sigma$. For the normalization and the average, we need to show that

$$\int_{-\infty}^{\infty} G(x; X, \sigma)dx = 1 \ , \tag{2.11}$$

which is easily done by substituting $y = (x - X)/\sigma$ and looking up the resulting integral in an integral table such as [10]. To verify that the average value is indeed $X$ we need to show that

$$\int_{-\infty}^{\infty} xG(x; X, \sigma)dx = X \ , \tag{2.12}$$

which is even simpler, because the substitution $y = (x - X)/\sigma$ leads to an integral that can be reduced to an exponential by substitution. For the RMS sigma we need to show that

$$\int_{-\infty}^{\infty} (x - X)^2 G(x; X, \sigma)dx = \sigma^2 \ , \tag{2.13}$$

which can be achieved by a similar substitution and subsequent inspection of an integral table. In passing we note that all integrals of Gaussians with polynomials in $x$ can be solved by parametric differentiation with respect to $B$ of the following generating function

$$G_B(x; X, \sigma) = \int_{-\infty}^{\infty} G(x; X, \sigma)e^{Bx}dx = e^{XB + \sigma^2 B^2/2} \tag{2.14}$$

and subsequently setting $B = 0$. Each differentiation with respect to $B$ pulls one power of $x$ down. The integral in Equation 2.14 can be calculated by completing the square in the exponent, which leads to a integral similar to the one we encountered for the normalization. Note how the repeated differentiation of the generating function $G_B(x; X, \sigma)$ with respect to $B$ results in the *moments* of a distribution

$$\langle x^m \rangle = \int_{-\infty}^{\infty} G(x; X, \sigma)x^m dx = \left(\frac{\partial}{\partial B}\right)^m \int_{-\infty}^{\infty} G(x; X, \sigma)e^{Bx}dx \bigg|_{B=0} \tag{2.15}$$

after setting $B = 0$ at the end.

The purpose of sketching the mathematical manipulations to solve the integrals is to indicate how Gaussians are rather benign integrands, even when considering Gaussian distributions of many variables, so-called *multi-variate Gaussian distributions*, which are often good approximations for the distribution of particles in accelerators, expressed through their phase space coordinates $x, x', y, y', \tau, \delta$. With a small abuse of notation we denote the six phase space variables collectively by the symbol $\vec{x}$ with components $x_i$. This allows us to write the $n-$dimensional multi-variate Gaussian as

$$\Psi(\vec{x}; \vec{X}, \sigma) = \frac{1}{(2\pi)^{n/2}\sqrt{\det \sigma}} \exp\left(-\frac{1}{2}\sum_{i,j=1}^{n}(\sigma^{-1})_{ij}(x_i - X_i)(x_j - X_j)\right) \ , \tag{2.16}$$

which describes distributions with average values $\vec{X}$ and *covariance matrix* $\sigma_{ij}$. In the one-dimensional limit this definition reverts to Equation 2.10. In order to show that the parameters $\vec{X}$ and $\sigma$ have the same interpretation as before, we need to have

$$\int \Psi(\vec{x}; \vec{X}, \sigma)x_i d^n x = X_i \quad \text{and} \quad \int \Psi(\vec{x}; \vec{X}, \sigma)(x_i - X_i)(x_j - X_j)d^n x = \sigma_{ij} \ , \tag{2.17}$$

which we state without proof. The calculations are lengthy and involve a multi-variate generating function, the equivalent of $G_B$ in Equation 2.14, and repeated parametric differentiations. These tricks are used later on in the book.

Note that the averages $\vec{X}$ and the covariance matrix $\sigma$ uniquely specify the Gaussian distribution and we will later use these parameters as proxies to characterize beams of charged particles. The parameters have succinct physical interpretations. $X_1 = \langle x \rangle$ is the horizontal position of the center of mass of all particles. It is a quantity that is experimentally accessible with beam-position monitors that normally are not sensitive to the positions of individual particles; they only sense averages. Likewise is $X_3 = \langle y \rangle$ the average vertical position and $X_2$ and $X_4$ are the angles of propagation of the beam. Moreover, $X_5 = \langle \tau \rangle$ describes the arrival time with respect to the reference particle at a location and $X_6 = \langle \delta \rangle$ is the average momentum deviation of the beam. The parameters on the diagonal of the covariance matrix are the squared beam sizes in the respective dimensions, such that $\sigma_{11}$ is the horizontal RMS beam size squared and $\sigma_{22}$ is the angular divergence, also squared. In the same way $\sigma_{33}$ and $\sigma_{44}$ describe beam size and angular divergence in the vertical plane. The fifth and sixth diagonal elements describe its bunch length and its momentum spread, respectively. Note that a beam described by a six-dimensional distribution function describes an entity that is confined in its phase space dimensions and can be visualized as a package that travels along the accelerator. Such a package is commonly called a *bunch*. In many accelerators many bunches propagate largely independently and at the same time.

In order to visualize a multi-variate distribution we consider a two-dimensional example of a Gaussian that is centered at the origin $(\vec{X} = 0)$ and therefore has the covariance matrix

$$\sigma = \begin{pmatrix} \langle x^2 \rangle & \langle xy \rangle \\ \langle xy \rangle & \langle y^2 \rangle \end{pmatrix} = \begin{pmatrix} 2 & 1 \\ 1 & 1 \end{pmatrix} \tag{2.18}$$

where we choose some numerical values for definiteness sake. The MATLAB code to generate the visualizations shown in Figure 2.5 is the following

```
sigma=[2,1;1,1]; siginv=inv(sigma);
psi=@(x,y)exp(-0.5*(siginv(1,1)*x.^2-2*siginv(1,2).*x.*y ...
          +siginv(2,2)*y.^2))./(2*pi*sqrt(det(sigma)));
[XX,YY]=meshgrid(-5:0.1:5,-5:0.1:5); ZZ=psi(XX,YY);
subplot(2,2,1); contour(XX,YY,ZZ)); xlabel('x'); ylabel('y');
subplot(2,2,2); surfc(XX,YY,ZZ)); xlabel('x'); ylabel('y');
subplot(2,2,3); plot(-5:0.1:5,0.1*sum(ZZ,1),'k'); xlabel('x');
subplot(2,2,4); plot(-5:0.1:5,0.1*sum(ZZ,2),'k'); xlabel('y');
```

where we use the continuation command ... in order to break up the long definition of psi. At the top of the script we define the sigma matrix and its inverse and simply code the distribution function from Equation 2.16 as the variable psi. Then we define the meshgrid structure to represent the coordinates compatible with the use of contour and surf to generate the contour and 3D-surface plots in the top row of Figure 2.5.

The plots in the lower row of Figure 2.5 show the projections of the two-dimensional distribution onto the $x$ and the $y-$axis. We find that even the projections are Gaussian and their respective RMS-widths are given by $\sigma_x^2 =$sigma(1,1) and $\sigma_y^2 =$sigma(2,2). We easily verify this by analytically integrating over one of the variables in a two-dimensional Gaussian. Integrating or summing over variables in a distribution function corresponds to not paying attention to the integrated variables. The dependence on the remaining variables leaves the projection of the original distribution function onto the space of the remaining variables.

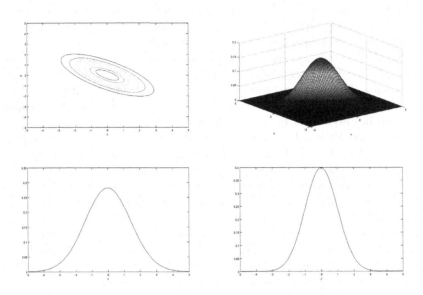

Figure 2.5 Contour (top left) and 3D-surface (top right) plots of a two-dimensional Gaussian distribution. The lower row shows the projections onto the $x$ and $y-$axes.

As an example, let us consider the distribution of a charged particle beam that depends on its six phase-space coordinates $x, x', y, y', \tau$, and $\delta$. If the beam impinges onto a fluorescent screen, we observe a two-dimensional image $\Phi(x, y)$ that depends on the two spatial coordinates $x$ and $y$, only. It corresponds to the projection of the six-dimensional distribution $\Psi(x, x', y, y', \tau, \delta)$ and can be determined by integrating over variables that are not observed

$$\Phi(x, y) = \int \Psi(x, x', y, y', \tau, \delta) dx' dy' d\tau d\delta \ . \tag{2.19}$$

The intensity of the image depends only on the number of particles that hit a particular location on the screen, irrespective of their angle or arrival time (to some limit related to the integration time of the camera) or momentum.

At this point we know where to put the magnets and where the reference trajectory and reference particle are. Moreover, we found a coordinate system whose origin "rides" on the reference particle and that we use to describe individual particles. Since there are many particles in a beam, we introduced distribution functions to describe large ensembles of particles. Since manipulating distribution functions often leads to complex mathematical manipulations we introduced the moments of the distribution as proxies for the essential characteristics (average position, beam width) of a charged particle distribution. Finally, we spent some time on Gaussian distributions which have the property of being uniquely specified by their average and their width, or for multi-variate distributions, by their averages, and their covariance matrix. All this effort provides us with a description of the particles and the distribution, the beam. But now is the time to find out how the beam propagates along the accelerator and determine how the accelerator components such as magnets and accelerating structure affect the state of the particles and consequently, also the beam.

## QUESTIONS AND EXERCISES

1. Build a ring with 12, 24, and 36 FODO cells respectively and prepare 2D plots and 3D images in OpenSCAD. Use the geometry defined in the array `fodo[]` on page 14, but adjust the deflection angle appropriately.

2. Build a racetrack ring with 12 equal FODO cells per arc and straight sections with 6 FODO cells each, based on the cells from the previous exercise. Prepare 2D plots and 3D images in OpenSCAD.

3. You need to design a beam line that takes the reference trajectory 100 m ahead and 10 m horizontally to the left. You have four dipole magnets available with a length of 2 m and quadrupoles with a length of 1 m, which should be spaced by approximately 5 m. (a) Sketch the geometry first and then implement it in a beam-line file; (b) generate the 2D plots; (c) generate the 3D model and load it in OpenSCAD.

4. You must build a transfer line to cross a road in an underground tunnel. Assume that the road is 10 m wide and you have four dipole magnets available, each 4 m long and capable of deflecting the beam by 15 degrees. Assume that the spacing of quadrupoles should be 5 m. Hint: the syntax for a coordinate rotation by angle `phi` in degrees is "20 1 0 phi".

5. Feel free to prepare nicer models for the magnets than the rectangular boxes.

6. Prepare a model of your favorite section of accelerator in your home institute and prepare 2D plots and 3D images. If you have access to a 3D-printer, make a 3D model.

7. Whenever you have difficulty visualizing a beam line later in this book, make a 3D model and have a look...

8. Derive the equations of motion for a particle in a field-free region.

9. Calculate the zeroth (normalization), the first and the second moment and the "central moment" (RMS width around the center) of the following distributions: (a) Triangular distribution: linearly rising from zero to unity for $c - w < x < c$, continuous at $c$ and linearly falling for $c < x < c + w$. It is zero everywhere else. Here $c$ is the center and $w$ the width of the box. (b) Lorentzian distribution $f(x) = 1/(w^2 + x^2)$ for all $x$; (c) Gaussian distribution $g(x) = e^{-x^2/2w^2}$. What percentage lies between $\pm w/2$ for the respective distributions?

10. Verify that the projections of the Gaussian with covariance matrix given by Equation 2.18 are Gaussians. What are the widths and how are they related to the elements in Equation 2.18. (a) Calculate the integrals numerically with MATLAB's `integral()` function. (b) Calculate the integral analytically.

11. Generate 27183 random numbers with the MATLAB function `random()` for a (a) Uniform; (b) Exponential; (c) Poisson with mean value 3; (d) Student's $t$–distribution with $\nu = 2$. Inspect the MATLAB `help` for the required parameters. Verify with the `hist()` function that the distributions behave as advertised. Note that `hist()` also returns arrays with the histogram data and the center positions of the bins.

12. Generate (a) 314; (b) 3142; (c) 31416 random numbers, sampled from a Gaussian (normal) distribution with center at zero and $\sigma = 1$. Then calculate the moments from the random numbers and check how well you can recover the input values for center and $\sigma$.

13. You can also use MATLAB's `fit()` function to fit parameters `a`, `b`, `c`, and `d` that parameterize a Gaussian and subsequently plot the result with the commands

```
gauss=@(a,b,c,d,x)a*exp(-((x-b).^2)./(2*c.^2))+d;
result=fit(x',y',fittype(gauss),'Start',[max(y),0.1,1,0])
plot(result,'k',x,y,'k*');
```

Note that you need to supply reasonable initial guesses for the four fit parameters. Use this to fit Gaussians to the data (a) from Exercise 12 and compare the fit parameters `b` and `c` with the moments you calculated there. (b) Analyze the distributions from Exercise 11 by also fitting Gaussians. (c) Implement fitting to a Lorentzian distribution, defined in Exercise 9.

14. (Central limit theorem) Generate $10^5$ random numbers sampled from a Gaussian distribution with $\mu = 0, \sigma = 1$ and calculate the sum of ten consecutive samples, such that there are $10^4$ samples left. Determine the zeroth, first, and second moments of the reduced set of samples. What is the average? The width? How are these values related to the original values?

15. (Central limit theorem) Repeat the previous exercise with a uniform distribution between $-1$ and $+1$.

16. (Failed central limit theorem, but Levy-stable) Repeat the previous exercise with a Lorentzian distribution, which in MATLAB is known as Student's $t$–distribution with $\nu = 1$.

# CHAPTER 3

# Transverse Beam Optics

In the previous chapter we discussed the description of particles and beams. In this chapter we describe how the elements in a beam line affect the beam. Manipulating beam parameters, mostly the beam position, size, or angular divergence in order to satisfy constraints coming from external requirements will be an important task to address. The requirements may come, for example, from an experiment. They may need particularly small beam sizes or particularly parallel beams. In order to do this we need to understand how the different components of an accelerator, in this chapter mostly magnets, influence the beam and how to combine them in a way to satisfy our requirements.

We start by breaking up a large accelerator into distinct elements and therefore need to understand the effect of each beam-line element on the particles. To simplify the discussion we start by considering a single particle only and how its phase-space coordinates $\vec{x} = (x, x', y, y', \tau, \delta)$ change from the entrance of the element to its exit. Thus, we seek to find a map $\mathcal{M}$ from the initial coordinates $\vec{x}_1$ to the those at the end $\vec{x}_2$ such that

$$\vec{x}_2 = \mathcal{M}\vec{x}_1 . \tag{3.1}$$

The entire accelerator is then represented by the concatenation of these maps. In the following we see that for many, and the arguably the most important, magnetic elements, the map $\mathcal{M}$ is linear and can be represented by a matrix.

In order to simplify the presentation further, we focus on the horizontal phase space with $x$ and $x'$ first, but later extend the discussion to comprise the vertical phase space and occasionally the longitudinal phase space when the need arises. The first element we discuss is the space between all other elements, where no external magnetic forces affect a particle. Such a section is commonly called *drift space* but is essentially an empty piece of beam pipe. In these regions the particle naturally travels on a straight line. If we assume that the drift has length $L$, we can express the coordinates at the end as

$$\begin{aligned} x_2 &= x_1 + Lx_1' \\ x_2' &= x_1' \end{aligned} \tag{3.2}$$

which is easy to understand from inspecting Figure 3.1. A particle comes from the right and travels towards the left. Initially it has a distance $x_1$ with respect to the reference trajectory and moves away from it with a positive angle $x_1'$. During the passage the angle does not change, but the distance to the $s-$axis increases linearly, according to the first of Equation 3.2. We see that the final coordinates, bearing a subscript 2, are linear combinations of the initial coordinates, bearing a subscript 1, and therefore we can write Equation 3.2 in

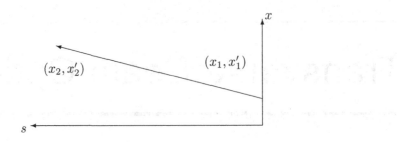

Figure 3.1 Particle trajectory in a drift space. The horizontal axis corresponds to the reference trajectory.

matrix form

$$\begin{pmatrix} x_2 \\ x_2' \end{pmatrix} = \begin{pmatrix} 1 & L \\ 0 & 1 \end{pmatrix} \begin{pmatrix} x_1 \\ x_1' \end{pmatrix}.$$ 

(3.3)

If we consider both horizontal and vertical planes simultaneously we find, by the same argument, that the map from initial to final coordinates is given by

$$\begin{pmatrix} x_2 \\ x_2' \\ y_2 \\ y_2' \\ \tau_2 \\ \delta_2 \end{pmatrix} = \begin{pmatrix} 1 & L & 0 & 0 & 0 & 0 \\ 0 & 1 & 0 & 0 & 0 & 0 \\ 0 & 0 & 1 & L & 0 & 0 \\ 0 & 0 & 0 & 1 & 0 & 0 \\ 0 & 0 & 0 & 0 & 1 & L/c\gamma_0^2 \\ 0 & 0 & 0 & 0 & 0 & 1 \end{pmatrix} \begin{pmatrix} x_1 \\ x_1' \\ y_1 \\ y_1' \\ \tau_1 \\ \delta_1 \end{pmatrix}$$

(3.4)

with copies of the $2 \times 2$ matrix from Equation 3.3 on the upper two (block-) diagonal entries. The $R_{56}$ accounts for the difference in arrival time for particles with different momenta and $\gamma_0$ is the particle's energy in units of its rest mass. It is easy to verify that the matrix for two consecutive drift spaces of length $L_1$ and $L_2$ can be obtained by multiplying the matrix for $L_1$ with that of $L_2$. Observe that in many pictures the particles propagate from the right to the left, which makes writing down the equivalent matrix equations easier, because matrices are usually multiplied from the left to a column vector that represents the particle.

In the following section we will determine the corresponding maps for quadrupole and dipole magnets. It turns out that most of them can be represented by matrices.

## 3.1 MAGNETS AND MATRICES

For a particular element, with potentials $\Phi$ and $\vec{A}$ given, we can expand the Hamiltonian $G$ from Equation 2.3, including the potentials $\Phi$ and $\vec{A}$ up to second order in the dynamical variables $x, x', y, y', \tau, \delta$. Hamilton's equations then lead to linear equations of motion, which can be integrated with elementary methods and the solutions can be converted to transfer matrices. They map the initial values of the dynamical variables of the particles to those at the end of the element. This is what the matrix from Equation 3.4 did for a drift space. However, instead of following the formal procedure, we use geometric reasoning for most beam-line elements, because this leads to a more intuitive picture of the dynamics.

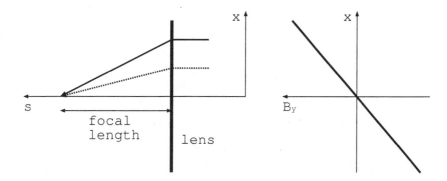

**Figure 3.2** A thin focusing lens (left) deflects parallel rays to a focal point a distance $f$ downstream of the lens. In a magnetic lens, a thin quadrupole, this demands the magnetic field to increase linearly with distance from the optical axis.

### 3.1.1   Thin quadrupoles

We start with a very short quadrupole, also called a *thin quadrupole*. It behaves in a similar way to a thin lens, known from light optics. We discuss long quadrupoles later, but we address thin quadrupoles first because they are very useful for estimating the beam-optical properties of beam lines. The defining property of a thin lens is that it kicks the particle—it changes its angle $x'$—proportional to the distance $x$ from the center of the lens. This behavior is illustrated on the left in Figure 3.2, where two particles come from the right and travel parallel to the axis. They are represented by the solid and the dotted lines. At the lens they receive a downward kick, proportional to their respective distance from the axis, which causes both particles to cross the axis at the same downstream location. The distance from the lens to the crossing point is, of course, the *focal length* of the lens. On the right-hand side in Figure 3.2, the linearly increasing magnetic field that causes the deflection is shown. It is zero on the reference trajectory and increases linearly with transverse position $x$ such that the force on a particle has the required dependence to cause parallel rays to cross the reference trajectory—equivalent to the optical axis—at the same point.

A matrix $Q$ that represents a transverse deflection $\Delta x'$ proportional to the transverse position $x$ is the following

$$\begin{pmatrix} x_2 \\ x'_2 \end{pmatrix} = \begin{pmatrix} 1 & 0 \\ -1/f & 1 \end{pmatrix} \begin{pmatrix} x_1 \\ x'_1 \end{pmatrix}. \tag{3.5}$$

The second equation reads $x'_2 = x'_1 - x_1/f$, which shows the required proportionality. The constant $f$ is the focal length. The choice of sign depends on the convention to assign positive focal length to focusing lenses. That the matrix $Q$ has the advertised property of directing all parallel rays to the focal point is easy to see by concatenating the matrix for the quadrupole with that for the subsequent drift space. In that case we have

$$\begin{aligned} \begin{pmatrix} x_3 \\ x'_3 \end{pmatrix} &= \begin{pmatrix} 1 & L \\ 0 & 1 \end{pmatrix} \begin{pmatrix} 1 & 0 \\ -1/f & 1 \end{pmatrix} \begin{pmatrix} x_1 \\ x'_1 \end{pmatrix} \\ &= \begin{pmatrix} 1 - L/f & L \\ -1/f & 1 \end{pmatrix} \begin{pmatrix} x_1 \\ x'_1 \end{pmatrix}, \end{aligned} \tag{3.6}$$

such that all parallel rays coming from the right, having $x'_1 = 0$, have the transverse position

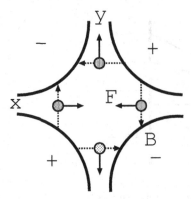

Figure 3.3 Forces in a quadrupole, if a positively charged particle moves into the plane. Note that the force points inwards in the horizontal plane and outwards in the vertical.

$x_3 = (1 - L/f)x_1$ which is zero, independent of the transverse position $x_1$, provided that $L = f$. In other words, all parallel rays cross the optical axis, the reference trajectory, at a distance equal to the focal length of the lens or the quadrupole. Similar to optical lenses, there are both focusing and defocusing quadrupoles, and they differ by the sign of the focal length. A defocusing lens has a negative sign, which is intuitively satisfying because the intersection with the axis lies before the lens for a defocusing lens, and after the lens for a focusing lens. We discuss the close analogy of light optics and charged-particle optics further in Appendix B.4.

Optical lenses are often round and the focal lengths in the horizontal and vertical planes are equal, even though there are exceptions, such as cylindrical lenses, which generate line foci. In magnetic quadrupole lenses, the deflection is generated by the magnetic field which has to obey Maxwell's equations, especially $\nabla \times \vec{B} = 0$. Inside the quadrupole we have $\partial B_y/\partial x = \partial B_x/\partial y$. This causes a magnetic field, linearly rising along the positive $x$−axis, to decrease along the positive $y$−axis. Another way of visualizing this behavior is by looking at the field lines (dotted) and the Lorenz force $F$ (solid) in the quadrupole shown in Figure 3.3. A particle on the horizontal axis is kicked towards the center of the quadrupole (focusing), whereas a particle on the vertical axis is deflected away from the quadrupole center (defocusing). In summary, a quadrupole that focuses in one plane, defocuses in the other plane. The $4 \times 4$ matrix for the two transverse planes that reflects this, is given by

$$\begin{pmatrix} x_2 \\ x_2' \\ y_2 \\ y_2' \end{pmatrix} = \begin{pmatrix} 1 & 0 & 0 & 0 \\ -\frac{1}{f} & 1 & 0 & 0 \\ 0 & 0 & 1 & 0 \\ 0 & 0 & \frac{1}{f} & 1 \end{pmatrix} \begin{pmatrix} x_1 \\ x_1' \\ y_1 \\ y_1' \end{pmatrix} . \tag{3.7}$$

By convention a quadrupole that focuses ($f > 0$) in the horizontal plane and defocuses in the vertical is called a *focusing quadrupole* and *defocusing* if it focuses vertically. Note also that the $4 \times 4$ matrix in Equation 3.7 contains two $2 \times 2$ blocks along the diagonal and $2 \times 2$ zero-matrices in the off-diagonal positions. The zeros in the off-diagonal $2 \times 2$ blocks imply that a normal quadrupole does not couple the horizontal motion in $x$ and the vertical motion in $y$.

### 3.1.2 Thick quadrupoles

The quadrupoles from the previous paragraphs were infinitesimally short, which is only an approximation, but real quadrupoles have a finite length. Like their thin counterparts, they have a vertical magnetic field $B_y$ that rises linearly with transverse distance $x$ from its center. Therefore, the field can be characterized by its gradient $g = \partial B_y / \partial x$ or after normalizing with the particle momentum $p/e$ by

$$k_1 = \frac{\partial B_y / \partial x}{p/e} = \frac{\partial B_y / \partial x}{B\rho} \; , \tag{3.8}$$

where we express the momentum $p$ in engineering units of $B\rho$ and Tm. We now need to work out the transfer matrix that maps the initial coordinates $x$ and $x'$ to those at at the end of the quadrupole. If we consider a thin longitudinal slice of length $\Delta s$ of such a quadrupole we can write for the change of deflection angle in the slice

$$\Delta x' = -k_1 \Delta s x \qquad \text{or} \qquad \frac{\Delta x'}{\Delta s} = -k_1 x \tag{3.9}$$

which, after taking the limit of $\Delta s \to 0$ we can write as

$$x'' + k_1 x = 0 \; . \tag{3.10}$$

This equation describes the motion for the horizontal phase-space coordinate of a particle. For $k_1 > 0$ this equation is solved by $\cos(\sqrt{k_1} s)$ and the corresponding sine function

$$x(s) = A_1 \cos(\sqrt{k_1} s) + A_2 \sin(\sqrt{k_1} s) \; . \tag{3.11}$$

The coefficients can be determined by matching to the initial values $x_1, x'_1$ and we obtain

$$x(s) = x_1 \cos(\sqrt{k_1} s) + \frac{x'_1}{\sqrt{k_1}} \sin(\sqrt{k_1} s) \; . \tag{3.12}$$

At the end of the quadrupole we have $s = l$ and can write for the $2 \times 2$ horizontal transfer-matrix

$$\begin{pmatrix} x_2 \\ x'_2 \end{pmatrix} = \begin{pmatrix} \cos(\sqrt{k_1} l) & \frac{1}{\sqrt{k_1}} \sin(\sqrt{k_1} l) \\ -\sqrt{k_1} \sin(\sqrt{k_1} l) & \cos(\sqrt{k_1} l) \end{pmatrix} \begin{pmatrix} x_1 \\ x'_1 \end{pmatrix} \tag{3.13}$$

that maps the initial coordinates $x_1, x'_1$ to those at the end of the quadrupole. Note that in the limit of a thin quadrupole with $l \to 0$, while keeping $k_1 l$ constant, the transfer matrix approaches that of a thin focusing quadrupole with focal length $k_1 l \to 1/f$.

In the vertical plane the quadrupole is defocusing and the sine of $k_1$ in Equation 3.10 is reversed. In that case we can solve the differential equation in terms of hyperbolic sines and cosines. The corresponding transfer matrix is then given by

$$\begin{pmatrix} y_2 \\ y'_2 \end{pmatrix} = \begin{pmatrix} \cosh(\sqrt{|k_1|} l) & \frac{1}{\sqrt{|k_1|}} \sinh(\sqrt{|k_1|} l) \\ \sqrt{|k_1|} \sinh(\sqrt{|k_1|} l) & \cosh(\sqrt{|k_1|} l) \end{pmatrix} \begin{pmatrix} y_1 \\ y'_1 \end{pmatrix} \tag{3.14}$$

and the $4 \times 4$ matrix can be built by placing the $2 \times 2$ matrices on the diagonal and $2 \times 2$ zero-matrices on the off-diagonal places to arrive at

$$\begin{pmatrix} x_2 \\ x'_2 \\ y_2 \\ y'_2 \end{pmatrix} = \begin{pmatrix} Q_f & 0_2 \\ 0_2 & Q_d \end{pmatrix} \begin{pmatrix} x_1 \\ x'_1 \\ y_1 \\ y'_1 \end{pmatrix} \; , \tag{3.15}$$

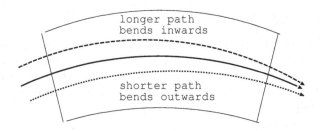

**Figure 3.4** Top-view onto a sector dipole magnet that illustrates weak focusing.

where $Q_f$ is the $2 \times 2$ matrix from Equation 3.13 and $Q_d$ from Equation 3.14. We denote the $2 \times 2$ matrix containing zeros only by $0_2$. If the quadrupole is defocusing, we need to exchange $Q_f$ and $Q_d$ in Equation 3.15.

Now we could already build straight beam-lines consisting of drift spaces and quadrupoles by constructing the matrices and seeing where particles are, depending on their initial conditions and on the strengths of the quadrupoles, but we are confined to straight beam-lines. In order to remedy this deficiency, we consider how dipole magnets affect the phase-space variables of the particles.

### 3.1.3 Sector dipole

The main task of the dipoles is to define the reference orbit as we have seen in the previous chapter, but they also have a small effect of the phase-space variables of the particle-motion *relative to* the reference trajectory.

In Chapter 2 we encountered a sector dipole in which the particles enter the magnet at a right angle with respect to the entrance and exit face of the magnet as shown in Figure 3.4. This implies that the entrance and exit faces are *not* parallel to each other and that the magnet looks like a slice of a pie, as is indicated in Figure 3.4, which shows the top view of a horizontal sector bending magnet with the reference trajectory, shown by a solid line. The upper trajectory, shown by a dashed line, enters the dipole further on the outside of the reference trajectory and experiences a longer magnet. It is therefore deflected more, as is shown in Figure 3.4. The converse is true for the trajectory on the inside, shown by a dotted line. The effect of the dipole is that a particle that deviates from the reference trajectory experiences a kick $\Delta x'$ proportional to its distance from the reference trajectory, which is similar to what happens in a focusing quadrupole and this effect is indeed called *weak focusing*. In order to determine the magnitude of this focusing effect, we consider the difference in bending angle that a particle with distance $x$ from the reference trajectory experiences in an infinitesimally short dipole of length $\Delta s$

$$\Delta x' = \phi(0) - \phi(x) = \frac{\Delta s}{\rho} - \frac{\Delta s}{\rho} \frac{\rho + x}{\rho} = -\frac{\Delta s}{\rho^2} x \; , \tag{3.16}$$

which can be rewritten as

$$x''(s) + \frac{1}{\rho^2} x = 0 \; , \tag{3.17}$$

resulting in an equation similar to that describing a quadrupole, except that $k_1$ is replaced by $1/\rho^2$. In the top left $2 \times 2$ block of the transfer matrix, which describes the horizontal motion, we can therefore use the same transfer matrix as in Equation 3.13 with $k_1$ replaced by $1/\rho^2$. Note that the focusing effect is proportional $\rho^2$ and is therefore independent of

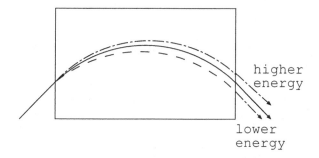

Figure 3.5 Effect of a dipole on particle with different energies.

the sign of the magnetic field $B$. This is particularly important when considering the weak focusing of magnets with alternating magnetic field, such as wigglers, which does *not* cancel on average. Since the vertical magnetic field $B_y$ in an ideal dipole is independent of the vertical position of the particle in the magnet, there is no additional focusing effect in the vertical plane and the transfer matrix for the lower right $2 \times 2$ is that of a drift space as given in Equation 3.3.

Dipole magnets with length $L$ deflect the beam by an angle given by $\phi_0 = BL/(p_0/e)$. Only particles with momentum $p_0$ experience this nominal deflection angle. If either the momentum differs by a small amount, or there is a small deviation $\Delta B$ in the dipole field, the angle is also different

$$\phi(\delta, \Delta B) = \frac{e(B + \Delta B)}{p_0(1 + \delta)} \approx \phi_0 \left(1 + \frac{\Delta B}{B}\right)(1 - \delta) \approx \phi_0 - \phi_0 \left(\delta - \frac{\Delta B}{B}\right) \qquad (3.18)$$

which implies that particles get a small additional kick, either due to a dipole error $\phi_0 \Delta B / B$ or a momentum deviation $-\phi_0 \delta$ with respect to the reference trajectory. We observe that the kick originates from a discrepancy of the momentum and the dipole field. For the time being, however, we focus on the relative momentum error $\delta = \Delta p / p$. Receiving a deflection angle, dependent on the beam energy is just the effect that is desired in a spectrometer, where the particles are sorted transversely, depending on their momentum.

Conceptually, *dispersion* is the trajectory of a particle with momentum offset $\delta$. Let us assume that a number of particles with different momenta are placed on the reference orbit at the start of the beam line. As the particles propagate along the beam line, they stay together on the reference orbit until they enter a dipole magnet, where they start diverging and are being sorted according to their energy. In the subsequent sections of the beam line the off-energy particles will perform oscillations around the reference trajectory. In the dipole magnets, however, they will receive small additional kicks that will change these oscillations with respect to the reference trajectory.

Since the dispersion is "generated in the dipoles," we now look closer at the detailed trajectory inside the dipole and will calculate an extended transfer matrix for the dipole that takes energy offset into account. The additional kick that a particle with momentum offset $\delta$ receives in an infinitesimally short dipole magnet is

$$\Delta x' = \delta \frac{\Delta s}{\rho} , \qquad (3.19)$$

which is the momentum-dependent effect that we need to add to the differential equation that describes the motion in the bending-plane of a dipole Equation 3.17 and we arrive at

$$x'' + \frac{1}{\rho^2}x = \frac{1}{\rho}\delta \ . \tag{3.20}$$

Since this is an ordinary linear differential equation we can assume that the solution is proportional to the momentum offset $\delta$ and introduce the dispersion function $D(s)$ with $x = D\delta$. The differential equation then simplifies to

$$D'' + \frac{1}{\rho^2}D = \frac{1}{\rho} \tag{3.21}$$

which, as can be easily verified, is solved by the Ansatz

$$D(s) = A\cos(s/\rho) + B\sin(s/\rho) + \rho \tag{3.22}$$

with integration constants $A$ and $B$. If the dispersion function and its derivative at the entrance of the dipole are denoted by $D_0$ and $D_0'$ we can express the integration constants in terms of the initial values and write the trajectory of the off-momentum particle as $x = D\delta$ in terms of its initial values as

$$D(s) = D_0\cos(s/\rho) + \rho D_0'\sin(s/\rho) + \rho[1 - \cos(s/\rho)] \tag{3.23}$$

and an equation for $D'(s)$ that can be calculated easily by differentiating Equation 3.23 with respect to $s$.

Since the dispersion function $D(s)$ describes the position of off-momentum particles in the accelerator, it is important in the design of accelerators and, since it stems from a *linear* differential equation, is easily calculated by transfer-matrices. The complete $6 \times 6$ transfer matrix for a sector bend is then given by

$$R = \begin{pmatrix} \cos\phi & \rho\sin\phi & 0 & 0 & 0 & \rho(1-\cos\phi) \\ -\sin(\phi)/\rho & \cos\phi & 0 & 0 & 0 & \sin\phi \\ 0 & 0 & 1 & l & 0 & 0 \\ 0 & 0 & 1 & 1 & 0 & 0 \\ -\sin\phi & -\rho(1-\cos\phi) & 0 & 0 & 1 & L/c\gamma_0^2 \\ 0 & 0 & 0 & 0 & 0 & 1 \end{pmatrix} \tag{3.24}$$

with the bending angle $\phi = L/\rho$. So far we have not motivated the entries in the fifth row, which describe the dependence of the arrival time (or equivalently, longitudinal position in the bunch) on the dispersion at the entrance of the dipole. Clearly, if the incoming dispersion is such that the initial angle $D_0'$ is pointing outwards in the dipole, the trajectory inside the dipole is longer and the particle arrives later, or further back in the bunch. A similar argument holds for $D_0$. The matrix element $R_{56}$ describes the effect that particles with different energies have slightly different speeds, despite being ultra-relativistic, and therefore have different arrival times.

### 3.1.4 Combined function dipole

Apart from the horizontally weak-focusing dipole we just encountered, sector-bends can have a quadrupole-like gradient added. This gradient can, for example, be generated by tilted pole faces, such as those shown in Figure 3.6, which shows an example of a combined function magnet. It is called a combined-function dipole because it serves two purposes:

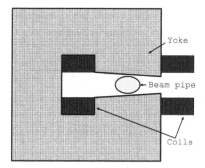

**Figure 3.6** Side view of a combined function dipole. The inclined pole faces in the magnet gap are responsible for a larger field on to the right compared to the left, thus causing an additional gradient on top of the dipole field.

it bends and it focuses similar to a quadrupole. The differential equations that govern the dynamics in a combined dipole is thus described by the following equations

$$x''(s) + \left(k_1 + \frac{1}{\rho^2}\right) x = 0$$

$$y''(s) - k_1 y = 0 \ . \tag{3.25}$$

The $2 \times 2$ transfer matrices that describe focusing in the respective planes of such a bend are given by Equation 3.13 and Equation 3.14, depending on the sign of $k_1 + 1/\rho^2$ and $k_1$.

### 3.1.5 Rectangular dipole

Apart from sector bending magnets, such as that shown in the top-view in Figure 3.4 there are also rectangular bends, so-called RBENDs, which have parallel entrance and exit faces. In such a magnet, the length of the trajectory does not depend on the horizontal offset, as was the case for sector bends, discussed earlier in this section. Therefore, there is no horizontal focusing in a rectangular bend. Of course, a quadrupole gradient can be added by shaping the pole face (see Figure 3.6) or other means such as additional coils.

In a rectangular bend the particles enter the fringe field with a horizontal angle and can interact with the longitudinal component of the $B$–field, which is present in the fringe-field region, because the vertical component $B_y$ varies with $s$ and due to $\partial B_y/\partial s = \partial B_s/\partial y$ also the longitudinal component $B_s$ varies with the vertical distance $y$ to the center of the magnet. A particle, crossing the fringe-field region with a horizontal angle, will therefore experience a vertical component of the Lorentz force. Thus, we find a vertical force that depends on the vertical position.

We will now approximately calculate the magnitude of this effect by first observing that

$$\frac{\partial B_s}{\partial y} = \frac{\partial B_y}{\partial s} \approx \frac{B_y}{g} \tag{3.26}$$

where $g$ is the full gap height. We thus assume that the full vertical field inside the magnet decays to zero over the longitudinal distance of one gap height. Moreover, here we assume that this decay is linear, which is only a crude approximation that is convenient for the calculations. Inside the fringe-field region the longitudinal component therefore can be approximated by $B_s \approx yB_y/g$ and points towards the magnet at its entrance face. The vertical force that a particle experiences is given by the vertical component of the Lorenz-force equation

$$\frac{dp_y}{dt} = ev_x B_s \approx -\frac{\phi}{2}c\frac{B_y}{g}y \, , \tag{3.27}$$

where the angle between the particle trajectory and the $B_s$ is negative and half the deflection angle $\phi$. Changing the time derivative $dt$ to the derivative along the beam line by $ds \approx c\,dt$, we obtain, after integrating over the longitudinal extent of the fringe-field,

$$\Delta p_y = -\frac{\phi}{2}B_y y \, . \tag{3.28}$$

Normalizing by the total momentum $p/e = B\rho$ results in

$$\Delta y' = \frac{\Delta p_y}{p} = -\frac{\phi}{2}\frac{y}{\rho} \, . \tag{3.29}$$

Rewriting this in terms of a transfer matrix, we find

$$\begin{pmatrix} y_2 \\ y_2' \end{pmatrix} = \begin{pmatrix} 1 & 0 \\ -\frac{\tan(\phi/2)}{2\rho} & 1 \end{pmatrix} \begin{pmatrix} y_1 \\ y_1' \end{pmatrix} \, , \tag{3.30}$$

where we replaced the approximate value of the deflection angle $\phi/2$ by $\tan(\phi/2)$, which follows from a more careful treatment that is, for example, shown in [11]. The matrix in Equation 3.30 is the map from just outside the fringe field to just inside the magnet. The same effect will affect the particle on its way out of the rectangular bend and therefore the combined effect of a rectangular bend with length $l$ in the non-deflecting plane is given by

$$R_y = \begin{pmatrix} 1 & 0 \\ -\frac{\tan(\phi/2)}{2\rho} & 1 \end{pmatrix} \begin{pmatrix} 1 & l \\ 0 & 1 \end{pmatrix} \begin{pmatrix} 1 & 0 \\ -\frac{\tan(\phi/2)}{2\rho} & 1 \end{pmatrix} \, , \tag{3.31}$$

which is used in beam optics codes.

### 3.1.6 Coordinate rotation

The next, this time, "virtual element" is a *coordinate rotation* around the $s-$axis that we use, for example, to create a vertically bending dipole magnet from a horizontally bending

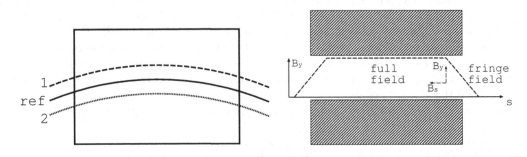

**Figure 3.7** Top-view (left) of a rectangular bending magnet and three horizontal trajectories. Side-view (right) of a rectangular bending magnet and the vertical magnetic field profile. Note the fringe field.

one. We can also create a so-called *skew-quadrupole* by rotating a normal quadrupole by 45 degrees, or $\pi/4$ in radian, around the longitudinal $s-$axis. Rotating in the $xy-$plane with an angle $\phi$ is simply achieved by a normal coordinate rotation

$$\bar{x} = x\cos\phi + y\sin\phi \quad \text{and} \quad \bar{y} = x(-\sin\phi) + y\cos\phi \tag{3.32}$$

and differentiating with respect to $s$ we find

$$\bar{x}' = x'\cos\phi + y'\sin\phi \quad \text{and} \quad \bar{y}' = x'(-\sin\phi) + y'\cos\phi \tag{3.33}$$

which leads to the transfer matrix

$$\begin{pmatrix} x_2 \\ x_2' \\ y_2 \\ y_2' \end{pmatrix} = \begin{pmatrix} \cos\phi & 0 & \sin\phi & 0 \\ 0 & \cos\phi & 0 & \sin\phi \\ -\sin\phi & 0 & \cos\phi & 0 \\ 0 & -\sin\phi & 0 & \cos\phi \end{pmatrix} \begin{pmatrix} x_1 \\ x_1' \\ y_1 \\ y_1' \end{pmatrix}. \tag{3.34}$$

If we denote the $4 \times 4$ transfer matrix by $R_r(\phi_r)$, the transfer matrix $Q_s$ for a skew quadrupole is given by

$$Q_s = R_r(-\pi/4) \begin{pmatrix} Q_f & 0_2 \\ 0_2 & Q_d \end{pmatrix} R_r(\pi/4) \tag{3.35}$$

where $Q_f$ and $Q_d$ are the $2 \times 2$ matrices from Equations 3.13 and 3.14. Vertically deflecting dipoles are described by sandwiching the matrix for a horizontally deflecting dipole between coordinate rotations for angles of $\pi/2$ and $-\pi/2$. Note that the angle $\phi_r$ is often referred to as a *roll-angle*.

### 3.1.7  Solenoid

In detectors for nuclear or high-energy physics experiments a longitudinal magnetic field is used to determine of the momenta of the collision products. This magnet, with longitudinal field $B_s$, is referred to as a *solenoid*. It also affects the beam particles, which, if they enter the solenoid at an angle, will follow helical trajectories inside the magnet. The reduction of the longitudinal field component $B_s$ near the magnet ends causes transverse field components $B_r$ to appear that focus the beam particles. The combined effect of the ends and the bulk of the magnet can be described [12] by a $4 \times 4$–transfer matrix that is the product of a matrix that describes focusing with strength $k_s = B_s/B\rho$ in both transverse planes and coordinate rotation $R_r(\phi_s)$ with the magnet length $L$ and $\phi_s = k_s L/2$. For the matrix we then find

$$R_s = R_r(\phi_s) \begin{pmatrix} Q_s & 0 \\ 0 & Q_s \end{pmatrix} \quad \text{with} \quad Q_s = \begin{pmatrix} \cos(\phi_s) & 2\sin(\phi_s)/k_s \\ -k_s\sin(\phi_s)/2 & \cos(\phi_s) \end{pmatrix}. \tag{3.36}$$

Note that the matrices in the previous equation commute and their order does not matter. Moreover, we observe that solenoids couple the transverse planes, just as skew quadrupoles or coordinate rotations do.

### 3.1.8  Non-linear elements

Apart from the magnets that can be represented by matrices, there are magnetic fields that have a *non-linear* dependence on the transverse coordinates $x$ and $y$. Some of these fields are due to errors in other magnets and others are due to magnets deliberately installed in the accelerator in order to correct some perturbation. For example, in Chapter 8 we will use sextupoles to correct the chromaticity. Here we will only consider magnetic fields that can be represented by a non-linear kick and in Section 4.2 we will see that they can be represented by $B_y + iB_x = B_0 \sum_{m=1}^{\infty} (b_m + ia_m)((x + iy)/R_0)^{m-1}$ with some reference field $B_0$, reference radius $R_0$, and multipole coefficients: $b_m$ for upright magnets and $a_m$ for skew magnets. Here $m = 2$ characterizes upright and skew thin-lens quadrupoles and $m = 3$ characterizes sextupoles, both upright ($b_3$) and skew ($a_3$). Octupoles, decapoles and higher multipoles follow the same scheme. When a beam particle traverses such a short magnet, it receives a transverse kick due to the Lorentz force given by $\Delta x' - i\Delta y' = \pm(B_y + iB_x)L/B\rho$, where $\pm$ represents the ambiguity originating from the charge of the particle; the same field kicks electrons and protons in opposite directions. This ambiguity is resolved by requiring the kick $\Delta x' < 0$ in a *horizontally focusing* quadrupole to be negative for positions with $x > 0$. The minus signs on the left-hand side indicate that a magnetic field in the positive horizontal direction kicks the particles downwards, $B\rho$ denotes the momentum of the particle, and $(B_y + iB_x)L$ is the field integrated over the length $L$ of the magnet. The kick that a particle receives is usually parameterized by a quantity $k_n L = -(\partial^n B_y/\partial x^n + i\partial^n B_x/\partial x^n)L/B\rho$ that can be related to Equation 4.13 and we find

$$\Delta x' - i\Delta y' = -\sum_{n=0}^{\infty} \frac{k_n L}{n!}(x + iy)^n = \frac{B_0 L}{B\rho} \sum_{m=1}^{\infty} (b_m + ia_m)\left(\frac{x + iy}{R_0}\right)^{m-1} . \tag{3.37}$$

Comparing coefficients we obtain

$$\frac{k_n L}{n!} = -\frac{(B_0/R_0^n)L}{B\rho}(b_{n+1} + ia_{n+1}) , \tag{3.38}$$

which facilitates translating between the different conventions to characterize non-linear magnetic fields. We will later use it in Chapter 11 to investigate the effect of the non-linear elements on the beam dynamics.

Now we have a well-filled toolbox of maps for non-linearities and, in particular matrices to describe the most common elements in an accelerator and it is time use it.

## 3.2 PROPAGATING PARTICLES AND BEAMS

In Chapter 2 we found that well-behaved distribution functions can be efficiently described by their first few moments, namely the zeroth moment, the particle number; the first moments, the centroids; and the second moments, the beam sizes. Here we discuss how the moments of the beam distribution propagate through a beam line. Once we can do this we have reasonably complete information about the behavior of the beam everywhere in the accelerator.

We start by considering how a single particle propagates through a single element or through an entire beam line, as described by a transfer matrix $R$, and then calculate the moments of the beam distribution at the end of the beam line by averaging the final coordinates over the initial distribution. To clarify this approach, we work this out in detail and describe the initial phase-space vector by $\vec{x} = (x_1, \cdots, x_n)$, where $n$ can be any number, but most often it will be 2, 4, or 6. The individual particle is assumed to propagate according to

$$\bar{x}_i = \sum_{j=1}^{n} R_{ij} x_j \tag{3.39}$$

if written in component form. We mark the particle coordinates at the end of the beam line with a bar. For averaging over the initial distribution function we use angle brackets. Averaging Equation 3.39 leads to

$$\bar{X}_i = \langle \bar{x}_i \rangle = \left\langle \sum_{j=1}^{n} R_{ij} x_j \right\rangle = \sum_{j=1}^{n} R_{ij} \langle x_j \rangle = \sum_{j=1}^{n} R_{ij} X_j \; . \tag{3.40}$$

The first equality is the definition of $\bar{X}_i$ as the ensemble average of the final coordinates $\bar{x}_i$. In the second equality we express it through the transfer matrix $R$ and initial coordinates $x_j$. Since the transfer matrix is the same for all particles and the summing is just a linear operation, we can pull the sum and $R$ from the average, such that only the average over the initial coordinates $x_j$ is left. In summary, Equation 3.40 states that the centroids $X_i$ propagate in the same way individual particles do, which is convenient, because we can use the single particle dynamics to describe the behavior of averages of a large ensemble of particles. In particular, the beam position monitors, which we will discuss further later in Chapter 7, are sensitive to the centroid of the beam motion, and we can model these measurements using the transfer matrices that were originally derived to describe the motion of single particles.

We now turn to the second moments and how they propagate in a beam line defined by transfer matrix $R$. The sigma matrix is in general defined by the *central second moments* of the distribution. "Central" in this context means that the centroid motion is subtracted. The sigma- or beam-matrix is then given by

$$\sigma_{ij} = \langle (x_i - X_i)(x_j - X_j) \rangle \; , \tag{3.41}$$

which is consistent with Equation 2.17 on page 22. In the remainder of this section we will, for the sake of simplifying the equations, assume that the centroid of the distribution is located on the reference trajectory, i.e., $X_i = 0$. The sigma matrix at the end of the beam line $\bar{\sigma}$ is then given by

$$\bar{\sigma}_{ij} = \langle \bar{x}_i \bar{x}_j \rangle = \left\langle \sum_{k=1}^{n} R_{ik} x_k \sum_{l=1}^{n} R_{jl} x_l \right\rangle = \sum_{k=1}^{n} R_{ik} \sum_{l=1}^{n} R_{jl} \langle x_k x_l \rangle = \sum_{k=1}^{n} \sum_{l=1}^{n} R_{ik} R_{jl} \sigma_{kl} \tag{3.42}$$

in terms of the initial sigma matrix $\sigma$ and the transfer matrix $R$. The first equality is just the definition of $\bar{\sigma}_{ij}$ and we exploit the fact that the transfer matrix is the same for all particles and the sums are linear to pull them out of the average. Note that Equation 3.42 is given in component form. Written in matrix form we find

$$\bar{\sigma} = R \sigma R^T \; , \tag{3.43}$$

where $R^T$ denotes the transpose of matrix $R$. In the calculation we have, strictly speaking, only shown that the sigma matrix propagates with Equation 3.43 if $\vec{X} = 0$, but with a little more effort it is straightforward to show that Equation 3.43 also holds when using the full definition of the sigma matrix from Equation 3.41.

Equations 3.40 and 3.43 enable us to propagate a beam, which is characterized by its first and second moments, through any beam line that is defined by the transfer matrices for all its elements. This method is implemented in many beam transport codes, starting from TRANSPORT [4] and MADX [3] to many others. We stress the importance of the sigma matrix, because it carries all the information about the beam properties such as beam size $\sigma_{11} = \sigma_x^2$ or angular divergence $\sigma_{22} = \sigma_{x'}^2$ throughout the accelerator.

After we know how to describe beam-line elements by matrices and the particles and beams by vectors and matrices, we are ready explore how these quantities move in an accelerator.

## 3.3 TWO-DIMENSIONAL

In this section, we confine ourselves to the horizontal transverse dimension and to very simple elements. We illustrate all calculations with examples coded in MATLAB or Octave.

### 3.3.1 Beam optics in MATLAB

We start by analyzing a simple beam line that consists of twenty straight FODO cells, similar to those we used in Chapter 2, but without dipole magnets. In order to simplify the calculations, we use thin quadrupoles instead of long ones. In that case we only need $2 \times 2$ matrices for drift spaces and for thin quadrupoles. We therefore write MATLAB functions that return the respective transfer matrices. The file for the drift space, we call it DD.m, is particularly simple

```
% DD.m, drift space
function out=DD(L)
out=[1,L;0,1];
```

It only receives one parameter, the length of the drift space L, as input and returns the $2 \times 2$ transfer-matrix for the element as parameter out. The function that returns the matrix for a thin quadrupole, named Q.m, is not much more difficult.

```
% Q.m, thin quadrupole
function out=Q(F)
out=eye(2);
if abs(F)<1e-8 return; end
out=[1,0;-1/F,1];
```

It works very similar to the one for the drift space, except that it receives the focal length F as input and returns the thin lens matrix for a quadrupole in the variable out unless the focal length is too small. In that case the unit matrix is returned.

Based on these functions for the transfer matrices we are ready to build a first beam transport code. We give it the name beamoptics1.m.

```
% beamoptics1.m, V. Ziemann, 181017
clear all; close all
ndim=2;   % 2 for 2x2 matrices
F=2.1;    % focal length of the quadrupoles
fodo=[ 1,  5,  0.2,  0;    % 5*0.2 m
       2,  1,  0.0, -F;    % QD
       1, 10,  0.2,  0;    % 10*0.2 m
       2,  1,  0.0,  F;    % QF/2
       1,  5,  0.2,  0];   % 5*0.2 m
beamline=repmat(fodo,20,1);     % name must be 'beamline'
nlines=size(beamline,1);        % number of lines in beamline
nmat=sum(beamline(:,ndim))+1;   % sum over repeat-count in column 2
Racc=zeros(ndim,ndim,nmat);     % matrices from start to element-end
Racc(:,:,1)=eye(ndim);          % initialize first with unit matrix
spos=zeros(nmat,1);             % longitudinal position
ic=1;                           % element counter
for line=1:nlines               % loop over input elements
   for seg=1:beamline(line,2)   % loop over repeat-count
```

```
    ic=ic+1;                    % next element
    Rcurr=eye(2);               % matrix in next element
    switch beamline(line,1)
      case 1   % drift
        Rcurr=DD(beamline(line,3));
      case 2   % thin quadrupole
        Rcurr=Q(beamline(line,4));
      otherwise
        disp('unsupported code')
    end
    Racc(:,:,ic)=Rcurr*Racc(:,:,ic-1);    % concatenate
    spos(ic)=spos(ic-1)+beamline(line,3); % position of element
  end
end
x0=[0.001;0];               % 1 mm offset at start
data=zeros(1,nmat);         % allocate memory
for k=1:nmat
  x=Racc(:,:,k)*x0;
  data(k)=x(1);             % store the position
end
plot(spos,data,'k'); xlabel('s [m]'); ylabel(' x [m]');
```

At the top of the script we clear the workspace, close all graphics windows, and define a parameter `ndim` indicating that we work with $2 \times 2$ matrices. It is used to create arrays with the right dimensionality. Next we define a parameter `F` that we use to set the focal length of the thin quadrupole and define the lattice `fodo` using the same syntax we used in Chapter 2, where we used the element code 2 to represent the thin quadrupole. It must have length zero and the focal length is specified in the fourth column. Note the comments following the % after each element description. MATLAB ignores everything following the percent sign.

Following the definition of `fodo` we make 20 consecutive copies of it with the `repmat()` function that concatenates 20 copies of `fodo` and copies it to the variable `beamline`. Once the complete `beamline` is assembled, we need to allocate arrays to hold all quantities we will calculate. For this we first determine how many elements the `beamline` contains and store that in the variables `nlines`. Likewise, taking the repeat-count of the elements into account, we determine the total number of matrices `nmap`. The array `Racc` will contain all transfer-matrices from the start to after each element, such that `R(:,:,1)` holds the transfer matrix from just before the start to itself and is initialized with the unit matrix, `eye(ndim)`. For example, `R(:,:,6)` holds the matrix from the start of the `beamline` to just after the element number 5, which is immediately upstream of the first quadrupole. Note that the first drift-space is subdivided in five subsections such that it counts as 5. The array `spos` holds the distance from the start of `beamline` to immediately after the respective element. For example, `spos(5)` returns 0.8 m.

After initializing the element counter `ic,` we are ready to step through the beam-line by iterating over the `lines` and over the segments `seg` that represent the repeat-count of the elements. Inside the loop, we first increment the element-counter `ic` and then `switch` according to the element-code stored in the first column of the beam-line description `beamline`. If it is 1, we use the previously defined function `DD.m` for a drift-space with length defined in the third column of the current `line` and assign the transfer-matrix to `Rcurr`. Likewise, if the element-code is 2, we set `Rcurr` to the transfer-matrix for a thin quadrupole with

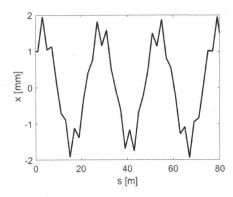 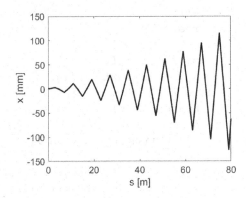

Figure 3.8 The transverse position of a particle that oscillates along the beam-line. On the left we have a stable beam-line with proper oscillations and on the right the oscillations are increasing, which indicates an unstable beam-line.

the focal length specified in the fourth column of `beamline`. If a code is `unknown,` a short message is printed. Later we will extend the list of known elements to comprise all those discussed in the previous section and will also extend the number of dimensions. At this point in the code, we know the transfer matrix for the current segment and we use it to update `Racc`, the array of accumulated transfer-matrices. Here we left-multiply the previously accumulated transfer-matrix in `Racc(:,:,ic-1)` with the current transfer-matrix `Rcurr`. In the same way, we fill the next position in the array `spos` by adding the length of the current segment.

As a first example we simply launch a particle with a position offset of 1 mm (`x0=[0.001;0]`), allocate an array in which to store the `data`, and loop over all positions. Finally, we plot the positions after each location, and annotate the axes. The code is shown in the last few lines of the example. Executing the script produces the plot shown on the left-hand side in Figure 3.8. We observe an oscillation along the 20 FODO-cells with an amplitude of up to 2 mm, which is larger than the initial starting amplitude x0 because the first quadrupole is defocusing and kicks the particle to larger amplitudes before it is bent back towards the reference orbit in the following focusing quadrupole. In this way the particle receives one kick after the other in each quadrupole it traverses and oscillates around the reference orbit until the end of the beam line.

On the left-hand side of Figure 3.8 the oscillation is stable, but if we decrease the focal length in the example code from 2.1 m to 0.999 m and run the script once again we find the oscillation with increasing amplitudes, shown on the right-hand side in Figure 3.8, which indicates an unstable beam line. The lesson we pick up from this exercise is that it is possible to set the quadrupoles in such a way that an accelerator is unstable and this normally leads to beam loss. Exploring possible values for the focal lengths $f$ further, we find that values with $f < -1$ m and $f > 1$ m lead to stable oscillations.

### 3.3.2 Poincarè section and tune

Let us continue to explore stable oscillations by observing the phase-space variables $x$ and $x'$ after every traversal through the FODO cell. This leads to a stroboscopic view of the particle motion, called a *Poincarè section* or phase-space portrait. To do this we change the

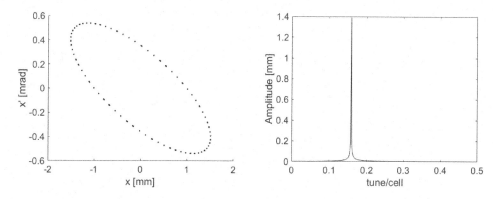

Figure 3.9 The phase-space of a particle and the FFT.

example code to only use a single FODO cell, instead of 20, by changing the definition of beamline near the top of the script to beamline=fodo; and replace the last three lines in the code by the following

```
hold on; xx=[0.001;0];  % 1 mm starting position
for k=1:100
  xx=Rturn*xx;
  plot(1e3*xx(1),1e3*xx(2),'.');
end
xlabel('x [mm]'); ylabel('x'' [mrad]');
```

The command hold on allows us to add multiple data points in the same plot without overwriting old ones. Then we define the initial conditions for the particle in the array xx[]. In the loop over k we map the phase-space state of the particle, as represented by xx, from one turn to the next for 100 iterations. In each iteration we place a dot on the plot and in the end, label the axes appropriately. We multiply the axes by 1000 in order to use the more appropriate scales mm and mrad. After the loop, we correspondingly label the axes and obtain the plot shown on the left-hand side in Figure 3.9, where we observe that the phase-space coordinates of the particle all lie on an ellipse and this is a strong indication that the dynamics of particles in accelerators is very similar to that of harmonic oscillators, well-known from elementary mechanics, whose phase-space portraits are also ellipses.

Since ellipses will play an important role further along in the discussion, we briefly digress and discuss transforming them into a canonical form. Ellipses are conic sections that can be represented by quadratic forms in the two phase-space coordinates $x$ and $x'$, such that we can write

$$2J = \gamma x^2 + 2\alpha x x' + \beta x'^2 \qquad (3.44)$$

with, at present, arbitrary coefficients $\alpha, \beta$, and $\gamma$ to describe the shape and orientation of the ellipse and a parameter $2J$ that describes its magnitude. Of course we choose a nomenclature that is consistent with convention. We now try to find a linear transformation with a unit determinant that transforms the phase-space coordinates $x$ and $x'$ to $\tilde{x}$ and $\tilde{x}'$, which transform the equation for the ellipse to that of a circle in the new variables. For the transformation between the variables we assume the form

$$\begin{pmatrix} x \\ x' \end{pmatrix} = \begin{pmatrix} a & 0 \\ b & c \end{pmatrix} \begin{pmatrix} \tilde{x} \\ \tilde{x}' \end{pmatrix} \qquad (3.45)$$

with determinant $ac = 1$. Inserting into Equation 3.44 leads to

$$2J = \left[\gamma a^2 + 2\alpha ab + \beta b^2\right] \tilde{x}^2 + 2\left[\alpha ac + \beta bc\right] \tilde{x}\tilde{x}' + \left[\beta c^2\right] \tilde{x}'^2 . \tag{3.46}$$

The requirement that this equation describes a circle amounts to the three conditions

$$1 = \gamma a^2 + 2\alpha ab + \beta b^2 , \quad 0 = \alpha ac + \beta bc , \quad \text{and} \quad 1 = \beta c^2 . \tag{3.47}$$

Solving these equations for $a, b,$ and $c$ leads to $a = \sqrt{\beta}$, $b = -\alpha/\sqrt{\beta}$, and $c = 1/\sqrt{\beta}$ and the consistency condition $\gamma = (1 + \alpha^2)/\beta$. The transformation matrix from Equation 3.45 that transforms the ellipses to circles then assumes the form

$$\begin{pmatrix} x \\ x' \end{pmatrix} = \begin{pmatrix} \sqrt{\beta} & 0 \\ -\alpha/\sqrt{\beta} & 1/\sqrt{\beta} \end{pmatrix} \begin{pmatrix} \tilde{x} \\ \tilde{x}' \end{pmatrix} . \tag{3.48}$$

The parameters appearing in Equation 3.44 are the well-known Twiss-parameters $\alpha, \beta, \gamma$ and the action, or Courant-Snyder invariant, $J$. Here the former describe the shape of the ellipse that a particle traces out, and the latter describes the size. The coordinate system in which the motion is confined to a circle is called *normalized phase-space,* a notion we will frequently encounter in the following sections. But back to numerically investigating the motion.

Since harmonic oscillators are characterized by their frequency, it is prudent to determine the oscillation frequency of our accelerator as well and we do so by extending the loop over N=1024 iterations and saving the position value xx(1) after every iteration in the array xpos. After the loop we calculate the Fourier-transform and plot the result with

```
y=2*abs(fft(xpos))/N;
plot((1:N/2)/N,y(1:N/2),'k');
```

which leads to the plot shown on the right-hand side in Figure 3.9, where a peak with amplitude of $1.4 \times 10^{-3}$ is visible near 0.16. This indicates that a fraction of about 0.16 of a full oscillation happens within one traversal of the FODO cell. The amplitude coincides with the maximum excursion in $x$ in the phase-space plot on the left-hand side in Figure 3.9.

We point out that harmonic oscillations are characteristic of slightly perturbed stable systems in equilibrium. "Equilibrium" implies that all external forces are balanced and add up to zero. If we consider an equivalent mechanical system, this means that the system is characterized by a potential function with zero first derivatives. The first non-zero term of the Taylor-series expansion of the potential can only be quadratic. This, in turn, implies that the motion derived from the potential is harmonic with system-specific eigenfrequencies. In accelerators these frequencies are called *tunes.*

### 3.3.3 FODO cell and beta functions

We continue the investigation by calculating the full-turn matrix $\hat{R}$ =Rturn for our simple system by hand from the matrices, such that we have

$$\hat{R} = D(L/2)Q(f)D(L)Q(-f)D(L/2) \tag{3.49}$$

where $D(L)$ is the matrix for a drift-space from Equation 3.3 and $Q(f)$ for a thin quadrupole from Equation 3.5. The same matrices are used in the scripts DD.m and Q.m, respectively. Here we use $L$ to represent the space between quadrupoles, such that the length of one FODO cell is $2L$. Inserting the matrices and evaluating the multiplications leads to

$$\hat{R} = \begin{pmatrix} 1 - \frac{L}{f} - \frac{L^2}{2f^2} & 2L - \frac{L^3}{4f^2} \\ -\frac{L}{f^2} & 1 + \frac{L}{f} - \frac{L^2}{2f^2} \end{pmatrix} , \tag{3.50}$$

which gives us a useful representation of the transfer-matrix for one cell in terms of geometrical quantities, the length $L$ and focal length $f$, which will later aid the discussion. Note that the determinant of the matrix is unity, because the matrices $D$ and $Q$ have a unit determinant.

The observation that there is an oscillation at the bottom suggests to decompose the full-turn transfer-matrix $R$ =Rturn into a matrix $\mathcal{A}$ that stretches and twists the coordinate-axes of the phase-space, a rotation matrix $\mathcal{O}$ and the inverse of $\mathcal{A}$ such that we write $\hat{R} = \mathcal{A}^{-1}\mathcal{O}\mathcal{A}$, where all matrices have a unit determinant as a consequence that $\hat{R}$ has a unit determinant. Therefore, only three of the four matrix-elements are independent. Since the rotation matrix $\mathcal{O}$ depends on one parameter, the rotation angle $\mu$, the matrix $\mathcal{A}$ depends on two more. Here we chose a form of the matrix that is inspired by the discussion that led to Equation 3.48. That matrix already maps a circle to an ellipse. Here we require the inverse operation, namely to map from the regular phase space $x$ and $x'$ to normalized phase space, where the motion is represented by a circle. Therefore we use the inverse matrix from Equation 3.48 on the right and the matrix itself on the left of the following equation

$$\hat{R} = \begin{pmatrix} \sqrt{\beta} & 0 \\ -\frac{\alpha}{\sqrt{\beta}} & \frac{1}{\sqrt{\beta}} \end{pmatrix} \begin{pmatrix} \cos\mu & \sin\mu \\ -\sin\mu & \cos\mu \end{pmatrix} \begin{pmatrix} \frac{1}{\sqrt{\beta}} & 0 \\ \frac{\alpha}{\sqrt{\beta}} & \sqrt{\beta} \end{pmatrix}$$

$$= \begin{pmatrix} \cos\mu + \alpha\sin\mu & \beta\sin\mu \\ -\frac{1+\alpha^2}{\beta}\sin\mu & \cos\mu - \alpha\sin\mu \end{pmatrix} . \tag{3.51}$$

with the definitions

$$\mathcal{O} = \begin{pmatrix} \cos\mu & \sin\mu \\ -\sin\mu & \cos\mu \end{pmatrix} \quad \text{and} \quad \mathcal{A} = \begin{pmatrix} \frac{1}{\sqrt{\beta}} & 0 \\ \frac{\alpha}{\sqrt{\beta}} & \sqrt{\beta} \end{pmatrix} . \tag{3.52}$$

The parameterization shown in the first line of Equation 3.51 has a simple interpretation. First we apply a coordinate transformation by the matrix with $\alpha$ and $\beta$, then we apply a rotation, followed by the inverse coordinate transformation. The matrix $\mathcal{A}$ is thus just an affine transformation that rescales and changes the angle of the coordinate axis. Since the motion in the new coordinate system is a circle, it is common to call this procedure "transforming into normalized phase space." Here $\beta$ and $\alpha$ are the commonly used *Twiss-parameters* and $\mu$ is sometimes called the *phase-advance*. In particular, $\beta$ is the ubiquitous beta function. It is a function of the position $s$, because it has a different value, depending on where we define the start of the FODO-cell. In this example we start in the middle of a drift space, but if we start immediately after a quadrupole, its value will be different.

Now we have two ways to express the transfer-matrix through one FODO-cell, either by using the "hardware" parameters $L$ and $f$ in Equation 3.50 or by using the parameterization using $\mu, \beta$, and $\alpha$. So, it is possible to express the latter parameters in terms of $L$ and $f$. Equating the matrices we get

$$\begin{pmatrix} 1 - \frac{L}{f} - \frac{L^2}{2f^2} & 2L - \frac{L^3}{4f^2} \\ -\frac{L}{f^2} & 1 + \frac{L}{f} - \frac{L^2}{2f^2} \end{pmatrix} = \begin{pmatrix} \cos\mu + \alpha\sin\mu & \beta\sin\mu \\ -\frac{1+\alpha^2}{\beta}\sin\mu & \cos\mu - \alpha\sin\mu \end{pmatrix} \tag{3.53}$$

and from adding the diagonal elements we find

$$\cos\mu = 1 - \frac{L^2}{2f^2} \tag{3.54}$$

and using the trigonometric equation $1 - \cos\mu = 2\sin^2(\mu/2)$ we obtain

$$\sin(\mu/2) = \frac{L}{2f} . \tag{3.55}$$

From the difference of the diagonal elements we obtain $\alpha$

$$2\alpha \sin \mu = -\frac{2L}{f} \quad \text{or} \quad \alpha = -\frac{L}{f \sin \mu} \ . \tag{3.56}$$

After some manipulations involving trigonometric functions, we find

$$\alpha = -\frac{1}{\sqrt{1 - L^2/4f^2}} \ . \tag{3.57}$$

From comparing the 12-elements of the transfer-matrices, we obtain

$$\beta \sin \mu = 2L - \frac{L^3}{4f^2} \tag{3.58}$$

which leads to

$$\beta = f\frac{2 - L^2/4f^2}{\sqrt{1 - L^2/4f^2}} \ . \tag{3.59}$$

Note that $\alpha$ and $\beta$ are calculated at the starting-point, which also equals the end-point of the periodic cell, as given in Equation 3.49. This description is also consistent with the one given by the array fodo in the MATLAB example.

In case we consider not only a single cell, but an entire circular accelerator, the full-turn transfer matrix is composed of many individual matrices for the elements of the ring, but we still have a full-turn matrix $\hat{R}$ available and can perform the analysis to determine $\alpha, \beta$ and the phase advance $\mu$ from $\hat{R}$. First, we calculate $\mu$ from the trace of the matrix $\hat{R}$, then $\beta$ from $\hat{R}_{12}$, and finally $\alpha$ from the difference of the diagonal elements as

$$\mu = \arccos\left(\frac{\hat{R}_{11} + \hat{R}_{22}}{2}\right) \ , \quad \beta = \frac{\hat{R}_{12}}{\sin \mu} \ , \quad \text{and} \quad \alpha = \frac{\hat{R}_{11} - \hat{R}_{22}}{2 \sin \mu} \ . \tag{3.60}$$

When $\hat{R}$ describes an entire ring the phase advance $\mu$ in units of $2\pi$ is called the *tune* $Q = \mu/2\pi$ of the ring. It describes the number of transverse oscillations performed by a particle on its journey around the ring. Note, however, that we can only determine the fractional part of the tune in this way because we only have information about the position of particles once per turn. Observing at a single point in the ring, we cannot know the integer number of oscillations the particle performed. The fractional part, however is crucial for the stability and robustness of operating a ring. We will return to this topic later.

Next, we relate these calculations to earlier observations from our numerical experiments. We see that the beta-function grows without bounds with increasing focal length which describes increasingly weaker quadrupoles. Moreover, as $f$ is smaller than $L/2$, the cosine in Equation 3.54 is less than its smallest permissible value of $-1$ and the square root in the denominator of Equation 3.57 and 3.59 become imaginary. This observation is consistent with the finding that the oscillations become unstable if F becomes less than 1 in the script and that is half the distance between quadrupoles in the beam-line definition of fodo. We thus conclude that a beam-line is stable if the sum of the diagonal elements, the trace of the full-turn matrix, lies between $-2$ and $2$, which is referred to as the *stability criterion* for transfer matrices describing rings.

It remains to compare the phase advance $\mu$ to the numerical examples where we found that the phase advance was around 0.16 if the focal length is $f = 2.1$ m. Inserting this value, together with $L = 2$ m in Equation 3.55 we obtain

$$\mu = 2 \arcsin(L/2f) = 2\pi \times 0.158 \tag{3.61}$$

which nicely shows the consistency of this analysis with the numerical experiments.

### 3.3.4 A complementary look at beta functions

We can exploit the view that there is "a harmonic oscillator at the bottom of every stable system" by realizing the similarity of focusing in a periodic beam-line as represented by its focusing strength $k_1(s) = k_1(s + L)$ with a harmonic oscillator that has a periodically varying spring-constant. Guided by this analogy we may search for quasi-periodic solutions. To do so, we start from the equation of motion for a single particle

$$x'' + k_1(s)x = 0 \, , \tag{3.62}$$

where $k_1(s)$ is entirely determined by the magnetic setup, the lattice. We then make a quasi-periodic Ansatz with functions $u(s)$ and $\psi(s)$ and integration constants $A$ and $\phi_0$

$$x(s) = Au(s)\cos(\psi(s) + \phi_0) \tag{3.63}$$

and calculate the derivatives

$$
\begin{aligned}
x' &= Au'\cos(\psi(s) + \phi_0) - Au\psi'\sin(\psi(s) + \phi_0) \\
x'' &= A\cos(\psi(s) + \phi_0)\left[u'' - u\psi'^2\right] - A\sin(\psi(s) + \phi_0)\left[2u'\psi' + u\psi''\right]
\end{aligned}
\tag{3.64}
$$

and, after inserting in the equation of motion, we collect terms in front of the sine and cosine

$$
\begin{aligned}
0 &= A\cos(\psi(s) + \phi_0)\left[u'' - u\psi'^2 + k_1(s)u\right] \\
&\quad - A\sin(\psi(s) + \phi_0)\left[2u'\psi' + u\psi''\right]
\end{aligned}
\tag{3.65}
$$

which implies the following conditions among the functions $u(s)$ and $\psi(s)$

$$0 = u'' - u\psi'^2 + k_1(s)u \qquad \text{and} \qquad 0 = 2u'\psi' + u\psi'' \, . \tag{3.66}$$

The second equation leads to

$$\frac{\psi''}{\psi'} = -2\frac{u'}{u} \tag{3.67}$$

and integrating once leads to

$$\ln\psi' = -2\ln u = \ln(1/u^2) \quad \text{or} \quad \psi' = \frac{1}{u^2} \, . \tag{3.68}$$

Historically, $u^2(s) = \beta(s)$ is called the beta function, but it is essentially the amplitude of the quasi-periodic oscillation which depends on the longitudinal position $s$ in the beam line. Moreover, instead of the constant $A$, often the *Courant-Snyder invariant* $J = A^2/2$ is used. We also note that the relation between $u$ and $\psi$ from Equation 3.68 can be written as

$$\psi(s) = \int_0^s \frac{ds'}{\beta(s')} \, . \tag{3.69}$$

It implies that the phase advances by a lot at locations where $\beta(s)$ is small. Now rewrite the equations for the trajectory $x$ and $x'$ in terms of the beta function

$$
\begin{aligned}
x &= \sqrt{2J\beta(s)}\cos(\psi(s) + \phi_0) \\
x' &= -\sqrt{\frac{2J}{\beta}}\left[\frac{\beta'}{2}\cos(\psi(s) + \phi_0) + \sin(\psi(s) + \phi_0)\right] \\
&= -\sqrt{\frac{2J}{\beta}}\left[\alpha(s)\cos(\psi(s) + \phi_0) + \sin(\psi(s) + \phi_0)\right] \, ,
\end{aligned}
\tag{3.70}
$$

where the second equation follows by differentiating the first. The fact that the motion is quasi-periodic and resembles a harmonic oscillator can be made explicit by solving the previous equations for cos and sin with the result

$$\cos(\psi(s) + \phi_0) = \frac{x}{\sqrt{2J\beta}} \quad \text{and} \quad \sin(\psi(s) + \phi_0) = -\frac{1}{\sqrt{2J\beta}} [\beta x' + \alpha x] . \tag{3.71}$$

Using the trigonometric identity

$$\cos^2(\psi(s) + \phi_0) + \sin^2(\psi(s) + \phi_0) = 1 \tag{3.72}$$

this leads to the expression

$$\begin{aligned} 2J &= \beta x'^2 + 2\alpha x x' + \frac{1+\alpha^2}{\beta} x^2 \\ &= \beta(s)x'^2 + 2\alpha(s)xx' + \gamma(s)x^2 \end{aligned} \tag{3.73}$$

which makes it obvious that the phase-portrait (Poincaré-plot) of $x$ and $x'$ is an ellipse that is parameterized in a way, consistent with Equation 3.44. The orientation of the ellipse varies along the lattice, because $\beta, \alpha = \beta'/2$, and $\gamma = (1+\alpha^2)/\beta$ depend on the longitudinal position $s$.

Now, we re-invent transfer matrices by expressing the integration constants $J$ and $\phi_0$ in terms of the initial coordinates $x_0$ and $x_0'$. At $s = 0$ we have

$$\begin{aligned} x_0 &= \sqrt{2J\beta_0} \cos\phi_0 \\ x_0' &= -\sqrt{\frac{2J}{\beta_0}} [\alpha_0 \cos\phi_0 + \sin\phi_0] , \end{aligned} \tag{3.74}$$

which we solve for $\cos\phi_0$ and $\sin\phi_0$ and replace the trigonometric functions in the general equation for $x$ and $x'$. We find

$$x = \sqrt{\frac{\beta}{\beta_0}} [\cos\psi + \alpha_0 \sin\psi] x_0 + \sqrt{\beta\beta_0} \sin\psi x_0' \tag{3.75}$$

$$x' = \frac{1}{\sqrt{\beta\beta_0}} [(\alpha_0 - \alpha)\cos\psi - (1 + \alpha\alpha_0)\sin\psi] x_0 + \sqrt{\frac{\beta_0}{\beta}} [\cos\psi - \alpha\sin\psi] x_0' ,$$

which is linear in the initial values $x_0$ and $x_0'$ and can be written as the following matrix equation

$$\begin{pmatrix} x \\ x' \end{pmatrix} = R \begin{pmatrix} x_0 \\ x_0' \end{pmatrix} \tag{3.76}$$

where the matrix $R$ can be written in the following form

$$R = \begin{pmatrix} \sqrt{\beta} & 0 \\ -\alpha/\sqrt{\beta} & 1/\sqrt{\beta} \end{pmatrix} \begin{pmatrix} \cos\psi & \sin\psi \\ -\sin\psi & \cos\psi \end{pmatrix} \begin{pmatrix} 1/\sqrt{\beta_0} & 0 \\ \alpha_0/\sqrt{\beta_0} & \sqrt{\beta_0} \end{pmatrix} , \tag{3.77}$$

which makes the dynamics obvious. First the particle is mapped into normalized phase space using $\beta_0$ and $\alpha_0$, followed by an oscillation. Finally it is mapped from normalized back to real space using $\beta$ and $\alpha$. If we calculate the matrix $R$ at the end of the period $s = L$, we recover the expression for the full turn matrix $\hat{R}$ in Equation 3.51 with $\psi = \mu$, $\alpha_0 = \alpha$, and $\beta_0 = \beta$.

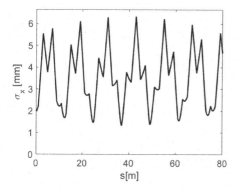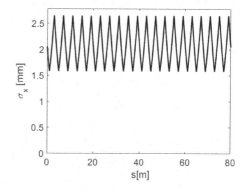

**Figure 3.10** The beam size $\sigma_x = \sqrt{\sigma_{11}}$ as a function of position $s$ along the accelerator for a mis-matched (left) and a matched (right) beam.

### 3.3.5 Beam size and emittance

So far, we only analyzed the propagation of single particles, but now we will move on to consider how a beam, as characterized by its centroids $\vec{X}$ and its beam-matrix $\sigma$, propagates along a beam-line. From general considerations, we already know that the averages, the beam-centroids, behave just like a single particle, see Equation 3.40, whereas the beam-matrix follows Equation 3.43 and our next task is to add the functionality to propagate beam-matrices to the MATLAB script. The following few lines at the end of the MATLAB script use Equation 3.43 to move the beam at the start of the beam line `sigma0`, here chosen arbitrarily, to each position in the beam-line and plot the beam size $\sigma_x = \sqrt{\sigma_{11}}$ at each position.

```
sigma0=[4,0;0,1];
for k=1:nmat
  sigma=Racc(:,:,k)*sigma0*Racc(:,:,k)';  % eq. 3.43
  s11(k)=sigma(1,1);   % record values for plotting
end
plot(spos,sqrt(s11)); xlabel(' s[m]'); ylabel('\sigma_x [mm]')
```

This code snippet is all that is needed to propagate a beam with known initial beam matrix `sigma0` along a known beam line. Of course, normally we also want to know the other, vertical beam sizes and the effect of momentum spread. We will address these points in due time. For now we consider the plot for the above example. It is shown on the left-hand side in Figure 3.10, where we observe that the beam size varies very irregularly and is at some places much larger than the initial beam size of 2 mm. This triggers the question whether it is possible to find an initial beam size that propagates in a more regular pattern along the beam line.

In order to find an initial beam-matrix `sigma0` that leads to a regularly oscillating beam size, we observe that we can use the matrix $\mathcal{A}$ from Equation 3.52 to construct such a $2 \times 2$ beam matrix $\sigma_0$ in the following form

$$\sigma_0 = \varepsilon_x \mathcal{A}^{-1} \left( \mathcal{A}^{-1} \right)^T = \varepsilon_x \begin{pmatrix} \beta & -\alpha \\ -\alpha & \gamma \end{pmatrix} \tag{3.78}$$

with $\gamma = (1 + \alpha^2)/\beta$ and an undetermined parameter $\varepsilon_x$. The rationale behind this construction is two-fold. First, all transfer-matrices $R$ have unit determinant $\det R = 1$. This implies that the determinant of beam matrices does not change, which follows from

$$\det \bar{\sigma} = \det(R\sigma R^T) = \det(R)\det(\sigma)\det(R^T) = \det(\sigma) \ . \tag{3.79}$$

It is therefore reasonable to give the conserved quantity a name, here *emittance* squared, or $\det \sigma = \varepsilon_x^2$. Since the determinant of $\mathcal{A}$ is unity, the determinant of $\sigma_0$ in Equation 3.78 is chosen to be $\varepsilon_x^2$. The second reason to choose $\sigma_0$ in that form is that it reproduces after one cell. This is easy to see by using the parameterization of the transfer-matrix as $R = \mathcal{A}^{-1}\mathcal{O}\mathcal{A}$, given in Equation 3.51. Calculating the beam-matrix $\bar{\sigma}$ after one cell, we find

$$\begin{aligned} \bar{\sigma} &= R\sigma_0 R^T \\ &= \mathcal{A}^{-1}\mathcal{O}\mathcal{A}\varepsilon_x\mathcal{A}^{-1}\left(\mathcal{A}^{-1}\right)^T\left(\mathcal{A}^{-1}\mathcal{O}\mathcal{A}\right)^T \\ &= \varepsilon_x\mathcal{A}^{-1}\mathcal{O}\mathcal{A}\mathcal{A}^{-1}\left(\mathcal{A}^{-1}\right)^T\mathcal{A}^T\mathcal{O}^T\left(\mathcal{A}^{-1}\right)^T \\ &= \varepsilon_x\mathcal{A}^{-1}\left(\mathcal{A}^{-1}\right)^T \\ &= \sigma_0 \end{aligned} \tag{3.80}$$

where we cancel terms whenever a matrix meets its inverse and we note that the inverse of an orthogonal rotation matrix $\mathcal{O}$ equals its transpose, which allows us to cancel the rotations. We thus find that the beam-matrix $\sigma_0$, as defined in Equation 3.78, reproduces itself after one cell. We immediately try this out and use Equation 3.60 to extract $\alpha$ and $\beta$ from a transfer matrix and encapsulate the procedure in a function R2beta() that receives a matrix $R$ and returns the tune $Q$, and the Twiss parameters $\alpha, \beta$, and $\gamma$.

```
% R2beta.m, V. Ziemann, 181017
function [Q,alpha,beta,gamma]=R2beta(R)
mu=acos(0.5*(R(1,1)+R(2,2)));
if (R(1,2)<0) mu=2*pi-mu; end
Q=mu/(2*pi);
beta=R(1,2)/sin(mu);
alpha=(0.5*(R(1,1)-R(2,2)))/sin(mu);
gamma=(1+alpha^2)/beta;
```

From the Twiss parameters we then construct the beam matrix according to Equation 3.78

```
eps0=1; sigma0=eps0*[beta0, -alpha0; -alpha0,gamma0]
```

where we simply set the emittance to unity such that the 11-element of the sigma matrix equals the beta function. Using this sigma0 and repeating the calculation of the beam sizes along the beam-lines results in the plot shown on the right-hand side in Figure 3.10, where we observe a regularly oscillating beam size along the beam line, with much smaller excursions, compared to the plot on its left. A beam-matrix that repeats itself after one cell is called a *matched beam*. On the other hand, sending an un-matched beam into a beam-line results in the very irregular beam sizes visible on the left in Figure 3.10. This is commonly called *beta-beating*.

Apparently there are two types of beta functions, matched and un-matched. The former is defined by the requirement for periodicity for one cell, or more generally, a section of a beam-line, and the latter depends on the beam matrix. In a straight beam-line the two types of beta functions do not necessarily agree and this results in beta-beating. In a circular

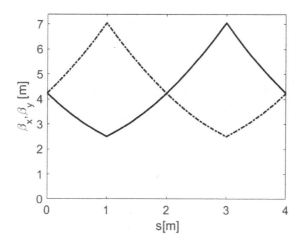

**Figure 3.11** The solid line shows the horizontal beta function in one FODO cell as a function along the position in the cell. The quadrupoles are located at 1 and 3 m where the beta function changes significantly. The dashed line shows the beta function after reversing the polarity of both quadrupoles, which corresponds to the situation in the vertical plane.

accelerator there is a natural periodicity requirement and the beta-function is unique and given by the calculations leading to Equation 3.60 on page 46. The two types of beta functions can also be categorized whether they depend on the hardware of the beam line, which determines the periodic beta-function, or whether they depend on the beam matrix $\sigma$ and we determine it from first calculating the emittance from $\varepsilon_x^2 = \det \sigma$ and then the beam's beta function from $\beta = \sigma_{11}/\varepsilon_x$ and $\alpha = -\sigma_{12}/\varepsilon_x$. This beta function we denote the beam beta function. It only provides a convenient parameterization of the beam matrix. The two types of beta-functions agree if the beam is matched to the lattice.

We also reiterate that matrix elements of the beam-matrix are related to the beam size by $\sigma_x^2 = \varepsilon_x\beta$ and angular divergence $\sigma_{x'}^2 = \varepsilon_x\gamma$. The parameter $\alpha$ describes whether the beam is convergent, if $\alpha > 0$, or divergent, if $\alpha < 0$. This is easy to see by considering the definition of the off-diagonal element of the beam-matrix $\sigma_{12} = -\varepsilon_x\alpha = \langle xx' \rangle$ where the second equation describes the matrix element as the average of the product of position and angle over all particles in the beam. If the particles in the beam with positive position $x$ point downwards, which means $x' < 0$, the particle trajectory points downwards and converges towards the reference trajectory. Conversely, particles with $x < 0$ and positive angle $x' > 0$, also move towards the axis. Thus if the average of the product $xx'$ is negative, most particles converge to the axis and the beam is convergent. The minus-sign in the definition of $\alpha$ therefore causes it to be positive for convergent beams.

When we design accelerators we normally build small cells that are repeated many times, with many magnets of the same type, which makes their production less expensive. These small cells, such as our FODO cell, are then characterized by the beta-functions, which indicates the beam sizes and one often sees beta function plots of the beam optics, as the beta functions are commonly called, displayed. Figure 3.11 shows this for our FODO cell. The solid line shows $\beta_x$ and the dashed line $\beta_y$ the beta-function in the vertical plane where

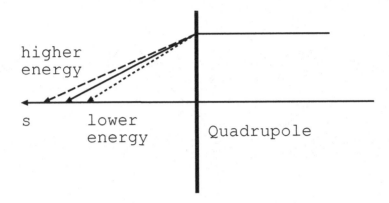

Figure 3.12 Effect of a quadrupole on particles with different energies. The focal point of particles with higher energy is further downstream than that of the reference particle and particles with lower energy.

the polarity of the quadrupoles is reversed. We find that the horizontal beta-function $\beta_x$ is minimum in the first quadrupole at $s = 1\,\mathrm{m}$ which is a defocusing quadrupole and it is maximum in the focusing quadrupole at $s = 3\,\mathrm{m}$. The focusing quadrupole "bends the beta-function back" just as it is about to go through the roof. In the vertical plane the situation is the converse. The horizontally defocusing quadrupole focuses in the vertical plane and the maximum of the vertical beta-function appears in the first quadrupole. Note that with the beta-function also the vertical beam size is maximum in the first quadrupole, where the horizontal beam size is minimal as indicated by the minimum of the solid line in Figure 3.11.

What have we achieved so far? We started with a description of a beam line and explored the motion of a single particle, or equivalently, the beam centroid and found that there is a stability criterion for beam-lines. Then we explored how to propagate beam-matrices along the beam-line and found parameterizations of the transfer-matrix and the beam-matrix in terms of beta-functions. This framework will give us a solid base to describe and design beam-lines and accelerators. But an extra ingredient that is missing is the effect of momentum deviations of the beam, so-called chromatic effects, and that is what we have to look at next.

## 3.4   CHROMATICITY AND DISPERSION

Here we consider two chromatic effects. First the effect of momentum-dependent focusing in quadrupoles, called *chromaticity* and then the effect of momentum-dependent bending of dipoles, which is called *dispersion*.

### 3.4.1   Chromaticity

The first effect we discuss depends on the momentum-dependence of the focal length of quadrupoles, the chromaticity. Recall that the inverse focal length is given by

$$\frac{1}{f} = \frac{e}{p}\frac{\partial B_y}{\partial x} = \frac{e}{p_0(1+\delta)}\frac{\partial B_y}{\partial x} = \frac{1}{f_0(1+\delta)} \approx \frac{1}{f_0} - \frac{\delta}{f_0}, \qquad (3.81)$$

where $\delta = (p - p_0)/p_0$ is the relative momentum offset of a particle with respect to the reference particle. Apparently, the momentum error has the same effect as an additional quadrupole with focal length $-f_0/\delta$. Consequently the focusing is momentum dependent and particles with different energies experience different focal lengths. This affects the longitudinal position of a focal point in a linear accelerator or the phase-advance per cell and thus the tune in a circular accelerator. Here we will consider the latter and explore the effect of a single additional quadrupole of strength $\hat{f} = -f_0/\delta$ at a location where the beta function is known to be $\hat{\beta}$. We assume $\alpha = 0$ to make the calculation more traceable, but if taking it along it will drop out of the calculation in the end.

Adding an additional quadrupole to a ring that has tune $Q$ starting at a location with beta-function $\hat{\beta}$ can be described by multiplying the transfer-matrix of the quadrupole to that of a ring with tune $Q$ and is given by

$$\hat{R} = \begin{pmatrix} 1 & 0 \\ -1/\hat{f} & 1 \end{pmatrix} \begin{pmatrix} \cos(2\pi Q) & \hat{\beta}\sin(2\pi Q) \\ -\sin(2\pi Q)/\hat{\beta} & \cos(2\pi Q) \end{pmatrix} \tag{3.82}$$

$$= \begin{pmatrix} \cos(2\pi Q) & \hat{\beta}\sin(2\pi Q) \\ -\sin(2\pi Q)/\hat{\beta} - \cos(2\pi Q)/\hat{f} & \cos(2\pi Q) - \hat{\beta}\sin(2\pi Q)/\hat{f} \end{pmatrix}$$

and we can now use Equation 3.60 to determine the new tune $Q + \Delta Q$ from $\hat{R}$

$$2\cos(2\pi(Q + \Delta Q)) = \hat{R}_{11} + \hat{R}_{22} = 2\cos(2\pi Q) - \frac{\hat{\beta}}{\hat{f}}\sin(2\pi Q) . \tag{3.83}$$

Since we assume that the extra quadrupole is weak, we expect a small change in the tune $\Delta Q$ and can rewrite cosine on the left-hand side as

$$\cos(2\pi(Q + \Delta Q)) \approx \cos(2\pi Q) - 2\pi\Delta Q \sin(2\pi Q) , \tag{3.84}$$

where we use that $\cos(2\pi\Delta Q) \approx 1$ and $\sin(2\pi\Delta Q) \approx 2\pi\Delta Q$. Solving for $\Delta Q$, we obtain

$$\Delta Q = \frac{\hat{\beta}}{4\pi\hat{f}} , \tag{3.85}$$

which is a well-known (in accelerator circles) equation to describe the tune-shift $\Delta Q$ as a consequence of a quadrupole error of magnitude $1/\hat{f}$. We notice that $\Delta Q$ is proportional to the beta-function $\hat{\beta}$ at the location of the error. This indicates that the beta-function not only describes the beam size, but also the *sensitivity to quadrupole errors*. Conversely, it is important to pay special attention to the manufacturing tolerances of quadrupoles located at positions where the beta functions are very large.

The tune-shift $\Delta Q$ due to a momentum error $\delta$ is found from realizing that $\hat{f} = -f_0/\delta$ and we find

$$\Delta Q = -\frac{\hat{\beta}}{4\pi f_0}\delta \tag{3.86}$$

which is the contribution of a single quadrupole with focal length $f_0$ to the chromaticity. The chromaticity $Q'_x = \Delta Q/\delta$ for the entire ring is consequently given by summing over the contributions of all quadrupoles

$$Q'_x = \Delta Q/\delta = -\sum_i \frac{\beta_i}{4\pi f_i} , \tag{3.87}$$

where the sum extends over all quadrupoles.

The quadrupole in the previous example is a thin quadrupole, but we can easily extend the range of validity of Equation 3.85 to comprise long quadrupoles, which have an effective focal length of $1/f \approx \hat{k}_1 l$. Thus we arrive at the tune-shift from a long quadrupole as

$$\Delta Q = \frac{1}{4\pi} \int_0^l \beta(s) \hat{k}_1 ds \ , \tag{3.88}$$

where $k_1 = (\partial B_y/\partial x)/B\rho$ is the normalized gradient of the quadrupole. For the chromaticity due to long quadrupoles we have $\hat{k}_1 = -k_1/\delta$ and we obtain in the same way as before

$$Q'_x = -\oint \beta(s) k_1(s) ds \ , \tag{3.89}$$

where the integral extends over the entire ring and picks up only contributions where $k_1(s)$ differs from zero and that is in the quadrupoles.

For our simple FODO cell with thin quadrupoles, it is very easy to implement the calculation of the chromaticity according to Equation 3.87 in the MATLAB script. We assume that the calculation of the matched beam-matrix and the beta functions is done previously and they are stored in an array beta(). Then, we only have to loop over all elements and add the contributions from the thin quadrupoles in the following code snippet.

```
xi=0;
ic=1;
for line=1:nlines
  for seg=1:beamline(line,2)
    ic=ic+1;
    if beamline(line,1)==2
      xi=xi-beta(ic)/(4*pi*beamline(line,4));
    end
  end
end
disp(['Chromaticity = ' num2str(xi,4)])
```

In the end, the numerical value of the chromaticity is displayed. In case we have to deal with long quadrupoles, we need to sum over their respective $k_1$−values weighted by the beta functions, but we leave that as an exercise.

A large value of the chromaticity is undesirable because it causes the tunes to vary significantly with the momentum of the beam particles and this has an important influence on the stability of the beam. To alleviate this dependence we thus need to compensate the chromaticity, but defer this task to Sections 8.4.4 and 8.5.4. Instead, we now turn to the momentum-dependence of dipole magnets.

## 3.4.2 Dispersion

We already considered the momentum dependence of the deflection angle of dipoles in Figure 3.5, which led to the transfer-matrix for a dipole given in Equation 3.24 on page 34. We see that a dipole magnet has the property of giving a particle a small horizontal change in position and offset, which is proportional to the momentum offset $\delta$ and is represented by the sixth column of the transfer matrix in Equation 3.24. Even if the particle was initially on-axis, it starts to deviate from the reference trajectory, and, provided the beam-line is stable, starts to oscillate. The amplitude of this oscillation is proportional to $\delta$ and the trajectory, normalized to $\delta$, is called the *dispersion*, often denoted by $D$. In this way the transverse offset $\bar{x}$ of the trajectory, due to the momentum offset, is $\bar{x} = D\delta$. It thus describes by how much the trajectories are separated due to their momentum and we may call it the *spectrometer function*. Though this name is not commonly used, it describes the function of the dispersion rather well.

We can treat this effect in our MATLAB simulation by including the momentum dependence in the transfer matrix. In order to keep the discussion transparent, we only consider the horizontal plane $x$ and $x'$ and assume that all dipoles have horizontal deflection angles and affect the horizontal motion, only. We therefore can model this by extending the previous simulation to $3 \times 3$ matrices where we place the entries in the sixth column of Equation 3.24 into the third column of our reduced model. The transfer-matrix for the horizontally deflecting sector dipole magnet is thus

$$R = \begin{pmatrix} \cos\phi & \rho\sin\phi & \rho(1-\cos\phi) \\ -\sin(\phi)/\rho & \cos\phi & \sin\phi) \\ 0 & 0 & 1 \end{pmatrix} \tag{3.90}$$

with the deflection angle $\phi = eBL/p$ and the radius of curvature $\rho = p/eB$. The MATLAB function that implements this is the following

```
% SB.m, sector bend
function out=SB(L,rho);
phi=L/rho;
out=eye(3);
if abs(phi)<1e-8
  out(1,2)=L;
else
  out(1:2,1:3)=[cos(phi),rho*sin(phi),rho*(1-cos(phi)); ...
          -sin(phi)/rho,cos(phi),sin(phi)];
end
```

where we see that the matrix is a straight translation of the matrix in Equation 3.90. In the same spirit we update the transfer-matrix for the drift space to $3 \times 3$ format by adding a third row and column with unity on the diagonal.

```
% DD.m, drift space
function out=DD(L)
out=eye(3);
out(1,2)=L;
```

Likewise we introduce the $3 \times 3$ transfer matrix for a thick quadrupole.

```
% QQ.m, thick quadrupole
function out=QQ(k,L)
ksq=sqrt(abs(k));
out=eye(3);
if abs(k) < 1e-6
  out(1,2)=L;
elseif k>0
  out(1:2,1:2)=[cos(ksq*L),sin(ksq*L)/ksq; ...
```

```
                     -ksq*sin(ksq*L),cos(ksq*L)];
    else
      out(1:2,1:2)=[cosh(ksq*L),sinh(ksq*L)/ksq; ...
                    ksq*sinh(ksq*L),cosh(ksq*L)];
    end
```

Note that only the transfer-matrix for the dipole has non-zero entries in the third column. This implies that only dipoles can change a zero dispersion to a non-zero value.

In the function `calcmat.m` that calculates the transfer-matrices we account for the sector-dipole and the thick quadrupole by adding the appropriate `case` statements and call the functions `SB()` and `QQ()`, respectively, as indicated in the following code-snippet.

```
switch beamline(line,1)
      :
   case 4    % sector dipole
     phi=beamline(line,4)*pi/180;  % convert to radians
     rho=beamline(line,3)/phi
     Rcurr=SB(beamline(line,3),rho);
   case 5    % thick quadrupole
     Rcurr=QQ(beamline(line,3),beamline(line,4));
      :
```

After these updates we are ready to explore what happens to a particle with a non-zero momentum offset $\delta$. We use the following beam-line description for our exercise. It is based on the example in Chapter 2 but with dipoles bending by 20 degrees only. In order to obtain smoother graphics we subdivide the elements in short segments, and the entire beam-line consists of three FODO-cells.

```
fodo=[ 1  5  0.5  0      ;
       5  5  0.2 -0.1799;
       1  3  0.5  0      ;
       4 10  0.2  20/10  ;
       1  3  0.5  0      ;
       5  5  0.2  0.1799;
       1  5  0.5  0      ];
beamline=repmat(fodo,3,1);
```

We explore the dispersion generated by the dipole by launching a particle with initial coordinates $(x, x', \delta) = (0, 0, 10^{-3})$ and display its horizontal position along the beam-line in Figure 3.13. We observe that the particle, launched at the left stays on-axis until it encounters the first dipole magnet at $s = 5$ m whence it starts deviating from the axis. Next it encounters a focusing quadrupole at $s = 7.5$ m that slightly bends the particle back towards the axis, but not enough. The particle increases its distance from the axis, even more so after passing the defocusing quadrupole at $s = 15$ m, but eventually it is bent back significantly by the focusing quadrupole at $s = 22.5$ m and so forth. But it is apparent that the orbit is non-periodic.

Note that we added a graphical representation of the magnets in the beam-line lattice at the bottom of the plot in order to aid interpretations of data on the plot. The function `drawmag()` that accomplishes this is discussed in Appendix B.5. It simply loops through the magnet lattice and draws rectangles of the magnets at the correct positions. We add input parameters `vpos` and `height` to place the drawing vertically on the plot so that it looks good.

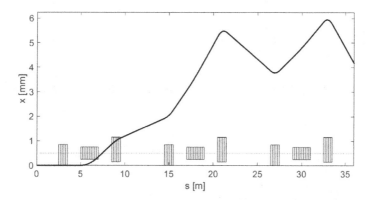

Figure 3.13 The trajectory of a particle with momentum offset $\delta = 10^{-3}$ in a FODO lattice with bending magnets.

But back to the dispersion. The question arises whether we can also find launch conditions for the dispersion that causes it to be periodic. For this we inspect the $3 \times 3$ transfer-matrix through a single FODO cell which for the above lattice file is

```
Rturn=
     1.1499      8.5130      1.5190
    -0.2279     -0.8177      0.1619
          0           0      1.0000
```

This matrix has the following structure

$$\hat{R} = \begin{pmatrix} \hat{R}_{11} & \hat{R}_{12} & \hat{R}_{16} \\ \hat{R}_{21} & \hat{R}_{22} & \hat{R}_{26} \\ 0 & 0 & 1 \end{pmatrix} \tag{3.91}$$

and we ask ourselves whether we can find a launch vector $(x, x', \delta)$ in which we also require $x$ and $x'$ to be proportional to $\delta$ that reproduces after one period. Thus we want to find a vector $(x, x', \delta) = (D\delta, D'\delta, \delta)$ that fulfills the following periodicity condition

$$\begin{pmatrix} D \\ D' \\ 1 \end{pmatrix} = \begin{pmatrix} \hat{R}_{11} & \hat{R}_{12} & \hat{R}_{16} \\ \hat{R}_{21} & \hat{R}_{22} & \hat{R}_{26} \\ 0 & 0 & 1 \end{pmatrix} \begin{pmatrix} D \\ D' \\ 1 \end{pmatrix}. \tag{3.92}$$

The first two lines can be rearranged and lead to

$$\begin{pmatrix} 1 - \hat{R}_{11} & \hat{R}_{12} \\ \hat{R}_{21} & 1 - \hat{R}_{22} \end{pmatrix} \begin{pmatrix} D \\ D' \end{pmatrix} = \begin{pmatrix} \hat{R}_{16} \\ \hat{R}_{26} \end{pmatrix}. \tag{3.93}$$

Solving for the dispersion $(D, D')$ we find

$$\begin{pmatrix} D \\ D' \end{pmatrix} = \frac{1}{(1 - \hat{R}_{11})(1 - \hat{R}_{22}) - \hat{R}_{12}\hat{R}_{21}} \begin{pmatrix} 1 - \hat{R}_{22} & -\hat{R}_{12} \\ -\hat{R}_{21} & 1 - \hat{R}_{11} \end{pmatrix} \begin{pmatrix} \hat{R}_{16} \\ \hat{R}_{26} \end{pmatrix}, \tag{3.94}$$

which denotes the periodic, or equilibrium, dispersion values at the start of the periodic beam line that is described by $\hat{R}$. These manipulations are easy to implement in MATLAB as shown in the following snippet

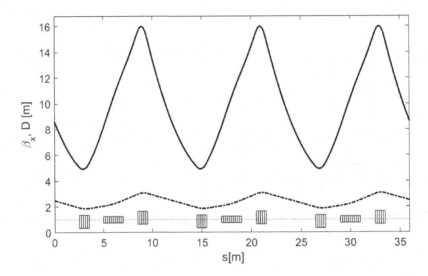

Figure 3.14  Matched dispersion (dashed) and horizontal beta function (solid).

```
D=(eye(2)-Rturn(1:2,1:2))\Rturn(1:2,3)
dd0=[D;1];  % initial periodic dispersion
for k=1:nmat
  x(k)=Racc(1,:,k)*dd0;
end
plot(spos,x,'k-.');
```

We show the plot of the periodic dispersion function $D$, together with the horizontal beta function, in Figure 3.14. Particles with energy error $\delta$ and travel on periodic trajectories have horizontal positions $D\delta$ along the accelerator, where $D$, the dispersion, is shown as the dashed trace in Figure 3.14.

The finite dispersion can have a major influence on the beam quality. Particularly in electron rings it determines the emittance by a process we now turn to.

### 3.4.3  Emittance generation

In the previous sections we found that the dispersion trajectory $D(s)$ is the closed orbit of a particle with momentum offset $\delta$. If we assume that a particle with energy offset $\delta_1$ is initially on its equilibrium orbit, the dispersion trajectory $D\delta_1$ is as shown in Figure 3.15. If that particle loses energy by emitting a photon and has the new energy offset $\delta_2 < \delta_1$ at a position with non-zero dispersion, it will stay at transverse position $D\delta_1$ but has energy $\delta_2$. The equilibrium orbit of the particle with momentum $\delta_2$ is, however, $D\delta_2$ and the particle finds itself away from its new equilibrium orbit and will therefore start oscillating around the new equilibrium orbit. In summary: initially the particle is on its equilibrium orbit, but through the energy loss, the equilibrium orbit has jumped away and the particle starts betatron oscillations around the new equilibrium orbit. The same argument holds for $D'$, the derivative of the dispersion. Note that this process is the dominant mechanism that determines the emittance in electron storage rings and synchrotron radiation sources. We will discuss this topic in depth in Section 10.1.

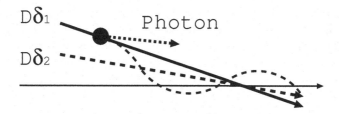

Figure 3.15 The mechanism that causes the excitation of betatron oscillations by a random energy loss.

The change in betatron state vector $(x, x')$ that the particle receives through a relative momentum loss $\delta$ at a position with dispersion $D$, is given by

$$\begin{pmatrix} x \\ x' \end{pmatrix} = -\delta \begin{pmatrix} D \\ D' \end{pmatrix} \tag{3.95}$$

or $\vec{x} = -\delta \vec{D}$. In the $m$−th turn the particle experiences the momentum change $\delta_m$ with corresponding betatron state change $\vec{x} = (x_m, x'_m) = (D, D')\delta_m$. After $n$ turns we then have to add all $\vec{x}_m$ over the previous turns with $m \leq n$

$$\begin{pmatrix} x_n \\ x'_n \end{pmatrix} = -\sum_{m=1}^{n} \delta_m R^{n-m} \begin{pmatrix} D \\ D' \end{pmatrix} . \tag{3.96}$$

The Courant-Snyder invariant $2J_n$ after $n$ turns, as defined in Equation 3.44, is given by $2J_n = \gamma x_n^2 + 2\alpha x_n x'_n + \beta x_n'^2$, where $\alpha, \beta$, and $\gamma$ are the Twiss parameters at the location where the momentum kicks $\delta_m$ are applied. Expressing the transfer matrix $R$ in Equation 3.96 by the representation from Equation 3.51 and after some algebra, we find for $2J_n$

$$2J_n = (\gamma D^2 + 2\alpha DD' + \beta D'^2) \sum_{k,l}^{n} \delta_k \delta_l \cos(2\pi(k-l)Q) . \tag{3.97}$$

The emission of photons is a random process and we can therefore assume that the $\delta_m$ from different turns $m$ are statistically independent. For the time being, we denote their rms value by $\delta_{rms}$, but will later calculate it from the spectrum of the emitted synchrotron radiation in Section 10.1. For the sum in Equation 3.97, we then obtain

$$\sum_{k,l}^{n} \delta_k \delta_l \cos(2\pi(k-l)Q) = n\delta_{rms}^2 \delta_{kl} , \tag{3.98}$$

where $\delta_{kl}$ is the Kronecker delta, which is unity for $k = l$, and zero otherwise. If we furthermore assume that we take the average over a large number of particles we use the ensemble average $\varepsilon = \langle J \rangle$ over the Courant-Snyder invariants, which is the emittance $\varepsilon$ of the ensemble. For the final result, we find

$$\frac{d\varepsilon}{dt} = \frac{\gamma D^2 + 2\alpha DD' + \beta D'^2}{T} \delta_{rms}^2 , \tag{3.99}$$

where $T$ is the revolution time. We find that the emittance growth $d\varepsilon/dt$ is proportional to

$$\mathcal{H} = \gamma D^2 + 2\alpha DD' + \beta D'^2 \tag{3.100}$$

and we immediately see that zero dispersion is desirable at locations, where the particles change their momenta, as is the case in accelerating cavities and, especially important for synchrotron light sources, in dipole magnets, where the beam loses energy due to synchrotron radiation. The design of small emittance lattices is focused on designing magnet configurations that have small beta functions and small dispersion, in particular in the dipoles. We shall use this as a guideline when discussing different types of lattices in Section 3.7.5 and return to the details of the emission of synchrotron radiation in Section 10.1.

### 3.4.4 Momentum compaction factor

The dispersion trajectory that we calculated in the previous section describes the orbit of a particle with momentum offset $\delta$. If such a particle traverses a dipole magnet it will lie further on the out- or inside, depending on the sign of the dispersion function at that location and the sign of $\delta$. If it lies further outside it will have a longer path to travel and therefore will arrive later at the exit of the dipole. The change in length $\Delta l$ with momentum variation of the path in a single dipole with bending radius $\rho$ is given by

$$\frac{\Delta l}{\delta} = \int_0^l \frac{D(s)}{\rho} ds \ . \tag{3.101}$$

In the case of a circular accelerator we can sum the path-length changes over all dipoles and normalize to the circumference $C$ of the ring to obtain the so-called *momentum-compaction* factor $\alpha$ defined by

$$\alpha = \frac{\Delta C}{C\delta} = \frac{1}{C} \int_{\text{all dipoles}} \frac{D(s)}{\rho} ds \tag{3.102}$$

which gives the fractional change of the circumference normalized to the energy $(\Delta C/C)/\delta$. This quantity plays a central role for the stability of the longitudinal motion of the particles in a storage ring as we shall see in Chapter 5.

So far we have covered the motion of particles in one transverse plane and the momentum dependence, but in general the motion in the all three planes, horizontal, vertical, and longitudinal, may be coupled and we address the motion in the two transverse planes in the following section.

## 3.5 FOUR-DIMENSIONAL AND COUPLING

The first thing we have to do in order to consider the simultaneous effect of both transverse planes simultaneously is to update the transfer-matrices to four dimensions. This is rather straightforward and we do not show the code here, but refer the reader to Appendix B.5.

Note that all $4 \times 4$ transfer-matrices we encountered so far were block-diagonal and this causes the motion in the two planes to be independent, or *un-coupled* in the sense that a particle launched on a horizontal oscillation will always stay in the horizontal plane. Conversely, a particle with a pure vertical oscillation stays in the vertical plane. In this way we can treat un-coupled systems as two independent $2 \times 2$ systems and can apply the analysis-methods from Section 3.3, once for the top left $2 \times 2$ block of the transfer matrix and once for the block on the lower right. We can therefore apply the R2beta function to calculate tunes and beta-functions for systems in the same way as before. We just use the top left or lower right $2 \times 2$ transfer matrix as input to R2beta.

This situation changes dramatically once we introduce coupling elements, such as quadrupoles that are installed with roll-angles. The matrix for a roll angle is given in Equation 3.34 and the MATLAB function that returns it is the following

```
% ROLL.m, roll-angle around s-axis
function out=ROLL(phi)   % phi in degree
c=cos(phi*pi/180); s=sin(phi*pi/180);
out=zeros(4);
out(1,1)=c; out(1,3)=s; out(2,2)=c; out(2,4)=s;
out(3,1)=-s; out(3,3)=c; out(4,2)=-s; out(4,4)=c;
```

which is self-explanatory. We refer to this transfer-matrix by the code 20 in the lattice file and place the roll-angle $\phi_r$ in degrees in the fourth column of the lattice description. Once the function for the transfer-matrix is written, we need to make the matrix-calculation routine calcmat.m aware of it by adding the following two lines to the switch statement

```
case 20   % coordinate roll
  Rcurr=ROLL(beamline(line,4));
```

such that it is automatically included in the calculation of the accumulated transfer-matrices stored in the array Racc. We illustrate the effect of *coupling* the planes by sandwiching the focusing quadrupole in the fodo lattice between elements of opposite roll-angle. This introduces a rolled coordinate system for the traversal of the quadrupole. The lattice file for the FODO cell is thus modified to look like

```
 1  5  0.2    0
 2  1    0  -2.1
 1  1  0.2    0
20  1    0    5
 2  1    0   2.1
20  1    0   -5
 1  5  0.2    0
```

We store this lattice description in a file named fodoroll.bl and load it with

```
fodor=dlmread('fodoroll.bl')
```

Calculating the transfer matrix from start to end of this file, we find that the matrix is no longer block-diagonal, but has non-zero elements in the top right and lower left $2 \times 2$ blocks.

Let us explore what happens in a long beam-line of eight FODO cells, where the focusing quadrupole in the second cell is rolled by 5 degrees, but we enter the system with a beam matrix that is matched to the unperturbed FODO cells. The script display_beta_beating.m that accomplishes this can be found in the online documentation. In this script we first load the unperturbed FODO-cell and determine the matched beta functions with R2beta and build the matched beam-matrix sigma0. Then we change the beam-line to consist of an unperturbed FODO cell, one perturbed with the rolled quadrupole and six more normal ones before updating the transfer-matrices for the long beam-line and plotting the beam sizes $\sigma_x$ and $\sigma_y$ along the beam-line and show the result in Figure 3.16. Here we see the sizes initially follow a regular pattern until they encounter the rolled quadrupole at $s = 7$ m, at which point the beam-sizes start to deviate from their regular oscillations. The rolled quadrupole apparently causes a mismatch.

We explore this mismatch further by rolling both quadrupoles by the same angle and also find that the off-diagonal blocks are populated with non-zero-elements. So, the question arises: how do we calculate these quantities in a coupled beam-line? Luckily, this problem was solved before, and we discuss the method first introduced by Edwards and Teng in [13].

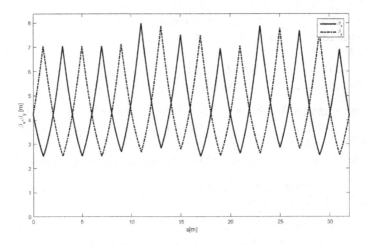

**Figure 3.16** The beta functions $\beta_x$ and $\beta_y$ along a beam line with a quadrupole in the second FODO cell that is rolled by 5 degrees. It causes the beta functions to beat downstream.

They devised an algorithm to find a parameterization of the transfer-matrix $\hat{R}$ that brings it into block-diagonal form such that we can write

$$\hat{R} = T^{-1} \begin{pmatrix} \hat{R}_x & 0_2 \\ 0_2 & \hat{R}_y \end{pmatrix} T \qquad (3.103)$$

where $0_2$ are $2 \times 2$ matrices containing only zeros and the $2 \times 2$ matrices $\hat{R}_x$ and $\hat{R}_y$ are the transfer-matrices of two eigenmodes that can be written in the form given in Equation 3.51 such that we have $\hat{R}_x = \mathcal{A}_x^{-1} \mathcal{O}_x \mathcal{A}_x$ and a similar expression for $\hat{R}_y$. The magic of [13] is the construction of the $4 \times 4$-matrix $T$ that achieves the de-coupling. Here we do not go into the details, but in Appendix B.5 provide the MATLAB function `edteng.m` that implements the algorithm described in [13]. The function receives the $4 \times 4$ full-turn matrix as input and returns $4 \times 4$ matrices $\mathcal{A}, \mathcal{O}$, and $T$ as well as an array of parameters such that we have

$$\hat{R} = T^{-1} \mathcal{A}^{-1} \mathcal{O} \mathcal{A} T \qquad (3.104)$$

with

$$\mathcal{A} = \begin{pmatrix} \mathcal{A}_x & 0_2 \\ 0_2 & \mathcal{A}_y \end{pmatrix} \quad \text{and} \quad \mathcal{O} = \begin{pmatrix} \mathcal{O}_x & 0_2 \\ 0_2 & \mathcal{O}_y \end{pmatrix} \qquad (3.105)$$

and the additional array makes the raw parameters such as eigen tunes, eigen beta functions, and coupling parameters available to the calling program. For the order of the parameters see Appendix B.5 but the first six parameters $Q_1, \alpha_1, \beta_1, Q_2, \alpha_2$, and $\beta_2$ are the eigen tunes and the eigen beta functions for the two eigenmodes.

We immediately use this to investigate the beam-line with the rolled quadrupole that led to the beam sizes shown in Figure 3.16 and assume that it represents a periodic lattice such that it makes sense to display periodic eigen beta functions. For this purpose we need to calculate the full-turn matrix at every point along the beam-line and then perform the Edwards-Teng decomposition to extract the eigen beta functions and plot them. This is accomplished by the following MATLAB script.

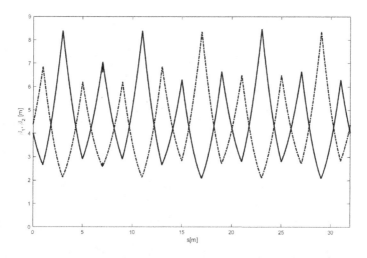

**Figure 3.17** The periodic eigen beta functions along the beam-line with one quadrupole rolled by 5 degrees.

```
% display_edteng_betas.m, V. Ziemann
clear all; close all
fodo=dlmread('fodo.bl'); fodor=dlmread('fodoroll.bl');
beamline=[fodo;fodor;repmat(fodo,6,1)];
[Racc,spos,nmat,nlines]=calcmat(beamline);
Rturn=Racc(:,:,end);   % full-turn-matrix at start
beta1=zeros(1,nmat); beta2=beta1;  % allocate space
for k=1:nmat
  R=Racc(:,:,k)*Rturn*inv(Racc(:,:,k));  % move FTM to point k
  [O,A,T,p]=edteng(R); beta1(k)=p(3); beta2(k)=p(6);
end
plot(spos,beta1,'k',spos,beta2,'k-.');
xlabel(' s[m]'); ylabel('\beta_1, \beta_2 [m]'); xlim([0, spos(end)])
```

After loading the files with the beamline descriptions we determine the full-turn matrix `Rturn` at the start of the beamline. After allocating space for the beta functions `beta1` and `beta2` we loop over all position `k` along the beam line and calculate the full-turn matrix starting at that position by moving to the start of the beam line first by applying the inverse of `Racc(:,:,k)`, moving forward through the entire beam-line and on to position `k` in the following turn. Then we pass this "moved" full-turn matrix to `edteng` and calculate the eigen tunes and beta functions. Those we save in `beta1` and `beta2` and later plot them in Figure 3.17. There we see that now the beta functions are periodic, but they show a significant beating. The constraint to maintain periodicity actually amplifies the effects of the rolled quadrupole compared to simply plotting of beam sizes along the beam line with initial beam-matrix held fixed, which was earlier shown in Figure 3.16.

As a final example we consider the effect of a rolled quadrupole on the eigen tunes and how well we can compensate the tunes with another upright quadrupole; a procedure that is known as the *closest-tune* coupling measurement. This method is based on the idea that

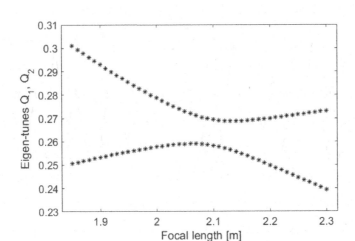

**Figure 3.18** The eigen tunes as a function of the focal length of the first focusing quadrupole while the second focusing quadrupole is rolled by 1 degree. Note that there is a minimum separation of the eigen tunes that can be achieved.

the eigenfrequencies of two coupled oscillators cannot coincide if they are coupled and the closest difference between the eigen tunes is proportional to the coupling constant. This idea is easy to verify by calculating the eigenfrequencies of a system made of two pendula coupled with an additional spring. We build an analogous system with the beam-line we used before and roll the quadrupole by only 1 degree instead of the 5 degrees we used earlier. The tunes move from $(Q_1, Q_2) = (0.2639, 0.2639)$ to $(0.2583, 0.2694)$, and we now scan the focusing quadrupole in the first FODO-cell and record the eigen tunes for all quadrupole settings. The following MATLAB script

```
fodo=dlmread('fodo.bl'); fodor=dlmread('fodoroll_1deg.bl');
beamline=[fodo;fodor;repmat(fodo,6,1)];
k=0;
for ff=1.85:0.01:2.3
  k=k+1; xval(k)=ff;
  beamline(4,4)=ff;  % QF is 4th element and F is 4th column
  [Racc,spos,nmat,nlines]=calcmat(beamline);  % update lattice
  [0,A,T,p]=edteng(Racc(:,:,end)); Q1(k)=p(1); Q2(k)=p(4);
end
plot(xval,Q1,'k*',xval,Q2,'k*'); xlim([1.83,2.32]);
xlabel(' Focal length [m]'); ylabel('Eigen tunes Q_1, Q_2')
```

produces Figure 3.18. In the script, we load the beam-line as before, and loop over the desired range of focal length values, update the correct slot in the `beamline`, re-calculate the transfer-matrices with `calcmat` and determine the eigen tunes with `edteng` before storing the values in the arrays `Q1` and `Q2`. Finally we plot the values and annotate the axes. On Figure 3.18 we clearly see the tunes varying with the quadrupole strength, but they are unable to come closer than $|Q_1 - Q_2| \approx 0.013$. It turns out that this smallest achievable difference can be used as a measurement for the "strength" of the coupling and is commonly used to adjust other skew-quadrupole correctors in order to minimize the coupling or the "closest tune." We will return to this topic in Chapter 8, where we will also discuss a method to compensate the coupling.

But now we move on to discuss how to tailor beam lines to serve specific purposes, such as having a certain phase advance or beam size at specific points.

## 3.6 MATCHING

In the previous sections we used beam-line description files that were defined before-hand without further discussion. But this is not really the normal situation. Often we have to find quadrupole settings or values of other elements that define a beam line such that it achieves a certain purpose. This is actually the task we set out to address in the first paragraph of this chapter on page 27. A typical example is adjusting the phase advance of a FODO or other cell to certain values or to find quadrupole values that cause the beam to have a certain beam-size at an experimental station. There the experiment may either require a beam waist, a focus, or maybe a particularly parallel beam. The procedure of adjusting magnet values in order to achieve constraints is commonly called *matching* and we explore a few simple examples in this section before using it more extensively later on.

### 3.6.1 Matching the phase advance

The first example is to adjust the phase advance of a FODO-cell to certain values, say $Q_x = 1/6$ and $Q_y = 0.25$. Since $1/6$ of a full circle corresponds to 60 degrees, such a cell is often referred to as a 60-90 degree FODO-cell. Since we have two constraints, the phase advances, to satisfy we need at least two parameters to vary and the obvious choices are the two quadrupoles in a FODO-cell. For convenience we just reproduce their beam-line description here

```
1  5  0.2   0
2  1    0 -2.1
1  1  0.2   0
2  1    0  2.1
1  5  0.2   0
```

which is stored in the file `fodo.bl`. The parameters we want to vary are the focal lengths of the quadrupoles, which reside in rows 2 and 4 and in column 4 in the beam-line description. Basically we need a program to adjust these values until the phase advances are those we want. Luckily, we can use the MATLAB function `fminsearch()`, which has the purpose of minimizing a function, to achieve this. All we have to do is to define a function that tells `fminsearch()` how close we are to the desired values. That function we call `chisq_tunes()` and it is defined by the following code:

```
% chisq_tunes.m, find focal length to set Qx and Qy of cell
function chisq=chisq_tunes(x)
global beamline      % need info about the beamline
beamline(2,4)=x(1);  % change quadrupole excitations
beamline(4,4)=x(2);
[Racc,spos]=calcmat(beamline,4); Rturn=Racc(:,:,end);
[Qx,alpha0x,beta0x,gamma0x]=R2beta(Rturn(1:2,1:2));
[Qy,alpha0y,beta0y,gamma0y]=R2beta(Rturn(3:4,3:4));
chisq=(Qx-0.166666)^2+(Qy-0.25)^2; % desired phase advance
```

This function receives an array x with focal length values for the two quadrupoles and plugs them into the proper slot in the beam-line description `beamline` that we pass to the `chisq_tunes` function as a `global` variable. There are other, more robust, ways to pass extra parameters to this function, but using `global` is the most transparent and least clumsy. Once `beamline` is updated with the new values, we update call `calcmat` to update the transfer-matrices and determine the phase advances with `R2beta`. This works well with an uncoupled beam-line, but in coupled beam lines we need to use `edteng` to calculate the phase advances. Finally, we return the squared difference of the calculated to the desired tunes as variable `chisq`.

We use the function `chisq_tunes` in the following example

```
% match_phase_advances.m, V. Ziemann
clear all; close all
global beamline          % make accessible to chisq_tunes.m
beamline=dlmread('fodo.bl');           % load beamline
x0=[-3,3.];                            % starting guesses
[x,fval]=fminsearch(@chisq_tunes,x0)   % matching
dlmwrite('fodo6090.bl',beamline,'\t'); % save to file
```

After clearing the workspace and declaring the `beamline` to be a `global` object, we define starting values for the focal lengths of the quadrupoles to be used in the search and launch the minimizer `fminsearch` to minimize the function `chisq_tunes`. It returns the final quadrupole values in the variable `x` and the achieved minimum values in `fval`. If the matching was successful, `fval` should be very small, say $10^{-8}$ or less. Finally we save the lattice description `beamline`, which now contains the updated values, to a file such that we can retrieve it later to use in other contexts. At this point we can also use the MATLAB function `plot_beta()` to plot the beam sizes or the beta functions in the matched FODO-cell. We do not reproduce the code here, but move on to find quadrupole settings to make small beams for experiments.

### 3.6.2  Match beta functions to a waist

In this example, we use three FODO cells with 60- and 90-degree phase-advance from the previous example and adjust the last four quadrupoles to create a double beam waist $\beta_x = \beta_y = 1\,\text{m}$ after an additional $1\,\text{m}$ long drift-space at the end of the beam-line. We need four quadrupoles in order to fulfill the requirements $\beta_x = \beta_y = 1\,\text{m}$ and $\alpha_x = \alpha_y = 0$. The latter requirement defines the waist and the requirement for the beta functions at the waist. The MATLAB function used to define the beam-line and to call from the minimizer `fminsearch` is the following:

```
% match_to_waist.m, V. Ziemann
clear all; close all
global beamline sigma0
beamline=dlmread('fodo6090.bl');
[Racc,spos,nmat,nlines]=calcmat(beamline); Rturn=Racc(:,:,end);
sigma0=periodic_beammatrix(Rturn,1,1);   % epsx=espy=1
extra_drift=[1,5,0.2,0];
beamline=[repmat(beamline,3,1);extra_drift];
x0=[-1,2.,-2,2];
[x,fval]=fminsearch(@chisq_waist,x0)
plot_betas(beamline,sigma0)
```

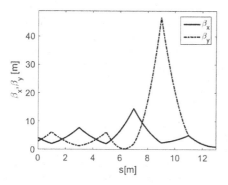
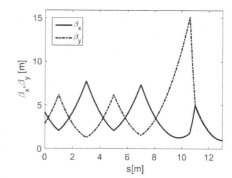

Figure 3.19 Matching a beam waist of $\beta_x = \beta_y = 1$ m at the end of the beam line. The solid lines are the horizontal beta function and the dashed the vertical. The left plot uses the regular quadrupole spacing of the FODO cells from the previous example and on the right plot the fifth quadrupole is moved closer to the final quadrupole, which relaxes the vertical beta function significantly.

First we clear the workspace and define the beamline description `beamline` and the initial beam matrix `sigma0` as global variables because we will need them inside the $\chi^2$−function and this is the easiest way to make them accessible. Then we load the beam-line description of the FODO-cell with a phase advance of 60 and 90 degrees that we saved in the previous example and calculate the matrices for one FODO-cell with `calcmat`. Passing `Rturn` and the horizontal and vertical emittances to the function `periodic_beammatrix` returns the periodic beam matrix that is constructed from the two $2 \times 2$ blocks on the diagonal of `Rturn` using `R2beta`. See Appendix B.5 for the code and further explanations. After having prepared the initial beam matrix, we define an extra drift-space that we want to add to the end of the beam line, because that is where our experimental station is located. Now we increase the beamline to contain three FODO-cells and add the extra drift space at the end before defining starting values for the four quadrupoles we want to vary. Finally, we call the minimizer and plot the beta functions with the function `plot_betas`, which encapsulates the plotting and axis annotations in a separate function and is also explained in Appendix B.5.

The $\chi^2$−function that we pass to the minimizer and that will find the quadrupole values is the following

```
% chisq_waist.m, find focal length to set waist at end
function chisq=chisq_waist(x)
global beamline sigma0      % need info about the beamline
beamline(7,4)=x(1);         % set quadrupole focal lengths
beamline(9,4)=x(2);
beamline(12,4)=x(3);
beamline(14,4)=x(4);
[Racc,spos,nmat,nlines]=calcmat(beamline); Rend=Racc(:,:,end);
sigma=Rend*sigma0*Rend';
chisq=(sigma(1,1)-1)^2+(sigma(3,3)-1)^2+sigma(1,2)^2+sigma(3,4)^2;
```

As input parameters, we receive the vector x with the four quadrupole values and assign them to the strength parameter, here the focal lengths, of the quadrupoles. The position of the quadrupoles are easily determined from inspecting the array beamline. We pick the lines with a code for a quadrupole. After the quadrupoles have their new values, we calculate the beam-line matrixes with calcmat and propagate the beam matrix at the start of the beam line sigma0 to the end and finally calculate the chisq as the sum of squares of the difference of the beam-matrix elements to their desired values which is also returned to the calling program.

Running the MATLAB script match_to_waist.m will produce the plot on the left-hand side in Figure 3.19. It shows that both beta-functions at the end are very small and the $\chi^2$ at the end fval is on the order of $10^{-9}$ which indicates that we actually fulfill the requirements with the quadrupole settings x= -0.9315, 1.5834, -1.3585, 1.6292 albeit at the expense of a rather large vertical beta function in the fifth quadrupole.

We explore to move quadrupoles around by moving the fifth quadrupole to be placed only 0.6 m upstream of the sixth to create a quadrupole-doublet. We simply add the instructions

```
beamline(11,2)=13;  % change quadrupole positions
beamline(13,2)=2;
```

just before calling fminsearch(), which changes the repeat-count of the drift-spaces such that the first becomes a little longer and the second, in between the quadrupoles, correspondingly shorter. Running the minimizer with the changed geometry once again results in the beta-functions shown on the right of Figure 3.19 with significantly relaxed vertical beta-function.

### 3.6.3 Point-to-point focusing

In the next example, we consider a simple imaging system where we require to image one focus to a second focus, which is commonly called *point-to-point* focusing, as illustrated in Figure 3.20. The system is given by a lens with focal length $f$ sandwiched between two drift spaces of length $b = 2$ m and $g = 1$ m and we want to determine the required focal length $f$. The point-to-point requirement implies that all rays starting in the middle of the beam pipe with $x = 0$ but having a non-zero angle $x'$ to cross the center of the beam-pipe after the second drift space. This is illustrated by the solid ray starting at the foot of the original image on the right which ends at the foot of the inverted image at the end of the beam-line. The requirement for point-to-point focusing is thus that all angles at the start of the beam-line are mapped to the same point on the optical axis at its end. But the matrix element that maps angles, having index number 2 in the transfer-matrix, to a position, having number 1, is the $R_{12}$.

The $\chi^2$-function to minimize therefore only contains the square of the $R_{12}$. The following code simply assigns the new focal length f to the entry in the beamline description that holds the focal length of the thin quadrupole, calculates the transfer-matrices and assigns the square of the $R_{12}$ at the end to the returned value chisq.

```
% chisq_R12.m, fit for matrix element
function chisq=chisq_R12(f)
global beamline
beamline(2,4)=f;
[Racc,spos,nmat,nlines]=calcmat(beamline);
chisq=Racc(1,2,end)^2;
```

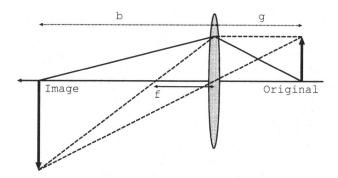

**Figure 3.20** Point-to-point imaging with a lens.

We minimize this cost-function, named `chisq_R12()`, in the following MATLAB-script, where we first define the `beamline` to be global in order to make it easily accessible inside `chisq_R12()` before defining the beam-line. After assigning a starting guess for the minimizer `fminsearch`, we call it with the `chisq_R12()`-function and the starting guess as argument.

```
% match_point_to_point.m, match the R12
global beamline
beamline=[1,  1,  1,  0;    % g
          2,  1,  0,  3;    % f
          1,  1,  2,  0];   % b
f0=3;  % starting guess
[f,fval]=fminsearch(@chisq_R12,f0)
```

Since we omitted the semicolon after the call to `fminsearch`, the final value for the focal length `f` and the final value of the cost-function `fval` are directly returned. For the focal length we obtain `f=0.6667` which is close to 2/3.

This example was deliberately chosen to be very simple and we can also solve the problem, quasi by hand, by explicitly calculating the matrix element $R_{12}$ and solving for the focal length that makes it zero. For the matrix $R$ at the end of the beam line we have

$$R = \begin{pmatrix} 1 & b \\ 0 & 1 \end{pmatrix} \begin{pmatrix} 1 & 0 \\ -1/f & 1 \end{pmatrix} \begin{pmatrix} 1 & g \\ 0 & 1 \end{pmatrix} = \begin{pmatrix} 1-b/f & b+g-bg/f \\ 1/f & 1-g/f \end{pmatrix} \tag{3.106}$$

and the requirement that the $R_{12}$ matrix element must be zero yields

$$b + g - \frac{bg}{f} = 0 \quad \text{or} \quad \frac{1}{b} + \frac{1}{g} - \frac{1}{f} = 0 \tag{3.107}$$

which returns the well-known imaging-equation for lenses from light optics. Solving for the focal length $f$ and inserting values for $b$ and $g$ we, again, find $f = 2/3$. Thus, we find that the requirement for point-to-point imaging leads to the well-known imaging-equation. In this simple system we can perform the optimization analytically. On the other hand, if we have to deal with a more complex beam-line, of course this becomes rather clumsy and we must resort to numerical solutions, which is much more convenient, especially if several constraints have to be fulfilled.

In this section we see how we formulate requirements as cost-functions and use

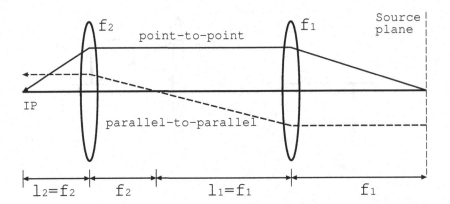

**Figure 3.21** A telescope images the source plane to the image plane at the interaction point (IP) with demagnification $-f_2/f_1$.

`fminsearch` to satisfy these requirements by varying a number of parameters such as quadrupole values or, in fact, any parameter in the `beamline` description. This works nicely for phase advances, for beta-functions, and also for individual matrix elements. Generalizing from these simple examples is straightforward and we will use that as a tool in the next section, where we discuss a number of beam-optical modules that are frequently used as the building blocks to construct larger systems, up to entire accelerators.

## 3.7 BEAM-OPTICAL SYSTEMS

In this section we address several of the often-used beam-optical modules. One example, we encountered before, the FODO-cell, is a module that is frequently used to cover long distances with a simple magnet lattice. The arcs of the LHC consist of FODO-cells with additional dipole magnets to force the particles on their circular path along the 27 km long beam pipe. Quadrupoles are used to focus deviating particles back towards their design orbit and thereby ensure stable operation. While the FODO-cells serve to transport the beam through the approximately 3 km long octants, the beams need to be focused to extremely small sizes inside the detectors, such as ATLAS and CMS, that are located in straight sections between the octants. In order to demagnify the beams to sizes of tens of microns, other optical modules are used, in this special case the *telescopes,* we consider next.

### 3.7.1 Telescopes

We start by utilizing the imaging equation from Section 3.6.3 and build telescopes, which are often used to obtain small beam-sizes, as is the case in the collision- or interaction-points (IP) of colliding beam accelerators such as the LHC.

Using two lenses with a drift space between them it is possible to build an optical telescope that has the desired optical properties. It creates a point-to-point image of the image plane (near the arc) to the IP, which means that the $R_{12}$ and the $R_{34}$ are made zero. Moreover, we also require that the imaging is parallel-to-parallel, which means that the $R_{21}$ and $R_{43}$ are zero. Consider a simple one-dimensional optical system as indicated in Figure 3.21, where the beam comes from the right and first passes through a lens with focal length $f_1$ and then through a drift space of length $f_1 + f_2$, whence the beam passes the second lens with focal length $f_2$, and after another drift space of length $f_2$, it arrives at the IP. We can easily write down the transfer matrix for the first drift-lens-drift module with index 1 with the result

$$R_1 = \begin{pmatrix} 1 & l_1 \\ 0 & 1 \end{pmatrix} \begin{pmatrix} 1 & 0 \\ -1/f_1 & 1 \end{pmatrix} \begin{pmatrix} 1 & l_1 \\ 0 & 1 \end{pmatrix} = \begin{pmatrix} 1 - l_1/f_1 & 2l_1 - l_1^2/f_1 \\ -1/f_1 & 1 - l_1/f_1 \end{pmatrix} . \tag{3.108}$$

Since we have chosen $l_1 = f_1$ the matrix simplifies to

$$R_1 = \begin{pmatrix} 0 & f_1 \\ -1/f_1 & 0 \end{pmatrix} \tag{3.109}$$

and we have a similar matrix $R_2$ for the second drift-quadrupole-drift system with index 2. Multiplying the two matrices with index 1 and 2, we arrive at the following transfer matrix $R$ that represents the beam optical system between the source plane and the IP

$$R = R_2 R_1 = \begin{pmatrix} -f_2/f_1 & 0 \\ 0 & -f_1/f_2 \end{pmatrix} . \tag{3.110}$$

Here we see that it describes a system that demagnifies the $x$ coordinate by the factor $M = -f_2/f_1$, the ratio of the focal lengths. The minus sign describes the inversion of a picture that is commonly encountered in normal telescopes. In practice, we now have to realize such a system with quadrupole magnets and the matching module discussed in earlier sections is very handy to find the quadrupole values once the geometry, i.e., the magnet lengths and distances, are known.

A small problem arises, because quadrupoles do not act like optical lenses that focus in both planes. As discussed in Section 3.1.1, quadrupoles focus the beam in one plane and they defocus in the other plane. We therefore have to combine several quadrupoles to obtain a system that focuses in both planes. We address such a system, a quadrupole triplet, in the next section.

### 3.7.2 Triplets

The telescope from the previous subsection represents a one-dimensional, or circularly symmetric, optical system. Quadrupoles, however, focus in one plane and defocus in the other. We therefore have to combine quadrupoles in such a way that their behavior resembles that of spherical lenses—they should focus in both planes. One such combination of quadrupoles is a *triplet* that consists of three closely spaced quadrupoles, where the central quadrupole has twice the strength and opposite polarity of the two equally powered outer quadrupoles.

The following lattice file `triplet.bl` describes such a beam line. It consists of a 5 m long drift-space, the three quadrupoles, separated by 1 m long drift spaces, and another 5 m long drift space. The outer quadrupoles have about twice the focal length of the inner quadrupole, therefore the inner quadrupole is twice as strong as the outer ones. Moreover, it has the opposite sign.

| | | | |
|---|---|---|---|
| 1 | 10 | 0.5 | 0 |
| 2 | 1 | 0 | 3.219 |
| 1 | 5 | 0.2 | 0 |
| 2 | 1 | 0 | -1.739 |
| 1 | 5 | 0.2 | 0 |
| 2 | 1 | 0 | 3.219 |
| 1 | 10 | 0.5 | 0 |

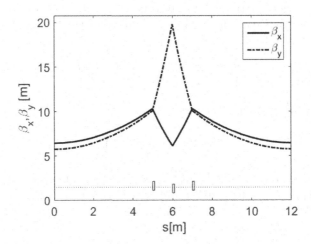

Figure 3.22 Beta functions for a triplet cell. Here the phase advance for the cell is adjusted to $\mu/2\pi = 0.25$ in both planes.

These quadrupole values were calculated with the function `match_triplet_tune.m`. It uses the $\chi^2$–function `chisq_tune_triplet.m`, which is similar to the one from Section 3.6.1 and matches the phase advances of this triplet cell to 0.25 in both planes. Finally, it calculates the transfer matrices with `calcmat`, determines the periodic beta functions and then plots them, which results in Figure 3.22.

We see that the beta-functions are periodic and long sequences of triplet cells are used in case equal beam sizes in long sections are required, such as long accelerating structures with narrow apertures. In that case it is beneficial to use a triplet lattice, which provides almost round beams with a rather constant and equal width in both planes. Only inside the triplet, and especially inside the central quadrupole the beta-functions, and thus also the beam sizes, differ significantly. From their appearance, the triplet cell maps an almost round beam from the entrance to an almost round beam at the exit of the cell and we may use triplets to construct telescopes to demagnify the beta functions in order to achieve very small and round beam spots for experiments.

We therefore use a system of three consecutive triplet cells. The first is only used to define the input beta functions and the second and third triplets form a telescope that we will adjust to obtain a small and round spot at the end of the beamline. To encode this requirement, we define a $\chi^2$–function that is very similar to the one from Section 3.6.2. It requires $\alpha_x$ and $\alpha_y$ to be zero and requires $\beta_x$ and $\beta_y$ to a small value. To fulfill these requirements, we vary the quadrupoles in the last two triplets, but always power the outer quadrupole pairs equally. The function `chisq_waist_triplet` achieves this.

```
% chisq_waist.m, find focal length to set waist at end
function chisq=chisq_waist_triplet(x)
global beamline sigma0      % need info about the beamline
beamline(9,4)=x(1); beamline(11,4)=x(2); beamline(13,4)=x(1);
beamline(16,4)=x(3); beamline(18,4)=x(4); beamline(20,4)=x(3);
[Racc,spos]=calcmat(beamline); Rend=Racc(:,:,end);
sigma=Rend*sigma0*Rend';
```

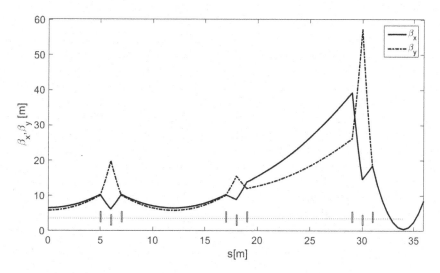

Figure 3.23 Beta functions for a telescope made of triplets.

```
chisq=(sigma(1,1)-0.5)^2+(sigma(3,3)-0.5)^2+sigma(1,2)^2+sigma(3,4)^2;
```

It is almost a straight copy of the function `chisq_waist` from Section 3.6.2, only this time we assign the four values to the six quadrupoles. The positions 9,11,13 of the quadrupoles in the second triplet and 16,18,20 in the third can easily be found by inspecting the `beamline` array. Running the matching code, we find that we have difficulties to achieve beta-functions at the end of the beam-line much smaller than 1 m. A way to achieve smaller spots is to make the final beam waist—the focus—closer to the final quadrupole. Basically, we need to shorten the beamline, for example, by reducing the repeat-count of the last drift in the `beamline` array from 10 to 6. Another option is to add a drift-space with a negative length to the end of the beamline. This has the added bonus that we actually see the minimum of the beta-functions. The following code illustrates this method.

```
% match_waist_triplet.m, V. Ziemann
clear all; close all
global beamline sigma0
t1=dlmread('triplet_25.bl'); beamline=t1;
[Racc,spos,nmat,nlines]=calcmat(beamline); Rturn=Racc(:,:,end);
sigma0=periodic_beammatrix(Rturn,1,1);
negdrift=[1,4,-0.5,0];
beamline=[repmat(t1,3,1);negdrift];
f0=[3.6,-1.8,3.6,-1.8]; % starting guess
[f,fval]=fminsearch(@chisq_waist_triplet,f0)
plot_betas(beamline,sigma0); drawmag(beamline,2,3)
```

This MATLAB script is based on the matching example from Section 3.6.2 but we also define the drift space with the negative length `negdrift` to contain 4 segments of $-0.5$ m length and add it to the end of the `beamline` that already contains three triplet-cells. Then we provide starting-guesses for `fminsearch` and call it to minimize the $\chi^2$−function chisq_waist_triplet. Finally, we plot the beta functions and add the magnet lattice to identify the positions of the quadrupoles.

Figure 3.23 shows the resulting plot. We see the three triplet-cells, of which the last two are detuned in order to achieve the small, and equal to 0.5 m, beta-functions. The values for the focal lengths of the quadrupoles are given by f=[6.5836,-3.1531,2.2944,-1.3211], where we, again, find the pattern that the excitations of inner quadrupoles of each triplet (second and fourth values) have a negative sign and have approximately half the focal lengths of the outer quadrupoles. The $\chi^2$ after the match is below $10^{-9}$ indicating that the match completed successfully. By increasing the number of negative drift space segments to 9 and thus moving the $s-$position of the waist to 0.5 m after the last quadrupole, we can even achieve beta-functions at the waist of 0.1 m in both planes.

Since the triplets treat both planes approximately equally, they can be considered as the equivalent of spherical lenses and are often a good choice if both planes need to be treated in a similar way, for example, if we require round spots at the experiment. If we, on the other hand, deal with beams that have large aspect ratios $\beta_y/\beta_x$, we can use quadrupole doublets.

### 3.7.3 Doublets

A doublet consists of two closely spaced quadrupoles with opposite polarity. The transfer matrix of such a system is given by

$$R_D = \begin{pmatrix} 1 & 0 \\ -1/f & 1 \end{pmatrix} \begin{pmatrix} 1 & l \\ 0 & 1 \end{pmatrix} \begin{pmatrix} 1 & 0 \\ 1/f & 1 \end{pmatrix} = \begin{pmatrix} 1+l/f & l \\ -l/f^2 & 1-l/f \end{pmatrix}, \qquad (3.111)$$

where $f$ is the focal length of the quadrupoles and $l$ the distance between them. The matrix element $R_{21} = -l/f^2$ that translates initial positions of final angles is responsible for focusing. It is always negative, irrespective of the sign of the focal length $f$, which implies that the doublet also focuses if we reverse the polarities of the quadrupoles. But this describes the situation in the other plane.

The periodic lattice with a phase advance of 0.2 in both planes per doublet cell is shown on the left-hand side in Figure 3.24, where we depict two consecutive cells such that the total phase advance in both planes is 0.4. We find that the beam-size variation is rather modest, and they vary by a factor of two from 7 to 14 m. Moreover, the beta-functions in one plane are the left-to-right mirror-image of those in the other plane. The large space between the doublets makes this type of lattice suitable for systems that require free space for installation of, for example, acceleration structures. Instead of using a sequence where focusing and defocusing quadrupoles alternate we can also use a sequence where two quadrupoles of a kind follow one another. We show such a beamline and the corresponding beta functions on the right-hand side in Figure 3.24. The total phase advance of the displayed sections with two doublets is again 0.4 in both planes, but this time one beta function is large in the space between the doublets while the other one is small. The role of the planes alternates in consecutive long drift spaces. Such a lattice is suitable, if we require a particularly small beam size in one plane as is the case if we want the synchrotron radiation emitted by electrons to be diffraction-limited in one plane.

We can also use doublets in telescopes, but have to keep in mind that the focusing in the two planes is different. We can use them if the beam at the IP has a large aspect-ratio, as is the case for the International Linear Collider (ILC), which uses a doublet as the final focusing lens. The asymmetry between the planes is easy to understand by considering the point-to-point focusing properties of a doublet which is important for using it in a telescope,

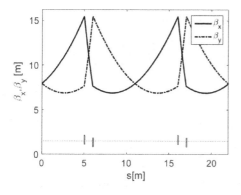
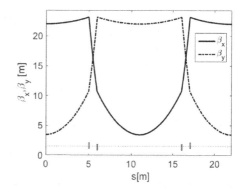

Figure 3.24 Beta functions for two doublet cells. On the left, the cells are periodic with alternating focusing and defocusing quadrupoles. In the right, two focusing and two defocusing quadrupoles follow each other. In the configurations, the phase advance in both planes is $\mu/2\pi = 0.4$.

see Section 3.7.1. We calculate the transfer matrix for a beam-line that starts with a drift space of length $5L$, followed by a doublet with focal length $f$ and distance $L$ between the quadrupoles, followed by another drift space of length $5L$. The resulting transfer matrix is

$$R = \begin{pmatrix} 1 + L/f - 5L^2/f^2 & 11L - 25L^3/f^2 \\ -L/f^2 & 1 - L/f - 5L^2/f^2 \end{pmatrix} \qquad (3.112)$$

where we see that the focal length $\hat{f} = \sqrt{25/11}L$ makes the $R_{12}$ matrix-element zero, and results in the transfer-matrix

$$R = \begin{pmatrix} -0.5367 & 0 \\ -0.44/L & -1.8633 \end{pmatrix}. \qquad (3.113)$$

The transfer-matrix for the other plane has the diagonal elements exchanged. And there we have the case that the doublet demagnifies, here by $-0.5367$ in one plane but amplifies by $-1.8633$ in the other plane. This is not surprising by considering the propagation of rays in the doublet. The other plane has focusing and defocusing quadrupoles reversed, but that corresponds to the rays going in the opposite direction through the doublet. If it is demagnifying in one plane, it does the opposite when going the other direction, or equivalently, propagating in the other plane.

After the first optical building blocks that deal predominantly with imaging in the transverse planes, we consider a system that is used to cancel the dispersion generated when deflecting the beam with dipole magnets. Such systems are called *achromats*.

### 3.7.4 Achromats

Dipole or bending-magnets deflect particles with different energies by different deflection angles and thereby sort the particles according to their energy. In other words they behave as a spectrometer and generate dispersion, where dispersion $D$ is defined through the trajectory $x = D\delta$ that a particle with non-zero momentum offset $\delta = (p - p_0)/p_0$ follows. Since a single dipole spreads the trajectories, we need at least a second dipole to collect the particles

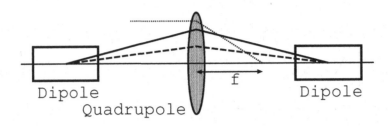

Figure 3.25 Achromat with dipoles deflecting in the same direction.

again and put them back on the reference trajectory. The simplest system we can make achromatic consists of two dipoles and a focusing quadrupole half-way between the dipoles. Figure 3.25 illustrates this configuration. The first dipole on the left spreads the particles according to their momentum such that they are sorted according to their momentum when they arrive at the quadrupole, but the sorting is linear in the momentum offset and the linearly rising off-axis field of the focusing quadrupoles deflects the particles with the largest excursion the most. In this way all particles arrive at the second dipole on the reference trajectory and by receiving the same momentum-dependent deflection as in the first dipole, are deflected back onto the reference trajectory. Note that this cancellation of momentum-dependent deflections in the two dipoles only works if the quadrupole provides point-to-point imaging between the centers of the dipoles. Here the focal length is a quarter of the distance between the dipole centers. Note that point-to-point focusing corresponds to a betatron phase advance of 180 degrees and the particles arrive at the second dipole with the opposite phase, but receive the same deflection as in the first dipole such that the deflections cancel.

The previous achromat works well with dipoles deflecting in the same direction, but how many quadrupoles do we need and how do we excite them if the dipoles deflect in opposite directions? This is, for example, the case, if the dipoles have to provide a parallel displacement of the reference trajectory in order to deliver the beams to multiple experimental areas, located side-by-side in a beam switchyard. Or to lift the reference trajectory from the basement, where a cyclotron is located, to an upper level, where the experiment areas are located. In the latter case, the dipoles are vertically deflecting and we have to employ the coordinate-rotations from Section 3.1.6 to describe the vertically deflecting dipoles. But to simplify the discussion we stay with horizontally deflecting dipoles of opposite polarity. In order to cancel the dispersion generated by the two dipoles we would love to have them at the same place because their effect would cancel locally. The next best thing is to have

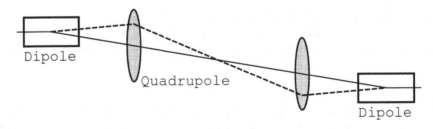

Figure 3.26 Achromat with dipoles deflecting in opposite directions.

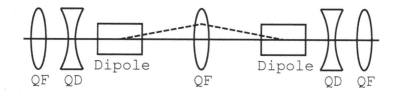

Figure 3.27 Double-bend achromat cell where the dashed line indicates the dispersion.

them separated by a phase-advance of 360 degrees. This can be accomplished by using two point-to-point focusing quadrupoles, which is the configuration shown in Figure 3.26 where the dipoles deflect in opposite directions and the two quadrupoles cause the dispersive orbit, denoted by a dashed line to perform a full 360-degree oscillation around the solid line that denotes the reference trajectory. The angle between the reference trajectory and the dispersive orbits in the two dipoles have equal magnitude, but opposite sign, whereas in the previous example from Figure 3.25 both magnitude and sign are equal. These two examples should serve as an intuitive guide how to build achromatic systems, but in general more constraints have to be fulfilled and more quadrupoles are part of the beam lines.

In those, more complex, cases the qualitative argument from the two previous paragraphs can be generalized by observing that the momentum-dependent offset and angle are given by the transfer-matrix elements $R_{16}$ and $R_{26}$, such that we can use these quantities in a fitting routine. The constraint is given by the requirement to make $R_{16}$ and $R_{26}$ zero by varying suitable quadrupoles within the beam-line section under consideration.

Systems with two dipole magnets are the simplest achromatic systems but more complex ones are frequently found in synchrotron-light sources.

### 3.7.5  Multi-bend achromats

One of the important figures-of-merit for a synchrotron-light source is the achievable small emittance. Moreover, they need moderately long straight sections with zero dispersion to place the specialty magnets—undulators-and wigglers—that produce the light. We thus need sections, also called cells, that have zero dispersion at the entrance and exit of the cell as well as the property to minimize the emittance. As discussed in Section 3.4.3 the emittance is determined by the emission of radiation inside the dipole magnets and proportional to $\mathcal{H}/|\rho|^3 \approx D^2/|\rho|^3$ where $\mathcal{H}$ is defined in Equation 3.100 and the inverse dependence on the bending radius $\rho$ will be discussed in Section 10.1. We thus need to ensure that the dispersion $D$ inside the dipoles is small and that is the case in the achromat shown in Figure 3.25 because the dispersion orbit starts in one dipole from zero and is returned to

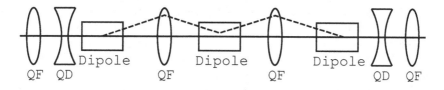

Figure 3.28 Triple-bend achromat cell where the dashed line indicates the dispersion.

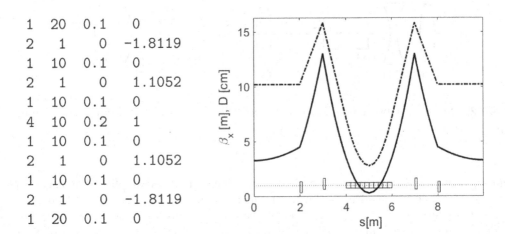

| 1 | 20 | 0.1 | 0 |
|---|----|-----|---|
| 2 | 1 | 0 | -1.8119 |
| 1 | 10 | 0.1 | 0 |
| 2 | 1 | 0 | 1.1052 |
| 1 | 10 | 0.1 | 0 |
| 4 | 10 | 0.2 | 1 |
| 1 | 10 | 0.1 | 0 |
| 2 | 1 | 0 | 1.1052 |
| 1 | 10 | 0.1 | 0 |
| 2 | 1 | 0 | -1.8119 |
| 1 | 20 | 0.1 | 0 |

**Figure 3.29** Left: the lattice file `tme.bl`. Right: the corresponding horizontal beta function (solid) and the dispersion (dashes) for one TME cell.

zero in the next dipole. Adding quadrupole doublets before the first and after the second dipole allows us to control the beta-functions in both planes as well. The layout of such a double-bend achromat (DBA), or Chasman-Green, cell is shown in Figure 3.27. DBA-cells are used in many synchrotron-light sources, most notably the NSLS in Brookhaven, Elettra in Trieste, and the ESRF in Grenoble.

The dependence of the emittance on the inverse bending radius $\rho$ indicates that it is advantageous to make the rings larger by increasing $\rho$ and building achromats consisting of weaker, but more dipoles. An example that uses three dipoles is the triple-bend achromats (TBA). A sketch of the layout of one cell is shown in Figure 3.28. Examples of light sources using this type of cell are the ALS in Berkeley, the TLS in Taiwan, and the first version of the PLS in Korea. Making the rings very large by increasing the circumference allows the combination of multiple dipoles to *multi-bend achromats*. For example, MAX-IV in Sweden uses seven dipoles in a single achromatic cell.

### 3.7.6 TME cell

If we relax the requirement for zero dispersion outside the cell, but emphasize the requirement for the smallest possible emittance, we arrive at the *theoretical minimum emittance* (TME) cell. Since we do not have to cancel the dispersion, we use a single dipole magnet and place quadrupoles in such a way that the dispersion and $\mathcal{H}$ is minimum inside the dipole magnet. See Figure 3.29 for the lattice file and optical functions. The two focusing quadrupoles adjacent to the dipole focus the dispersion down to the smallest possible value at the expense of having non-zero dispersion outside the cell because we can only manipulate a non-zero dispersion with quadrupoles alone. Moreover, the quadrupoles are often very strong. TME cells are sometimes considered for damping rings for linear colliders. Their sole requirement is to provide the smallest possible emittance without concern for dispersion-free straight sections to place undulators or wigglers.

Figure 3.29 shows the lattice file and a plot of the dispersion and the horizontal beta function for a TME-cell with a 10-degree dipole that has phase advances of 240 and 120 degrees

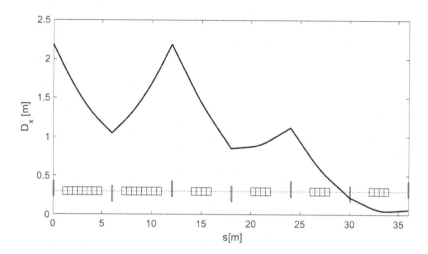

Figure 3.30 A dispersion suppressor in a 90-degree FODO lattice.

in the horizontal and vertical planes, respectively. Note the very small dispersion with scale given in cm, rather than meters. We see that the lattice is similar to the doublet lattice shown on the right of Figure 3.24 with a dipole sandwiched between the doublets with focusing quadrupoles that have the purpose to minimize dispersion inside the dipole.

### 3.7.7 Dispersion suppressor

The dispersion in the interaction regions of storage-ring colliders must also be zero in order to prevent the momentum spread $\sigma_\delta$ to increase of the beam size at the interaction point. The dispersion that is inevitably generated by the dipole magnets in the arcs therefore has to be canceled at the ends of the arcs. And this is the purpose of a dispersion suppressor which is often implemented by reducing the excitation of the dipoles at the ends of the arcs. In [14] Helm determines patterns of dipole excitations that cancel the dispersion. Provided that the phase-advance in a FODO-cell is a sub-multiple of 180 degrees horizontal phase advance, it is possible to replace all dipoles within the last 180 degrees of the arc by dipoles of half their normal length. This creates an interference pattern, opposite in phase to the periodically oscillating dispersion in the arcs, which cancels the dispersion at the end of the arc. Conceptually, this has a similar purpose as the "nose" at the bow of large ships, which also creates an interference pattern that reduces the creation of waves. In the dispersion suppressor the dipoles with reduced excitation cancel the dispersion wave.

Figure 3.30 shows an example with a dispersion suppressor in a FODO arc with a phase-advance of 90 degrees per cell in both planes. On the left of the figure the dispersion, shown as the solid line, starts with the periodic values for the arc and the first cell contains two long dipoles. The following two cells, covering a phase-advance of 180 degrees, have the length of the dipoles halved and we see that the periodic dispersion coming from the arcs is reduced to almost zero. Close observation shows that the dispersion is not exactly zero, because the weak focusing of the shorter dipoles is different and slightly changes the horizontal phase advance in the suppressor-cells. This, however, is a small effect, proportional to $1/\rho^2$ where $\rho$ is the bending radius of the dipoles, and is only visible in this example, because we chose dipoles with small bending radii. In most circumstances, in large storage rings with large bending radii the discrepancy is not important.

A dispersion suppressor is normally used to interface arcs of a storage ring to straight sections, often called interaction regions, where the experiments with their detectors are located.

### 3.7.8 Interaction region

The $\mu$m–scale beam sizes at the interaction point (IP) in the SLAC Linear Collider (SLC) in the 1990s and the nm–scale sizes in future linear colliders such as the International Linear Collider (ILC) or the Compact LInear Collider (CLIC) require extremely strong quadrupoles very close to the IP. If the focusing is very strong and the beta functions $\beta^*$ at the IP are very small, it is easy to show that at a distance $s$ before or after the IP they assume the value $\beta(s) = \beta^* + s^2/\beta^*$. For very small values of $\beta^*$, say mm or less, the beta functions at the closest quadrupoles, which typically are a few meters away, are on the order of several km. This is indicated by the dashed line, labeled $\beta$ near the IP in Figure 3.31. Moreover, since the focal lengths of quadrupoles depend on the momentum of the particles the finite momentum spread of the beam will increase the beam size at the IP, because particles with different momenta have their waist at different longitudinal positions. This longitudinal dependence of the focal point is also called *chromaticity*, just like the chromaticity in a circular accelerator we discussed in Section 3.4.1, but here the interpretation is different. But despite this difference the magnitude is also determined by the product of quadrupole strength $k_1$ and the beta function and the dominant contribution comes from the quadrupoles closest to the IP. In order to compensate this effect we need another momentum-dependent source of focusing, preferably close to the final focus quadrupoles. Thus all linear collider final focus systems have to deal with correcting the chromaticity and in the following we discuss the solution that was implemented at the SLC.

The conceptual layout of the SLC final focus is shown in Figure 3.31. The beam comes from the right and moves towards the left. It first passes through a matching section, which contains a dispersion suppressor and several quadrupoles and skew quadrupoles in order to match the beam matrix to the design values in the final focus. If a beam with design parameters arrives, the transfer matrix through the matching section is a negative unit matrix and will leave the sigma matrix untouched. The next section contains a telescope that demagnifies the beam size by a factor $M_x = 8.5$ in the horizontal and $M_y = 3.1$ in the vertical plane. Weak dipoles in the chromatic correction section (CCS) generate some dispersion, indicated by the dotted line. Sextupoles placed in this section act as momentum-dependent quadrupoles and are used to compensate the chromaticity by the mechanism we later discuss in Section 8.5.4. By placing equally powered sextupoles with phase advances of 180 degrees apart it is possible to cancel unwanted nonlinear aberrations, a concept we return to in Section 11.3. Two independently powered families of sextupoles, indicated by the solid and the hashed rhombs, are used to adjust the horizontal and the vertical chromaticity independently. Finally, a second telescope, constructed using two triplets, demagnifies by a factor of $M = 4$ in both planes, before the beam meets the other, counter-propagating beam at the IP.

The correction of the chromaticity with a dedicated CCS with its four-dipole achromat makes the entire final focus system extremely long, especially at high beam momenta because the dipoles create very little dispersion in the sextupoles, needed to correct the chromaticity. This problem can be alleviated by realizing that we only need to make the dispersion $D$ zero at the IP, but not necessarily the derivative of the dispersion $D'$. Thus

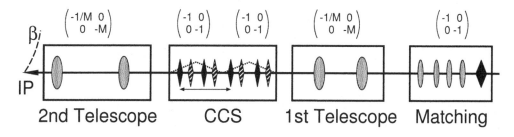

**Figure 3.31** The conceptual layout of the SLC final focus. The beam travels from right towards the IP at the left. Above the boxes are the nominal transfer matrices for the respective sections displayed. $M$ denotes the demagnification factor for the telescopes.

we can have significant dispersion near the IP and place sextupoles next to the final focus quadrupoles and cancel the chromaticity where it is created. A highly flexible system can be designed around this concept as is shown by Raimondi and Seryi in [15]. Their design will be used in future linear collider final focus systems because it shortens the length of the system by kilometers besides having many other advantages discussed in [15].

The systems discussed so far dealt mostly with transverse properties of the beams. In some cases, however, we need to address the longitudinal properties, for example to create extremely short bunches. Examples are linear colliders and accelerators that drive free-electron lasers such as LCLS, SACLA, or the European XFEL.

### 3.7.9 Bunch compressors

In a bunch compressor we seek to reduce the bunch length at the expense of the momentum spread. This is achieved by accelerating the bunch off-crest in an accelerating cavity. In this way the head of the bunch can be made to receive a lower energy than the tail of the bunch. We now have to produce a device that translates momentum difference into arrival time difference. A chicane is such a device, where three dipole magnets with bending angles $\phi$, $-2\phi$, and $\phi$, respectively, are arranged as is shown in Figure 3.32. The idea is to give particles with different energies different path lengths. In particular, a particle with higher energy will be deflected less in the dipoles and will take a shortcut on the inside of the chicane, resulting in a shorter path length. This shortening can be calculated by first considering the length of the unperturbed path

$$l = \frac{2L}{\cos\phi} \approx \frac{2L}{1 - \phi^2/2} \approx 2L\left(1 + \frac{\phi^2}{2}\right) . \tag{3.114}$$

If the momentum offset is $\delta$ the bending angle will be reduced by $\phi \to \phi/(1+\delta)$, we find

$$l(\delta) = 2L\left(1 + \frac{\phi^2}{2(1+\delta)^2}\right) \approx 2L\left(1 + \frac{\phi^2}{2}\right) - 2L\phi^2\delta , \tag{3.115}$$

where we observe that the first term is equal to $l(0)$. The $R_{56}$, which describes the path length change as a function of momentum offset $\delta$, is thus given by

$$R_{56} = \frac{l(\delta) - l(0)}{\delta} \approx -2L\phi^2 \tag{3.116}$$

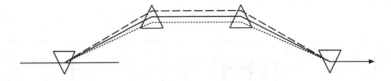

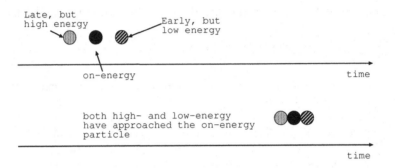

**Figure 3.32** Layout of a bunch compressor. A particle with higher momentum (dotted) than the reference particle (solid) is deflected less and takes a shorter path through the chicane and arrives earlier. A particle with lower momentum (dashed) takes a longer path and therefore arrives later.

**Figure 3.33** Three particles with different momenta have a larger distance between them before the chicane (upper). After the chicane the late high-momentum particle has caught up with the reference particles because it takes a short-cut and, conversely, the early low-momentum particle arrives later because its path is longer. This leads to the reduced distance between the particles.

which is proportional to the total length of the chicane and the square of the bending angle $\phi$.

In accelerators operating with low-energy beams the energy can be modulated by exciting a short, isolated section of beam pipe by a time-varying voltage with respect to the adjacent beam pipe. This produces an energy and velocity modulation of non-relativistic beams, such that the fast beam particles catch up with their slower companions in a drift space and cause the beams to be bunched after some distance. This method is called *velocity bunching*.

All beam optics modules we discussed so far rely on magnets to deflect the trajectory and focus the beams and now is the time to have a closer look at how to design and build these magnets.

## QUESTIONS AND EXERCISES

1. Consider one transverse dimension only, such that you can use the 2D-version of the software for this exercise. Prepare a FODO cell with thin-lens quadrupoles, having the same absolute value of their focal lengths, that starts in the middle of a 5 m long drift space. Adjust the quadrupoles such that the phase advance is (a) 60 degrees; (b) 90 degrees; (c) 77 degrees.

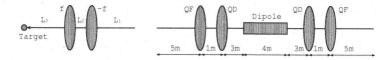

Figure 3.34  The geometry of the beam line for Exercises 6 (left) and 7 (right).

2. Use the 60-degree FODO cell from the previous exercise and replace the thin quadrupoles by long quadrupoles with a length of $0.2, 0.4, 0.8\,\text{m}$. Adjust their strength such that the phase advance per cell remains 60 degrees. By how much does the periodic beta function at the start of the cell change? Express the change in percent.

3. Use the cells you prepared in Exercise 1 and (a) prepare phase-space plots by plotting $x$ versus $x'$ once per turn for 314 turns. Select a few different starting positions. Discuss what you observe. (b) Build beam lines of 20 cells for each of the three phase advances and display the position along the beam line when launching a particle with an angle of $x_0' = 1\,\text{mrad}$. (c) Unless you had already subdivided the drift spaces in short segments, say 10, do so now and replot the orbit.

4. Still using the 2D software: (a) build a beam line of two 60-degree cells and one 90-degree cell, then adjust the quadrupoles in the middle cell to match the Twiss parameters at the end of the first 60-degree cell to those of the 90-degree cell. (b) Replace the 90-degree cell by the 77-degree cell and repeat the matching from part (a).

5. Use the FODO cells with the geometry from Exercise 1 and (a) determine their limit of stability, i.e., for given cell half-length $L$, what values of $f$ permit a periodic solution? (b) What phase-advance per cell $\mu$ corresponds to the limiting cases? (c) Calculate the maximum $\beta_{\text{max}}$ and minimum beta function $\beta_{\text{min}}$ within a cell as a function of the phase-advance per cell $\mu$. (d) Generate a plot of the $\beta_{\text{max/min}}$ versus $\mu$.

6. You are responsible for a short beam line which has the layout shown on the left-hand side in Figure 3.34 with $L_1 = L_3 = 2\,\text{m}$ and $L_2 = 1\,\text{m}$ and a quadrupole doublet with two magnets of equal strength but opposite polarity. The first one is horizontally defocusing and the second is focusing. The colleague who works on the accelerator upstream promises to provide a beam at the entrance of your section which has a horizontal waist there and an rms width of $5\,\text{mm}$ and rms angular divergence of $2\,\text{mrad}$. (a) What is the sigma-matrix at the entrance of your section? (b) What is the transfer matrix for your beam line? (c) What beam size at the end of the beam line do you find for $f = 1\,\text{m}$? (d) Your friends from the experimental group want a small horizontal beam size at their target position at the left of your beam line. To what strength do you need to adjust the quadrupoles? (e) What is the minimum achievable beam size? (f) Assume that the incoming beam has the same vertical sigma matrix as the horizontal. What is the vertical beam size on the target?

7. Use the doublet cell shown on the right-hand sides in Figures 3.24 and 3.34 and (a) place one $4\,\text{m}$ long sector dipole with a bending angle of 12 degrees in the middle of the drift space. Calculate the periodic dispersion and plot it. (b) Now you want to find out whether it is better to place the dipole between the focusing or between the defocusing quadrupoles, you analyze the configuration with a quadrupole polarities reversed and calculate the dispersion once again. (c) If this example describes a section of an electron ring, in which case do you expect a smaller emittance? (d) Numerically

evaluate $\mathcal{H}$ from Equation 3.100 (inside the dipoles) for each of the different variants and find out which one will have the lower emittance.

8. Now use the 4D software, and prepare FODO cells with 60, 90 and 77 degrees in both planes. (a) Then build a beam line to match the Twiss parameters from the 60-degree cell to the 90-degree cell. How many intermediate cells do you need and explain why? Then repeat part (a) by matching into a final 77-degree cell.

9. Use FODO cells with the same geometry as in the previous exercise and find the range of focal lengths of the QF and the QD that permit stable periodic Twiss parameters. The easiest way is to scan the quadrupole strengths $1/f$ for both quadrupoles within reasonable limits and prepare a graph with the two $1/f$ on the axes and mark the spot on the graph with an asterisk if the combination of focal lengths results in a stable periodic lattice.

10. Design a ring that consists of eighteen FODO cells with a length of 10 m having a phase advance of 90 and 60 degrees in the horizontal and plane, respectively. You can use thin lens quadrupoles and place one 2 m-long sector dipole per cell in between the quadrupoles. (a) Find quadrupole values to adjust the tunes to $Q_x = 4.27$ and $Q_y = 3.38$, respectively. (b) Prepare plots of the beta functions in one FODO cell, and for the entire ring. (c) Calculate the chromaticities.

11. Repeat the previous exercise, but use the lattice with doublet cells, shown on the right-hand side in Figure 3.24, instead. Use the doublet cell from Exercise 7 with the 4 m long dipole. (a) Adjust the tune to $Q_x = 7.27$ and $Q_y = 7.38$ and (b) plot the beta functions for one segment and for the entire ring. (c) Track particles, starting with $x_0 = y_0 = 1\,\mathrm{mm}$, for 1024 turns, record the positions, Fourier transform them, and verify that the fractional values of the tunes are correct.

12. Your beam line needs to cross a highway in an underground tunnel. In order to transport the beam downwards by approximately 4 m, as shown in Figure 3.26, you use two 1 m long sector dipole magnets with a bending angle of 30 degrees, each. In between the dipoles you place two quadrupoles. Adjust their focal lengths, such that the beam line from the entrance to the first dipole to the exit of the second is achromatic. You can ignore the Twiss parameters in this exercise.

13. Two ultra-relativistic electrons, initially traveling together, but one of them having a momentum 1 % higher than the other one, pass a bunch compressor with four dipole magnets, spaced by 1 m. If each dipole deflects the electrons by 3 degrees, by how much are they longitudinally separated after the bunch compressor?

14. In the ring with tunes $Q_y = 4.27$ and $Q_y = 3.38$, which you prepared in Exercise 10, one of the horizontally focusing quadrupoles was accidentally mounted, such that it is rolled by 5 degrees around the beam axis. (a) Track particles with 1 mm initial offsets, record turn-by-turn positions for 1024 turns, and make an FFT to determine the tunes. By how much do they differ from the design values? (b) By changing the excitation of all QF, try to make the tunes as equal as possible. You can use the edteng() function to determine the tunes directly from the transfer matrices. For what focal length are the tunes closest? (c) Set the focal lengths of all QF to this value and verify the tune with tracking and FFT. Make sure that the values agree reasonably well.

# CHAPTER 4

# Magnets

In Chapters 2 and 3 we saw that dipole magnets define the reference trajectory in an accelerator and that quadrupole magnets ensure that particles stay close to it. In this chapter we will address topics regarding the design and construction of these magnets.

## 4.1 MAXWELL'S EQUATIONS AND BOUNDARY CONDITIONS

The spatial an temporal dependence of electro-magnetic fields in general, and of the magnets in particular, are governed by Maxwell's equations [16]

$$\vec{\nabla} \cdot \vec{D} = \rho \quad , \qquad \vec{\nabla} \times \vec{E} = -\frac{\partial \vec{B}}{\partial t} \quad ,$$

$$\vec{\nabla} \cdot \vec{B} = 0 \quad , \qquad \vec{\nabla} \times \vec{H} = \vec{J} + \frac{\partial \vec{D}}{\partial t} \quad . \tag{4.1}$$

The two equations in the first line describe the dynamics of electric fields and the two equations in the second line that of magnetic fields. Besides the above equations there are relations among the four fields $\vec{D}, \vec{E}, \vec{B}$, and $\vec{H}$. In particular, the magnetic flux density $\vec{B}$ is related to the magnetic field $\vec{H}$ by $\vec{B} = \mu_r \mu_0 \vec{H}$ and the electric field $\vec{E}$ is related to the displacement field $\vec{D}$ by $\vec{D} = \varepsilon_r \varepsilon_0 \vec{E}$. Here $\mu_0$ is the permeability and $\varepsilon_0$ the permittivity of vacuum, whereas $\mu_r$ and $\varepsilon_r$ describe material properties that differ from those in vacuum. The fields are excited by the charge density $\rho$ and the current density $\vec{J}$.

In this chapter, however, we will neglect the electric fields $\vec{E}$ and $\vec{D}$, and focus on the equations for the magnetic fields $\vec{B}$ and $\vec{H}$ and how they are excited by currents. To understand how the currents excite magnetic fields we consider the fourth equation $\vec{\nabla} \times \vec{H} = \vec{J}$, where we can ignore the displacement field $\partial \vec{D}/\partial t$ for constant fields. We now use Stokes's theorem to convert the equation to integral form

$$\int_S \vec{J} \cdot d\vec{S} = \int_S \left( \vec{\nabla} \times \vec{H} \right) \cdot d\vec{S} = \int_{\partial S} \vec{H} \cdot d\vec{l} \quad . \tag{4.2}$$

Here the first identity follows from integrating the fourth of Equation 4.1 over a surface area $d\vec{S}$ and the second identity follows from Stokes's theorem. It allows us to express the integral over $S$ as a line integral of the field $\vec{H}$ along the perimeter $\partial S$ around $S$. An immediate application is the calculation of the field around a wire. It is given by the enclosed current $I = \int_S \vec{J} \cdot d\vec{S}$, where $S$ is a circle of radius $r$. Inserting in Equation 4.2, we recover Ampere's law $I = 2\pi r H(r)$ or $B(r) = \mu_0 I/2\pi r$, which describes the magnetic fields caused by an infinitely long wire. In the derivation we implicitly assumed that the space around the wire does not contain material with relative permeability $\mu_r$ different from unity.

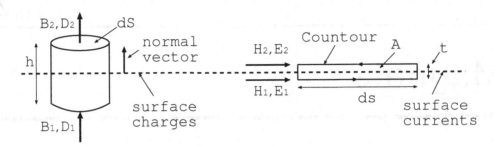

Figure 4.1 Boundary conditions for Maxwell's equations. See the text for explanations.

If, on the other hand, materials with different $\mu_r$ and $\varepsilon_r$ have a common interface, we need to discuss the behavior of the fields $\vec{D}, \vec{B}, \vec{E}$, and $\vec{H}$ at those boundaries. If we use Gauss's law [16] to convert the two Maxwell equations for the divergence into integral form and consider a small cylinder, as shown on the left-hand side in Figure 4.1, we find that the normal component of the displacement current $\vec{D}$ flux density and the $\vec{B}$ have to obey

$$\int \vec{n} \cdot \vec{D}_2 dS - \int \vec{n} \cdot \vec{D}_1 dS = \int \rho dV = \sigma_s dS \quad \text{and} \quad \int \vec{n} \cdot \vec{B}_2 dS - \int \vec{n} \cdot \vec{B}_1 dS = 0 \ , \quad (4.3)$$

which leads to

$$\vec{n} \cdot \left( \vec{D}_2 - \vec{D}_1 \right) = \sigma_s \quad \text{and} \quad \vec{n} \cdot \left( \vec{B}_2 - \vec{B}_1 \right) = 0 \tag{4.4}$$

in the limit of a vanishingly small cylinder with height $h$ and cross sectional area $dS$. Here $\sigma_s$ is the surface charge density. We then use Stokes's theorem on the other two of Maxwell's equations and calculate the line integral of the tangential component of $\vec{H}$ and $\vec{E}$ along the contour shown on the right-hand side in Figure 4.1. This integral is equal to the surface currents enclosed in the contour and we find [16]

$$\vec{n} \times \left( \vec{E}_2 - \vec{E}_1 \right) = 0 \quad \text{and} \quad \vec{n} \times \left( \vec{H}_2 - \vec{H}_1 \right) = j_s \ . \tag{4.5}$$

To summarize the discussion about the boundary conditions, we denote the normal components by a subscript $\perp$, such that $\vec{n} \cdot \vec{B} = B_\perp$ and the tangential by a subscript $\parallel$, such that $\vec{n} \times \vec{E} = E_\parallel$. Using this notation $B_\perp$ and $E_\parallel$ are continuous across the boundary and we have $\Delta D_\perp = \sigma_s$ and $\Delta H_\parallel = j_s$. If one of the materials is metallic with infinite conductivity, we additionally find $E_\parallel = 0$ and $B_\perp = 0$.

After understanding how to generate magnetic fields with currents and how the fields behave at boundaries, we need to understand its dynamics in materials. As before, we consider magnetostatic fields only and can therefore assume that $\vec{E}$ and $\vec{D}$ are zero. Maxwell's equations for the flux density $\vec{B}$ then reduce to

$$\vec{\nabla} \cdot \vec{B} = 0 \quad \text{and} \quad \vec{\nabla} \times \vec{B} = 0 \ . \tag{4.6}$$

The first of these equations allows us to express the flux density $\vec{B}$ as the curl of a vector potential $\vec{A}$, such that $\vec{B} = \vec{\nabla} \times \vec{A}$. The second of these equations permits a complementary description of the fields in vacuum and is given in terms of the gradient of a potential $\Phi$ through $\vec{B} = -\vec{\nabla}\Phi$. For the characterization of the fields it is most convenient to use the second description and inserting in the first equation leads to

$$\triangle \Phi = 0 \ , \tag{4.7}$$

which implies that the magnetic potential $\Phi$ has to fulfill the Laplace equation. Finding the magnetic fields therefore reduces to solving the Laplace equation subject to suitable boundary conditions.

After these general results that are valid in three dimensions, we will simplify the situation further by considering long magnets. Their fields can be derived by studying the two-dimensional transverse geometry, only.

## 4.2 2D-GEOMETRIES AND MULTIPOLES

For long magnets, it suffices to only consider transverse components of the magnetic flux density $\vec{B}$ in the two transverse directions $x$ and $y$, only. Here we denote them by $u = B_x$ and $v = -B_y$. For these components the equations in Equation 4.6 reduce to

$$\frac{\partial u}{\partial x} = \frac{\partial v}{\partial y} \quad \text{and} \quad \frac{\partial u}{\partial y} = -\frac{\partial v}{\partial x} . \tag{4.8}$$

These equations we recognize as the *Cauchy-Riemann equations* for a complex function $w(z) = u(x,y) + iv(x,y) = B_x(x,y) - iB_y(x,y)$ of a complex variable $z = x + iy$. Since any complex function $w(z)$ can be expressed as the derivative $w(z) = i\,dF/dz$ of another complex function

$$F(z) = A(x,y) + iV(x,y), \tag{4.9}$$

we find

$$B_x = -\frac{\partial V}{\partial x} = \frac{\partial A}{\partial y} \quad \text{and} \quad B_y = -\frac{\partial V}{\partial y} = -\frac{\partial A}{\partial x} , \tag{4.10}$$

where we identify $V(x,y)$ as the two-dimensional analogons of the three-dimensional potential $\Phi$ and $A(x,y)$ as the out-of-plane component of the vector potential $\vec{A}$. This is the same component that was denoted by $A_s$ in Equation 2.2. Realizing that Maxwell's equation in two dimensions are closely related to the theory of complex functions makes it possible to employ powerful methods such as conformal mapping and the existence of all derivatives of analytic functions, namely, those that obey the Cauchy-Riemann equations.

A special potential is that of a filament that carries a current $I$, which is given by $F(z) = (-\mu_0 I/2\pi)\log(z)$. The magnetic flux density derived from it is

$$B_x - iB_y = \frac{\mu_0 I}{2\pi i z} \tag{4.11}$$

with $z = x + iy$. We observe that by taking the modulus on both sides we recover Ampere's law. The existence of all derivatives permits us to write the complex potential $F(z)$ as a Taylor series in $z = x + iy$

$$F(z) = -B_0 R_0 \sum_{m=1}^{\infty} \frac{b_m + ia_m}{m} \left(\frac{z}{R_0}\right)^m \tag{4.12}$$

where $B_0$ is a suitably chosen reference field and $R_0$ a reference radius. The coefficients $b_m$ and $a_m$ are commonly called *multipole coefficients* and are defined to be consistent with [12] and lead to

$$iw(z) = B_y + iB_x = -\frac{dF}{dz} = B_0 \sum_{m=1}^{\infty} (b_m + ia_m) \left(\frac{z}{R_0}\right)^{m-1} . \tag{4.13}$$

The first non-trivial coefficient is the dipole coefficient with $m = 1$ and we consider the power

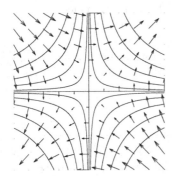 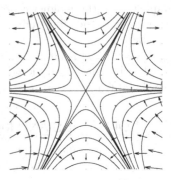

Figure 4.2 Equipotential lines and field vectors for an upright quadrupole (left) and an upright sextupole (right).

series that truncates after the first term, which results in $F_D(z) = -B_0(b_1 + ia_1)(x + iy)$. Using Equation 4.10, we find the components of the flux density to be $B_x = a_1B_0$ and $B_y = b_1B_0$. We can thus describe a dipole with a purely ($b_1 = 1, a_1 = 0$) vertical field component $B_0$ by the complex potential $F_D = -B_0(x + iy)$. According to Equation 4.9 the imaginary part of $F_D$ defines the potential $V(x, y) = -B_0y$ and we immediately see that the equipotential lines are given by $V(x, y) = V_0$ with some constant $V_0$ or equivalently $V_0 = B_0y$ which implies that the equipotential lines are given by $y = V_0/B_0$ which is constant and are thus parallel to the horizontal axis. Since the pole faces of magnets are equipotential lines we find the expected result that a dipole with a vertical field has horizontal pole faces.

The second non-trivial coefficient with $m = 2$ is called the quadrupole coefficient and the potential is given by $F_Q(z) = -(B_0/2R_0)(b_2 + ia_2)(x + iy)^2$ from which we derive the magnetic flux densities with the help of Equation 4.10 and obtain

$$B_x = (B_0/R_0)(a_2x + b_2y) \qquad \text{and} \qquad B_y = (B_0/R_0)(b_2x - a_2y) \qquad (4.14)$$

where we see that the components of the flux densities $B_x$ and $B_y$ grow linearly with $x$ and $y$, just as we required for a quadrupole in Chapter 3. In particular, for an upright quadrupole ($b_2 = 1, a_2 = 0$) the gradient is given by $g = \partial B_y/\partial x = B_0/R_0$. The potential then becomes $F_Q = -(B_0/2R_0)(x + iy)^2$ and the equipotential lines—they also define the pole faces—are given by $V_0xy$ with $V_0$ being the value of the potential. Apparently, here the choice of the reference field $B_0$ is the field on the pole tip and $R_0$ is the pole-tip radius. We show the equipotential lines as well as the flux-density vectors for the upright quadrupole on the left-hand side in Figure 4.2. Note that the field along the horizontal axis grows linearly. For the upright quadrupole only $b_2$ is non-zero. If, on the other hand, only $a_2$ is non-zero, the equipotential lines are rotated by 45 degrees and the corresponding magnet is called a skew quadrupole.

The next multipole with $m = 3$ is a sextupole. It is characterized by a complex potential $F_S(z) = -(B_0/3R_0^2)(b_3 + ia_3)(x + iy)^3$. If only $a_3$ is non-zero, it is called skew-sextupole. If only $b_3$ is non-zero, the magnet is called "upright" and its equipotential lines are given by $V_0 = (B_0b_3/(3R_0^2))(y^3 - 3x^2y)$, according to Equation 4.9. They are shown, together with the flux-density vectors, on the right-hand side in Figure 4.2. The vertical component $B_y$ along the horizontal axis points upwards on either side of the origin, consistent with a quadratic dependence on $x$.

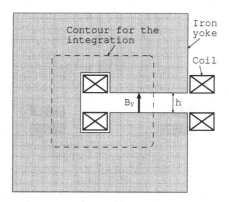

Figure 4.3  Geometry of a C-shaped dipole.

In the same manner, the equipotential lines for higher multipoles, such as octupoles, decapoles, and do-decapoles are defined by their complex potentials. Occasionally, it is desirable to combine several multipoles in one magnet. The most common combination is a dipole magnet with a quadrupolar component added in order to provide additional focusing, Figure 3.6 in Section 3.1.4 shows an example. The shape of the pole faces for such magnets is determined, as before, by the complex potentials. In this case, we need to add the potentials for a dipole with field $B_0$ and for a quadrupole with gradient $g$, which yields $V(x, y) = -B_0 y - gxy$. Note, that in combined function magnets the quadrupolar component is rigidly linked to the dipole field and cannot be adjusted independently. One is not limited to combining dipoles and quadrupoles. In the arcs of the SLC the very tight bending radius made it necessary to combine even sextupoles for chromatic correction on top of the dipole and quadrupolar fields, because there was not enough space available for separate magnets.

So far, we addressed the admissible potentials and fields. Next we need to address how to excite the fields and first we consider iron-dominated magnets that are excited by driving currents through coils.

## 4.3  IRON-DOMINATED MAGNETS

The magnetic fields in iron-dominated magnets are predominantly defined by the shape of the pole faces, which define the equipotential lines of a multipole, or combination of multipoles. These magnetic poles are excited by coils that are wound around them and our task is to determine the magnitude of the fields as a function of the current in the coils.

### 4.3.1  Simple analytical methods

First we consider a C-shaped dipole magnet as shown in Figure 4.3 that is driven by two pairs of coils which consist of $N$ turns that each carry a current $I$. We now use Stokes's theorem from Equation 4.2 to relate the enclosed current in the contour, shown as a dashed line in Figure 4.3, to the line integral of the magnetic field $\vec{H}$ along the perimeter, which is the contour itself. For the total current we find $\int_S \vec{J} \cdot d\vec{S} = 2NI$ where we have to add $NI$

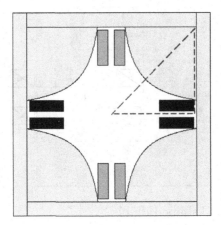

Figure 4.4 The contour used to determine the excitation of a quadrupole.

from the upper and $NI$ from the lower coil. For the line integral along the contour, we have

$$2NI = \int_{\text{gap}} H_y dy + \int_{\text{iron}} H_{\text{iron}} dl = \int_{\text{gap}} \frac{B_y}{\mu_0} dy + \int_{\text{iron}} \frac{B_{\text{iron}}}{\mu_0 \mu_r} dl \approx \frac{B_y h}{\mu_0} \ . \tag{4.15}$$

Here we split the contour into one part across the magnet gap and a second part through the iron where the relative permeability is much larger than unity $\mu_r \gg 1$, such that in the limit $\mu_r \to \infty$ we can neglect this contribution to the integral, which remains the integral of the magnetic field $H_y = B_y/\mu_0$ across the gap of height $h$. Solving for the flux density $B_y$, we find

$$B_y = \frac{\mu_0 (2NI)}{h} \ . \tag{4.16}$$

As an example, we consider a dipole with a gap of $h = 0.1\,\text{m}$ and that is excited by coils with $N = 40$ windings that are driven by $I = 1\,\text{kA}$. This results in a magnetic flux density $B_y \approx 1\,\text{T}$. We observe that $B_y$ is inversely proportional to the gap height $h$ such that magnets with large gaps either require large currents $I$ or many turns $N$. Essentially, dipole magnets with a large bore require many Ampere-turns $NI$ to achieve a high field. Conversely, in order to operate economically, the magnet designer should strive to design magnets with the smallest gap that is compatible with other constraints. After the coarse design of a dipole let us move on to the design of a quadrupole magnet.

Normal-conducting quadrupole magnets are excited by coils that are wound around their poles. The top right pole shown in Figure 4.4 is excited by the coil in which current flows into the paper through the right conductor on the top and return through the upper coil on the right. In the same way the other poles are excited and the resulting field pattern is shown on the left in Figure 4.2. If we now apply Stokes's theorem to the contour shown as a dashed triangle in Figure 4.4, we see that the enclosed current is given by the number of Ampere-turns $NI$ in the coil and the line integral has three contributions; the horizontal path from the center of the magnet to the right edge of the magnet, the path inside the iron, and the path from the center of the magnet to the pole face radius $a$.

$$NI = \int_{\text{horiz}} \frac{B dl}{\mu_0} + \int_{\text{iron}} \frac{B dl}{\mu_0 \mu_r} + \int_0^a \frac{gr dr}{\mu_0} \approx \frac{ga^2}{2\mu_0} \tag{4.17}$$

The first integral is zero, because the magnetic flux only has a vertical component on the center line and the second integral vanishes in the limit of infinite permeability $\mu_r \to \infty$. Along the path from the center of the quadrupole to the pole face, Figures 4.2 and 4.4 illustrate that the field is parallel to the path of integration and increases linearly with the radius. Evaluating the integral, and solving for $g$, we arrive at

$$g = \frac{\partial B_y}{\partial x} = \frac{2\mu_0 NI}{a^2} . \tag{4.18}$$

We realize that the gradient is inversely proportional to the square of the pole-face radius $a$. This implies that large currents are needed to power strong quadrupoles with a large bore, such as those closest to the interaction points of colliders, where the beta functions are very large, as discussed in Section 3.7.8.

The relation between exciting current and the resulting field for other multipoles can be derived in the same way. We choose an integration path around one coil that has two contributions that vanish. On the horizontal leg the field is perpendicular to the integration path and the leg that passes through the iron vanishes in the limit $\mu_r \to \infty$. The only non-zero contribution comes from the leg between the center of the multipole and the pole tip at radius $a$. Along this path the field is parallel to the integration path and grows as $r^n$. The resulting field gradient thus becomes

$$\frac{\partial^{n-1} B_y}{\partial x^{n-1}} = \frac{n\mu_0 NI}{a^n} . \tag{4.19}$$

Here we see that the multipole gradient scales inversely with the $n$th power of the pole-tip radius and high gradients require large currents $I$ or many winding turns $N$. Both quantities are limited by the available space, the current carrying capability of copper and the ability to remove the heat that is generated due to the finite resistance of the coils.

The equations in this section often serve as a first estimate of design parameters for multipole magnets. Note, however, that the equations are only valid in the limit of infinite permeability $\mu_r$ and under the assumption that the pole faces are perfect and extend infinitely. To improve the magnet design we have to resort to numerical methods and for this task we employ the MATLAB PDE-toolbox to solve the partial differential equations.

### 4.3.2 Using the MATLAB PDE toolbox

Here we use numerical methods to address the following topics: finite permeability $\mu_r$, finite size pole faces, and saturation of magnetic flux in the iron. As tool we employ the MATLAB PDE toolbox, which uses a finite-element algorithm to solve Maxwell's equations in discretized form on a mesh. Since the magnetic flux density $\vec{B}$ is divergence-free, we can derive it from a vector potential $\vec{A}$ by $\vec{B} = \vec{\nabla} \times \vec{A}$ and inserting in Maxwell's Equations 4.1 we find

$$\vec{\nabla} \times \left( \frac{1}{\mu_r} \vec{\nabla} \times \vec{A} \right) = \mu_0 \vec{J} . \tag{4.20}$$

If we use the Lorentz gauge $\vec{\nabla} \cdot \vec{A} = 0$ and we confine ourselves to two-dimensional problems, this implies that we only have one component of the vector potential $A_z$ and one component of the current density $J_z$. The equation for $A_z$ then becomes

$$-\vec{\nabla} \cdot \left( \frac{1}{\mu_r} \vec{\nabla} A_z \right) = \mu_0 J_z , \tag{4.21}$$

which reduces to the two-dimensional Poisson equation if $\mu_r$ is constant. It is this equation that the PDE toolbox solves after we provide the permeability $\mu_r$ and the current $J_z$ in various sub-domains of our problem. Once MATLAB has solved the previous equation we can determine the transverse components of the magnetic flux density $B_x$ and $B_y$ from $B_x = \partial A_z/\partial y$ and $B_y = -\partial A_z/\partial x$ in much the same way as in Equation 4.10. Apart from specifying the material properties $\mu_r$ and $J_z$ in the sub-domains, we need to specify the boundary condition on the outer boundaries of the integration volume. We can specify the value of the potential, which is called a *Dirichlet* boundary condition. A common example is setting the value to zero far away from the regions with currents. The other alternative is to specify the normal component of the potential, which is called the *von Neumann* boundary condition. A common example is setting the normal derivative to zero, which implies that the tangential component of the magnetic field $\vec{H} = \vec{\nabla} A_z/\mu_r$ vanishes. We frequently encounter this at symmetry planes.

Equation 4.21 is a PDE and MATLAB refers to it using the generic notation

$$m\frac{\partial^2 u}{\partial t^2} + d\frac{\partial u}{\partial t} - \vec{\nabla} \cdot \left(c\vec{\nabla}u\right) + au = f \tag{4.22}$$

and uses the generic names $m, d, c, a$, and $f$ to define the coefficients of the PDE and $u$ to refer to the solution itself and to formulate boundary conditions. Generalized von Neumann conditions are of the form

$$\vec{n} \cdot (c\vec{\nabla}u) + qu = g \tag{4.23}$$

where $\vec{n}$ is the normal vector on the boundary and we need to specify $q$ and $g$. Dirichlet boundary conditions have the form

$$hu = r \tag{4.24}$$

and we specify either $h$ and $r$ or set $u$ to a fixed value. Comparing with Equation 4.21 we see that $u$ corresponds to $A_z$, that $c$ corresponds to $1/\mu_r$, and that $f$ corresponds to $\mu_0 J_z$, while $m, d$, and $a$ are zero.

In order to use the PDE solver we have to follow a number of steps that are common to most numerical solvers:

1. define the geometry;
2. define boundary conditions;
3. discretize the geometry on a mesh;
4. specify material properties;
5. solve the differential equations;
6. post-process the solution to extract physically relevant properties.

Let us start with the definition of the geometry. MATLAB allows us to define basic shapes such as circles, ellipses, rectangles, and polygons from which we build our model of the magnet. We first define the basic shapes in the form of a column vector and then define the geometry as sums, differences, or intersections of the basic shapes. In order to illustrate the procedure, we consider a simple example: the C-shape dipole magnet discussed in the previous section. We assume that the magnet yoke has width and height of 1.6 m. The gap is $h = 0.1$ m, and the coils are rectangular with a width of 0.3 m and height of 0.2 m. The first component is the yoke, which we define as a polygon. It is advisable to first sketch the shape on a piece of paper and note the edges of the polygon; the $x$-values in the first row and the $y$-values immediately below. Once this table is complete we transfer the data to the MATLAB file and define the yoke as a column vector whose entry in the first row is the code, here 2, for a polygon, followed by number of points, here 13. Then follows the list of 13 $x$-values and then the 13 $y$-values.

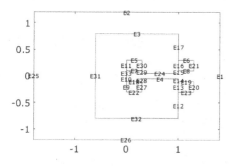 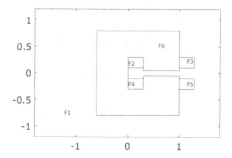

Figure 4.5  The sub-domains and edges of the C-shaped dipole magnet.

```
yoke=[2; 13; ...
  0;0;0.3;0.3;1;1;-0.6;-0.6;1;1;0.3;0.3;0; ...
  0;0.3;0.3;0.05;0.05;0.8;0.8;-0.8;-0.8;-0.05;-0.05;-0.3;-0.3];
```

After the yoke is defined, we enter the definitions of the coils C1 to C4 in the same way and finally the enclosing World that envelopes the whole integration volume

```
C1=[3;4;0;0;0.3;0.3;  0.1;0.3;0.3;0.1;zeros(18,1)];
    :
World=[3;4;-1.8;1.8;1.8;-1.8;-1.2;-1.2;1.2;1.2;zeros(18,1)];
```

Here we use the code 3 to define a rectangle. The meaning of the other entries is the same as for the polygon. Note that the definitions for the geometric shapes all need to have the same number of rows, which explains their padding with zeros. Note that the shapes are plain MATLAB arrays and, instead of using numbers, we can also use variables or functions to generate the various shapes. This makes parametric studies very easy to implement.

Once the basic shapes are defined, we can assemble the model, give names to the shapes, and define their relation with the following commands:

```
gd=[World,yoke,C1,C2,C3,C4];                  % assemble geometry
ns=char('World','yoke','C1','C2','C3','C4')'; % names of the shapes
sf='World+yoke+C1+C2+C3+C4';                  % relation
g=decsg(gd,sf,ns);
```

First we assemble the column vectors for each of the shapes into one matrix gd for "geometry descriptor" before assigning names to the shapes in the variables ns. The use of the char() function ensures the names all have the same length and that ns becomes an array of characters. In the next line we define sf to hold the relation of the different shapes. Here we only use the + operator, but we can also use − to define, for example, holes, such that R-C describes a circular hole defined by C in a rectangle R. Using * allows us to define regions that are part of two shapes. For example R*C describes the region that is both in R *and* in C. But for the C-shaped dipole we only need to add the shapes. Finally, we are ready to assemble the full geometry description with the decsg() function in the variable g. Admittedly, this way of entering the geometry is somewhat arduous, but it gives the highest flexibility in defining the geometry and later in post-processing the results.

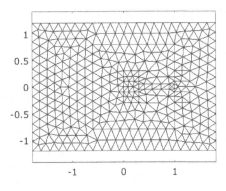 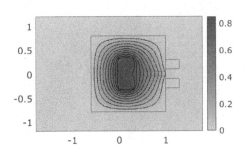

Figure 4.6 The mesh (left) with the default mesh size that is refined by adjusting Hmax to 0.05 to generate the solution (right).

With the definition of the geometry completed, we add it to the model, which is a data structure that holds all information about the simulation. We create an empty model with the createpde() function and since we only have a single variable $A_z$ in the PDE Equation 4.21, we use 1 as the argument and we add the geometry description g to the model in the call to the geometryFromEdges() function.

```
model=createpde(1);
geometryFromEdges(model,g);
pdegplot(model,'EdgeLabels','on')
figure; pdegplot(model,'SubDomainLabels','on');
```

And then we are ready to inspect the geometry with the pdegplot() function; once to show the EdgeLabels and once to show the SubDomainLabels. The corresponding plots are shown in Figure 4.5. The most important information is the labels of the outer boundaries, here 1,2,25, and 26. We need them to define the boundary conditions with the following call

```
applyBoundaryCondition(model,'Edge',[1,2,25,26],'u',0);
```

which is fairly self-explanatory; on edges with labels 1,2,25, and 26 we require the value of the solution u to be zero.

The next task is to discretize the geometry with a call to the generateMesh() function and to show the result with the pdemesh() function.

```
generateMesh(model);  % or generateMesh(model,'Hmax',0.05);
pdemesh(model);
```

The resulting mesh is shown on the left of Figure 4.6. In the figure MATLAB chooses the default mesh size, but we can also require a mesh with smaller triangles by specifying the maximum edge of a triangle with the Hmax argument. In the remainder, we set Hmax equal to 0.05.

With the mesh in place, we are ready to specify the material properties in the sub-domains. We refer to the right image in Figure 4.5, where we find the labels for the sub-domains. In particular, 1 labels the outside of the magnet, 2 to 5 the coils and 6 the magnet yoke. Next, we enter the values m, d, c, a and f for the respective sub-domains in multiple

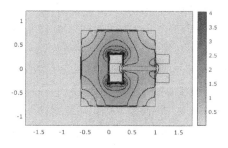 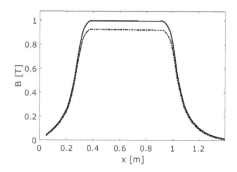

Figure 4.7 The magnetic flux density distribution (left) and the values in the mid-plane of the gap (right) for $\mu_r = 5000$ (solid) and $\mu_r = 500$ (dashes).

calls to the `specifyCoefficients()` function. We remember that $c = 1/\mu_r$ and therefore we specify $c = 1$ everywhere, except the yoke, where we set $c = 1/\mu_r = 1/5000$. Likewise, $a$ is zero everywhere, but we assume the coils to carry $NI = 40\,\text{kA-turns}$. This causes $f$ to be $f = \mu_0 NI/A$ where $A = 0.2 \times 0.3\,\text{m}^2$ is the cross section of the coils. For $f$ we thus find the numerical value 0.8378, which we assign to the left coils with one sign, and the other coils with the opposite sign.

```
specifyCoefficients(model,'m',0,'d',0,'c',1,'a',0,'f',0,'Face',1);
specifyCoefficients(model,'m',0,'d',0,'c',1,'a',0,'f',0.8378,'Face',[2,4]);
specifyCoefficients(model,'m',0,'d',0,'c',1,'a',0,'f',-0.8378,'Face',[3,5]);
specifyCoefficients(model,'m',0,'d',0,'c',1/5000,'a',0,'f',0,'Face',6);
result=solvepde(model);
figure; pdeplot(model,'XYdata',result.NodalSolution,'contour','on');
hold on; pdegplot(model);
```

Then we solve the PDE with a call to `solvepde()`, which sets up the system of equations, solves them, and returns the solution as `result`. We refer to the solution as `u=result.NodalSolution` and immediately plot it on the right-hand side in Figure 4.6 to inspect the solution. Note that we plot both `u` and the contour lines. Here `hold on` and the call to `pdegplot()` superimposes the geometry on the plot.

With the solution $u = A_z$ available, we can proceed to extract other physical quantities, such as the magnetic flux density $(B_x, B_y) = (\partial u/\partial y, -\partial u/\partial x)$. Both gradients are already provided in `result` as `ux=results.XGradients` and `uy=results.YGradients` and we can therefore proceed to calculate the magnitude of the field with the `hypot()` function. A call to the `pdeplot()` function then displays the magnitude of the magnetic flux density `Bn` with contour lines and geometry superimposed.

```
Bn=hypot(result.XGradients,result.YGradients);
figure; pdeplot(model,'xydata',Bn,'contour','on');
hold on; pdegplot(model);
```

The left-hand side in Figure 4.7 shows this plot. We immediately observe on the color bar that values above 3 T occur in the iron, especially close to the coils. This causes saturation of the iron and we will later improve the solution by taking magnet saturation into account. But we first explore the present solution further by plotting the flux density on the midplane

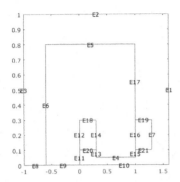 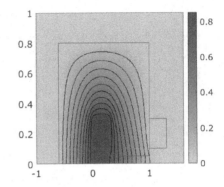

Figure 4.8 The geometry and solution of the magnet when exploiting the mid-plane symmetry of the C-shaped dipole magnet.

of the magnet gap. First, we create arrays x and y to describe the path along which we seek to calculate the magnetic field. Then we employ the `evaluateGradient()` function to determine the gradients along this path, use Equation 4.10 to ensure the fields have the correct sign, and calculate the magnitude of the magnetic flux density B.

```
x=0.05:0.01:1.4; y=zeros(1,length(x));
[dAx,dAy]=evaluateGradient(result,x,y); Bx=dAy; By=-dAx; B=hypot(Bx,By);
plot(x,B,'k'); xlabel('x [m]'); ylabel('B [T]')
```

The solid line in the plot on the right-hand side in Figure 4.7 displays this absolute value of the magnetic flux density $B$. We note that the field in the middle of the gap is around 1 T, in agreement with the estimate using Equation 4.16. But we see that the field rolls off at both ends of the magnetic gap near $x = 0.3\,\text{m}$ and $x = 1\,\text{m}$, an effect we will consider more closely below.

But first we investigate the effect of bad iron with a much lower permeability of $\mu_r = 500$ instead of 5000, which is as easy as changing the definition of the parameter $c$ inside the yoke (Face 6) in a call to the `specifyCoefficients()` function, before solving the PDE with `solvepde()`. The resulting flux density in the gap is shown as the dashed line on the right-hand side in Figure 4.7. Apparently, the field in the gap is much lower.

In the first example we simulated the field in the entire magnet, which is redundant, because the magnet has a midplane symmetry and it is sufficient to model only the upper half and adjust the boundary conditions accordingly. All we have to do is to change the polygon that describes the yoke to

```
Cmag=[2; 8; ...
   0;0;0.3;0.3;1;1;-0.6;-0.6; ...
   0;0.3;0.3;0.05;0.05;0.8;0.8;0];
```

and only define the two upper coils before defining the boundary conditions with the following code

```
applyBoundaryCondition(model,'Edge',[1:3],'u',0);
applyBoundaryCondition(model,'Edge',[8:10],'q',0,'g',0);
```

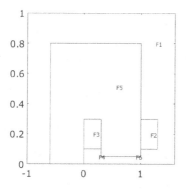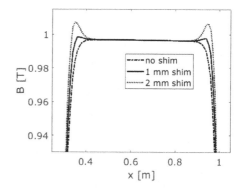

Figure 4.9 On the left the shims are added to the magnet geometry as 5 mm wide iron pieces with a height of 1 or 2 mm. On the right the magnetic flux density in the gap is plotted for a magnet without shims (dot-dashed), with a 1 mm high shim (solid) and with 2 mm shims (dotted).

where the first line defines the Dirichlet boundary conditions on the left, right, and upper outer boundary. The second call defines von Neumann boundary conditions on the edges on the midplane. The numbers we obtain by inspecting the `EdgeLabels` that are shown on the left-hand side in Figure 4.8. Moreover, we only have to specify the coefficients `m`, `d`, `c`, `a`, and `f` in four sub-domains: the yoke, two coils, and the rest of the integration volume. On the right-hand side in Figure 4.8, we show the solution $u$ returned by `solvepde()` that corresponds to the right-hand side in Figure 4.6. It is easy to verify that the magnetic field in the gap equals that calculated with the full model we used before. Note that the number of nodes for `Hmax=0.05` is only 2362 as opposed to 7940 that were used before when simulating the magnet without symmetries taken into account. Normally, it is advisable to exploit symmetries, because fewer resources are required. This often results in faster execution time and the ability to decrease the mesh size in order to solve more complex problems.

With this more efficient simulation model, we investigate the roll-off of the magnetic flux density at the ends of the magnet gap. A common way to improve the field in this region is called *shimming*. It is based on adding small pieces of iron near the end of the poles. In the model, we simply add small, 5 mm-wide rectangles just below the pole ends, as shown on the left in Figure 4.9, where the labels 4 and 6 indicate the position of the shims on either end of the upper pole face. The magnetic flux density in the midplane of the magnet gap is shown on the right for shim heights of 0, 1, and 2 mm as the dot-dashed, solid and dotted line, respectively. We find that the 1 mm shims clearly extend the good-field region by about 0.05 m on both sides of the magnet gap while a 2 mm shim obviously causes a significant overshoot at the ends. We point out that the shim height depends on the magnet gap $h$.

So far, we did not take saturation of iron into account, which starts around 1.5 to 2 T, depending on the quality of the iron or steel used when manufacturing the magnet. Adapting the simulation to taking saturation into account is as easy as defining the relative permeability $\mu_r$ to depend on the magnetic flux density. Here we will use the following dependence

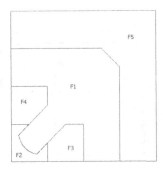 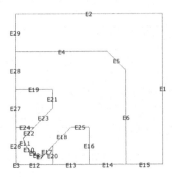

Figure 4.10 The domains (left) and edges (right) for a quarter of a quadrupole.

for the relative permeability

$$\mu_r = \frac{5000}{1+0.05B^2} + 200 \, , \qquad (4.25)$$

which is taken from the example for a magnetostatic problem in the documentation of the MATLAB PDE toolbox. All we have to do is to change two lines; first we define the function mufun() that implements Equation 4.25 and then we have to refer to it when defining the parameter $c$ in the call to specifyCoefficients() pertaining to the domains of the model that are affected by saturation.

```
mufun=@(location,state)1./(200+5000./(1+0.05*(state.ux.^2+state.uy.^2)));
specifyCoefficients(model,'m',0,'d',0,'c',mufun,'a',0,'f',0,'Face',4);
```

Here state.ux and state.uy are the derivatives of the potential and the sum of their squares gives the flux density squared $B^2$. Though not used in this example, we briefly note that state.u refers to the potential itself and, for example, location.x to the $x$–position in the domain of integration.[1] The rest of the script remains the same in our case. In the present example, however, the effect of taking the non-linearity of the iron into account, is very small and hardly visible when plotting the midplane magnetic flux density such as shown on the right-hand side in Figure 4.7.

With these tools at hand it is possible to explore many other dipole geometries, such as H-magnets, but we leave that as an exercise and explore quadrupoles instead.

### 4.3.3 Quadrupoles

Quadrupoles can be treated in much the same way as dipoles. We realize, however, that quadrupoles have a four-fold symmetry and we therefore only need to model one quadrant, if we treat the symmetry axes suitably by specifying von Neumann boundary conditions. The quadrupole has width and height 0.6 m and a pole-tip radius of 53 mm. We show the geometry with the sub-domains specified on the top left plot in Figure 4.10. We see that the iron yoke, labeled by index 1, the coils by indices 3 and 4, and the extra space by indices 2 and 5. The iron yoke and the coils are specified by polygons and the edges or segments of the polygons are shown on the right plot at the top in Figure 4.10. Note that we subdivided the

---

[1]Executing help FunctionCoefficientFormat in MATLAB displays the available information.

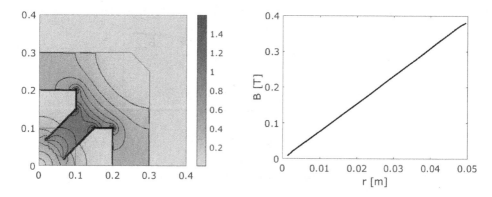

Figure 4.11 The magnetic flux density in the entire simulation volume (left) and along a line from the origin to the pole-tip center (right).

hyperbolic pole tip into a number of shorter segments and we also added a short segment, labeled 3, in the origin. This was necessary to avoid an ambiguity with the von Neumann boundary conditions on the left boundary where the tangential component is vertical and on the bottom boundary, where it is horizontal. We simply added a short diagonal segment near the origin where we enforce zero Dirichlet boundary conditions. We set the current density in the coils with cross section $87.5\,\mathrm{cm}^2$ to be $J_z = 10^6\,\mathrm{A/m}^2$ which results in a total current of 8750 A-turns.

The rest of the simulation code follows the previous examples and after specifying the boundary conditions, we generate the mesh, specify the coefficients $m, d, c, a,$ and $f$, and solve the equations with `solvepde()`, which returns the vector potential $u = A_z$ from which we calculate the magnetic field components $B_x$ and $B_y$. Their modulus $B = \sqrt{B_x^2 + B_y^2}$ is shown on the left-hand side in Figure 4.11 where the strongest field appears near the neck of the pole tip. There the iron will saturate first, if we increase the current in the coils further. In the present scenario, the field on the pole tip is approximately 0.4 T. We read this value off of the right plot in Figure 4.11, which shows $B$ along the diagonal from the origin towards the pole-tip center. We determine the gradient from a linear fit to be 7.7 T/m, which we compare to the estimate from Equation 4.18 which results in gradient of 7.8 T/m. After these initial simulations of the base performance of the quadrupole, we can start optimizing it by including saturation of the iron in our modeling, change the shape of the pole-tip, add shims, or round the corners near the neck of the pole-tip, where we observe high magnetic flux densities. We may also investigate iron or steel with different permeabilities $\mu_r$. And this brings us to the technological aspects of iron-dominated magnets.

### 4.3.4 Technological aspects

One particular aspect of iron-dominated magnets is the finite permeability of the iron and the fact that it decreases with increasing magnetic fields, which is a consequence of the hysteresis, illustrated in Figure 4.12. The slope near the origin is given by the permeability $\mu_0 \mu_r$ of the material. We observe that the magnetic flux density saturates at some value. The level at which the iron saturates and the relative permeability $\mu_r$ are material dependent and vary significantly. For example, we used $\mu_r = 5000$, which is a typical value for good-

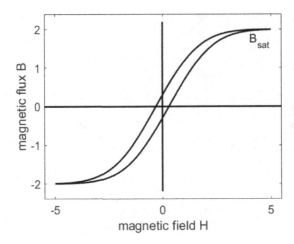

Figure 4.12 The magnetic flux $B$ as a function of the magnetic field $H$ illustrating the hysteresis of iron.

quality iron. Mu-metal, used for shielding magnetic fields, has ten times higher values. The saturation fields range from 1.5 to somewhat over 2 T, but in any case the data sheets of the manufacturers must be consulted to obtain accurate descriptions.

Note that in large parts of the previous sections we assumed a one-to-one relationship between the magnetic field $H$ and the flux density $B$, but from Figure 4.12 it is obvious that this relationship is multi-valued and normally also depends on the history of excitation levels of a magnet. In practice, this implies that occasionally the magnets have to be *demagnetized* or, as it is often called, *standardized*. This is achieved by cycling the power supplies of the magnet, initially between the maximum and minimum values, and then repeating the procedure with successively decreasing amplitudes. In this way, the magnet iron is forced to follow a gradually diminishing hysteresis curve until it cycles around zero excitation, at which point the magnet has "forgotten" its history and almost no residual magnetization remains. This procedure results in a well-defined initial state from which the desired magnetic flux density can be reached in a reproducible way.

One of the reasons why the excitation of iron-dominated magnets depends on their history are eddy-currents that are excited by quickly changing currents. This normally happens in accelerator magnets used in rapidly cycling synchrotrons. They often cycle their magnetic field several times per second and a common way to avoid excessive eddy-currents is to assemble the magnets from *laminated* sheets of metal, in the same way transformers are built. Building magnets from stacks of laminations is particularly advantageous for large series of equal magnets, because the stencil to stamp out the laminations from the sheet metal only needs to be paid for once, and then a large number of laminations can be stamped at moderate cost. Moreover, in this way, only moderately light-weight laminations need to be assembled, rather than large blocks of iron.

The magnets are excited by driving a current through the coils and the ohmic losses due to the finite resistance of the conductor limit the maximum current. A common limit for the current density is about $10\,A/mm^2$, provided that the coils are water cooled and the pressure drop across the coils is sufficiently large to ensure the flow to be turbulent [17]. In these circumstances the conductors are insulated by epoxy and have a hole in their center

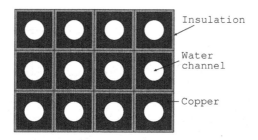

Figure 4.13  Sandwich of conductor coils with water-cooling channels.

through which the water flows. Figure 4.13 illustrates the cross section through such a "sandwich" of $3 \times 4$ conductors with embedded water channels.

In small magnets, such as those used as steering correctors, only moderate currents are needed to excite the magnet. If the current density in the coils stays below approximately $1 \, \text{A/mm}^2$, cooling by air-convection is sufficient and we can avoid the additional complexity of installing a water-cooling system.

Iron-dominated magnets are the most common magnets in accelerators and their design and manufacturing is well-understood. This makes them the first choice unless one wants to use permanent magnets to avoid electricity costs or magnetic flux densities above 2 T are required. In the latter, the high-field regime, super-conducting magnets are the only choice.

## 4.4  SUPER-CONDUCTING MAGNETS

In super-conducting magnets, the flux densities are almost entirely determined by the distribution of currents in the system, and in the basic configuration, no iron is used to shape the field. In this way, the achievable flux densities are not limited by saturation of the iron, and superconductors permit extremely large current densities, because there are no ohmic losses. The disadvantage is, however, that super-conducting magnets need to be cooled to temperatures close to absolute zero by embedding them in a bath of liquid Helium at 4.2 K, and in special cases even to 1.9 K, where Helium becomes super-fluid.

The wires used to power the magnets are made of superconducting material that exhibits a vanishing electrical resistance to direct currents near zero absolute temperature. In certain materials, two electrons interact through phonons, which are lattice vibrations of the crystal, and form bound states, so-called Cooper-pairs, which can travel unimpeded through the material [18]. When exposed to a magnetic fields, *Type-I* superconductors expel the magnetic fields, provided it is below a critical field strength $H_c$. Unfortunately, their $H_c$ is too low to use this type of material for magnets. Instead, one uses *Type-II* superconductors. They completely expel magnetic fields below the first critical temperature $H_{c1}$ and partially expel it in a range between $H_{c1}$ and the second critical field $H_{c2}$, before completely loosing superconductivity at fields above $H_{c2}$. The field $H_{c2}$ depends on the material and notably on the temperature to which the material is cooled. It is higher for lower temperatures, which explains why high-field magnets are often cooled with superfluid Helium at 1.9 K, rather than the more easily accessible liquid Helium at 4.2 K. The ability of Type-II superconductors to support superconducting currents makes them the material of choice for the wires.

The most common material for the wires is Nb-Ti, an alloy with approximately equally

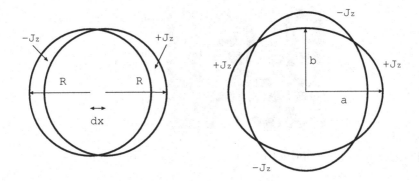

Figure 4.14 Two intersecting cylinders cause a dipolar field in the inner region (left) and intersecting ellipses cause a quadrupolar field (right).

shared weight of niobium and titanium. It consists of filaments of Nb-Ti, extruded to a diameter of several microns. A few thousands of the filaments are embedded in a copper matrix to form strands with a diameter of about 1 mm. The copper matrix is needed to carry the large currents and dissipate heat that is generated when the wire loses superconductivity and quenches. Several tens of the strands are assembled to form the wires that are finally used to wind the coils. Nb-Ti is widely used, because it can carry current densities between 1000 and 2000 A/mm² and it has good mechanical properties. For example, it can be easily extruded. Other materials, such as Nb₃Sn can be used to reach higher fields, but is very brittle and the manufacturing process is very complex and expensive.

These superconducting wires form the base for the current-dominated dipole and quadrupole magnets, which we will consider in the following section.

### 4.4.1 Simple analytical methods

If we apply Equation 4.2 to the inside of a cylinder with a homogeneous current density $J_z$, we find

$$\pi r^2 J_z = 2\pi r H(r) \qquad \text{or} \qquad B(r) = \mu_0 H(r) = \frac{\mu_0 J_z}{2} r \ . \tag{4.26}$$

The field grows linearly with radius $r$. Since the flux density has only an azimuthal component we obtain for the components in the horizontal and vertical direction

$$B_x = -\frac{\mu_0 J_z}{2} y \qquad \text{and} \qquad B_y = \frac{\mu_0 J_z}{2} x \ . \tag{4.27}$$

If we superimpose two cylinders separated horizontally by a distance $dx$, as shown on the left in Figure 4.14, the currents with opposite current density simply superimpose and cancel in the inner region. The fields superimpose as well, such that $\bar{B}_x = 0$ and

$$\bar{B}_y = \frac{\mu_0(-J_z)}{2}(-dx/2) + \frac{\mu_0 J_z}{2}(dx/2) = \frac{\mu_0 J_z dx}{2} \ , \tag{4.28}$$

which is purely vertical and constant in the inner region, just what is required for a dipole.

We can treat quadrupolar fields in a similar way if we observe that the field components in an elliptic cylinder with homogeneous current density are given by [19]

$$B_x = -\frac{\mu_0 J_z a}{a+b} y \qquad \text{and} \qquad B_y = \frac{\mu_0 J_z b}{a+b} x \ , \tag{4.29}$$

where $a$ and $b$ are major and minor axes of the ellipse. Summing contributions from two ellipses with opposite current densities and rotated by 90 degrees, we obtain for the flux densities

$$\bar{B}_x = \mu_0 J_z \frac{b-a}{b+a} y \qquad \text{and} \qquad \bar{B}_y = \mu_0 J_z \frac{b-a}{b+a} x , \tag{4.30}$$

which grow linearly with distance from the origin, just as a quadrupolar field should.

Since the magnetic flux densities are generated by the distribution of currents, it is prudent to investigate the contribution of a current filament at position $z_0 = x_0 + i y_0$ to the multipoles at position $z = x + iy$. We already know from the discussion at the end of Section 4.2 that the potential of a current filament can be written as $F(z) = (\mu_0 I / 2\pi) \log(z - z_0)$, and that the corresponding flux density is given by

$$\hat{\underline{B}}^* = \hat{B}_x - i\hat{B}_y = \frac{\mu_0 I}{2\pi} \frac{i}{z - z_0} = \frac{-i\mu_0 I}{2\pi z_0} \sum_{n_0}^{\infty} \left(\frac{z}{z_0}\right)^n . \tag{4.31}$$

Here we introduce the abbreviation $\underline{B} = B_x + iB_y$ and use the asterisk to denote the complex conjugate. In the last equality, we first extract $z_0$ from the denominator and then write the resulting expression as a power series in $z/z_0$, which is permissible for $|z/z_0| < 1$, the region inside the current distribution. Comparing this expression with Equation 4.13, we see that $n = 0$ corresponds to the contribution to the dipole component and $n = 1$ to the quadrupole component.

The question of how to place the current filaments in such a way that the combined effect of all filaments only produces a single multipole can be addressed by introducing polar coordinates with $z = re^{-\phi}$ and $z_0 = r_0 e^{-\phi_0}$ which allows us to rewrite the previous equation as

$$\hat{\underline{B}}^*(\phi_0) = \frac{-i\mu_0 I}{2\pi r_0} \sum_{n_0}^{\infty} \left(\frac{r}{r_0}\right)^n e^{in\phi} e^{-i(n+1)\phi_0} \tag{4.32}$$

where we added the argument $\phi_0$ to the left-hand side in order to make the dependence of the magnetic flux density $\hat{B}$ on the location of the current filament obvious. Now, if the current filaments have an azimuthal current distribution that follows $dI/d\phi_0 = \hat{I} \cos(m\phi_0 + \hat{\phi})$, we see that integrating over all current filaments

$$\underline{B}^* = \frac{-i\mu_0 \hat{I}}{2\pi r_0} \left(\frac{r}{r_0}\right)^n e^{in\phi} \int_0^{2\pi} e^{-i(n+1)\phi_0} \cos(m\phi_0 + \hat{\phi}) d\phi_0 \tag{4.33}$$

extracts a single Fourier harmonic with $m = n + 1$. Here $\hat{\phi}$ describes the orientation and distinguishes between upright and skew multipoles. We thus find that an azimuthal current distribution with a $\cos(m\phi_0)$-dependence creates as pure multipole of order $n+1$. In particular, a $\cos(\phi_0)$-dependence results in a dipolar field distribution and a $\cos(2\phi_0)$-dependence in a quadrupolar field distribution.

After the basic layout we use the MATLAB PDE toolbox to verify the design and then progress to construct the coils from more convenient rectangular current leads, rather than the crescents shown in Figure 4.14 or filaments with a current distribution with a $\cos(m\phi_0)$ dependence.

### 4.4.2  PDE toolbox

We first verify that intersecting ellipses with homogeneous current densities generate a dipole field as given by Equation 4.28. Since the geometry is particularly simple, we use the

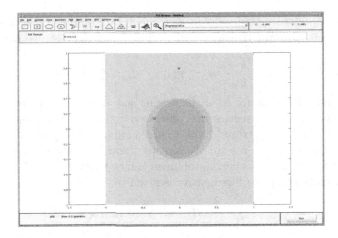

Figure 4.15 The interactive PDF toolbox interface with the quadrupole geometry.

interactive mode of the PDE toolbox, and define the geometry with the following sequence of commands

```
pderect([-1,1,-1,1],'W')
pdecirc(0.05,0,0.4,'C1')
pdecirc(-0.05,0,0.4,'C2')
```

which causes the window with the `pdeModeler`, shown in Figure 4.15, automatically to appear on the screen. The large rectangle, labeled `W`, denotes the integration volume and the two circles, labeled `C1` and `C2` define the coils of the dipole magnet. First we change the type of problem in the middle of the toolbar from "Generic Scalar" to "Magnetostatics," then we select the boundary condition mode by pressing the button with $\partial\Omega$. This changes the display to highlight the boundaries and double-clicking the boundaries opens a window in which to enter values. In this problem we use Dirichlet boundary conditions with $u = 0$ on the boundary, which is the default, and we do not have to change anything. Next we select "PDE Mode" and "Show Sub-domain Labels" from the "PDE" menu. This causes the display to show the labels of the sub-domains and double-clicking on the number opens a dialog box in which we enter $\mu_r$ and $\mu_0 J_z$ for the respective sub-domains. In this case $\mu_r = 1$ everywhere, and the right crescents has positive current, say $\mu_0 J_z = 1$, and the left crescent has $\mu_0 J_z = -1$. The other sub-domains carry no current and we set $\mu_0 J_z = 0$. Once the geometry, boundary, and material properties are defined, we create a mesh by pressing on the button with the triangle and then refine it by pressing the button immediately to its right. Pressing the equal sign solves the system, and, after a short while, displays the solution, here the potential, in the same window. The button to the right of the equal sign opens a post-processing dialog, where we select to display the "magnetic flux density" with "Arrows." The display then changes to reflect this choice. If we want to manipulate the data further we can export the data structures to the MATLAB workspace for the mesh `p`, `e`, and `t` and the solution `u` from the "Mesh" and "Solve" menu, respectively. Then we can use the commands we used before to plot the solution along a line or other sub-domains.

In practice it is very expensive to manufacture crescent-shaped coils and basing the design of the magnet instead on coils with simple geometries, such as a square blocks, is highly desirable. We already know from the discussion in the previous section that a cosine-like azimuthal current distribution results in a dipolar field distribution. We therefore

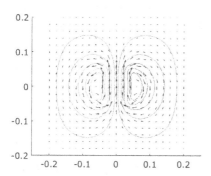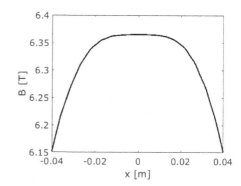

Figure 4.16 The geometry, equipotential lines, and flux density for a superconducting dipole (left), and the magnitude of the flux density along a horizontal line in the midplane of the magnet (right).

investigate, whether we can find simple current distributions that can generate a dipole field near the center of the magnet. We start by approximating the cosine by a distribution with constant current density over a given azimuthal range, where we have positive current density for $\phi = \pm\alpha$ and negative density for $\phi = 180^o \pm \alpha$. Experimenting with varying $\alpha$ it quickly becomes obvious that $\alpha = 60^o$ produces the most homogeneous flux density near the center of the dipole.

The left-hand side in Figure 4.16 shows the geometry, the equipotential lines and the magnetic flux density as arrows. We approximate the square blocks by a polygon that closely follows two concentric circular arcs of radius 5 and 7 cm. The current density we use is $500 \, \mathrm{A/mm^2}$. Using the same functions as in previous sections to calculate the flux density along a line, we find that this configuration results in a fairly homogeneous flux density with magnitude $B = 6.4 \, \mathrm{T}$, shown on the right-hand side in Figure 4.16. Of course, we can try to improve the quality of the field by distributing the current blocks in different ways and increase the field by adding a second layer of current blocks at larger radii, but that is left as an exercise. Instead we consider a current density distribution that results in a quadrupolar flux density.

Stimulated by the success of the simple current distribution that resulted in a fairly homogeneous dipole field, we try an equally simple distribution to generate quadrupolar fields. A little experimenting with the angular width $\pm\alpha$ of the current blocks results in $\alpha = 30^o$ that produces a rather linear increase of the vertical flux density component $B_y$ along the horizontal axis with $y = 0$. Figure 4.17 shows the geometry, the equipotential lines, and the flux density, shown as arrows. From a fit of a straight line to the plot of $B_y$ versus horizontal position $x$ we find the gradient to be $g = 114.8 \, \mathrm{T/m}$. Even here we can improve the current distribution by placing the current blocks in different ways and increase the gradient by adding layers of current blocks, but leave that as an exercise, as well.

Superconducting magnets are the only magnets to reach flux densities of several Teslas, but occasionally, there are other requirements to fulfill, such as moderately high fields in very tight spaces, or the need to avoid power supplies that drive currents through coils altogether. This is the realm of permanent magnets.

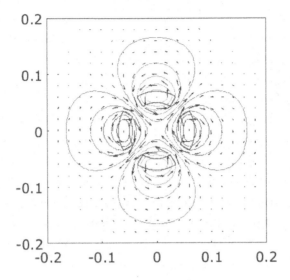

Figure 4.17 The geometry, equipotential lines, and flux density for a superconducting quadrupole.

## 4.5 PERMANENT MAGNETS

Permanent magnets are blocks of magnetized material that act as "flux pumps" for magnetic field lines without the need for external excitation by electric currents. Moreover, the flux density they provide is independent of their size, which makes it possible to use them in situations where moderately high flux densities, on the order of one Tesla, are needed in very tight spaces. Today they are frequently found in electric motors, on physicist's whiteboards, and in magnets used to guide the beams in particle accelerators, where they are heavily used in specialty magnets, undulators and wigglers, to produce synchrotron radiation and in strongly focusing machines, such as CBETA [20]. The latter uses magnets that are based on design ideas presented by Halbach in [21] on which we also base the following discussion.

There are several types of permanent magnets available, often made of Samarium-Cobalt or Neodym-Iron. The process in which they are produced consists of rapidly cooling a molten mixture of the ingredients and then grinding the cold material to a very fine powder. The powder is subsequently exposed to a strong magnetic field under high pressure, which aligns the grains to their preferred magnetic orientation. In a second step the blocks are heated and compressed further in a process called *sintering,* before machined to their final form as blocks or disks. In a last step, the material is subjected to an even higher magnetic field than before. This process imprints the large remanent magnetic flux density $B_r$ on the magnet, which can exceed 1 T for some materials. The direction in which the field is imprinted is commonly called the *easy axis* of the magnet.

The main properties of permanent magnets are summarized in the relation of an externally generated magnetic field $H$ to the flux density $B$

$$B_{||} = \mu_r \mu_0 H_{||} + B_r \qquad \text{and} \qquad B_\perp = \mu_r \mu_0 H_\perp \, , \qquad (4.34)$$

where $B_{||}$ denotes the flux density parallel to and $B_\perp$ perpendicular to the easy axis. $B_r$ is the remanent field due to the magnetization of the material. It only has a non-zero component

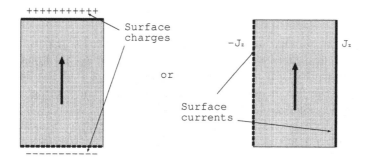

**Figure 4.18** Homogeneous permanent magnets can be described by either surface charges (Equation 4.35) or by surface currents (Equation 4.36) perpendicular to the paper.

$B_r$ along its easy axis. Moreover, $\mu_r$ is very close to unity, such that we will assume $\mu_r = 1$ in the following. This has the remarkable property that the fields generated by permanent magnets linearly superimpose, provided no other materials with large permeabilities are nearby. The permanent magnets are transparent to external fields $H$, but they contribute by virtue of their remanent field $B_r$.

Equation 4.34 can be written in vectorized form as $\vec{B} = \mu_0 \vec{H} + \vec{B}_r$ where $\vec{B}_r$ points along the easy axis. The magnetic flux density $\vec{B}$ has to fulfill Maxwell's Equations 4.1, with $\vec{\nabla} \cdot \vec{B} = 0$ and $\vec{\nabla} \times \vec{H} = 0$ in the absence of external currents or electric fields. The first equation leads to

$$\mu_0 \vec{\nabla} \cdot \vec{H} = -\vec{\nabla} \cdot \vec{B}_r = \rho_r \tag{4.35}$$

where we observe that the divergence of the remanent field $B_r$ behaves just like a source-term $\rho_r$ for the magnetic field $\vec{H}$. If we derive the magnetic field from a scalar potential $\vec{H} = -\vec{\nabla}V$ we obtain the Poisson equation for the potential $\triangle V = -\rho_r/\mu_0$. The second relation leads to

$$\vec{\nabla} \times \vec{B} = \vec{\nabla} \times \vec{B}_r = \mu_0 \vec{J}_r \tag{4.36}$$

such that the curl of the remanent field behaves analogously to a source current $\vec{J}_r$. This allows us to use Equation 4.11 to calculate the magnetic flux for two-dimensional geometries, to which we confine ourselves in the remainder.

If the permanent magnet blocks are homogeneous, the curl or divergence of the remanent field in Equations 4.35 and 4.36 only change on the surface of the material. This immediately leads to a description of the permanent magnets in terms of surface charges or currents, as shown in Figure 4.18. We can therefore model their behavior in the MATLAB PDE toolbox by adding, for example, thin layers of width $w$ with current density $J_z = B_r/w\mu_0$ to the sides of the magnets that are parallel to their easy axis. The corresponding currents are enormous; the current density in a 0.1 mm layer of a magnet with $B_r = 1$ T is about $8 \, 10^8$ A/m$^2$ and the total current $I = J_z w h$ in a block with height $h = 50$ mm corresponds to 40 kA! A description, complementary to the one using currents, is based on using magnetic surface charges and is also useful in some circumstances. Note that this does not imply the existence of magnetic monopoles; it is just a convenient description of the string of dipoles, aligned with the easy axis in the material, that show one end on the surface. Here we do not pursue numerical simulations further, but use the special properties of the permanent magnet's material ($\mu_r \approx 1$, linear superposition of fields) to calculate the fields by analytical methods instead.

Equation 4.36 implies that we can determine the flux density $B$ from a distribution of sources that are equivalent to electrical currents. Therefore, we use Equation 4.11, which describes the flux density of a current filament, and the fact that the fields superimpose linearly, to obtain the flux density due to all currents by convoluting with the current density. In two-dimensional geometries this yields

$$\underline{B}^*(\hat{z}) = \frac{\mu_0}{2\pi i} \int_\Omega \frac{J_r}{\hat{z} - z} dxdy = \frac{1}{2\pi i} \int_\Omega \left[ \frac{\partial(B_r)_y/\partial x}{\hat{z} - z} - \frac{\partial(B_r)_x/\partial y}{\hat{z} - z} \right] dxdy \qquad (4.37)$$

where the integration extends over a volume $\Omega$ containing the permanent magnet material with $z = x + iy$. The first expression in the square bracket can be transformed with the help of partial integration over the variable $x$

$$\int_\Omega \frac{\partial(B_r)_y/\partial x}{\hat{z} - z} dxdy = \int_{\partial\Omega} \frac{(B_r)_y}{\hat{z} - z} dy - \int_\Omega \frac{(B_r)_y}{(\hat{z} - z)^2} dxdy \ . \qquad (4.38)$$

The first integral vanishes, because it extends over a region just outside the permanent magnet material where $B_r$ is zero. The second integrand in Equation 4.37 can be treated in a similar way and after assembling the contributions, we finally obtain for the field outside the permanent magnet material

$$\underline{B}^*(\hat{z}) = \frac{1}{2\pi} \int_\Omega \frac{B_r}{(\hat{z} - z)^2} dxdy \ . \qquad (4.39)$$

The right-hand side does not contain derivatives of $\underline{B}_r$ but instead $(\hat{z} - z)^2$ in the denominator, which is the same that appears in the Green's function for elementary dipoles. Equation 4.39 thus tells us that the field outside the permanent magnet material at point $\hat{z}$ is given as the distribution of dipoles at points $z = x + iy$, convoluted with the Green's function for dipoles.

### 4.5.1 Multipoles

We note the resemblance of Equation 4.39 with the second equality in Equation 4.31, where we calculated the contributions of each current filament to the multipole coefficients. Here we do the same, but instead of distributions of filaments, we have to deal with distributions of magnetic dipoles and instead of a Green's function $1/(\hat{z} - z)$ in Equation 4.31, we have to use the dipolar Green's function $1/(\hat{z} - z)^2$. For positions $\hat{z} < z$ we have

$$\frac{1}{(\hat{z} - z)^2} = \frac{d}{dz}\left( \frac{1}{\hat{z} - z} \right) = -\frac{d}{dz}\left( \sum_{n=1}^\infty \frac{\hat{z}^{n-1}}{z^n} \right) = \sum_{n=1}^\infty \frac{n\hat{z}^{n-1}}{z^{n+1}} \ . \qquad (4.40)$$

Inserting in Equation 4.39 we finally obtain an equation that describes the contribution of $B_r$ in the permanent magnet to the multipole coefficient of order $m = n - 1$

$$\underline{B}^*(\hat{z}) = \sum_{m=0}^\infty \left[ \frac{m+1}{2\pi} \int_\Omega \frac{B_r}{z^{m+2}} dxdy \right] \hat{z}^m \qquad (4.41)$$

as the expression in the square bracket. In particular $m = 0$ describes the contribution to the dipole and $m = 1$ to the quadrupole component. In Section 4.4.1, we sought current

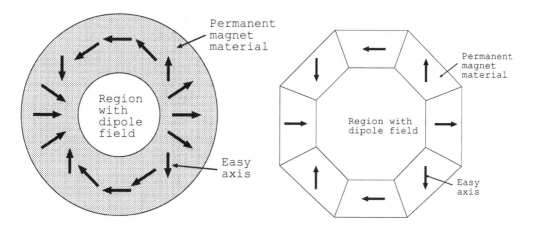

Figure 4.19 Illustration of the easy axis of the permanent magnet material continuously rotating in order to provide a horizontal dipole field in its center (left) and a rendition with discrete trapezoidal blocks (right).

distributions that led to a single non-zero multipole coefficient and found the $\cos(n\phi)$ distribution for the currents. Here we seek an angular distribution of remanent fields that results in a single multipole coefficient as well.

First, we consider a dipole magnet with a horizontal field in which the permanent magnet material is distributed in a ring around a region in which we desire the dipole field with $m = 0$. Figure 4.19 shows the geometry, where we assume that the beam moves into the paper inside the inner region. It is easy to understand that the easy axis of the permanent magnet material, both on the left and on the right, point towards the right, because that will pump field lines from the left to the right through the inner region. Likewise, the easy axis on the top and bottom should point right to left, because that will pump the field lines back to the left again. In between these regions we interpolate. In this way the easy axis appears to "tumble around" twice if we go around the ring once. Let us therefore assume that the angle of the easy axis with respect to the horizontal axis $\alpha$ is given by $\alpha = 2\phi$ where $\phi$ is the azimuthal position around the ring, such that we can describe each position in the ring by $z = x + iy = re^{i\phi}$. We thus have

$$\underline{B}_r = B_r e^{i\alpha} = B_r e^{2i\phi} \ . \tag{4.42}$$

Inserting this expression into Equation 4.41, rewriting $dxdy = rdrd\phi$ in cylinder coordinates, and integrating from inner radius $r_i$ to outer radius $r_o$ we find

$$
\begin{aligned}
\underline{B}^*(\hat{z}) &= \sum_{m=0}^{\infty} \left[ \frac{m+1}{2\pi} \int_0^{2\pi} d\phi \int_{r_i}^{r_o} rdr B_r e^{2i\phi} \frac{1}{r^{m+2}} e^{-i(m+2)\phi} \right] \hat{z}^m \\
&= B_r \sum_{m=0}^{\infty} \left[ \frac{m+1}{2\pi} \int_{r_i}^{r_o} \frac{dr}{r^{m+1}} \int_0^{2\pi} e^{2i\phi - i(m+2)\phi} d\phi \right] \hat{z}^m \\
&= B_r \log(r_o/r_i) \ ,
\end{aligned}
\tag{4.43}
$$

where we use that the integral over $d\phi$ is $2\pi$ if $m = 0$ and is zero otherwise. Apparently, the rule that the easy axis rotates twice $\alpha = 2\phi$ leads to a configuration in which only the

$m = 0$ Fourier-harmonic is non-zero and the flux density in the inner region is that of a dipole with strength $B_r \log(r_o/r_i)$, pointing towards the positive horizontal axis.

Encouraged by the success with the dipole we generalize the "tumbling law," by how many times $k$ the easy axis rotates, while it moves around the ring once, to

$$\alpha = k\phi \qquad \text{such that} \qquad \underline{B}_r = B_r e^{ik\phi} \tag{4.44}$$

and insert this expression into Equation 4.41, whence we obtain

$$
\begin{aligned}
\underline{B}^*(\hat{z}) &= B_r \sum_{m=0}^{\infty} \left[ \frac{m+1}{2\pi} \int_{r_i}^{r_o} \frac{dr}{r^{m+1}} \int_0^{2\pi} e^{ik\phi - i(m+2)\phi} d\phi \right] \hat{z}^m \\
&= \frac{m+1}{m} \frac{B_r}{r_i^m} \left( 1 - \frac{r_i^m}{r_o^m} \right) \hat{z}^m \quad \text{for} \quad k = m+2 .
\end{aligned} \tag{4.45}
$$

Again, the integral over the angle $\phi$ is $2\pi$ only for $k = m+2$ and zero otherwise. In order to obtain a multiple of order $m$ the easy axis needs to tumble $m+2$ times around the ring. For a quadrupole with $m = 1$ we therefore need 3 tumbling rotations of the easy axis. It is noteworthy that the pole-tip field at radius $|\hat{z}| = r_i$ can exceed $B_r$. If the outside radius of the ring $r_o$ becomes very large, $r_o \gg r_i$, the ratio by which it exceeds $B_r$ approaches $(m+1)/m$.

## 4.5.2 Segmented multipoles

We have shown that the "tumbling law" from Equation 4.44 results in multipolar fields of order $m$, but it required the easy axis to rotate continuously as a function of angle $\phi$, which is very difficult to manufacture. It is therefore prudent to investigate configurations where the ring is made of a number of trapezoidal segments that have a constant easy axis within each segment. On the right-hand side of Figure 4.19 we show a magnet that is made of eight trapezoidal segments and approximates the magnet with the continuous rotation of the easy axis shown in the left-hand side. In the following, we assume that the number of segments $M$ is arbitrary.

To calculate the contribution of all segments, we start by calculating the contribution of a single segment, shown to be located symmetrically around the $x$-axis in Figure 4.20, and then determine how its contribution changes when rotated by an angle $\phi$. Note that the easy axis for the different segments also rotates, albeit determined by the desired multipolarity of the magnets according to Equation 4.44. But let us first calculate the contribution to the magnetic flux density $\tilde{B}$ around the origin of the horizontally placed segment by using Equation 4.41 and use the fact that the easy axis is constant throughout the integration volume $\Omega$, shown as the lower shaded region

$$\underline{\tilde{B}}^*(\hat{z}) = \underline{B}_r \sum_{m=0}^{\infty} \left[ \frac{m+1}{2\pi} \int_\Omega \frac{dxdy}{z^{m+2}} \right] \hat{z}^m . \tag{4.46}$$

We now evaluate the integral only over $\Omega$ with $z = x + iy$ and obtain

$$
\begin{aligned}
\int_\Omega \frac{dxdy}{(x+iy)^{m+2}} &= \int_{r_i}^{r_o} dx \int_{-x\tan\alpha}^{x\tan\alpha} \frac{dy}{(x+iy)^{m+2}} \\
&= \frac{i}{m+1} \left[ \frac{1}{(1+i\tan\alpha)^{m+1}} - \frac{1}{(1-i\tan\alpha)^{m+1}} \right] \int_{r_i}^{r_o} \frac{dx}{x^{m+1}} \\
&= \frac{2}{(m+1)^2} (\cos\alpha)^{m+1} \sin((m+1)\alpha) F_m(r_i, r_o) ,
\end{aligned} \tag{4.47}
$$

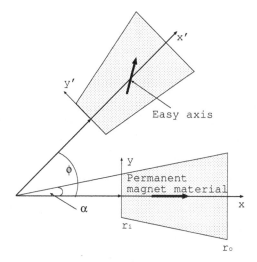

**Figure 4.20** Two trapezoidal segments of permanent magnet material.

where we introduce the abbreviation $\alpha = \pi/M$ for half of the opening angle of the trapezoid and then use it to parameterize the upper and lower boundary of the integration of $dy$ by $y = \pm x \tan \alpha$. For the integral over $dx$ we introduce the abbreviation

$$F_m(r_i, r_o) = (m + 1) \int_{r_i}^{r_o} \frac{dx}{x^{m+1}} = \begin{cases} \log(r_o/r_i) & \text{for } m = 0 \\ \frac{m+1}{m} \left( \frac{1}{r_i^m} - \frac{1}{r_o^m} \right) & \text{otherwise.} \end{cases} \tag{4.48}$$

Inserting in Equation 4.46, yields

$$\underline{\tilde{B}}^*(\hat{z}) = \underline{B}_r \sum_{m=0}^{\infty} \left[ \frac{(\cos \alpha)^{m+1} \sin((m+1)\alpha)}{(m+1)\pi} F_m(r_i, r_o) \right] \hat{z}^m . \tag{4.49}$$

The flux density $\tilde{B}$ is the contribution from the single horizontally placed segment. The contribution from a second segment that is rotated by an angle $\phi$ with respect to the first segment, is shown in Figure 4.20. Here we can proceed in the same way as for the first segment, but need to take into account that the easy axis has rotated by $k\phi$ according to Equation 4.44 and that the integration must be done over the coordinates $x'$ and $y'$. Those, however, are related to the coordinates $x$ and $y$ by $x' + iy' = e^{i\phi}(x + iy)$. Since there is a factor $(x' + iy')^{m+2}$ in the denominator of the integrand the contribution of the second segment is given by

$$\underline{\tilde{B}}^*(\hat{z}') = \underline{\tilde{B}}^*(\hat{z}) e^{ik\phi} e^{-i(m+2)\phi} \tag{4.50}$$

and we observe that the contribution from the second segment is only phase shifted by the angle $(k - m - 2)\phi$ with respect to the contribution from the first segment. Therefore the flux density $\hat{B}$ from $M$ segments, rotated by an angle $\phi = 2\pi/M$ with respect to each other, is given by

$$\underline{\hat{B}}^* = \underline{\tilde{B}}^* \sum_{j=0}^{M-1} e^{2\pi i(k-m-2)j/M} , \tag{4.51}$$

with $\underline{\tilde{B}}^*$ given by Equation 4.49. The sum is only non-zero if $(k - m - 2)/M$ is a positive or negative integer $\nu$, in which case it equals $M$. We therefore only have contributions to harmonics $m$ that fulfill $m = k - 2 + \nu M$. For the segmented dipole shown on the right of Figure 4.19 we have $k = 2$ and $M = 8$, such that this dipole also has harmonics $m = 8, 16, \ldots$ other than the wanted dipole harmonic $m = 0$.

Finally we collect all terms and find for the flux density of a segmented multipole

$$\hat{\underline{B}}^*(\hat{z}) = \underline{B}_r \sum_{\nu=0}^{\infty} \left[ \cos(\pi/M)^{m+1} \frac{\sin((m+1)\pi/M)}{(m+1)\pi/M} \right] \times F_m(r_i, r_o)\hat{z}^m \delta_{m,k-2+\nu M} \quad (4.52)$$

where the $\delta_{m,k-2+\nu M}$ describes the harmonics that are caused by the segmentation. The difference from the perfect multipoles described by Equation 4.43 and 4.45 are the additional harmonics and a reduction of the excitation, which is described by the factor in the square brackets. Appendix A.3 adapts the calculations to cubic magnetic blocks, and describes the construction of small magnets by inserting the magnet cubes in 3D-printed frames.

### 4.5.3 Undulators and wigglers

Synchrotron radiation sources and free-electron lasers rely heavily on undulator and wiggler magnets to provide transversely oscillating trajectories for the charged particles to emit synchrotron radiation. In order to emit large intensities of short-wavelength radiation, these magnets must reach very high magnetic fields and very short oscillation periods. These requirements can be fulfilled by using permanent magnets to construct the undulators and wigglers.

It turns out that the magnetic fields of undulators are closely related to those of multipole magnets with high multipolarity as is apparent from considering Figure 4.21, which shows a weakly curved segment of a multipole with very large inner radius $r_i$ and four segments per azimuthal distance $\lambda$. If we consider the flux density along the dashed line a distance $g$ below the magnets, we immediately see that the radial field component oscillates between pointing towards the magnets and in the reverse direction with period $\lambda$. It is easy to understand that increasing the radius $r_i$, while also increasing the multipolarity $m$ in such a way that the period length $\lambda$ is kept fixed, will, in the limiting case, result in a planar undulator. To formalize this idea, we introduce the scaling variable $p$, which scales the number of periods of length $\lambda$ around the circumference of length according to $2\pi r_i = 2p\lambda$. When applying this scaling to the segmented dipole on the right-hand side in Figure 4.19 we find that it consists of $p = 2$ four-segment periods with a total number of $M = 4p$ segments while the multipolarity is given by $m + 1 = 2p$. Using the dipole as a reference, we calculate the flux density on the dashed line that lies a distance $g$ below the segments with the help of Equation 4.52. For convenience, we summarize the scaling of the variables with $p$ here

$$m + 1 = 2p, \qquad r_i = p\lambda/\pi, \qquad M = 4p, \qquad \hat{z} = i(r_i - g) + z \quad (4.53)$$

and proceed to address the parts of Equation 4.52 one at a time. Inserting these scaling relations into the expression in the square brackets and taking the limit of $p \to \infty$ we find that the square bracket reduces to the constant $\sin(\pi/4)/(\pi/4) \approx 0.9$. The expression $F_m(r_i, r_o)\hat{z}^m$ we rewrite as

$$F_m(r_i, r_o)\hat{z}^m = \frac{m+1}{m}\left(1 - \frac{r_i^m}{r_o^m}\right)\frac{\hat{z}^m}{r_i^m} \quad (4.54)$$

and treat each of the factors independently. The factor $(m+1)/m = 2p/(2p-1)$ obviously

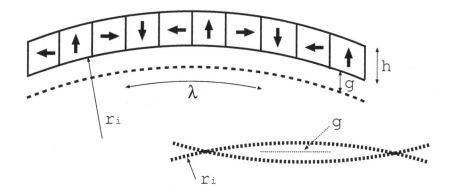

Figure 4.21  The geometry of an undulator in the Halbach-geometry. In the limit of $r_i \to \infty$ the parameter $g$ is the half-gap and the $h$ is the height of the permanent magnet-blocks.

approaches unity in the limit $p \to \infty$. The second factor becomes

$$
\begin{aligned}
\left(1 - \frac{r_i^{2p-1}}{r_o^{2p-1}}\right) &= 1 - \left(\frac{p\lambda/\pi}{p\lambda/\pi + h}\right)^{2p-1} = 1 - \left(\frac{1}{1 + \pi h/p\lambda}\right)^{2p} \\
&\approx 1 - (1 - \pi h/p\lambda)^{2p} \longrightarrow 1 - e^{-2\pi h/\lambda}
\end{aligned}
\tag{4.55}
$$

where we used $r_o = r_i + h$. In the last step, we use the representation $(1 - x/p)^p \to e^{-x}$ for the exponential function and see that the expression approaches $1 - e^{-2\pi h/\lambda}$ in the limit $p \to \infty$. In a similar fashion we obtain for the last factor in Equation 4.54

$$
\frac{\hat{z}^m}{r_i^m} = \frac{i(r_i - g) + z}{r_i} = i^{2p-1}\left(1 - \frac{\pi(g - iz)}{p\lambda}\right)^{2p-1} \longrightarrow i\, e^{-2\pi(g - iz)/\lambda}
\tag{4.56}
$$

where we tacitly assume that $p$ is even to fix the sign before the imaginary unit. Finally we realize that the effect of the second assembly of permanent magnets, the lower yoke, can be visualized as lying on a similar circle with radius increasing with $p$. The insert on the lower right of Figure 4.21 illustrates this. We need to note, however, that the variable $z$ has the opposite sign on the second circle. Since the fields from the permanent magnets on both circles superimpose linearly, the flux density around the dashed mid-gap line is given by

$$
\begin{aligned}
\hat{\underline{B}}^*(\hat{z}) &= i\underline{B}_r \frac{\sin(\pi/4)}{\pi/4}\left(1 - e^{-2\pi h/\lambda}\right)\left(e^{-2\pi(g-iz)/\lambda} + e^{-2\pi(g+iz)/\lambda}\right) \\
&= 2i\underline{B}_r \frac{\sin(\pi/4)}{\pi/4}\left(1 - e^{-2\pi h/\lambda}\right)e^{-2\pi g/\lambda}\cos(2\pi z/\lambda) ,
\end{aligned}
\tag{4.57}
$$

where we only took the lowest harmonic into account. Note that in the mid-plane of the gap the field oscillates with period length $\lambda$ and it is purely vertical. The field decreases exponentially with increasing half-gap $g$. Increasing the height $h$ of the permanent magnets beyond $\lambda/2$ increases the peak field only marginally. In that case, the field is given by $\hat{\underline{B}}^*(\hat{z}) \approx 1.723i\underline{B}_r e^{-2\pi g/\lambda}\cos(2\pi z/\lambda)$. Equation 4.57 is commonly used to design permanent magnet undulators in the described permanent magnet configuration, which is referred to as the Halbach-configuration.

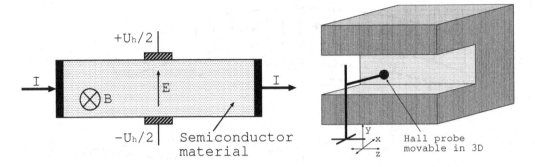

Figure 4.22 Sketches to illustrate a Hall sensor (left) and its use on a magnet measuring bench (right).

The manufacture of the magnet assembly with the very strong permanent magnets requires special care, because the magnet blocks have the tendency to align with their easy axes aligned, rather than following the pattern with easy axes rotating by 90 degrees from block to block. Often they are fixed in a rigid aluminum frame. The frames for upper and lower magnet assembly are often movable in the vertical direction in order to adjust $B_y$ in the mid-plane due to the exponential dependency on the half-gap $g$ in Equation 4.57. We refer to Appendix A.3 for the construction of undulators from magnetic cubes inserted in 3D-printed frames.

At this point, we have the tools available to make a first design of iron-dominated, super-conducting, and permanent magnet-based systems. Of course, once the magnets are built, they will differ from the design, either due to the used approximations or due to manufacturing tolerances. In either case, however, we need to verify their performance, before installing them in an accelerator. And this brings us to measuring the magnets.

## 4.6  MAGNET MEASUREMENTS

When measuring magnets we distinguish methods that locally measure the magnetic flux density at one or several points or methods that determine global quantities such as field integrals or multipole coefficients. Here, we first discuss local measurements using sensors based on the Hall-effect.

### 4.6.1  Hall probe

*Hall sensors* are based on semiconductor materials, such as GaAs or InSb, that have a high mobility of the charge carriers. The Lorentz-force due to a magnetic flux density, perpendicular to the direction of the current flow through the semiconductor, causes the charge carriers to deflect sideways and accumulate on electrodes. The schematics on the left in Figure 4.22 illustrates the geometry. The accumulated charges cause a transverse electric field that compensates the Lorentz-force such that subsequent charge carriers can travel unimpeded. The potential difference between the two electrodes $U_h$ is proportional to the magnetic flux density $B$ and can be amplified and measured with a voltmeter, resulting in a signal that is proportional to $B$. Note that only the flux density perpendicular to the plane of the sensor can be detected. Three sensors, mounted in orthogonal directions are, however, capable of measuring all three components of the flux density, albeit not exactly at the same

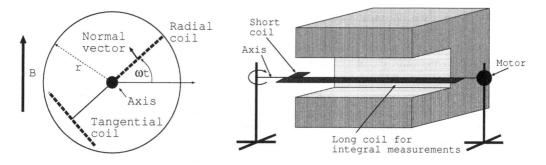

Figure 4.23 A sketch of a rotating coil (left) with positions of radial and tangential coils shown as dashed lines. The sketch on the right shows operating the rotating coil on a measuring bench.

location, because the center positions of the sensors are separated by small distances that depend on the design of the device. For precision measurements, one has to keep in mind that the ambient temperature affects the semiconductor sensors and precautions need to be taken to compensate temperature variations electronically or by keeping the measurement setup in a temperature-controlled environment. A package of one or several sensors, with appropriate temperature compensation, is often called a Hall probe.

A typical setup to move the Hall-probe around in a measuring volume, here a C-shaped dipole magnet, is shown on the right-hand side in Figure 4.22. The Hall-probe can consist of one or several individual sensors for the three components of the magnetic flux density and is mounted on a translation stage that allows the probe to be moved around in the measurement volume, for example the gap of the dipole in Figure 4.22. For obvious reasons special care must be taken to avoid vibrations when designing the mechanical translation stage. The output of such a measurement is a table of the magnetic flux densities for each position where the probe is read out. This data can subsequently be analyzed further in a spreadsheet or in MATLAB. In Appendix A.4, we describe a simple measuring bench with a Hall sensor mounted on a translation stage and use it to measure the magnets built in Appendix A.3. Special care must be taken when using the measured field map in beam-transport calculations. This may lead to unphysical results, because the three components of the flux density are not measured at the same position and even the interpolated fields do not necessarily satisfy Maxwell's equations. Normally, one should try to fit the field-map to analytical expressions that implicitly fulfill Maxwell's equations.

While Hall-probes are the standard device to perform local field measurements, rotating coil measurements are the normal method to determine integral quantities of magnets.

## 4.6.2 Rotating coil

Rotating a coil with cross section $A$ and normal vector $\vec{n}_A$ in a magnetic field the flux $\Phi = A\vec{n}_A \cdot \vec{B}$ enclosed by the coil changes with time and induces a voltage $U = -d\Phi/dt$ in the coil. The basic geometry is shown on the left-hand side in Figure 4.23, where we indicate that both radial and tangential coils can be used. It is easy to see that in the field of a dipole the coil, rotating with angular frequency $\omega$, causes both flux $\Phi$ and voltage $U$ to oscillate with the same frequency $\omega$. On the other hand, in a quadrupolar field, positions 180 degrees apart, will cause the same $\Phi$ and the resulting voltage will oscillate with $2\omega$.

Sampling the voltage at a rate much higher than $\omega$ and subsequently Fourier-transforming the signal will reveal harmonics that are related to the multipolarity. For example, a coil that is not properly centered in a quadrupole will also show a first harmonic and reducing it allows us to find the magnetic center of the quadrupole. The voltage caused by each multipole component of the field will depend on the magnetic flux density integrated over the coil but is straightforward to calculate given the magnetic potential in Equation 4.12 and the fields resulting from it, but we leave this as an exercise.

The mechanical setup is shown on the right-hand side of Figure 4.23, where we see that the motor on the right will turn the axis with a long and a short coil attached to it. The long coil provides information about the integral multipole components of the magnet, including those coming from the fringe-fields, whereas the short coil will allow us to probe a subsection of the magnet and probe for inhomogeneities of the field. The practical difficulties of performing rotating-coil measurements are knowing the area of the coil, limited mechanical rigidity of the coil which might cause it to sag, and sliding contacts to bring the electrical signals from the coil, which rotates, to the digitizer, which normally stands on the floor.

Despite these difficulties rotating coil measurement systems are widely used to rapidly measure and characterize large numbers of conventional multipole magnets. Undulators, however, require additional considerations.

### 4.6.3  Undulator measurements

Undulators are straight devices and they should only produce a wiggling particle motion inside the undulator, but they should change neither angle nor position of the beam particles at the exit. We only consider the horizontal motion here and the deflection angle is proportional to the vertical field $B_y$ the particle experience. The total deflection angle for the entire device is thus proportional to the integral of $B_y$ along the undulator, and this quantity $I_{1x} = \int_0^L B_y ds$ is called the *first field integral*. The integral extends over the entire device, including the end-fields. The position at the exit can be calculated by integrating the angle at longitudinal location $s$, which is proportional to $\int_0^s B_y(s')ds'$ along the whole device. This quantity $I_{2x} = \int_0^L ds \int_0^s B_y(s')ds'$ is called the *second field integral*. In order to verify that an undulator magnet does not perturb the orbit of an accelerator we must measure the two field integrals. In most undulators the first integral is close to zero, but for a pure sine or cosine-like field there is a systematic non-zero contribution due to the fields at the ends of an undulator. They can, however, be approximately compensated by tapering the periods at the entrance and exit of an undulator with an excitation pattern of $1/4, -3/4, 1, \ldots$.

It is of course possible to measure the vertical flux density $B_y$ along an undulator with a Hall probe and integrate numerically, but we can also use the *stretched wire* method [22]. It is based on a wire that is mounted on translation stages on either end of the undulator and the coil is completed with a wire in the field-free region. When moving both translation stages in parallel the enclosed flux of the coil changes proportional to $B_y$ along the wire, which is precisely the first field integral. By moving the translation stages in opposite directions, the flux changes in a way that is related to the second field integral.

There are other methods as well, but we leave them to the specialized literature. We move on to address the problem of how to accelerate particles in linear and circular accelerators.

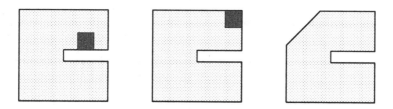

Figure 4.24  Sketches of the dipoles with the bad iron and the knocked-out corner.

## QUESTIONS AND EXERCISES

1. Calculate the Ampere-turns $NI$ for a dipole with a field of 1.5 T. The dimensions of the coil are $15 \times 8$ cm and the gap is 12 cm. Do you need to use water cooling, or does air cooling suffice?

2. Determine the number of Ampere-turns for a quadrupole with pole-tip radius 5 cm and a gradient of $\partial B_y/\partial x = 20$ T/m. If your power supplies are limited to operate below 100 A, how many turns in the magnet coils are suitable.

3. It turns out that the dipole magnet, shown in Figure 4.5 was built with sub-standard iron. There is a square region, 20 cm $\times$ 20 cm, located directly above the gap in which $\mu_r = 50$. See Figure 4.24 for an illustration. (a) Model the geometry and determine the fields and, in particular, the field in the mid-plane of the gap, as shown on the right-hand side in Figure 4.9; (b) try to salvage the magnet by restoring the field quality with shims. Document your success by showing the field in the gap; (c) find out whether saturation of the iron changes this result; (d) how bad would the magnet be, if the bad iron has $\mu_r = 1$?

4. In a second magnet the area with $\mu_r = 50$ is located in the top right corner of the magnet, as shown in the middle of Figure 4.24. Explore, by how much the field in the mid-plane of the gap is affected.

5. Somebody mishandled a crane and badly bumped with a second magnet into your dipole. The impact knocked out a triangular wedge that extends 50 cm in the vertical and horizontal directions at the top left corner, as shown on the right-hand side in Figure 4.24. Analyze the problem and figure out by how much the vertical field in the gap are affected.

6. The other magnet that was hanging on the crane is of the same type and has a similar wedge knocked out from its lower right corner. Analyze this magnet as well.

7. It turned out that the quadrupole, shown in Figures 4.10 and 4.11 is too large for your project. The width of the magnet must be reduced to 0.5 m. You therefore need to redesign the quadrupole, such that the pole-tip radius remains 5 cm, but its half-width is reduced from 0.3 to 0.25 m. Prepare a plot, similar to the one on the right-hand side in Figure 4.11 to document how well this is possible. Can you reduce the size of the magnet even further without sacrificing the achievable gradient? Investigate!

8. Design a sextupole magnet with a pole-tip radius of 5 cm and a maximum outer diameter of 40 cm. Determine the gradient $\partial^2 B_y/\partial x^2$ that is achievable with air-cooled coils.

9. Place four current filaments with a current of 1 kA at the corners of a square with 10 cm to a side, as shown in Figure 4.25. (a) Powering the two upper filaments with the same polarity produces a dipole field. Use Equation 4.11 to calculate $B_y$ in the midplane.

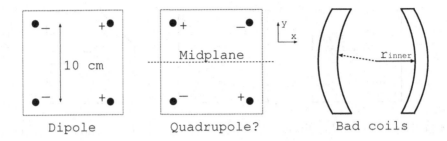

**Figure 4.25** The current filaments are arranged in a square and powered to generate dipole (left) and quadrupole (middle) field. A sketch to illustrate the bad coils from Exercises 10 and 11.

Apart from the dipole, what other multipoles appear? What is their magnitude? (b) Calculate the field gradients $\partial B_y/\partial x$, if the filaments are powered as shown in the middle image in Figure 4.25. What is wrong with this quadrupole? (c) Place the filaments in such a way that you obtain a quadrupole.

10. When the coils for the super-conducting dipole magnet shown in Figure 4.16 were delivered, you found out that the outer radius of the left coil is 1 mm larger than that of the right coil. Assume that the position of the inner radii of the coils are fixed. Moreover the coils are powered in series, such that the total current in both coils is equal. Determine the quadrupole gradient that this asymmetry causes.

11. You cannot move the left coil, but please explore if you can move the right coil further to the right to salvage the magnet. Can you reduce the gradient? Quantify the multipoles that arise in the process.

12. One of the segments in an M=8 dipole from Section 4.5.2 has only 90 % of the remanent field $B_r$ of the other magnets. How large is the ensuing gradient?

13. Design an permanent magnet undulator with a period of 8 cm and a peak field of 0.4 T on the beam axis. Assume that you can obtain good-quality magnets with $B_r = 1.2$ T. (a) plot $B_y$ and $B_x$ on the beam axis along the undulator; (b) plot $B_y$ and $B_x$ along a vertical line at a longitudinal position where the vertical field on-axis is maximum; (c) repeat this at a location where the vertical field is zero.

14. For the construction of an undulator you had ordered square blocks of permanent material. After their arrival you find out that the remanent field is only 70 % of the specified value. Can you recover field on the beam axis by redesigning the undulator and placing two magnets on top of each other?

15. Calculate the induced voltage when rotating the radial coil with radius $r$, shown on the left-hand side in Figure 4.23 with angular frequency $\omega$ in a (a) dipole; (b) quadrupole; and (c) quadrupole, horizontally displaced by a small amount $d_x$.

16. Repeat the calculations from the previous exercise for the tangential coil, also shown on the left-hand side in Figure 4.23. The end points of the coil lie on a circle with radius $r$.

# CHAPTER 5

# Longitudinal Dynamics and Acceleration

In the previous chapters we discussed magnets to guide and focus particles with a given momentum $p_0$ or energy. We noted that static magnetic fields do not change the energy of moving particles, only their direction of propagation. In order to accelerate the particles and change their energy, we need electric fields in the direction of propagation. The fields are given by the derivatives of the potential $\Phi$ in Equation 2.1. Changing the energy of the particles also changes their speed and thus their arrival time at a given location, at least at low, non-relativistic energies. At high energies the trajectory of the particles may change, and this may also change the arrival time of the particles as described by the momentum compaction factor $\alpha$ that we discussed in Section 3.4.4. If the electric field is time-varying, particles will gain different energies, depending on the arrival time. The interplay of energy gain and arrival time will determine the dynamics of their particles in the so-called *longitudinal phase space,* which is described by the arrival time $\tau$ and the relative momentum offset $\delta = (p - p_0)/p_0$, the latter is often used instead of the energy, as discussed in Sections 2.2 and 2.3.

In this chapter, we will discuss simple acceleration structures and the beam dynamics, both in circular and in linear accelerators. Various technological aspects, such as the generation, transport, and control of power are deferred to the next chapter. Here we start by considering methods to use electrical fields to transfer energy to the beam. Static fields beyond a few MV/m, such as those used in Van-de-Graaff and Tandem accelerators, lead to problems with the high-voltage insulation. Therefore electro-magnetic fields in the radio-frequency range are used to reach higher energies. The drift-tube linear accelerator, shown in Figure 1.2, is one example. It is based on "hiding" the particles in drift tubes, while the polarity of the accelerating field is reversed, such that the particle experiences a longitudinal field with the correct polarity, the next time it crosses the gap between drift tubes. In the limiting case, we may consider a single acceleration gap and regard the beam pipe as the drift tube. If the accelerator is circular, we often use individual accelerating gaps; and they are typically located in resonant structures, the *acceleration cavities.* Since cavities play such a central role in charged particle accelerators, we will look at a simple prototype—the *pill-box cavity*—more closely, and analyze the electro-magnetic fields that can oscillate inside the cavity in the following sections.

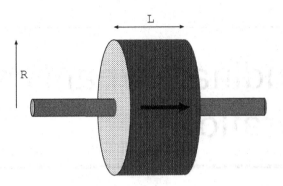

Figure 5.1 A pill-box cavity with the longitudinal electrical field vector indicated.

## 5.1 PILL-BOX CAVITY

The geometry of the pill-box cavity is shown in Figure 5.1 as the darker cylinder with length $L$ and radius $R$. The beam pipe extends from the left to the right-hand side and normally has a diameter much smaller than $R$. The dynamics of the electro-magnetic fields is governed by Maxwell's equations in vacuum with no charges or currents present. The general form of these equations, shown in Equation 4.1, in Chapter 4 then simplifies to

$$\vec{\nabla} \cdot \vec{\mathbf{E}} = 0 \quad , \qquad \vec{\nabla} \times \vec{\mathbf{E}} = -\mu_0 \frac{\partial \vec{\mathbf{H}}}{\partial t} \quad ,$$

$$\vec{\nabla} \cdot \vec{\mathbf{H}} = 0 \quad , \qquad \vec{\nabla} \times \vec{\mathbf{H}} = \varepsilon_0 \frac{\partial \vec{\mathbf{E}}}{\partial t} \quad . \tag{5.1}$$

The inside of the cavity is evacuated and both permeability and dielectric constants are those of vacuum. Furthermore, we assume that the metallic walls of the cavity have infinite conductivity, and, following the discussion in Section 4.1, this implies that the normal component of the flux density $\vec{n} \cdot \vec{\mathbf{H}}$ and the tangential component of the electric field $\vec{n} \times \vec{\mathbf{E}}$ are zero. Here $\vec{n}$ is a vector, normal to the surface.

We now have to solve the set of Equations 5.1 with the boundary conditions fulfilled on the metallic walls of the cylindrical pill-box cavity. Taking the curl on both sides of the second equation for $\vec{\mathbf{E}}$ in Equation 5.1 we obtain

$$\vec{\nabla} \left( \vec{\nabla} \cdot \vec{\mathbf{E}} \right) - \vec{\nabla}^2 \vec{\mathbf{E}} = -\mu_0 \varepsilon_0 \frac{\partial^2 \vec{\mathbf{E}}}{\partial t^2} \quad , \tag{5.2}$$

where we expressed the curl of $\vec{\mathbf{B}}$ by the temporal derivative of $\vec{\mathbf{E}}$. The first term on the left-hand side is zero, because the divergence of $\vec{\mathbf{E}}$ vanishes and this leaves us with

$$\vec{\nabla}^2 \vec{\mathbf{E}} - \mu_0 \varepsilon_0 \frac{\partial^2 \vec{\mathbf{E}}}{\partial t^2} = 0 \quad , \tag{5.3}$$

which is the wave equation. We thus find that the fields in the cavity propagate as waves and that the propagation speed is given by $c^2 = 1/\mu_0 \varepsilon_0$, the speed of light in vacuum. Since these waves will be reflected from the walls of the cavity we expect standing-wave patterns to emerge.

We are mostly interested in field patterns, or modes, that have an electric field component $E_z$ in the beam's direction of propagation. We assume that this can be described by a wave propagating in the $z$–direction with frequency $\omega$ and the wave vector, often also called *propagation constant* $k_z$

$$\vec{\mathbf{E}} = \vec{E}e^{i(\omega t - k_z z)} \qquad \text{and} \qquad \vec{\mathbf{H}} = \vec{H}e^{i(\omega t - k_z z)} \ , \tag{5.4}$$

where $\vec{E} = \vec{E}(x, y)$ and $\vec{H} = \vec{H}(x, y)$ depend on the transverse coordinates, only. Inserting into Equation 5.1 and sorting the eight independent components, we obtain

$$
\begin{aligned}
\frac{\partial E_x}{\partial x} + \frac{\partial E_y}{\partial y} &= ik_z E_z \\
\frac{\partial E_z}{\partial y} - \frac{\partial E_y}{\partial z} &= \frac{\partial E_z}{\partial y} + ik_z E_y = -i\omega\mu_0 H_x \\
\frac{\partial E_x}{\partial z} - \frac{\partial E_z}{\partial x} &= -ik_z E_x - \frac{\partial E_z}{\partial x} = -i\omega\mu_0 H_y \\
\frac{\partial E_y}{\partial x} - \frac{\partial E_x}{\partial y} &= -i\omega\mu_0 H_z \\
\frac{\partial H_x}{\partial x} + \frac{\partial H_y}{\partial y} &= ik_z H_z \\
\frac{\partial H_z}{\partial y} - \frac{\partial H_y}{\partial z} &= \frac{\partial H_z}{\partial y} + ik_z H_y = i\omega_0\varepsilon_0 E_x \\
\frac{\partial H_x}{\partial z} - \frac{\partial H_z}{\partial x} &= -ik_z H_x - \frac{\partial H_z}{\partial x} = i\omega_0\varepsilon_0 E_y \\
\frac{\partial H_y}{\partial x} - \frac{\partial H_x}{\partial y} &= i\omega_0\varepsilon_0 E_z \ .
\end{aligned}
\tag{5.5}
$$

It now turns out that we can express all transverse components of the electric and magnetic fields through the longitudinal component $E_z$ and $H_z$. For example, eliminating $E_y$ from the second and seventh equations, we get

$$\omega_0\varepsilon_0 \frac{\partial E_z}{\partial y} - \beta\frac{\partial H_z}{\partial x} = i\left(k_z^2 - \frac{\omega^2}{c^2}\right)H_x \ , \tag{5.6}$$

where we used $\mu_0\varepsilon_0 = 1/c^2$. Introducing $k_c^2 = \omega^2/c^2 - k_z^2$ and solving for $H_x$ leads to

$$H_x = \frac{i}{k_c^2}\left(\omega\varepsilon\frac{\partial E_z}{\partial y} - k_z\frac{\partial H_z}{\partial x}\right) \ . \tag{5.7}$$

We observe that the transverse component $H_x$ can be expressed through derivatives of the longitudinal component $E_z$ and $H_z$. Likewise, the other transverse components can be expressed in a similar way by eliminating the one transverse component without a derivative. We find

$$
\begin{aligned}
H_y &= \frac{-i}{k_c^2}\left(\omega\varepsilon\frac{\partial E_z}{\partial x} + k_z\frac{\partial H_z}{\partial y}\right) \\
E_x &= \frac{-i}{k_c^2}\left(k_z\frac{\partial E_z}{\partial x} + \omega\mu\frac{\partial H_z}{\partial y}\right) \\
E_y &= \frac{i}{k_c^2}\left(-k_z\frac{\partial E_z}{\partial y} + \omega\mu\frac{\partial H_z}{\partial x}\right) \ .
\end{aligned}
\tag{5.8}
$$

At this point we have expressed all transverse field components through the longitudinal ones, but we still have the freedom to specify either $E_z$, or $H_z$, or both. The special case with $E_z = 0$ describes *transverse electric*, or TE–waves, but they are obviously unsuitable to accelerate particles along the $z$–direction. Instead, we choose *transverse magnetic*, or TM–waves, by requiring $H_z$ to be zero, in which case the four equations depend on the longitudinal field components $E_z$ and from Equations 5.7 and 5.8, we obtain

$$H_x = \frac{i\omega\varepsilon}{k_c^2}\frac{\partial E_z}{\partial y} \ , \quad H_y = -\frac{i\omega\varepsilon}{k_c^2}\frac{\partial E_z}{\partial x} \ , \quad E_x = -\frac{ik_z}{k_c^2}\frac{\partial E_z}{\partial x} \ , \quad E_y = -\frac{ik_z}{k_c^2}\frac{\partial E_z}{\partial y} \ . \tag{5.9}$$

Now we know how to calculate the transverse field components from the longitudinal, but still have to find the longitudinal components $E_z$. And this we can do by inspecting Equation 5.3, which is actually three equations, one for each component of the vector $\vec{E}$ and in particular also for the longitudinal component.

$$\triangle E_z + \frac{\omega^2}{c^2}E_z = 0 \ , \tag{5.10}$$

where we assume a harmonic time-dependence proportional to $e^{i\omega t}$.

The cylindrical symmetry of the pill-box cavity suggests that using cylinder coordinates will later facilitate to satisfy the boundary conditions. This transforms Equation 5.10 into

$$\frac{1}{r}\frac{\partial}{\partial r}\left(r\frac{\partial E_z}{\partial r}\right) + \frac{1}{r^2}\frac{\partial^2 E_z}{\partial \phi^2} + \frac{\partial^2 E_z}{\partial z^2} + \frac{\omega^2}{c^2}E_z = 0 \ . \tag{5.11}$$

We attempt to solve this linear partial differential equation with the separation ansatz $E_z = f(r)g(\phi)h(z)$. Inserting it into Equation 5.11, we obtain

$$\frac{1}{rf}\frac{\partial(rf')}{\partial r} + \frac{1}{r^2}\frac{g''}{g} + \frac{h''}{h} + \frac{\omega^2}{c^2} = 0 \ . \tag{5.12}$$

Obviously, $h''/h$ only depends only on $z$ and all the other terms depend on $r, \phi$, or are constant. Therefore $h''/h$ must be a constant, which we call $-k_z^2$ in order to be consistent with Equation 5.4, and find

$$\frac{h''}{h} + k_z^2 = 0 \quad \text{or} \quad h''(z) + k_z^2 h(z) = 0 \ , \tag{5.13}$$

which is solved by

$$h(z) = E_0 e^{\pm ik_z z} \ . \tag{5.14}$$

The value of $-k_z^2$ will be determined later by the boundary conditions. The remaining equation still depends on $f$ and $g$ is given by

$$\frac{1}{rf}\frac{\partial(rf')}{\partial r} + \frac{1}{r^2}\frac{g''}{g} = -\left(\frac{\omega^2}{c^2} - k_z^2\right) = -k_c^2 \ . \tag{5.15}$$

This equation can be rewritten as

$$\frac{r}{f}\frac{\partial(rf')}{\partial r} + k_c^2 r^2 = -\frac{g''}{g} \ . \tag{5.16}$$

Again, since the left-hand side only depends on $r$ and the right-hand side only on $\phi$, each side must be independently constant and equal to some constant $k_r^2$. We thus obtain the two equations

$$r^2 f''(r) + rf'(r) + (k_c^2 r^2 - k_r^2)f(r) = 0 \quad \text{and} \quad g''(\phi) + k_r^2 g(\phi) = 0 \ . \tag{5.17}$$

The second equation is either solved by exponentials or by sine and cosine functions with argument $k_r \phi$. Periodicity in $\phi$ implies that $2\pi k_r = 2\pi m$ or, equivalently, that $k_r$ needs to be an integer. For $g(\phi)$ we then obtain

$$g(\phi) = e^{\pm im\phi} . \tag{5.18}$$

We now insert $k_r = m$ into the first equation and make a variable substitution $s = k_c r$. This leads to

$$s^2 f'' + s f' + (s^2 - m^2)f = 0 , \tag{5.19}$$

which is just the defining equation for the Bessel functions of integer order [23], denoted by $J_{\pm m}(s)$. For $f(r)$ we therefore obtain

$$f(r) = J_m(k_c r) , \tag{5.20}$$

where we have substituted back the original variable $r$. Collecting the solutions for $f, g$, and $h$ we find the solution for the longitudinal electric field component

$$E_z(r, \phi, z, t) = E_0 J_m(k_c r) e^{\pm im\phi} e^{\pm ik_z z} e^{i\omega t} \tag{5.21}$$

inside the cavity. Here $E_0$ is a constant amplitude.

If we assume that the cavity boundaries are perfectly conducting, we have $E_z = 0$ at $r = R$ and this implies that $J_m(k_c R) = 0$ and we conclude that the Bessel-function must have a zero on the surface of the cavity. If we denote the $n$-th zero of the Bessel function $J_m$ by $\gamma_{mn}$ we must have $k_c R = \gamma_{mn}$. For example, the mode with $m = 0$ therefore requires $k_c R = 2.405$ where 2.405 is approximately the first zero of $J_0$. The requirement that the electric field vanishes at $z = 0$ and at $z = l$ can be fulfilled by combining the two exponentials with argument $\pm ik_z z$ to a cosine with the same argument that vanishes at $z = 0$ and $l$ whereby we get $k_z l = p\pi$ with integer $p$. Collecting these constraints for $k_z$ and $k_c$ we obtain a dispersion relation for the resonance frequencies

$$\frac{\omega_{mnp}^2}{c^2} = \left(\frac{\gamma_{mn}}{R}\right)^2 + \left(\frac{p\pi}{l}\right)^2 \tag{5.22}$$

with integers $m, n$, and $p$. Obviously, if the geometry of the cavity is given through its radius $R$ and length $l$ the admissible frequencies are given by $f_{mnp} = \omega_{mnp}/2\pi$ in Equation 5.22. Only these frequencies satisfy the boundary condition of vanishing fields on the metallic boundaries.

We can turn the argument around and use Equation 5.22 to design a cavity that has the desired frequency that we deem useful for our accelerator. A value for the frequency that is often chosen is $500\,\text{MHz}$ and we now have to choose $R$ and $l$ suitably. We also require that $500\,\text{MHz}$ is the fundamental mode, or the lowest possible eigenfrequency. We therefore pick $m = 0$ and $n = 1$, the first zero of the zeroth Bessel function. Solving Equation 5.22 for $R$, we obtain

$$R = \frac{2.405\lambda}{2\pi\sqrt{1 - (p\lambda/2l)^2}} , \tag{5.23}$$

where we introduced the wave length $\lambda = c/f = 0.6\,\text{m}$. If we select $p = 0$, we find the relation between the radius and the wavelength

$$R = \frac{2.405\lambda}{2\pi} = 0.23\,\text{m} \tag{5.24}$$

for the mode characterized by $(mnp) = (010)$. It is usually denoted the $\text{TM}_{010}$-mode. It has a non-zero longitudinal field component $E_z$, which is used to accelerate particles. The longitudinal magnetic field-component is zero and the other components can be derived from Equation 5.9 with $E_z$ from Equation 5.21. Note that the dimensions are of similar magnitude to those of a cookie-jar, whose modes we will analyze in Appendix A.5.

In order to accelerate particles, we externally excite the $TM_{010}$–mode with a radio-frequency generator, but all the other modes are still present and it happens that the beam itself excites these modes, which may lead to instabilities, a topic we return to in Chapter 12. For the time being, we consider the cavity solely as a device that periodically provides a longitudinal electric field and will now focus on how it affects the beam. Other aspects, such as the generation of the radio-frequency power needed to excite the fields in the cavities is deferred to Chapter 6.

We continue by observing that the accelerating cavity has a finite length $l$ and the finite velocity of the particles implies that a given particle cannot receive the maximum acceleration voltage all the time during its traversal of the cavity. The effective voltage "seen" by the particle is reduced by the so-called transit time factor.

## 5.2 TRANSIT-TIME FACTOR

The energy gain $\Delta W$ of a particle with charge $q$ that arrives in the center of the cavity with phase $\phi$ is given by

$$\Delta W = q \int_{-l/2}^{l/2} \hat{E}_z \cos(\omega_{rf} t + \phi) dz \tag{5.25}$$

with $t(z) = z/\beta c$, where $\beta c$ is the velocity of the particle and $\hat{E}_z$ is the peak longitudinal electric field in the pill-box cavity. The integral boundaries extend from the position the particle enters the cavity on one side until its exits on the other. Evaluating the integral, we find

$$\Delta W = \left[ q\hat{E}_z l \cos(\phi) \right] \frac{\sin(\omega_{rf} l / 2\beta c)}{\omega_{rf} l / 2\beta c} . \tag{5.26}$$

The expression in the square brackets is the energy gain the particle had received during the traversal, had it been exposed to the peak field the whole time. Since the field varies during the traversal, it is reduced by the *transit time factor* $T(x) = \sin(x)/x$ with $x = \omega_{rf} l / 2\beta c$. Considering the 500 MHz cavity from Equation 5.24 with a length of $l = 0.2$ m, for relativistic electrons with $\beta$ close to unity, we find $T \approx 1$, whereas for low-energy protons with $\beta = 0.1$ we find $T \approx 0.6$. This inefficiency motivates the use of special accelerating cavities, such as half-wave resonators or spoke cavities, for low-energy particles. These special cavities are adapted to minimize the distance across which the particles experience the accelerating field—the accelerating gap.

We now turn to the dynamics of beams in circular accelerators, where the cavities affect the energy of the particles, but the travel time for one turn affects the arrival-phase of a particle in the cavity and it needs to be synchronized with the frequency of the cavity. If done correctly, the particles will perform stable oscillations, a subject we investigate in the following section.

## 5.3 PHASE STABILITY AND SYNCHROTRON OSCILLATIONS

The energy gain in an accelerating cavity $\Delta W$ depends on the arrival time of the particles or, equivalently, their arrival phase $\phi$. In a circular accelerator, such as a storage ring, the RF-frequency $f_{rf} = \omega_{rf}/2\pi$ of the cavity must be chosen to be an integer multiple of the

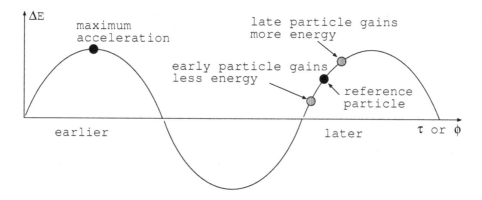

Figure 5.2 The energy $\Delta E$ gained by a particle in a cavity as a function of the arrival time $\tau$ or phase $\phi = \omega_{\mathrm{rf}}\tau$.

revolution frequency $f_0 = 1/T_0$, where $T_0$ is the revolution time for one turn. This integer $h = f_{\mathrm{rf}}/f_0$ is called the *harmonic number* of the RF-system.

The synchronous phase $\phi_s$ of the RF-cavity at which the reference particle should arrive is determined by external requirements, such as to achieve maximum acceleration, which is indicated by the particle located at the crest of the oscillation in Figure 5.2. In other accelerators the design phase may be determined by the requirement to replenish the losses from the emission of synchrotron radiation or the interaction with a target. Increasing the energy of the beam during an energy ramp is another example that we address in Section 5.6. To be specific, we consider a storage ring, where the energy $U_d$ must be delivered to the beam every turn and this needs to be provided by the accelerating cavity. The synchronous phase $\phi_s$, at which the beam receives this energy is then given by

$$U_d = q\hat{V} \sin \phi_s \, , \tag{5.27}$$

where $\hat{V}$ is the peak acceleration voltage that a particle can experience in the cavity. It is determined by the peak longitudinal electric field, the length of the cavity, and the transit-time factor as given in Equation 5.26. A second particle, arriving at a slightly different time $\tau$, will then experience the following energy gain in the cavity

$$\Delta E = q\hat{V} \left[\sin(\omega_{\mathrm{rf}}\tau) - \sin \phi_s\right] \, . \tag{5.28}$$

This situation is illustrated on the right-hand side in Figure 5.2, where the reference particle is located at the phase $\phi_s$ and an early and a late particle are also shown. They experience an energy kick with respect to the reference particle given by the previous equation. We note that it is convenient to use the phase variable $\phi$ instead of arrival time $\tau$. These variables are related by $\phi = \omega_{\mathrm{rf}}\tau$.

We relate the change in energy to the change in the phase-space variable $\delta = \Delta p/p$ with

$$\frac{\Delta E}{E} = \beta^2 \frac{\Delta p}{p} = \beta^2 \delta \, , \tag{5.29}$$

which follows from differentiating $E = \sqrt{m^2 c^4 + p^2 c^2}$, known from relativistic kinematics. In most circumstances, the change of energy $\Delta E$ is small compared to the total energy of the particle and we can replace the derivative with respect to time $d/dt$ by averaging the change over one revolution period $T$. For the temporal variation of $\delta$ we then find

$$\frac{d\delta}{dt} \approx \frac{1}{T_0\beta^2} \frac{\Delta E}{E} = \frac{e\hat{V}}{T_0\beta^2 E} \left[\sin\phi - \sin\phi_s\right] , \qquad (5.30)$$

which relates the change in relative momentum per revolution period $d\delta/dt$ to the arrival time $\tau$, here expressed in terms of the phase variable $\phi = \omega\tau$.

After analyzing what happens to the particles in the cavity we need to investigate how the travel-time through the arcs of the ring is affected by the momentum deviation $\delta$ and on the magnetic field in the dipoles $\Delta B/B$. The latter is relevant, because changing the field in the dipoles changes the length of the orbit such that the arrival phase of the particles is systematically shifted to one side, causing them to systematically gain or lose energy. This mechanism is used to accelerate the beams—just increase the dipole field and the beam moves to a new arrival phase in the cavity to receive the right energy to stay synchronous. This, of course, requires the motion to be stable and that we will show in later parts of this section.

The dependence of the arrival time at the cavity on the magnetic field in the dipoles is given in terms of the momentum compaction factor $\alpha$ by $\Delta\tau/T_0 = -\alpha\Delta B/B$ following the discussion after Equation 3.18 where we saw that the dispersion originates from a mismatch of the momentum of the particles to the dipole field. Therefore, we find for the dependence of the circumference on $\delta$ and $\Delta B/B$

$$\frac{\Delta C}{C} = \alpha \left(\delta - \frac{\Delta B}{B}\right) . \qquad (5.31)$$

The time for one revolution is given by $T_0 = C/v$ where $v$ is the speed of the particle. From $p = \beta\gamma mc$ we find $dv/v = d\beta/\beta = (1/\gamma^2)dp/p = \delta/\gamma^2$ and for the relative change of the revolution time $\Delta T/T_0$, we obtain

$$\frac{\Delta T}{T_0} = \frac{\Delta C}{C} - \frac{\Delta v}{v} = \left(\alpha - \frac{1}{\gamma^2}\right)\delta - \alpha\frac{\Delta B}{B} = \eta\delta - \alpha\frac{\Delta B}{B} , \qquad (5.32)$$

where we introduce the *phase-slip factor* $\eta = \alpha - 1/\gamma^2$. It describes the dependence of the revolution time, or equivalently revolution frequency, as a function of the momentum deviation. There are two contributions: the change of the circumference with momentum, as described by $\alpha$, and the change in the speed of the particle. The latter effect is mostly important in low-energy rings. Note, however, that $\eta$ can change sign as the particles are accelerated and $\gamma$ increases. The energy $\gamma_T = 1/\sqrt{\alpha}$, at which $\eta$ is zero, is called the *transition energy*, and it has an important influence on the stability of the particle's motion, as we shall shortly see.

The change in revolution time $\Delta T/T_0$ in Equation 5.32 can be interpreted as the change of arrival time in the cavity $d\tau/T_0 = \Delta T/T_0$ of a particle with momentum deviation $\delta$. The factor proportional to $\Delta B/B$ describes the variation of the phase with the field $B$ in the main dipoles of the ring. For the change in arrival phase we then obtain

$$\frac{d\phi}{dt} \approx \frac{\omega_{\rm rf}d\tau}{T_0} = \omega_{\rm rf}\eta\delta - \omega_{\rm rf}\alpha\frac{\Delta B}{B} . \qquad (5.33)$$

Equation 5.33 and 5.30 jointly describe the dynamics of particles in the longitudinal phase-space $\phi = \omega_{\rm rf}\tau$ and $\delta$. Equation 5.30 describes the effect of arrival phase $\phi$ on the change of momentum in the cavity $d\delta/dt$ and Equation 5.33 describes the change of arrival phase as a function of the momentum $\delta$. Differentiating the latter equation with respect to time and inserting in the first, results in

$$\ddot{\phi} - \frac{\omega_{\rm rf}\eta}{T_0\beta^2}\frac{q\hat{V}}{E}\left[\sin\phi - \sin\phi_s\right] = 0 \ . \tag{5.34}$$

For small phase deviations $\psi = \phi - \phi_s$ from the design phase $\phi_s$ Equation 5.34 can be rewritten by using $\sin\phi - \sin\phi_s \approx \psi\cos\phi_s$. This allows us to express it in the following form

$$\ddot{\psi} + \Omega_s^2\psi = 0 \qquad \text{with} \qquad \Omega_s^2 = -\frac{\omega_{\rm rf}\eta\cos\phi_s}{T_0\beta^2}\frac{e\hat{V}}{E} \ , \tag{5.35}$$

valid for small oscillations. Equation 5.35 is the differential equation for a harmonic oscillator and describes oscillations with the synchrotron frequency $\Omega_s$. These oscillations in the phase-momentum phase space are called *synchrotron oscillations*. The motion of the phase $\psi$ and momentum offset $\delta$ will be harmonic with $\psi = \hat{\psi}\sin\Omega_s t$ and $\delta = \hat{\delta}\cos\Omega_s t$ where $\hat{\psi}$ and $\hat{\delta}$ are the maximum amplitudes of the phase and the momentum offset. Note that Equation 5.33 for $\Delta B/B = 0$ implies that $\hat{\psi}$ and $\hat{\delta}$ are related by $\hat{\psi} = (\omega\eta/\Omega_s)\hat{\delta}$.

The oscillations described by Equation 5.35 are only stable if $\Omega_s^2$ is positive, which implies that the product of $\eta$ and $\cos\phi_s$ must be negative. Considering the definition of the phase slip factor $\eta = \alpha - 1/\gamma^2$ in terms of the momentum compaction factor $\alpha$ and the kinematic factor $\gamma$ we see that for low energies the $\gamma$-factor dominates and makes $\eta$ negative which implies that $\cos\phi_s$ must be positive. The transition energy where $\eta$ is zero separates the low-energy regime where a storage ring operates, colloquially speaking, below transition, from the high-energy regime, where rings operate above transition. In the latter case we find that $\cos\phi_s$ must be negative and consequently the phase must be close to 180 degrees, near the point where the RF voltage crosses zero from positive to negative voltages. This is the case for practically all electron accelerators. Normally, the particles will assemble at the design-phase, where the motion is stable: below transition around zero phase, where the slope of $\Delta E$ versus $\tau$ is rising and shown in Figure 5.2, and above transition around 180 degrees, where it is falling. Around these phases the particles perform stable oscillations.

We now return to Equation 5.34 and the motion of particles with large amplitudes, as might occur when injecting particles with too high energy into a ring.

## 5.4 LARGE-AMPLITUDE OSCILLATIONS

When injecting new particles into a ring, normally the energy and phase of the new particles should be adjusted to arrive near the point where stable oscillations occur. If, however, the timing or the energy are wrong, the particles are still governed by Equation 5.34, but the small-angle approximation, that led to Equation 5.35, is no longer valid. We therefore use Equation 5.34, but with the simplifying assumption $\phi_s = 0$, and arrive at the equation of a mathematical pendulum

$$\ddot{\phi} + \Omega_s^2\sin\phi = 0 \ . \tag{5.36}$$

By multiplying this equation with $\dot{\phi}$ we see that it gives rise to an integral of motion

$$\frac{d}{dt}\left[\frac{1}{2}\dot{\phi}^2 - \Omega_s^2\cos\phi\right] = 0 \qquad \text{or} \qquad \frac{1}{2}\dot{\phi}^2 - \Omega_s^2\cos\phi = A \ . \tag{5.37}$$

The constant $A$ can be expressed in terms of the maximum phase excursion $\hat{\phi}$ because at $\phi = \hat{\phi}$ we have $\dot{\phi} = 0$ and find $A = -\Omega_s^2\cos\hat{\phi}$. Inserting into the previous equation yields

$$\frac{1}{2}\dot{\phi}^2 + \Omega_s^2(\cos\hat{\phi} - \cos\phi) = 0 \ . \tag{5.38}$$

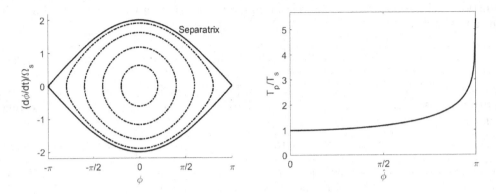

Figure 5.3 Phase portrait of the longitudinal phase space (left) and the oscillation period as a function of the amplitude $\hat{\phi}$.

Introducing the maximum amplitude $\hat{\phi}$ characterizes the trajectories in much the same way as we can distinguish different oscillations of a children's swing by their maximum amplitude. Note that another option would be to describe the trajectories by their total energy.

We can now solve Equation 5.38 for $\dot{\phi}$ and plot it as a function of $\phi$ and obtain

$$\dot{\phi} = \Omega_s \sqrt{2(\cos \phi - \cos \hat{\phi})} \tag{5.39}$$

and plot $\dot{\phi}$ as a function of $\phi$ for parameter values $\hat{\phi} = \pi/5, 2\pi/5, \dots, \pi$. We show the plots in the $\phi, \dot{\phi}$ phase plane on the left-hand side in Figure 5.3 and observe that for small $\hat{\phi}$ the trajectories follow ellipses, but as the amplitude increases, the phase space is increasingly distorted up to a limiting curve for $\hat{\phi} = \pi$. It is drawn as a solid line and is called the *separatrix*, because it separates stable oscillations from unbounded trajectories.

Note that $\dot{\phi}$ is related to the energy deviation $\delta = \Delta p/p$ by Equation 5.33, which implies that the vertical axis in Figure 5.3 is actually just a rescaled $\delta$ where the rescaling is given by $\dot{\phi} = \eta \omega_{\mathrm{rf}} \delta$. This implies that there is a *maximum momentum acceptance* $\delta_{max}$, namely the height of the separatrix, which is often called the *bucket half-height*. It is given by

$$\delta_{max} = \frac{2\Omega_s}{\eta \omega_{\mathrm{rf}}} = 2\sqrt{\frac{1}{\eta \omega_{\mathrm{rf}} \beta^2 T_0} \frac{e\hat{V}}{E}} , \tag{5.40}$$

which clearly shows that the momentum acceptance is proportional to the root of the cavity voltage $\hat{V}$. If a particle is badly injected or suffers a very large energy loss it can end up outside the separatrix, where it will not perform stable synchrotron oscillations.

We found that the particles with small phase angles $\phi$ perform harmonic synchrotron oscillations. This will, however, change as the amplitude increases and the motion becomes increasingly an-harmonic. To illustrate this we calculate the oscillation period $T_p$ as a function of the amplitude by rearranging Equation 5.39 in the following way

$$\frac{\Omega_s T_p}{4} = \Omega_s \int_0^{T_p/4} dt = \int_0^{\hat{\phi}} \frac{d\phi}{\sqrt{2(\cos \phi - \cos \hat{\phi})}} = \frac{1}{2} \int_0^{\hat{\phi}} \frac{d\phi}{\sqrt{(\sin^2(\hat{\phi}/2) - \sin^2(\phi/2))}} , \tag{5.41}$$

where we calculate the time to get from $\phi = 0$ to the extreme phase $\phi = \hat{\phi}$, which covers a quarter the oscillation period $T_p/4$. The integral on the right-hand side can be brought into the standard form of a complete elliptic integral $K(x)$, as defined in [23], by the substitution $\sin \psi = \sin(\phi/2)/\sin(\hat{\phi}/2)$, whence we arrive at

$$\frac{\Omega_s T_p}{2} = 2 \int_0^{\pi/2} \frac{d\psi}{\sqrt{1 - \sin^2(\hat{\phi}/2) \sin^2 \psi}} = 2K(\sin(\hat{\phi}/2)) . \tag{5.42}$$

Solving for the oscillation period $T_p$, we obtain

$$T_p = \frac{4}{\Omega_s} K(\sin \hat{\phi}/2) = T_s \left(\frac{2}{\pi}\right) K(\sin \hat{\phi}/2) \tag{5.43}$$

with the small-oscillation amplitude revolution period $T_s = 2\pi/\Omega_s$. The right-hand side in Figure 5.3 shows the revolution period $T_p$, normalized to the small amplitude period $T_s$, as a function of the amplitude $\hat{\phi}$ in the range $0 \leq \hat{\phi} \leq \pi$. The oscillation period diverges as the amplitude $\hat{\phi}$ approaches $\pi$ which corresponds to a starting phase close to one of the nodes of the separatrix.

In the following paragraphs we determine transfer maps [24], non-linear variants of the transfer matrices from Chapter 3, that allow us to map starting positions $\phi_0$ and $\dot{\phi}_0$ to final positions $\phi(t)$ and $\dot{\phi}(t)$ after an arbitrarily long time $t$. To find this map, we use the initial values of $\phi_0$ and $\dot{\phi}_0$ to express the integration constant $A$, which appears in Equation 5.37, instead of the maximum amplitude $\hat{\phi}$ to arrive at

$$\frac{1}{2}\dot{\phi}^2 - \Omega_s^2 \cos \phi = E = \frac{1}{2}\dot{\phi}_0^2 - \Omega_s^2 \cos \phi_0 . \tag{5.44}$$

Later on it will be convenient to express the cosine in terms of the square of a sine with half angle as argument. We therefore use the trigonometric identity $\cos y = 1 - 2\sin^2(y/2)$ and rewrite the previous equation as

$$\dot{\phi}^2 = 2[E + \Omega_s^2 \cos \phi] = \dot{\phi}_0^2 + 4\Omega_s^2 \sin^2(\phi_0/2) - 4\Omega_s^2 \sin^2(\phi/2) \tag{5.45}$$

and, after introducing the abbreviation $k^2 = (\dot{\phi}_0^2/2\Omega_s)^2 + \sin^2(\phi_0/2)$, we obtain

$$\dot{\phi}^2 = (2\Omega_s k)^2 \left[1 - \frac{1}{k^2} \sin^2(\phi/2)\right] , \tag{5.46}$$

which is the starting point for our further investigations.

First, we note that $k^2 = 1$ defines a boundary between a region where the sign of the expression in the square brackets may change ($k^2 < 1$) and where the sign is always the same ($k^2 > 1$). The boundary is the separatrix and given by

$$\dot{\phi} = 2\Omega_s \sqrt{1 - \sin^2(\phi/2)} = \pm 2\Omega_s \cos(\phi/2) , \tag{5.47}$$

which, after using some trigonometric identities, is the same as given in Equation 5.39 for $\hat{\phi} = \pi$.

If $k^2 > 1$ the expression in the square brackets in Equation 5.46 never changes sign and we solve the expression for $dt$

$$dt = \frac{1}{2\Omega_s k} \int_{\phi_0}^{\phi} \frac{d\phi'}{\sqrt{1 - (1/k^2)\sin^2(\phi'/2)}} . \tag{5.48}$$

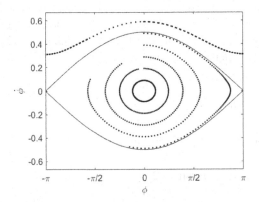

**Figure 5.4** Trajectories of six particles (dots) starting at $\phi = 0$ and $\dot\phi = 0.09, 0.19, \ldots, 0.59$ for one small-amplitude synchrotron period $T_s = 2\pi/\Omega_s$. Note that particles starting with larger amplitudes fall increasingly short of completing a full turn.

With the substitution $y = \phi'/2$ we transform the integral to the standard form for the incomplete elliptic integral of the first kind [23]

$$\Omega_s k t = \int_{\phi_0/2}^{\phi/2} \frac{dy}{\sqrt{1 - (1/k^2)\sin^2 y}} = F(\phi/2, 1/k) - F(\phi_0/2, 1/k) \ . \tag{5.49}$$

The elliptic integrals have inverses, which can be expressed in terms of the Jacobi elliptic functions sn, cn, and dn [23]. In particular, for $F(x, 1/k) = w$, the inverse is given by $\sin(x) = \mathrm{sn}(w, 1/k)$. Applied to the previous equation, we find

$$\sin(\phi/2) = \mathrm{sn}\left(\Omega_s k t + F(\phi_0/2, 1/k), 1/k\right) \ , \tag{5.50}$$

which gives us the phase of a particle with known initial coordinates $\phi_0$ and $\dot\phi_0$ after an arbitrary time $t$. The other phase-space coordinate $\dot\phi$ after time $t$ can be found from Equation 5.46. Inserting $\sin(\phi/2)$ from Equation 5.50 leads to

$$\dot\phi^2 = (2\Omega_s k)^2 \left(1 - \frac{1}{k^2}\,\mathrm{sn}(w, 1/k)^2\right) = (2\Omega_s k)^2\,\mathrm{dn}(w, 1/k)^2 \tag{5.51}$$

with the abbreviation $w = \Omega_s k t + F(\phi_0/2, 1/k)$. Taking the square root results in $\dot\phi = 2\Omega_s k\,\mathrm{dn}(w, 1/k)$, the other phase-space coordinate. To summarize, the two equations

$$\begin{aligned}
\sin(\phi/2) &= \mathrm{sn}\left(\Omega_s k t + F(\phi_0/2, 1/k), 1/k\right) \\
\dot\phi &= 2\Omega_s k\,\mathrm{dn}\left(\Omega_s k t + F(\phi_0/2, 1/k), 1/k\right)
\end{aligned} \tag{5.52}$$

with $k^2 = (\dot\phi_0^2/2\Omega_s)^2 + \sin^2(\phi_0/2)$ map the initial phase-space coordinates $\phi_0$ and $\dot\phi_0$ to those after an arbitrary time $t$. The Jacobi-elliptic functions sn and dn take the place of the trigonometric functions we normally encounter for small-amplitude oscillations. The case with $k^2 > 1$ corresponds to points inside the separatrix and next we consider points outside the separatrix.

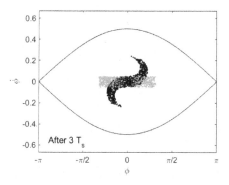
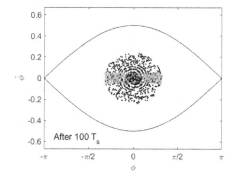

Figure 5.5 The distribution of 1000 injected particles (grey) after three synchrotron oscillation periods (left) and after 100 periods (right), where a homogeneous, or matched, distribution emerges.

For $k^2 < 1$ we transform Equation 5.46 to

$$\Omega_s t = \frac{1}{2} \int_{\phi_0}^{\phi} \frac{d\phi'}{\sqrt{k^2 - \sin^2(\phi'/2)}} = \int_{z_0}^{z} \frac{dz}{\sqrt{1 - k^2 \sin^2(z)}} = F(z, k) - F(z_0, k) , \qquad (5.53)$$

where the second equality follows from the substitution $\sin(\phi'/2) = k \sin z$, which transforms the integral to the standard form of the incomplete elliptic integral with suitably transformed integral boundaries $z$ and $z_0$ with $\sin z_0 = \sin(\phi_0/2)/k$. Using the inverse of the elliptic integral in terms of Jacobi elliptic functions, and using Equation 5.46 to calculate the other phase-space coordinate $\phi$, we find the map from initial to final coordinates for $k^2 < 1$

$$\begin{aligned}
\sin(\phi/2) &= k \operatorname{sn}(\Omega_s t + F(z_0, k), k) \\
\dot{\phi} &= 2\Omega_s k \operatorname{cn}(\Omega_s t + F(z_0, k), k) .
\end{aligned} \qquad (5.54)$$

Equations 5.52 and 5.54 are straightforward to code in MATLAB in the function named pendulumtracker.m, given in Appendix B.5. Given the small amplitude synchrotron frequency $\Omega_s$ and the time to integrate $t$ as well as the initial phase space coordinated $\phi_0$ and $\dot{\phi}_0$ it returns $\phi$ and $\dot{\phi}$ at the end of the integration time. We now use this function to explore the dynamics of particles inside and outside a radio-frequency bucket. For all simulations we choose the numerical value $\Omega_s = 2\pi/T_s = 0.25$.

In Figure 5.4 we follow six particles that start at $\phi = 0$ with $\dot{\phi} = 0.09, 0.19, \ldots, 0.59$ for 100 time steps, each having the duration of $T_s/100$. We observe that the particle with the smallest amplitude actually completes one period and the circle is closed. Particles with larger amplitudes fall short by an increasing degree. The particle closest to the separatrix only completes a little more than one half of a full turn. This behavior is a complementary view of the increasing oscillation amplitude shown on the right-hand side in Figure 5.3. Often, this is called *amplitude-dependent tune shift*. Moreover, the particle that starts outside the separatrix with starting amplitude $\dot{\phi} = 0.59$ does not even follow a periodic trajectory.

The MATLAB function to track particles for arbitrarily long times enables us to investigate the effect of, for example, injecting a bunch that is too long. On the left-hand side in Figure 5.5 we see the initial (gray) and final (black) distribution of 1000 particles injected with evenly distributed initial phase-space coordinates $\phi_0 = \pm 1$ and $\dot{\phi}_0 = \pm 0.05$.

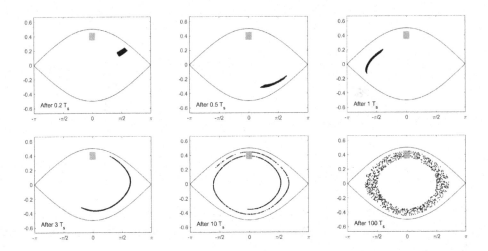

Figure 5.6 Filamentation after injection with too high energy. 1000 particles (grey) are followed for up to 100 synchrotron oscillation periods (dark).

After three oscillation periods distinct spiral arms have developed and the vertical size, proportional to the energy spread, has increased. After 100 oscillation periods a homogeneous circular distribution has evolved. Its shape will not change further and it signifies the *matched distribution* for the given synchrotron frequency $\Omega_s$. We note that, in order to maintain an unchanged, matched, beam distribution of the ensemble of particles, the size in the phase variable $\phi$, which is proportional to the bunch length $\sigma_\phi$ and the spread $\sigma_{\dot\phi}$ in the phase-space coordinate $\dot\phi$, must be related by $\sigma_\phi = \sigma_{\dot\phi}/\Omega_s$.

We continue by analyzing an incorrect energy of an injected bunch of particles in Figure 5.6. The initial phase space coordinates of the 1000 particles are evenly distributed with $\phi_0 = \pm 0.15$ and $\dot\phi_0 = 0.4 \pm 0.05$ and shown as gray dots. The plots on the top row in Figure 5.6 show the distribution after $0.2\,T_s, 0.5\,T_s$, and $T_s$. The initially square distribution is increasingly sheared and distorted, but the particles stay within a recognizable phase space area. The ensemble oscillates but gets wider in the process. The bottom row shows the distribution after $3\,T_s, 10\,T_s$, and $100\,T_s$. The particles all have different amplitudes and correspondingly different oscillation frequencies, which causes the particles to spread out more and more. After $10\,T_s$, a spiral that winds for one and a half turns is discernible, but after $100\,T_s$ no visible structure is left and the particles are smeared out along a band. Note, however, that each particle still oscillates independently with the amplitude it initially had, but the ensemble of all particles is smeared out across a wide range of phases. We conclude that the initial energy-mismatch led to a rather large spread of both phases and energies.

The left-hand plot in Figure 5.7 shows the average phase $\langle\phi\rangle$ (top) and energy $\langle\dot\phi\rangle$ (bottom) for the ensemble of particles as function of time for the same starting conditions used in Figure 5.6. We observe that there is an apparent damping of the averages. This effect, often called *Landau damping*, is not real damping in the sense that energy is dissipated, but a loss of coherence in the ensemble of particles. We mentioned in the previous paragraph that the oscillation amplitude of each individual particle is unaffected, but the particles get out of step and the coherent motion of the average is lost. Instead, the distribution of the ensemble becomes wider. The right-hand side in Figure 5.7 shows the initial and final projections of

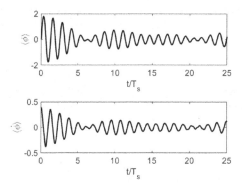
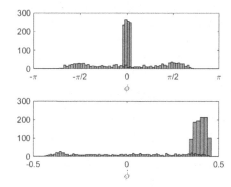

Figure 5.7  The average phase $\langle \phi \rangle$ and energy $\langle \dot{\phi} \rangle$ (left, top and bottom) as a function of time, which shows decoherence, often referred to as Landau damping. The projections of the initial distribution (narrow) and after $100\,T_s$ (wide), corresponding to the bottom right plot in Figure 5.6, are shown on the right.

the distribution of particles after $100\,T_s$, which corresponds to the bottom right plot in Figure 5.6. Initially the distributions are well localized, as shown by the histograms in lighter gray. After $100\,T_s$ the projections of the smeared out ring, visible in Figure 5.6, appear as spread-out projection in Figure 5.7, for the phase in the top plot and the energy $\eta$ in the bottom plot. Again, the loss of coherence leads to a widened final distribution. In order to maintain a small beam size after injecting into a ring one obviously has to pay special care to adjust the energy and also the timing, which would lead to the same widening of the distribution.

The upper plot on the right-hand side in Figure 5.7 shows the projection of the particles' phase-space distribution onto the phase axis. Often these projections can be recorded, for example, with a fast oscilloscope, and used to reconstruct the two-dimensional distribution in a process called *bunch tomography* [25]. Since each projection does not contain information about $\dot{\phi}$, the algorithm assumes that the distribution in $\dot{\phi}$ spreads evenly over a range of values and then propagates the resulting "vertical strips" backwards in time to a common reference. Superimposing these back-propagated strips from many projections will cause the original distribution to reappear, at least approximately. An implementation, using `pendulumtracker()` for the back-propagation, is discussed in Appendix B.5.

Once a beam is injected in a ring and has assumed a distribution that is matched to the given RF-voltage, it is possible to manipulate the bunch distribution by adjusting the parameters of the radio-frequency. This is called *RF gymnastics* [26] and is the topic of the next section.

## 5.5   RF GYMNASTICS

The longitudinal dynamics of the particles is governed by Equation 5.36, which also describes an equivalent Hamiltonian system, the mathematical pendulum. It can be shown [8] that the action variables in such Hamiltonian systems remain constant if a system parameter, such as the length of the pendulum, or $\hat{V}$ of the RF-system, *changes slowly.* In other words, the action variable is an *adiabatic invariant.* Here "slowly" means that $d\Omega_s/dt \ll \Omega_s^2$. Since the action variable of each particle remains constant, their average, the longitudinal emittance, also stays constant in such an operation. In the following, we assume that the bunch extensions are sufficiently small to stay in the linear regime of synchrotron oscillations.

An immediate consequence is that we can lengthen the bunch by slowly reducing $\hat{V}$ to a much smaller value $\tilde{V}$. At the same time, the momentum spread decreases, such that the distribution in phase space is very flat. In order to maintain adiabaticity, this process takes many synchrotron oscillations. But once the bunch is long, we can abruptly increase $\hat{V}$ back to its original value $\hat{V}$. The particles, all of a sudden, find themselves highly mismatched to the RF-voltage and start synchrotron oscillations with the original frequency $\Omega_s$. After a quarter of an oscillation period, the bunch length and momentum spread have exchanged roles, and the bunch is now short, but has a large momentum spread. If we extract the beam at this moment, we can deliver a short bunch to an experiment. This manipulation is called *bunch rotation* [26].

Occasionally, it is desirable to use one RF-frequency during injection to capture the incoming beam, but later deliver a larger number of shorter bunches to an experiment or to the next machine to accelerate further. In such circumstances one reduces the RF-voltage $\hat{V}$ to zero and lets the bunches spread out around the ring, albeit with very low momentum spread. Once the beam is thus "homogenized," a second RF-system, operating at a higher frequency, is turned on and its voltage is slowly increased to capture the particles. The particles then assemble in the RF-buckets corresponding to the higher RF-frequency. This method is called *debunching-rebunching* [26].

Instead of first turning one RF-voltage off before slowly increasing the amplitude of the second one, one can start ramping up the second voltage, which has twice the original RF-frequency, while the first one is still at its nominal value. The second RF-frequency causes the fixed point at the center of the separatrix to become unstable and forms two stable fixed points separated in phase. The particles in the bunch then start to assemble around these new fixed points and once the first RF-voltage is slowly decreased to zero, the particles have split in two buckets, each containing half the original particles. This operation is called *bunch splitting* [26].

If a storage ring is equipped with an RF-system that can operate in a wide range of frequencies, it is possible to compress a bunch train. This keeps the same number of particles per bunch and the same number of filled buckets, but creates additional empty buckets. This is achieved [26] by simultaneously decreasing the RF voltage of the first system, which operates at a harmonic number $h_1 = \omega/\omega_0$, and increasing the voltage of a system with a slightly large harmonic number, say $h_2 = h_1 + 2$. This operation is called *batch compression*.

After having analyzed the longitudinal motion of particles at fixed energy, we now turn to investigating their dynamics during acceleration.

## 5.6  ACCELERATION

During acceleration the radio-frequency system has to continuously transfer energy to the particles, which means, by virtue of Equation 5.27 that the design phase $\phi_s$ has to be different from zero or $\pi$. In a ring, changing the phase of the RF system, will only cause the arrival time of the beam to move synchronously with the changing phase, but does not accelerate the particles. On the other hand, by continuously increasing the field of the main dipoles $\Delta B/B$, we systematically shorten the circumference of the ring, as described by Equation 3.102, and thereby force the particles to systematically arrive earlier than the zero-crossing of the RF voltage. This effect is already included in Equation 5.33, which contains the shortening of the orbit due to $\Delta B/B$. When ramping the dipole field with a

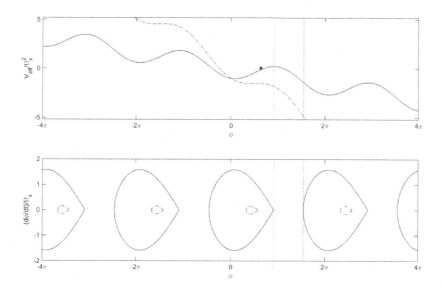

Figure 5.8 The effective potential $V_{\text{eff}}(\phi)$ (top) for a synchronous phase of $\phi_s = 15$ degrees (solid) and $\phi_s = 75$ degrees (dashed). The bottom plot shows the corresponding regions where stable oscillations are possible.

ramp linear in time and proportional to $\Delta B = \dot{B}t$ we can write

$$\frac{\Delta B}{B} = \frac{\rho \dot{B} t}{B\rho} = \frac{\rho \dot{B} t}{p} = \frac{\rho \dot{B} t}{\beta E/c} \ . \tag{5.55}$$

Inserting in Equation 5.33 gives us

$$\frac{d\phi}{dt} = \omega_{\text{rf}} \eta \delta - \frac{\omega_{\text{rf}} \alpha c \rho \dot{B}}{\beta E} t \tag{5.56}$$

and differentiating with respect to time and inserting $d\delta/dt$ from Equation 5.30 results in

$$\frac{d^2\phi}{dt^2} = \frac{\omega_{\text{rf}} \eta e \hat{V}}{T_0 \beta^2 E} [\sin\phi - \sin\phi_s] - \frac{\omega_{\text{rf}} \alpha c \rho \dot{B}}{\beta E} = \frac{\omega_{\text{rf}} \eta e \hat{V}}{T_0 \beta^2 E} [\sin\phi - \sin\phi_s - \sin\phi_a] \ , \tag{5.57}$$

where we express the term proportional to $\dot{B}$ as being due to an additional phase angle $\phi_a$ with

$$q\hat{V} \sin\phi_a = \frac{\alpha \beta c T_0 \rho \dot{B}}{\eta} = \frac{\alpha C}{\eta} \rho \dot{B} \ . \tag{5.58}$$

Here $C = \beta c T_0$ is the circumference of the ring. We see that a constant ramp rate of the main dipoles $\dot{B}$ causes the beam to assume a non-zero phase angle $\phi_a$, in addition to any other losses described by the phase angle $\phi_s$.

Thus the dynamics of particles during acceleration is described by Equation 5.34 with $\phi_a$ defining the synchronous phase $\phi_s$. Equation 5.34 naturally leads to an integral of motion

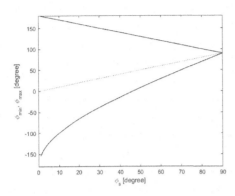
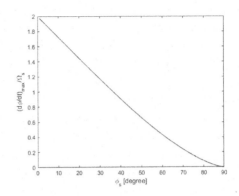

Figure 5.9 The stable-phase region (left) and the bucket half-height (right) as a function of the synchronous phase $\phi_s$.

by multiplying the equation with $\dot\phi$ and realizing that

$$\frac{d}{dt}\left[\frac{1}{2}\dot\phi^2 - \hat\Omega_s^2(\cos\phi + \phi\sin\phi_s)\right] = 0 \quad \text{and} \quad A = \frac{1}{2}\dot\phi^2 - \hat\Omega_s^2(\cos\phi + \phi\sin\phi_s) \,, \quad (5.59)$$

which implies that the expression in the square brackets is constant with value $A$. Here we introduced the abbreviation $\hat\Omega_s^2 = |\omega_{\mathrm{rf}}\eta q\hat V/T_0\beta^2 E|$, which is the synchrotron frequency $\Omega_s^2$ from Equation 5.35 for zero phase angle $\phi_s$. Equation 5.59 can be interpreted to describe the conservation of the energy, here represented by the constant $A$, of mechanical system of a particle with unit mass in the effective potential $V_{\mathrm{eff}}(\phi) = -\hat\Omega_s^2(\cos\phi + \phi\sin\phi_s)$, which is shown on the upper plot in Figure 5.8 for a synchronous phase $\phi_s = 15$ degrees. We see that there are potential wells in which the mass point can perform stable oscillations. A mass point is shown as a small dot near $\phi = 2\pi/3$ in Figure 5.8 as an illustration. But the range of stable oscillations is limited, because mass points launched from rest ($\dot\phi_0 = 0$) and with phases in the range denoted by the vertical dashed lines cannot be captured by the potential wells. We also observe that there are equilibrium or fixed points, defined by $dV_{\mathrm{eff}}/d\phi = 0$, where the mass point remains at rest. Solving for the fixed points we find that they must obey $\sin\phi = \sin\phi_s$. These points at $\phi_s + 2n\pi$ are those at the bottom of the well and are therefore stable. The points at $(2n+1)\pi - \phi_s$ correspond to the local maxima visible in the upper plot in Figure 5.8. They are unstable, because the smallest perturbation will cause the mass point to move away from this fixed point. For comparison, we also show the potential for a synchronous phase angle of $\phi_s = 75$ degrees as a dashed line. It is much steeper and the potential wells are much shallower, which already indicates that the range where stable oscillations are possible, is much reduced.

These regions, where stable oscillations are possible, are enclosed by the separatrix, shown on the bottom plot in Figure 5.8. They are characterized by values of the constant $A$ that correspond to the value of the effective potential at the unstable fixed point. It is given by

$$A = \hat\Omega_s^2\left(\cos\phi_s - [(2n+1)\pi - \phi_s]\sin\phi_s\right) \,. \quad (5.60)$$

Inserting in Equation 5.59 and solving for $\dot\phi$ we find the equation that defines the separatrix to be

$$\dot\phi = \pm\hat\Omega_s\sqrt{2(\cos\phi + \cos\phi_s - [(2n+1)\pi - \phi - \phi_s]\sin\phi_s)} \,, \quad (5.61)$$

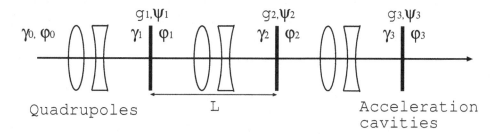

Figure 5.10 The layout of the model linear accelerator.

which is used to generate the plot in the bottom plot in Figure 5.8. Compared to the situation with $\phi_s = 0$ that we already encountered in Figure 5.3 and 5.4, the range of phases that allows stable oscillations is reduced. The second, much smaller, set of separatrices, shown as dashed lines in the bottom plot are the regions of stable oscillations if the phase angle is $\phi_s = 75$ degrees. They are indeed much smaller than those for 15 degrees; both the extent in phase and in the vertical direction, indicating a much reduced momentum acceptance.

The range of stable phases extends from the phase $\phi_{min}$ at which the expression under the root in Equation 5.61 is zero to the unstable fixed point at $\phi_{max} = (2n + 1)\pi - \phi_s$. On the left-hand side in Figure 5.9, we show the range of stable phases as the area between the solid lines. The dotted line denotes the synchronous phase $\phi_s$. We observe that for $\phi_s$ close to zero the range extends from $-180$ degrees to $180$ degrees, or $-\pi$ to $\pi$, which corresponds to the situation shown in Figure 5.3. With increasing synchronous phase the range of stable phases shrinks until it is reduced to zero at a synchronous phase angle of $\phi_s = 90$ degrees, which corresponds to acceleration on the crest of the radio-frequency voltage. On the right-hand side in Figure 5.9, the half-height of the separatrix is shown as a function of $\phi_s$. The maximum of the separatrix is given by the value of $\dot\phi$ in Equation 5.61, evaluated at $\phi = \phi_s$, because it is located directly above the stable fixed point. We observe that the height of the separatrix, which defines the momentum acceptance, diminishes as a function of $\phi_s$ and it vanishes at $\phi = 90$ degrees.

We conclude that the range of phases and the momentum acceptance is reduced at increasing synchronous phase angles $\phi_s$ and operating at maximum acceleration with $\phi_s = 90$ degrees, is only marginally feasible, because the stable region of synchrotron oscillations shrinks to a single point and stability of the longitudinal motion is lost.

## 5.7 A SIMPLE WORKED EXAMPLE

In the hands-on spirit of this book, we now consider a MATLAB model, described in detail in Appendix B.5, of the simple linear proton accelerator shown in Figure 5.10. The beam enters from the left with energy $\gamma_0$ and passes through a number of cells, say 30, with length $L$, each made of one cavity with design phase $\psi_i$ and maximum energy gain $g_i$ and a quadrupole doublet to provide transverse focusing. Just upstream of cavity number $i$ the phase of the beam is given by $\phi_i$ and the energy after the cavity is $\gamma_i$.

The equations that govern the dynamics in the longitudinal phase space are

$$\gamma_{i+1} = \gamma_i + g_i \sin(\phi_i - \psi_i) \qquad \text{and} \qquad \phi_{i+1} = \phi_i + \frac{\omega L/c}{\sqrt{1 - 1/\gamma_i^2}} , \qquad (5.62)$$

where the first equation describes the energy added to the particle in the cavity. Note that $g_i$

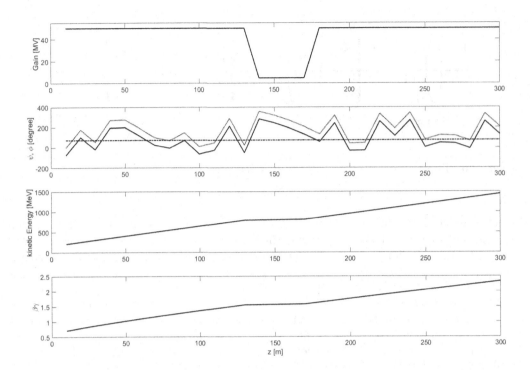

**Figure 5.11** The energy gain per module (top plot), the cavity phases (second), the energy profile (third), and the momentum (bottom), used to scale the quadrupole excitations.

is given in units of the particles rest mass that may be called gamma-units. The energy gain depends on the design phase $\psi_i$ of that cavity and the actual arrival phase $\phi_i$ of the particle. The second equation describes the change in arrival phase that depends on the distance $L$ between the cavities, the speed of the particle $v_i = c\sqrt{1 - 1/\gamma_i^2}$, and the frequency $\omega$ of the cavity. These equations correspond to Equation 5.30 and 5.33, which we encountered previously.

The transverse dynamics is defined by a doublet lattice, similar to the one shown on the left-hand side in Figure 3.24 with a cell length of 10 m and a nominal focal length for the quadrupoles of $f = \pm 2$ m. This leads to a phase advance per doublet cell of 97.2 degrees. The required currents to drive the coils of the quadrupoles increase proportionally to the desired momentum profile along the linac. Therefore, they depend on the phases and on the gradients in the cavities. We normally want to operate the linac despite a malfunctioning power generator with a cavity that does not accelerate the beam. Under these circumstances we can reconfigure the quadrupole gradients to achieve nominal transverse focusing and still operate the linac, albeit at lower energy.

Let us go through the necessary steps to set up the linac. The initial energy, given by $\gamma_0$ is known and we assume that the arrival phase $\phi_0$ is zero. The maximum achievable gradients $g_i$ are determined by external factors and we need to accept them as given. But we can set the operating phases $\psi_i$ of the cavities, based on a compromise to reach the maximum energy to which we want to accelerate the particles, balanced by the requirement for stability and robustness. The latter dictates to choose phase angles lower than 90 degrees

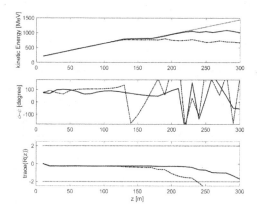

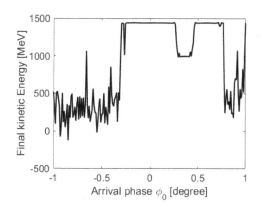

Figure 5.12 Variation of the initial (arrival) phase $\phi_0 = -0.31$ degree (solid) and $\phi_0 = 0.77$ degree (dashes).

in order to have an acceptable stable-phase range. We consult Figure 5.9 and settle on a synchronous phase for the cavities of $\psi_i = 75$ degrees. In a linac we pay the electricity bill to provide maximum power to the beam to accelerate to the highest possible energy, but then we are forced by beam dynamics to waste some of this power by operating at a phase angle below 90 degrees. Other constraints, for example, from operating feedback systems for the beam energy, may also require to operate below the maximum in order to allow actually increasing the energy delivered to the beam by changing a cavity phase. Knowing the gains and phases of the cavities, we can determine the design energy profile of the beam along the linac and this, in turn, determines the currents we need to excite the quadrupole magnets to provide the nominal focal length of $\pm 2$ m. This completes the setup of the linac and we can explore stability by varying parameters.

Figure 5.11 shows the energy gain $g_i m_p c^2$ in MV/module along the linac. Modules 14 to 17 only operate at 10 % of the nominal gradient and this is shown as the dip around $z = 150$ m. The second plot shows the arrival phase of the nominal beam with initial phase $\phi_0 = 0$ and kinetic energy $T_0 = 200$ MeV, which translates to an initial $\gamma_0 = 1 + T_0/m_p \approx 1.21$. The cavity phases $\psi_i$, shown by the solid line, follow the reference particle such that their difference remains at the nominal acceleration phase of 75 degrees. The latter is shown as the dashed line. With these gains and phases the kinetic energy $T = (\gamma - 1)m_p$ of the beam along the linac is shown in the third plot. Also here the "bad modules" show up with a reduced rate of acceleration near $z = 150$ m. The fourth plot shows the scaling of the quadrupole currents with the momentum $p/m_p c = \beta\gamma$. We see that $\beta\gamma$ covers the range from approximately 0.7 to 2.3, such that we need to excite the coils of the quadrupoles by currents that vary by a factor 3.5 in order to guarantee a constant nominal focal length. This completes the basic setup of the linac and next we explore its robustness by varying several basic parameters.

The criteria we use to judge operational robustness is the kinetic energy at the end of the linac and the transverse stability of the transfer matrices for each cell. The latter plays a role, because the quadrupole excitations are calculated for the nominal energy profile, as shown in Figure 5.11. If, however, the actual beam energy at a longitudinal position $z$ differs from the nominal, the focal length of the quadrupoles also differs from the nominal and that may lead to unstable transfer matrices. As a proxy for this effect we use the trace

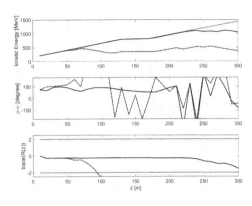
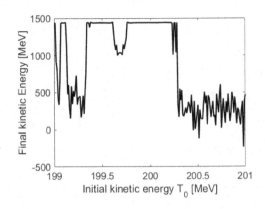

Figure 5.13  The final energy as a function of the initial energy with $T_0 = 200.27$ MeV (solid) and $T_0 = 198.97$ MeV (dashes).

of the transfer matrix for each cell, which should lie between $-2$ and $2$. For a focal length of $\pm 2$ m, the nominal value for the trace is $-0.25$.

In Figure 5.12, we show the effect of varying the initial arrival phase $\phi_0$. On the left-hand side, we display the energy profile (top), the difference of cavity phase and arrival phase of the beam (middle) and the trace of the cell transfer-matrix (bottom) along the linac. We show two examples with $\phi_0 = -0.31$ degree (solid) and $\phi_0 = 0.77$ degree (dashes). The chosen examples visualize particles losing synchronism in the middle section of the linac. At some point they arrive too late and therefore receive too little energy which makes them arrive even later in the next module. The middle plot, showing the phase difference between cavity and arrival phase of the beam, should be constant at $\phi_s = 75$ degrees, but the beam initially shows oscillations that last until about $z \approx 120$ m for one example (solid) and up to $z \approx 220$ m for the other example. The bottom plot shows that the trace of the transfer matrix is lowered, because the quadrupoles are too strong for the actual beam energy and this leads to overfocusing. Eventually this may lead to a much increased transverse beam size and particle loss. On the right-hand side in Figure 5.12, we display the final energy while scanning the input phase and we observe that the acceptable phase variation is on the order of $\pm 0.2$ degrees.

In Figure 5.13, we show the effect of varying the input energy $T_0$ instead of the phase $\phi_0$. Otherwise the plots show the same quantities as in Figure 5.12. On the left-hand side the chosen energies are $T_0 = 200.27$ MeV (solid) and $T_0 = 198.97$ MeV (dashes). We see that the examples lead to a loss of synchronism at different points along the linac, visible in the middle plot as large phase excursions. The mismatch of energies from that point onward leads to transverse over-focusing, as indicated by the trace of transfer matrix, shown on the bottom plot. On the right-hand side in Figure 5.13, we show the achievable final energy as a function of the input energy. The window for which we can expect to reliably reach the final energy is $\pm 0.2$ MeV or on the order of $10^{-3}$ of the energy.

As a final example we consider random variations of the phase $\psi_i$ in each module, so-called phase jitter. The right-hand side in Figure 5.14 shows the distribution of final energies at the end of the linac for 1000 different realizations of uniformly distributed phase errors $\pm 1$ degree. We see that the final energies vary in the range of $\pm 2$ MeV. A few realizations that lead to higher energies around 1444 MeV also show up. For a slightly increased range of phase errors of 1.2 degrees, most of the realizations lead to energies close to the nominal

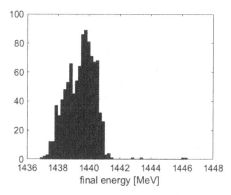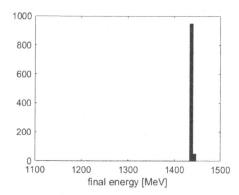

Figure 5.14 Distribution of final energies for random phase variations of 1 degree (left) and 1.2 degrees (right).

energy, as is shown on the corresponding plot on right-hand side in Figure 5.14. There are, however, a few realizations that lead to significantly reduced final energies, as is witnessed by the horizontal scale that extends to much lower values. We conclude that we should be able to control the phase of radio-frequency power delivered to the cavities in the modules should be stable to within a degree.

After these discussions of the longitudinal dynamics of the beam, in the next chapter, we will turn our attention to the technological aspects of generating, guiding, controlling, and coupling power to the beam.

## QUESTIONS AND EXERCISES

1. Explain, why TM-modes are used to accelerate the beam.

2. Can you use TE-modes to accelerate a beam?

3. Give a typical example of a TEM-mode. The example should be so obvious as to immediately strike your eye. Why are they unsuitable for direct acceleration?

4. (a) Calculate $E_z$ for a $TM_{010}$ mode and use MATLAB's surf() function to display it. (b) What is $H_z$ for this mode? (c) Calculate $H_x, H_y, E_x$, and $E_y$ and display.

5. Repeat the previous exercise (a) for a $TM_{020}$ mode; (b) for a $TM_{110}$ mode; (c) for any other $TM_{mnp}$ mode of your choice.

6. What is the size of a 100 MHz pillbox cavity used for electrons? Prepare a plot of the transit time factor as a function of the length of the cavity.

7. In the CELSIUS storage ring in Uppsala, protons were injected with a kinetic energy of 50 MeV and slowly ($\dot{B}/B \approx 0$) accelerated up to a kinetic energy of 1360 MeV. The ring had a circumference of 82 m, a momentum compaction factor of $\alpha = 0.123$, and typically used an acceleration voltage of $\hat{V} = 2$ kV. Plot (a) the synchrotron frequency and (b) the bucket half-height as a function of the kinetic energy for the relevant energy range.

8. For an electron storage ring, calculate the bucket half-height as a function of the ramp rate $\dot{B}/B$ and plot the result.

9. Simulate the bunch rotation, mentioned in Section 5.5, by tracking 1000 sample particles with the help of the `longitudinal_matching_example.m` script and `pendulumtracker()` MATLAB function, available from this book's web page. Prepare a matched beam with a five times smaller synchrotron frequency than used in the preparation of Figure 5.5. The bunch should have a long extension in $\phi$ but small height in $\dot{\phi}$. Then, instantaneously increase the synchrotron frequency back to the original value and track for 1/4 synchrotron period. How does the distribution look like now? In the real machine, instead of the synchrotron frequency, which parameter would you change?

10. Simulate de-bunching in the same way as in the previous exercise. Start from the situation shown on the right-hand side in Figure 5.5 and reduce the synchrotron frequency in 100 equal steps to zero. At each step, track for 100 synchrotron periods such that the distribution is matched to the new frequency. In the end the beam should cover all phases, but have a small momentum spread.

11. And now simulate the process in reverse by increasing the synchrotron frequency back to its original value. What do you observe?

12. Simulate rebunching on the (a) second and (b) third harmonic of the fundamental radio frequency.

# CHAPTER 6

# Radio-Frequency Systems

In this chapter, we cover the generation and control of the radio-frequency power that is used to accelerate the beams. We will follow the flow of electrical power from the wall plug through power amplifiers and through waveguides to the coupling antennas into the accelerating structures, where the microwaves accelerate the beam. If a beam contains large numbers of particles, it "consumes" a significant fraction of the power delivered to the structures, and appears as a load that varies in time. We will briefly discuss the diagnostics and control mechanism to stabilize the power delivered to the beam.

The radio-frequency signals with the desired frequency, amplitude, and phase start their journey towards the beam in a frequency generator as a low-power signal at a power level of mW to W, and must be amplified. And that is the topic of the next section.

## 6.1  POWER GENERATION AND CONTROL

The power difference between the input and the output of an amplifier must be provided by a power supply that converts the three-phase electricity from the power grid, the "wall-plug power," usually with the help of transformers and rectifier diodes, to a level suitable to operate the amplifier. If the accelerator, and consequently also the RF-amplifier, is operated in pulsed mode, we need to adapt the time structure of the power delivered to the amplifier with a *pulse-forming network* or a *modulator*. The former can only provide short, typically µs–long pulses, and is based on a coaxial cable, charged to high voltage, that is rapidly discharged through a high-power switch, often a *thyratron*. A modulator is a pulsed power-supply in which a continuously charged capacitor bank is periodically discharged through a high-power switch. This releases a high-voltage pulse to a pulse-transformer, which adapts the voltage and current—the impedance—to that of the amplifier. Filter circuits, an example is called *bouncer,* are often added in series to the primary side of the pulse transformer in order to limit the droop of the voltage due to the discharging capacitors.

Now that we have the power available to amplify the low-power RF signal, we turn to the different types of amplifiers. Most amplifiers are actually multi-stage devices that consist of several pre-amplifiers, each of which increases the power level typically by a factor of less than 100, or 20 dB(power) before the last, high-power, stage. The first stages often use *solid-state amplifiers,* based on MOSFET transistors (LDMOS), specially adapted for RF operation [27]. Matching networks adapt the input and output impedance of the transistor to that of the surrounding circuits. Modern transistors provide an output power on the order of 1 kW with an efficiency of about 70 %, which makes water-cooling of the substrate, onto which the transistor is mounted, necessary. These losses present a problem, because operating at high frequencies of several 100s of MHz requires the transistor to be small but operating at high power requires them to be large in order to limit the power density and the ability to cool the device. With the per-transistor power limited, we need to join the output of a very large number transistors in power-combiners in order to reach 100s of kW levels [28].

High-power RF amplifiers based on vacuum tube technology, often *tetrodes*, were used for most of the 20th century to transmit radio and TV signals in the frequency range of up to a few 100 MHz and even modern accelerators [29] use them, where suitable, even though the frequency and power ranges are limited. In tetrodes, electrons are emitted from a heated cathode and attracted to the anode at high positive voltage. The large current is modulated by a moderate voltage applied to a control grid close to cathode. A fourth electrode, the screen grid, shields the effect of surface charges, accumulated on the anode, and makes operation with high gain more reliable. The four electrodes are mounted radially in high-power tetrodes in order to efficiently cool the outermost lying anode, where the power of the impinging electrons is dissipated. The large anode current is used to drive an antenna in an output cavity, optimized for the operating frequency of the tetrode. As an example, we consider the system used to drive cavities in a section of the ESS, where an anode voltage of 16 kV and current of up to 18.7 A is used to operate a 352 MHz system that amplifies a 6.7 kW input signal to produce 200 kW output power with a 15 dB gain and an efficiency of 67 % [30]. In this system, the control-grid is negatively biased in order to only cause electrons to flow during the positive half-cycle of a sine. This improves the efficiency, because current only flows part of the time and this mode of operation is called class-B operation. In contrast, when operating in class-A, a constantly flowing current is modulated around its average value. This constant average current deteriorates the efficiency and therefore, often class-B mode operation is chosen instead.

In vacuum tubes, such as tetrodes, the anode serves a dual purpose: it extracts the power from the electrons by producing a large anode current at radio-frequencies, and, at the same time, stops the electrons. It can be beneficial to separate the two functions and use an resonating cavity to extract the RF power and, further downstream, an electron collector to stop the electrons. Devices, based on this technology, are called *inductive output tubes* (IOT). Since the RF output is produced directly, without first producing an anode current, the simpler design of the collector helps to improve the lifetime of the devices.

The capacitance of the control-grid to ground limits the frequencies to below the GHz range. In order to overcome this limiting factor, *klystrons* are used in order to produce high power-levels in the GHz range. RF-cavities are used to modulate the velocity of the electrons. Velocity bunching, where faster electrons catch up with the slower ones, causes the initially continuous electron current to bunch at the modulating frequency. The bunched high-current electron beam then excites modes in a second cavity, tuned to the modulating frequency, and its power is extracted at the high-power output port. In klystrons, the electrons are guided by a longitudinal magnetic field that is produced by solenoids. Their performance is normally optimized by adding additional cavities between the modulating cavity and output cavities. The intermediate cavities are not powered, but excited by the beam and they are tuned both to the fundamental and to higher harmonics. The fields created in these idle cavities are designed to enhance the bunching and increase the efficiency. At very high power levels (MW), space-charge repulsion among the electrons reduces the bunching, but using multiple electron beams side-by-side within the same klystron alleviates this limitation. These devices are called multi-beam klystrons and are used, for example, in the European XFEL in Hamburg. Typical anode voltages range up to 100 kV with a currents in the 10 to 100 A range.

The RF power produced by the tubes and klystrons needs to be transported to the accelerating structures, where it is used to accelerate the beam. Apart from transporting the high power, it may need to be split, or combined, diagnosed, and even dissipated in dummy-loads. Devices to achieve these operations are the topic of the next section.

## 6.2 POWER TRANSPORT: WAVEGUIDES AND TRANSMISSION LINES

In Section 5.1, we found that for electro-magnetic waves, defined by Equation 5.4, all transverse electric and field components can be expressed through the longitudinal components $E_z$ and $H_z$ with the help of Equation 5.7 and 5.8. If both $E_z = 0$ and $H_z = 0$ the waves are called TEM waves [31]. The waves with $H_z = 0$ and $E_z \neq 0$ are called transverse magnetic, or TM, waves. We sought fields of this type in the pill-box cavity in Section 5.1 in order to support a longitudinal electric field $E_z$ to accelerate the beam. A third type of waves, with $E_z = 0$ and $H_z \neq 0$, is called transverse electric, or TE, waves. All their transverse field components can be calculated from $H_z$ alone

$$H_x = -\frac{ik_z}{k_c^2}\frac{\partial H_z}{\partial x}, \quad H_y = -\frac{ik_z}{k_c^2}\frac{\partial H_z}{\partial y}, \quad E_x = -\frac{i\omega\mu_0}{k_c^2}\frac{\partial H_z}{\partial y}, \quad E_y = \frac{i\omega\mu_0}{k_c^2}\frac{\partial H_z}{\partial x} \ , \qquad (6.1)$$

which follows from Equation 5.7 and 5.8 by setting $E_z = 0$ and using $k_c^2 = \omega^2/c^2 - k_z^2$. Just as the electric field $\vec{E}$, the magnetic field $\vec{H}$ obeys Equation 5.3 and we find that all, and in particular, the longitudinal component of the magnetic fields fulfills $\triangle H_z + (\omega^2/c^2)H_z = 0$. Inserting the Ansatz for $H_z$ from Equation 5.4 we obtain the following equation for the TE modes

$$\left(\frac{\partial^2}{\partial x^2} + \frac{\partial^2}{\partial y^2}\right) H_z = -k_c^2 H_z = \left(k_z^2 - \frac{\omega^2}{c^2}\right) H_z \ , \qquad (6.2)$$

which is an eigenvalue equation for $H_z$. Its solutions depend on the boundary conditions for $H_z$, and since the electric field components $E_x$ and $E_y$ have to vanish on perfectly conducting metallic boundaries, Equation 6.1 dictates the respective derivatives $\partial H_z/\partial y$ and $\partial H_z/\partial x$ to vanish on the wave-guide walls as well. Thus, we are facing the problem of finding eigen solutions to the 2-dimensional Laplace operator with von-Neumann boundary conditions.

We start by considering the propagation of TE waves in a rectangular wave-guide. For the geometry we assume that the waveguide extends from $x = 0$ to $x = a$ in the horizontal direction from $y = 0$ to $y = b$ in the vertical. For $H_z$, we then make the Ansatz

$$H_z = B_{mn} \cos\left(\frac{m\pi x}{a}\right) \cos\left(\frac{n\pi y}{b}\right) \ , \qquad (6.3)$$

which obviously obeys $\partial H_z/\partial x = 0$ for $x = 0$ and $x = a$ and $\partial H_z/\partial x = 0$ for $y = 0$ and $y = b$. The $B_{mn}$ are the amplitudes of the respective modes labeled by the mode number $m$ and $n$. Inserting $H_z$ from Equation 6.3 into Equation 6.2 yields the dispersion relation for the propagation constant $k_z$ of the $TE_{mn}$ waves

$$-\left(\frac{m\pi}{a}\right)^2 - \left(\frac{n\pi}{b}\right)^2 = k_z^2 - \frac{\omega^2}{c^2} = -k_c^2 \quad \text{or} \quad k_z = \sqrt{\frac{\omega^2}{c^2} - \left(\frac{m\pi}{a}\right)^2 - \left(\frac{n\pi}{b}\right)^2} \ , \quad (6.4)$$

where the modes are labeled by two integers $m$ and $n$ of which at least one must be non-zero. The other components of the magnetic and the electric field can now be calculated from Equation 6.1. The *impedance* of an electro-magnetic wave is given by the ratio of $E_x$ and $H_y$ and for the TE-modes is given by $Z_{TE} = E_x/H_y = \omega\mu_0/k_z$ with the frequency-dependent propagation constant $k_z$ given by Equation 6.4.

Waveguides are used to transport RF-power and this property is described by the *Poynting vector,* given by $\vec{\mathbf{S}} = \vec{\mathbf{E}} \times \vec{\mathbf{H}}$, which points in the direction of propagation. The speed of delivery of a pulse of RF-power is described by the *group velocity* $v_g = d\omega/dk_z = k_z c/\sqrt{k_z^2 + k_c^2}$, which is always less than the speed of light, whereas the *phase velocity* is given by

$$v_p = \frac{\omega}{k_z} = c\frac{\sqrt{k_z^2 + k_c^2}}{k_z} \tag{6.5}$$

and it always exceeds the speed of light. The product of phase and group velocities is given by $v_p v_g = c^2$. Note that frequencies $\omega$ with $\omega^2/c^2 < \omega_c^2/c^2 = k_c^2 = (m\pi/a)^2 + (n\pi/b)^2$ lead to an imaginary propagation constant and that implies that waves with this frequency are exponentially damped and are called *evanescent.* The frequency $f_c = \omega_c/2\pi$ is called the *cutoff frequency* and separates the regime with propagating waves from that with evanescent waves. For the mode with $m = 1$ and $n = 0$ we see that the cutoff frequency is $f_c = c/2a$ and for the cutoff wavelength $\lambda_c$ we find $\lambda_c = c/f_c = 2a$, which implies that half a wavelength has to fit between the transverse walls and waves with longer wavelength will be attenuated and become evanescent.

Many commonly used waveguides have an aspect ratio of two, such that $a = 2b$, which ensures that the cutoff frequency of the $TE_{01}$ mode is twice that of $TE_{10}$ and therefore only a single mode can propagate in the frequency range between the respective cutoff frequencies. These waveguides are labeled by WR-xx, where xx is the magnitude of the larger dimension, here $a$, given in units of $1/100$ of an inch. For example, we use WR-2300 waveguides in our lab to transport 352 MHz RF-power from the amplifier to the cavity.

The many components of the electric and magnetic fields are difficult to visualize and we therefore employ the MATLAB PDE toolbox to determine them numerically and then display and analyze them further. Numerical methods not only help with visualization, but also provide a tool to analyze geometries that are difficult or impossible to handle analytically. First, we simulate the rectangular waveguide discussed in the previous paragraphs. The full MATLAB code is available in Appendix B.5, so here we only highlight the most important steps. After defining and inspecting the geometry, following the same procedure already used for the magnets in Chapter 4, we mesh the problem with a call to the `generateMesh()` function. Then we apply the von Neumann boundary conditions and specify the model coefficients. Since here we deal with the eigenvalue problem posed in Equation 6.2, a call to the `solvepdeeig()` function returns the eigenvalues $k_c^2$ and the eigen functions $H_z$ in the variable `result`.

```
applyBoundaryCondition(model,'Edge',[1:4],'q',0,'g',0); % Neumann
specifyCoefficients(model,'m',0,'d',1,'c',1,'a',0,'f',0,'Face',1);
result=solvepdeeig(model,[1,5000]);
eigenvalues=result.Eigenvalues; Hz=result.Eigenvectors;
[p,e,t]=meshToPet(model.Mesh);
[dHx,dHy]=pdegrad(p,t,Hz(:,1)); Hx=-dHx; Hy=-dHy; Ex=-dHy; Ey=dHx;
```

It is easy to verify that the eigenvalues returned are very close to the analytical values shown on the left-hand side in Equation 6.4. The call to the `meshToPet()` function creates an array to describe the points, edges, and triangles of the mesh, which is needed in the subsequent call to the `pdegrad()`, which returns the gradient of the solution and yields the transverse field components by using Equation 6.1. Here we do not include the fore-factors from Equation 6.1, because we are only interested in the field patterns. The transverse field patterns can then be displayed with a call to `pdeplot(model,'flowdata',[Hx;Hy])` and the like. In the MATLAB script from the appendix we also plot the electric field on axis and

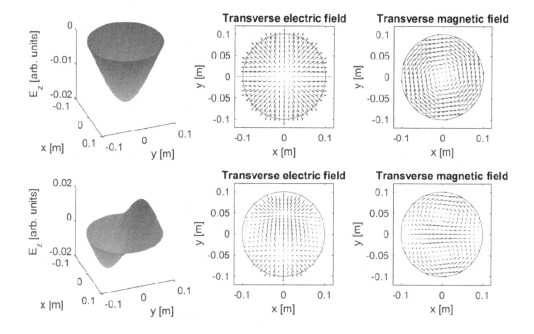

Figure 6.1   The field components of the lowest (upper row) and second lowest (bottom row) TM mode in a circular waveguide.

recover the sinusoidal dependence we expect from calculating fields for the lowest modes analytically.

As a second example we consider TM waves in a circular waveguide with 10 cm radius and show the resulting fields for the two modes with the lowest eigenvalues in Figure 6.1. The only changes with respect to the previous example is that, instead of solving Equation 6.1, we need to solve Equation 5.10 with Ansatz from Equation 5.4. This leads to an eigenvalue equation, equivalent to Equation 6.2, for the longitudinal electric field component $E_z$. Moreover, for TM waves the longitudinal component $E_z$ vanishes on the metallic boundaries and we need to implement Dirichlet boundary conditions here. The call to the MATLAB function thus reads

```
applyBoundaryCondition(model,'Edge',[1:4],'u',0); % Dirichlet
```

and we use the `solvepdeeig()` function to return the eigenvalues $k_c^2$ and longitudinal field component $E_z$. The remaining field components are given by Equation 5.9, which in MATLAB reads

```
[dEx,dEy]=pdegrad(p,t,Ez(:,mymode)); Hx=dEx; Hy=-dEy; Ex=-dEx; Ey=-dEy;
```

and several calls to the `pdesurf()` and `pdeplot()` functions generate the plots shown in Figure 6.1. Inspecting the eigenvalues from the call to `solvepdeeig()` we see that they correspond to those given for the pill-box cavity by Equation 5.22 with $p = 0$. This is no surprise, because the pill-box cavity from Section 5.1 is cylindrical waveguide, terminated by conducting plates, such that the waves are not propagating, but form a standing-wave pattern instead. Based on these simple examples it should be easy to change the geometry and, for example, analyze the effect of finite manufacturing tolerances.

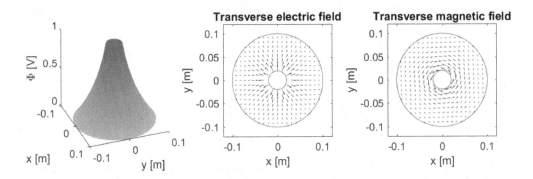

Figure 6.2 The potential $\Phi$ (left) and the field components in a coaxial waveguide.

For frequencies in the 100s MHz range the waveguides are rather large and it is advantageous to transport RF-power using *coaxial waveguides*, or, at low power, coaxial cables. They have concentric inner and an outer conductors with either air or a dielectric filling the intermediate space. The waves that propagate on coaxial lines are TEM waves which are characterized by $E_z = 0$ and $H_z = 0$. From Equation 5.5 we find that this leads to

$$\frac{\partial E_x}{\partial x} + \frac{\partial E_y}{\partial y} = 0, \quad \frac{\partial E_y}{\partial x} - \frac{\partial E_x}{\partial y} = 0, \quad \frac{\partial H_x}{\partial x} + \frac{\partial H_y}{\partial y} = 0, \quad \frac{\partial H_y}{\partial x} - \frac{\partial H_x}{\partial y} = 0 \qquad (6.6)$$

and to

$$ik_z E_y = -i\omega\mu_0 H_x, \quad -ik_z H_x = i\omega\varepsilon_0 E_y, \quad -ik_z E_x = -i\omega\mu_0 H_y, \quad ik_z H_y = i\omega\varepsilon_0 E_x . \qquad (6.7)$$

Consistency between the first and the second pair of the latter equations requires that $k_z^2 = \omega^2\mu_0\varepsilon_0 = \omega^2/c^2$, or $k_z = \omega/c$, which is the dispersion relation of a wave propagating in a coaxial wave guide. Comparing with the dispersion equation for TE-waves, Equation 6.4, we see that the cutoff wave-vector $k_c$ now is zero and all waves, independent of their frequency or wave vector, can propagate. We encountered equations, similar to those of Equation 6.6 before, when discussing purely transverse magnetic fields in Chapter 4, where we realized that the first and second pair of equations are equivalent to the Cauchy-Riemann equations. Here, we can therefore derive, for example, the electric field components $E_x$ and $E_y$ from a purely transverse scalar potential $\Phi(x, y)$ that fulfills the Laplace equation $\triangle\Phi = 0$ via the expressions $E_x = -\partial\Phi/\partial x$ and $E_y = -\partial\Phi/\partial y$. At this point we see that we can calculate the field pattern of TEM waves by solving an equivalent electro-static problem, namely solving the Laplace equation with constant voltages on metallic boundaries. For azimuthally symmetric field configurations and in cylindrical coordinates $(r, \phi)$, the Laplace equation is given by

$$\frac{1}{r}\frac{\partial}{\partial r}\left(r\frac{\partial\Phi}{\partial r}\right) = 0 \qquad (6.8)$$

such that $\Phi$ does not depend on the variable $\phi$. It is straightforward to verify that $\Phi(r) = A\ln r + B$ with arbitrary integration constants $A$ and $B$ satisfies Equation 6.8. Matching them to a given potential $\Phi(r_i) = V_0$ on the inner conductor with radius $r_i$ and $\Phi(r_o) = 0$ on the outer conductor with radius $r_o$ we find that the potential is given by $\Phi(r, \phi) = V_0 \ln(r_o/r)/\ln(r_o/r_i)$. Converting back to Cartesian coordinates and calculating the fields are left as a exercise. Instead, we use MATLAB to calculate them in the file TEMcoax.m in

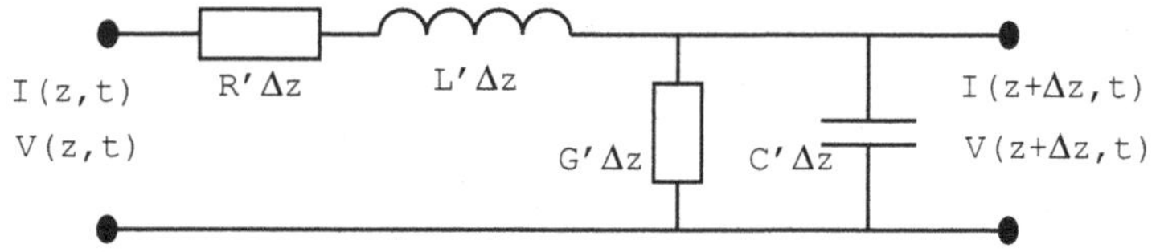

Figure 6.3 The lumped-elements model for a transmission line. See text for details.

Appendix B.5. After defining the geometry, meshing, and defining the boundary conditions we call the solver and then calculate the fields with

```
specifyCoefficients(model,'m',0,'d',0,'c',1,'a',0,'f',0,'Face',1);
result=solvepde(model); Phi=result.NodalSolution;
Ex=-result.XGradients; Ey=-result.YGradients; Hx=-Ey; Hy=Ex;
[p,e,t]=meshToPet(model.Mesh);
pdesurf(p,t,Phi);
```

Here m, d, c, a, and f are the coefficients of the PDE and solvepde solves the Laplace equation. We then store the solution as defined on the nodes of the mesh in the variable Phi. In the following line, we extract the gradients, which yields the electric field components. The magnetic field components are given by Equation 6.7, where we did not include the proportionality constant, because we only are interested in the field patterns. After the call to meshToPet() we call pdesurf() to generate the left-most plot in Figure 6.2. The electric and magnetic field patterns are shown in the two other plots. Here we see that the electric field only has a radial component and the magnetic field only an azimuthal component that encircles the inner conductor.

Waveguides and coax lines are two examples of *transmission lines* that are often characterized by current and voltage-waves traveling along them. The connection to the description in terms of electric and magnetic fields, we used in the previous paragraphs, is made by considering the fields in a coax line at a fixed position $z$ and realizing that the azimuthal magnetic field $H_\phi(r, z)$ is created by a current $I(z)$ traveling on the central conductor. Moreover, the integral of the radial electric field component yields the voltage difference $V(z)$ between the inner and outer conductor:

$$I(z) = \oint H_\phi(r, z) r d\phi \qquad \text{and} \qquad V(z) = \int_{r_i}^{r_o} E(r, z) dr \ . \qquad (6.9)$$

It can be shown [31] that the averaged power flow given by the integral of the Poynting vector over the cross section of the coax line equals the power flow given as the product of current and voltage.

Using the description in terms of voltages and currents permits us to use the lumped-element description of a short section of the transmission line with length $\Delta z$ shown in Figure 6.3. Here $L'$ represents the inductance per unit length, such that $L'\Delta z$ is the inductance of a short section of the conductor. Likewise, $C'\Delta z$ is the capacitance between inner and outer conductor. The series-resistance $R'$ and shunt-conductance $G'$ are introduced to describe losses in the conductors or the dielectric between the conductors. For harmonic currents and voltages with a $e^{i\omega t}$ dependence, Kirchhoff's rules lead, in the limit of infinitesimal short sections with $\Delta z \to dz$, to the *telegrapher's equations*

$$\frac{dV}{dz} = -(R' + i\omega L')I(z) \qquad \text{and} \qquad \frac{dI}{dz} = -(G' + i\omega C')V(z) \ . \qquad (6.10)$$

Inserting one of the equations into the other leads to $d^2V(z)/dz^2 - \gamma^2 V(z) = 0$ and $d^2I(z)/dz^2 - \gamma^2 I(z) = 0$ with the complex propagation constant $\gamma = \alpha + i\beta = \sqrt{(R' + i\omega L')(G' + i\omega C')}$. Solutions to these equations are waves traveling to the left and to the right. They are given by

$$V(z) = V_+ e^{-\gamma z} + V_- e^{\gamma z} \qquad \text{and} \qquad I(z) = I_+ e^{-\gamma z} + I_- e^{\gamma z} . \tag{6.11}$$

Inserting the expression for $V(z)$ into the left-hand side in Equation 6.10, we find $I(z) = \gamma (V_+ e^{-\gamma z} - V_- e^{\gamma z})/(R' + i\omega L')$. Comparing with $I(z)$ in Equation 6.11, we see that we can relate the current and the voltage by

$$I(z) = \frac{1}{Z_0} \left( V_+ e^{-\gamma z} - V_- e^{\gamma z} \right) \qquad \text{with} \qquad Z_0 = \sqrt{\frac{R' + i\omega L'}{G' + i\omega C'}} . \tag{6.12}$$

Here $Z_0$ has the units of Ohms and is the *characteristic impedance* of the transmission line. It is given by the ratio of voltage and current, analogously to the earlier definition in waveguides as the ratio of electric and magnetic fields. For a line without losses, $R' = 0$ and $G' = 0$, it is given by the ratio of the inductance and the capacitance $Z_0 = \sqrt{L'/C'}$, while the real part of the propagation constant $\alpha$ vanishes and the imaginary part is given by $\beta = \omega \sqrt{L'C'}$. Without proof [31] we mention that this propagation constant $\beta$ agrees with $k_z$, calculated from the analysis of the electro-magnetic fields earlier in this section.

The purpose of the transmission line is to move the RF-power from the amplifier discussed in Section 6.1 to a "consumer", for example an acceleration structure, and the latter can be represented by a load impedance $Z_L$ at the end of the transmission line. Let us choose the reference point, where the load is located, to be $z = 0$. There, the current and voltage are related by $V(0) = Z_L I(0)$ with $V(z)$ given by Equation 6.11 and $I(z)$ by Equation 6.12. Solving for the reflected voltage $V_-$ we find

$$V_- = \frac{Z_L - Z_0}{Z_L + Z_0} V_+ \quad \text{or} \quad V_- = \Gamma V_+ \quad \text{with} \quad \Gamma = \frac{Z_L - Z_0}{Z_L + Z_0} , \tag{6.13}$$

where $\Gamma$ is the reflection coefficient. We observe that matching the load impedance $Z_L$ to the impedance of the transmission line $Z_0$ minimizes the reflections and all power is "consumed" by the load.

If $Z_L = 0$, the transmission line is short-circuited and the wave is reflected with $\Gamma = -1$ and $V_- = -V_+$. This results in a counter-propagating wave and leads to standing waves along the transmission line. The voltage at position $z$ is $V(z) = V_+(e^{-i\beta z} - e^{i\beta z}) = -2iV_+ \sin(\beta z)$ and for the current we find $I(z) = V_+(e^{-i\beta z} + e^{i\beta z})/Z_0 = 2V_+ \cos(\beta z)/Z_0$. For the ratio of voltage and current a distance $z = -L$ to the left of the short we obtain $Z(-L) = iZ_0 \tan(\beta L)$, which is purely imaginary or reactive. We thus find that the impedance varies periodically with distance $L$. For an open circuit with $Z_L = \infty$ we have $V_- = V_+$ and showing that the impedance in that case is $Z(-L) = -iZ_0 \cot(\beta L)$ we leave as a exercise.

The transmission line shown in Figure 6.3 is an example of the more general device, a two-port network with current $I_1$ and voltage $V_1$ on the one side and the corresponding quantities with index 2 on the other side. Other examples are attenuators, filters, or amplifiers. Many of these devices are linear and their inner workings can be understood with the help of Kirchhoff's rules, just as we did earlier, when discussing the transmission line itself. The linearity makes it possible to prepare so-called ABCD matrices that relate the input current and voltage to the output voltages

$$\begin{pmatrix} V_1 \\ I_1 \end{pmatrix} = \begin{pmatrix} A & B \\ C & D \end{pmatrix} \begin{pmatrix} V_2 \\ I_2 \end{pmatrix} \tag{6.14}$$

in much the same way we earlier used transfer matrices for the phase-space coordinates. Likewise we can use the ABCD matrices for two-port components to cascade several of them. We will use them further on and list matrices of a few examples in the following table:

| | | |
|---|---|---|
| (series impedance $Z$, ports 1–2) | $\begin{pmatrix} 1 & Z \\ 0 & 1 \end{pmatrix}$ | (shunt impedance $Z$, ports 1–2) $\begin{pmatrix} 1 & 0 \\ 1/Z & 1 \end{pmatrix}$ |
| (transmission line $\beta, L, Z_0$) $\begin{pmatrix} \cos\beta L & iZ_0\sin\beta L \\ (i/Z_0)\sin\beta L & \cos\beta L \end{pmatrix}$ | | (transformer $1{:}n$) $\begin{pmatrix} 1/n & 0 \\ 0 & n \end{pmatrix}$ |

The top line shows the symbols and matrices for series and shunt impedances and the bottom row those for a loss-less transmission line with characteristic impedance $Z_0$, propagation constant $\beta$, and length $L$, as well as a transformer with winding ratio $n$.

Representing networks by matrices automatically takes Kirchhoff's rules into account and a useful application is *matching* load impedances $Z_L$ to line impedances $Z_0$. Two examples are shown in Figure 6.4. The matrix representation of the left-hand example, where a transformer is used, leads to the following equation

$$\begin{pmatrix} V_0 \\ I_0 \end{pmatrix} = \begin{pmatrix} 1/n & 0 \\ 0 & n \end{pmatrix}\begin{pmatrix} 1 & 0 \\ 1/Z_L & 1 \end{pmatrix}\begin{pmatrix} V_{Z_L} \\ 0 \end{pmatrix} \qquad \text{such that} \qquad Z_0 = \frac{V_0}{I_0} = \frac{1}{n^2}Z_L \quad (6.15)$$

where we used that no current enters or leaves the circuit at the end where $Z_L$ is located. We find that a transformer with winding ratio $n$ reduces the impedance $Z_L$ by $n^2$. This is suitable for $|Z_0| < |Z_L|$, but reversing the transformer works in the converse case, where $|Z_0| > |Z_L|$. Note, that a transformer cannot match a complex impedance $Z_L$ to a real $Z_0$, because both impedances have the same ratio of real to imaginary part. This deficiency is remedied by the circuit shown on the right-hand side in Figure 6.4, which uses two reactive impedances $iX$ and $iY$ to match the real and the imaginary part of the load impedance $Z_L$ to $Z_0$. Multiplying the matrices gives voltage $V_0$ and current $I_0$ and the condition to match the impedances is $Z_0 = V_0/I_0 = iX + 1/\left[1/iY + 1/(R_L + iX_L)\right]$. Separating real and imaginary parts yields two equations for the unknown values $X$ and $Y$ for the matching reactances. Solving the equations involves long, but standard, algebra that we do not repeat here. Instead, we refer to the specialized literature [31] for the solution that turns out to be suitable for $Z_0 < R_L$, and the converse solution that is suitable for $Z_0 < R_L$, and involves moving the line reactance $iX$ to a position adjacent of $Z_L$.

The analysis of matching networks so far only used lumped circuit elements, but is equally well applicable to transmission lines involving waveguides. Instead of resistors, capacitances, or inductors so-called *matching stubs* are used. They consist of either cylinders,

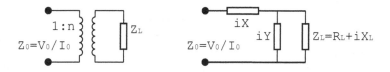

**Figure 6.4** Matching a load impedance $Z_L$ to a line impedance $Z_0$ with the help of a transformer (left) and with a two reactances (right).

often movable, that are inserted in the waveguide, or of alcoves to a waveguide. These elements create local interference patterns of the passing waves. They change the magnitude and phase of magnetic and electric fields components in different ways, and thereby change the impedance of the wave. The design of these elements requires electro-magnetic simulation tools that are beyond our scope.

The ABCD matrices are not the only matrix description for circuit networks and other descriptions of the same physics exist; for example, the *impedance matrix* that returns the voltages on the ports as a consequence of currents flowing into the ports. Its inverse is called the *admittance matrix*. The matrix elements of the different matrix-descriptions can be expressed in terms of each other, but we leave this to the specialized literature [31]. Instead we briefly discuss the commonly used characterization of network elements in terms of *S–parameters*, which have the big advantage of being experimentally accessible with network analyzers by exciting a port $n$ with a voltage $V_{n+}$ and recording the voltage response $V_{m-}$ exiting from the same port $n$ or from another port $m$. Thus, they are generalizations of reflections and transmission coefficients. It is important to realize that only a single port is excited and it is assumed that no reflections from other ports occur; they must be matched by their characteristic impedance. For a given port, say $n$, we can relate the forward and backward traveling voltages with the help of Equations 6.11 and 6.12 and the notation $V_n = V_{n+} + V_{n-}$ and $I_n = (V_{n+} - V_{n-})/Z_0$. Solving for $V_{n+}$ and $V_{n-}$ we obtain $V_{n+} = (V_n + I_n Z_0)/2$ and $V_{ni} = (V_n - I_n Z_0)/2$. In passing we mention that in the literature often the quantities $a_n = V_{n+}/\sqrt{Z_0}$ and $b_n = V_{n-}/\sqrt{Z_0}$ are used. The S–parameters are defined by

$$S_{11} = \frac{V_{1-}}{V_{1+}}\bigg|_{V_{2+}=0}, \quad S_{12} = \frac{V_{1-}}{V_{2+}}\bigg|_{V_{1+}=0}, \quad S_{21} = \frac{V_{2-}}{V_{1+}}\bigg|_{V_{2+}=0}, \quad S_{22} = \frac{V_{2-}}{V_{2+}}\bigg|_{V_{1+}=0}, \quad (6.16)$$

and since the $V_{n\pm}$ can be expressed in terms of currents and voltages, the S–parameters for two-port devices can be converted to ABCD and the other matrices.

S–parameters are, however, not limited to two-port devices and describe devices with different numbers of ports equally well and we will discuss a few examples that frequently appear in RF systems. A *load* is a one-port device whose purpose is to dissipate all incident power. It therefore must be matched to the transmission line feeding it. The reflections, given by its $S_{11}$, should be as small as possible. Typical values are $10^{-3} < S_{11} < 10^{-2}$. Loads are used in many installations during commissioning to ensure proper operation of subsystems and during later operation to dissipate undesired reflected signals. We already mentioned transmission lines, attenuators, filters, and amplifiers as typical examples for two-port networks and will not discuss them further. Transitions between different types of transmission lines, for example from a coaxial line to a waveguide, shown on the left-hand side in Figure 6.5, are further examples of two-port networks. Examples for three-port devices are 3 dB *power-splitters*. They divide the power incident on one port equally to two other ports. They are used, for example, to distribute the power from one klystron to two or several accelerating structures. *Circulators* are three-port devices that use magnetic materials, often ferrites, to produce an asymmetry between signals propagating one way and those in the reverse direction. In particular, they pass signals from the first port to the second, but power entering the second port is directed to the third port, often connected to a load to dissipate the power. Circulators are commonly used to protect klystrons or other amplifiers from the reflected power from mismatched accelerating structures. It is possible to eavesdrop on the signals in RF distribution systems with *directional couplers*. They are four-port devices that pick up a small fraction, typically $10^{-3}$ to $10^{-6}$, of the RF power that passes through them from port 1 to port 2. In their simplest incarnation, they consist

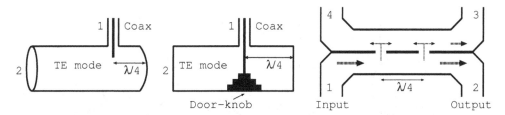

Figure 6.5 Coupling a coaxial line or a cable to a circular waveguide (left) and a transition from a coaxial line to a square waveguide with a door-knob (middle). The right-most figure shows a two-hole directional coupler.

of a parallel waveguide, coupled with two holes a distance $\lambda/4$ apart, to the power-carrying waveguide. In the upstream port 4 of the "diagnostic waveguide" the signals from the two coupling holes are $\lambda/2$ out-of-phase and cancel, as is illustrated on the right-hand side in Figure 6.5, while on the downstream port 3 they add to a non-vanishing contribution that can be coupled to coaxial cable and transported to analyzing electronics, such as an oscilloscope.

Using waveguides and other transmission lines enables us to transport the RF-power into the vicinity of the accelerating structures and cavities, but now we need to address the question of how to efficiently couple the power into the structures themselves and that is the topic of the next section.

## 6.3 COUPLERS AND ANTENNAS

The task of a power coupler is to transfer the RF-power from a coaxial line or waveguide to a resonant structure, such as a cavity. There are three basic methods [32] to accomplish this and we already mentioned two of them in conjunction with directional couplers: coupling through holes or slots, and coupling waveguides to coaxial cables. Figure 6.5 shows two examples of the latter, which both have a coaxial line coming from the top and only a short section of the central conductor extends into the circular waveguide and serves as a short antenna. Ensuring that the distance to the right-hand end of the waveguide is a quarter of the wavelength ensures the electric field, which must be zero at the end, to have a maximum at the location of the small antenna. In the example in the middle, the coupling to the waveguide is increased by extending the center conductor to the other side of the waveguide and terminating it in a so-called *door-knob*. The sinusoidal current flowing in the center conductor behaves similar to dipole antenna and radiate power into the modes that propagate in the waveguide. Note that in these transitions the signal can equally well flow from waveguide to the coaxial line or in the reverse direction and such devices are therefore called *reciprocal*. On the right-hand side in Figure 6.5, we show the directional coupler mentioned near the end in the previous section. Here power flows from port 1 to port 2, but part of the magnetic field, which is largest on the waveguide surface, can penetrate as an evanescent wave through the small coupling holes into the second waveguide, where it excites waves that propagate with the same frequency as the original one.

The two mechanisms of coax antennas and slot coupling are also used to couple power into the accelerating cavities. The left-hand side in Figure 6.6 shows a waveguide coupled with a coupling-hole to a pillbox resonator and in the middle a coaxial coupler, similar to those used to excite the superconducting cavities in LEP at CERN. On the right-hand

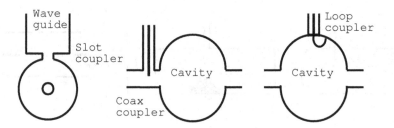

**Figure 6.6** Illustrations of three coupling methods to feed power into resonating acceleration structures: slot couplers, coaxial couplers, and loop couplers.

side, the third coupling mechanism is shown, where a small loop couples to the azimuthal magnetic field component and thereby excites the fields in the cavity. The design and optimization of these couplers requires numerical three-dimensional electro-magnetic field calculations and the interested reader is referred to [32] and the references given there.

A qualitative understanding of the process of filling a cavity can, however, be derived from the equivalent circuit model, shown in Figure 6.7, where the power source is represented by a matched current source $I_g$ that is connected through a transmission line to the power coupler, which is modeled by a transformer with a winding ratio $n$. The resonating cavity or accelerating structure is modeled as an $RLC$–resonator. The inductance $L$ and capacitance $C$ define the resonance frequency $\hat\omega$ and the resistance $R$ defines the losses in the cavity. Note that there are additional losses due to the source impedance of the generator $Z_0$, but transformed to the other, resonator, side of the transformer. Using ABCD matrices we can easily calculate the voltage $V_L$ and current $I_L$ at the reference plane, indicated by the vertical dashed line in Figure 6.7. The circuit then directly translates to

$$
\begin{pmatrix} V_L \\ I_L \end{pmatrix} = \begin{pmatrix} 1/n & 0 \\ 0 & n \end{pmatrix} \begin{pmatrix} 1 & 0 \\ 1/R & 1 \end{pmatrix} \begin{pmatrix} 1 & 0 \\ 1/i\omega L & 1 \end{pmatrix} \begin{pmatrix} 1 & 0 \\ i\omega C & 1 \end{pmatrix} \begin{pmatrix} V_c \\ I_c = 0 \end{pmatrix}
$$
$$
= \begin{pmatrix} 1/n & 0 \\ n/R + n\,(i\omega C + 1/i\omega L) & n \end{pmatrix} \begin{pmatrix} V_c \\ 0 \end{pmatrix} \tag{6.17}
$$

where each circuit element from right to left is replaced by the matrix for the corresponding shunt impedances. $V_c$ is the voltage in the cavity and since we consider operating the circuit

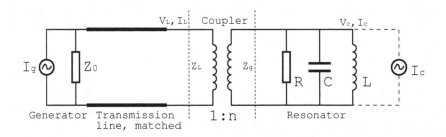

**Figure 6.7** Equivalent circuit with a current source, the generator, with current $I_g$ driving a resonant structure, such as a cavity. The second current source $I_c$ models the beam and will be discussed in Section 6.6.

without beam, the circuit is terminated at the right-hand side and the current $I_c$ is zero. The impedance $Z_L$, which is the load that the generator drives is then given by the ratio of voltage $V_L$ and current $I_L$ at the reference plane

$$Z_L = \frac{V_L}{I_L} = \frac{1/n}{n/R + n(i\omega C + 1/i\omega L)} = \frac{R/n^2}{1 + iR\sqrt{C/L}\left(\omega\sqrt{LC} - 1/\omega\sqrt{LC}\right)} \ . \tag{6.18}$$

Introducing the resonance frequency $\hat{\omega} = 1/\sqrt{LC}$ and loss-parameter $Q_0 = R\sqrt{C/L}$ we can write the previous equation as

$$Z_L = \frac{R/n^2}{1 + iQ_0\left(\frac{\omega}{\hat{\omega}} - \frac{\hat{\omega}}{\omega}\right)} \tag{6.19}$$

and realize that $Z_L$ is the load resistance to which the generator with source impedance $Z_0$ delivers the power and we can use Equation 6.13 to calculate the reflection coefficient $\Gamma = (Z_L - Z_0)/(Z_L + Z_0)$. Obviously, all power is delivered to the cavity, provided that $Z_L = Z_0$.

The power coupler does not only deliver power to the resonator, it also allows power to escape from the resonator, which is easily seen by realizing that the source impedance $Z_0$ is placed in parallel to the resistor $R$. We have, however, taken into account that the resonator "sees" $Z_0$ through the coupler, which steps up the voltage by a factor $n$ and steps down the current by the same factor, such that the impedance, being the ratio of voltage and current, is stepped up by $n^2$. We thus find $Z_g = n^2 Z_0$ and the total losses are given by the total resistance $1/R_t = 1/R + 1/Z_g = 1/R + 1/n^2 Z_0$. The ratio of internal losses, given by $R$ and external losses, given by $Z_g = n^2 Z_0$ is commonly called the *coupling factor* $\beta = R/n^2 Z_0$ and one associates a quantity $Q_E = n^2 Z_0 \sqrt{C/L}$, the external quality-factor, to the losses due to power escaping from the resonator through the coupler. The sum of the losses are then described by the *loaded-Q*, or $Q_L$. It is given by $1/Q_L = 1/Q_0 + 1/Q_E = (1 + \beta)/Q_0$ and simply describes the two contributions to the power loss from the resonator: $Q_0$ describes the power dissipated in the cavity walls and $Q_E$ the power escaping through the coupler.

The above definition of the coupling factor $\beta = R/n^2 Z_0$ and $Z_L$ from Equation 6.19 now allows us to express the reflection coefficient $\Gamma$ as

$$\Gamma = \frac{Z_L - Z_0}{Z_L + Z_0} = \frac{\beta - 1 - iQ_0\delta}{\beta + 1 + iQ_0\delta} \qquad \text{with} \qquad \delta = \frac{\omega}{\hat{\omega}} - \frac{\hat{\omega}}{\omega} \ , \tag{6.20}$$

which succinctly summarizes the coupling of an external generator to a resonator by two parameters: $Q_0$, which describes the internal losses of the resonator and the coupling factor $\beta$, which describes both the power transfer into the resonator, but also the power escaping from the resonator. The coupling factor $\beta$ can be adjusted by changing, for example, the length of the antenna, shown in Figure 6.6, or the angle of the loop with respect to the magnetic field lines. Tuning it to make the reflection coefficient $\Gamma$ zero for $\omega = \omega_0$ requires $\beta = 1$; this condition is called *critically coupled* and describes steady-state operating conditions with destructive interference of the power reflected by the coupler and that escaping from the resonator. The case with $\beta < 1$ is called *under-coupled* and describes the situation where the power is escaping from the resonator faster than it is replenished. The converse situation with $\beta > 1$ is called *over-coupled*.

To implement tunability of the coupling factor $\beta$ is often technically challenging, because the available space is very tight. Moreover, the design engineer has to accommodate vacuum barriers, so-called windows, because the waveguides are usually air-filled, while the resonating structures are evacuated. These windows must withstand the flow of up to MW of RF power with electric field levels in the multi-MV range. This may lead to harmful discharges that must be avoided, because they can destroy the vacuum windows. A further difficulty is *multipacting,* resonant avalanches of field-emitted electrons that bounce back and forth between surfaces and increase in numbers every time they impact on the coupler surface. This usually happens intermittently during power-conditioning and commissioning and only at certain power levels. Couplers for super-conducting cavities must include heat barriers to separate the waveguides at room temperature from the resonating structures at liquid Helium temperatures.

Instead of discussing the technical realization of couplers further, we turn to the resonating structures that are used to accelerate the beam. First we look at the internal losses in a cavity which determine, for example, $Q_0$ in Equations 6.19 and 6.20.

## 6.4   POWER TO THE BEAM: RESONATORS AND CAVITIES

In Section 5.1 we determined the resonating modes in a pill-box cavity. Moreover, in the previous section, we found that the internal losses, expressed by the $Q_0$ play an important role for filling the cavity. Therefore we will consider the origin of these losses closer and determine them for the pill-box cavity.

### 6.4.1   Losses and quality factor $Q_0$ of a pill-box cavity

In Section 5.1, we assumed that the walls of the cavity are perfectly conducting, but this assumption is only approximately fulfilled and the finite conductance will be responsible for the dissipation of some energy. The reason is the tangential component of the magnetic field, which is continuous at the interface between the inside of the cavity and the walls. Since it is time-varying, it will cause an electric field that in turn drives eddy currents in the cavity walls; just like in a transformer or induction stove, they will heat the metal through ohmic losses.

The quantitative description is based on Maxwell's equations in materials with finite conductance $\sigma$, where electric fields $\vec{\mathbf{E}}$ drive conduction currents $\vec{\mathbf{J}}$ via Ohm's law $\vec{\mathbf{J}} = \sigma\vec{\mathbf{E}}$. The two curl equations from Equation 5.1 then read

$$\vec{\nabla} \times \vec{\mathbf{E}} = -\mu_0 \frac{\partial \vec{\mathbf{H}}}{\partial t} \quad \text{and} \quad \vec{\nabla} \times \vec{\mathbf{H}} = \sigma\vec{\mathbf{E}} + \varepsilon_0 \frac{\partial \vec{\mathbf{E}}}{\partial t} \ . \tag{6.21}$$

Applying the curl operator to the second equation and assuming a harmonic time-dependence with $\vec{\mathbf{H}} = \vec{\mathbf{H}}'e^{i\omega t}$, we find, after using $\vec{\nabla} \cdot \vec{\mathbf{H}} = 0$, that

$$\triangle\vec{\mathbf{H}}' = (i\omega\mu_0\sigma - \mu_0\varepsilon_0\omega^2)\vec{\mathbf{H}}' = \gamma^2\vec{\mathbf{H}}' \tag{6.22}$$

with $\gamma^2 = i\omega\mu_0\sigma - \mu_0\varepsilon_0\omega^2$. Likewise, applying the curl operator to the first equation, we find $\triangle\vec{\mathbf{E}}' = \gamma^2\vec{\mathbf{E}}'$. Both $\vec{\mathbf{E}}$ and $\vec{\mathbf{H}}$ show exponential dependence on a local coordinate $z'$ that points into the metallic surface

$$\vec{\mathbf{E}}' = \vec{E}_0 e^{-\gamma z'} \tag{6.23}$$

with $\gamma \approx \sqrt{i\omega\mu_0\sigma} = (1+i)\sqrt{\omega\mu_0\sigma/2} = (1+i)/\lambda_s$. Here we use the fact that at high frequencies $\omega$, the first term in $\gamma^2$ is much larger than the second, because conduction currents are much larger than the displacement currents. Therefore, the fields decay exponentially into the metal with decay length $\lambda_s = \sqrt{\omega\mu_0\sigma/2}$, the *skin depth*. The factor $1+i$ accounts for a phase shift by $\arg(1+i) = \pi/4$.

We can now relate the magnetic field on the surface, which is continuous, to the electric field just inside the metal by exploiting $-i\omega\mu_0\vec{H}' = \vec{\nabla} \times \vec{E}'$, the first of Equations 6.21, and assume that the tangential component of the magnetic field $H_{0y'}$ on the interface points towards the local $y'$–direction, such that the only component of the electric field affected by curl-operator is the $E_{0x}$ and we get

$$-i\omega\mu_0 H_{0y'} e^{-\gamma z'} = -\frac{\partial}{\partial z'} E_{0x'} e^{-\gamma z'} , \qquad (6.24)$$

such that we arrive at the electric field just below the surface $E_{0x'} = -i(\omega\mu_0/\gamma)H_{0y'} \approx -(1+i)\sqrt{\omega\mu_0/2\sigma}H_{0y'}$, expressed in terms of the magnetic field which is continuous across the boundary.

Inside the metal, the electric fields $E_{x'} = E_{0x'}e^{-\gamma z'}$ drive conduction currents $J_{x'} = \sigma E_{x'}$, which will dissipate energy, and lead to a time-averaged power loss density $d^3P/dx'dy'dz' = J_{x'}E_{x'}^*/2 = J_{x'}J_{x'}^*/2\sigma$, where the asterisk denotes complex conjugate. Integrating over $dz'$, extending far into the metal, we obtain

$$\frac{d^2P}{dx'dy'} = \frac{\omega\mu_0}{2}H_{0y'}^2 \int_0^\infty e^{-2z'/\lambda_s}dz' = \frac{\omega\mu_0\lambda_s}{4}H_{0y'}^2 = \frac{1}{2\lambda_s\sigma}H_{0y'}^2 . \qquad (6.25)$$

Remembering that we selected the $y'$–direction to be parallel to the tangential component $H_\parallel$ of the magnetic field on the surface, we obtain the total power loss $P$ by integrating over the inner surface areas of the cavity

$$P = \frac{1}{2\lambda_s\sigma} \int H_\parallel^2 dA = \frac{R_s}{2} \int H_\parallel^2 dA \qquad (6.26)$$

where we introduce the sheet resistivity $R_s = 1/\lambda_s\sigma$.

The magnetic field of a TM$_{010}$ mode is purely azimuthal and is given by $H_\phi = -i(E_0/Z_0)J_1(p_1r/R)$ where $p_1 = 2.405$ is the first zero of the Bessel function $J_0$ with $J_0(p_1) = 0$. This field is tangential on the circumference of the pill box and on the two faces at the end. We calculate the losses on the circular circumference $P_c$ first. Here we have $r = R$ which makes the integral trivial to evaluate

$$P_c = \frac{R_s}{2} \int H_\phi^2 R dz d\phi = \frac{\pi R_s E_0^2}{Z_0^2} J_1^2(p_1)Rl \qquad (6.27)$$

and the losses $P_e$ on the two end faces are

$$P_e = 2\frac{R_s}{2} \int H_\phi^2 r d\phi dr = \frac{2\pi R_s E_0^2}{Z_0^2}R^2 \int_0^1 J_1^2(p_1t)tdt = \frac{\pi R_s E_0^2}{Z_0^2}J_1^2(p_1)R^2 , \qquad (6.28)$$

where we used the identity $\int_0^1 J_1^2(p_1t)tdt = J_1^2(p_1)/2$. The total losses $P_t$ for the fundamental mode of the pill-box cavity are the sum of $P_c$ and $P_e$

$$P_t = \frac{\pi R_s E_0^2}{Z_0^2}J_1^2(p_1)R(R+l) , \qquad (6.29)$$

where $J_1(p_1) \approx 0.52$.

We compare these losses to the time-averaged energy $U$ in the cavity, which is given by the integral of the energy density of the electric and magnetic fields over the volume $V$ of the cavity

$$U = \frac{1}{4} \int \left(\varepsilon_0\vec{E}'^2 + \mu_0\vec{H}'^2\right) dV . \qquad (6.30)$$

Since the two contributions from the electric and the magnetic fields are equal, it suffices to calculate twice the contribution from the electric field and since the only non-zero component is $E_z = E_0 J_0(p_1 r/R)$, we need to calculate

$$U = \varepsilon_0 E_0^2 \int J_0^2(p_1 r/R) r d\phi dr dz = \pi R^2 l \varepsilon_0 E_0^2 \int_0^1 J_0^2(p_1 t) t dt = \frac{\pi R^2 l \varepsilon_0 E_0^2}{2} J_1^2(p_1) \quad (6.31)$$

where we used the identity $\int_0^1 J_0^2(p_1 t) t dt = J_1^2(p_1)/2$.

These are the losses we previously described by the *quality factor* $Q_0$ in Equation 6.19, which is also given by the ratio of the stored energy $U$ and as $2\pi$ times the losses during one oscillation period $P_t/\omega_1$ of, here, the first mode with frequency $\omega_1/c = p_1/R$.

$$Q_0 = \frac{\omega_1 U}{P_t} = \frac{p_1 Z_0}{2R_s} \frac{1}{1 + R/l} = \frac{G}{R_s} \quad (6.32)$$

where we see that $Q_0$ on one hand depends on the *geometry constant* $G$ through geometric quantities: the radius $R$ and the length $l$. On the other hand, it is inversely proportional to the surface resistance $R_s$. Since the surface resistance depends on the penetration of the fields into the conductor through the skin effect, it is frequency dependent and proportional to $R_s \propto \sqrt{\omega_{\mathrm{rf}}}$. For copper at room temperature and at $f_{\mathrm{rf}} = \omega_{\mathrm{rf}}/2\pi = 1\,\mathrm{GHz}$ it is $R_s \approx 8\,\mathrm{m\Omega}$. For our pillbox cavity with $R = 0.23\,\mathrm{m}$ and assuming a length of $l = 0.2\,\mathrm{m}$, we find $Q_0 = 3.7 \times 10^4$. If we were to construct the same cavity from super-conducting niobium at $4\,\mathrm{K}$, where the surface resistance is on the order of $R_s \approx 100\,\mathrm{n\Omega}$, we obtain $Q_0 = 2.1 \times 10^9$. This indicates that the losses in the normal-conducting copper cavity are about five orders of magnitude larger than in a corresponding super-conducting cavity and explains why the latter are used in accelerators with continuously running radio-frequency systems that require large power delivered to the beam, such as LEP. Other examples are linear accelerators operating with long macro-pulses, often in the ms range, such as the European XFEL or the ESS.

When exciting a cavity with an external generator, we want to accelerate particles, but we do not want to dissipate power in the cavity walls. Using the symbols from Figure 6.7: we want $V_c$ to accelerate, but do not want to dissipate $P_t = \langle V_c^2 \rangle/R = \hat{V}_c^2/2R$ in the *shunt resistor* $R$. The angle brackets denote averaging over time such that we have $\langle V_c^2 \rangle = \hat{V}_c^2/2$. The voltage $\hat{V}_c$ is related to the peak electric field $E_0$ by $\hat{V}_c = TE_0 l$, where $T$ is the transit-time factor, defined in Section 5.2. For the shunt resistance $R$ we therefore find

$$R = \frac{\hat{V}_c^2}{2P_t} = \frac{Z_0^2}{\pi J_1^2(p_1) R_s} \times \frac{T^2 l^2}{R(R + l)} \quad (6.33)$$

for our pillbox cavity. The second factor contains only parameters that define the geometry of the cavity: the length $l$, the radius $R$, and the transit-time factor $T$. The first factor contains only numerical constants and the surface resistance $R_s$. We note that a small value of $R_s$, as found in super-conducting cavities, results in a large shunt resistance $R$.

Comparing Equations 6.32 and 6.33, we observe that both definitions are inversely proportional to the dissipated power $P_t$, which makes it natural to consider the ratio $R/Q_0$. It is given by

$$\frac{R}{Q_0} = \frac{\hat{V}_c^2}{2\omega_1 U} = \frac{Z_0 T^2}{\pi p_1 J_1^2(p_1)} \left( \frac{l}{R} \right) \quad (6.34)$$

and relates the voltage "seen" by the beam $\hat{V}_c$ to the stored energy $U$ in the cavity. It only depends on natural constants and parameters defining the geometry of the cavity. The art of designing cavities involves maximizing this quantity, because it maximizes the efficiency of the cavity geometry, but is independent of material properties.

We could analyze our simple pill-box cavity analytically, but for other geometries we need to employ the numerical methods we turn to in the next section.

## 6.4.2 General cavity geometry with the PDE toolbox

In most circumstances the geometry of cavities differs from the simple and analytically tractable models with high degrees of symmetry, such as the pill-box cavity and we have to employ numerical tools to find the eigenfrequencies and the fields. In order to illustrate the methods, we first investigate the influence of the beam pipe aperture on the fundamental mode and eigenfrequency of a pill-box cavity with a length $l = 0.3$ m and radius $R = 0.23$ m, such that the frequency of the fundamental mode, according Equation 5.23, should be close to 500 MHz. The pipe through which the beam enters and exits the cavity has a radius of 1 cm. The geometry is assumed to be azimuthally symmetric and is shown on the left-hand side in Figure 6.8. In the center, extending from $-0.15$ to $0.15$ m is the cavity visible with the beam pipe sticking out near to $\pm 0.4$ m.

The eigenfrequencies and fields of TM-modes are determined by Equation 5.10 subject to the boundary conditions that the tangential electric field components are zero on the metallic surfaces. Since we seek azimuthally symmetric solutions, we can omit the dependence on the azimuthal angle and turn Equation 5.10 to

$$\frac{\partial E_z}{\partial r} + r\frac{\partial^2 E_z}{\partial r^2} + r\frac{\partial^2 E_z}{\partial z^2} = -r\frac{\omega^2}{c^2}E_z \; . \tag{6.35}$$

As before, we will use the MATLAB PDE toolbox to solve this problem, but, since we will use cylindrical coordinates, we first have to adapt the Laplace operator in Equation 6.35 to a form that the toolbox can handle, namely we have to "put it into divergence form," which, according to the manual for the PDE toolbox, is written as $-\vec{\nabla}\cdot(c\vec{\nabla}E_z)+aE_z = \lambda dE_z$. Here $\vec{\nabla} = (\partial/\partial r, \partial/\partial z)$ is the gradient operator in cylindrical coordinates $r$ and $z$. Moreover, the $2 \times 2$ matrix $c$ and the functions $a$ and $d$ depend on $r$ and $z$. Comparing this form to Equation 6.35 we find that $a = 0, d = -r$, and $c$ is a matrix with $-r$ on the diagonal and zeros on the off-diagonal. These preparations enable us to solve azimuthally symmetric, sometimes called 2.5-dimensional, problems. We define the geometry of the problem with cavity and beam pipe sections attached and specify Dirichlet boundary conditions for boundaries parallel to the beam pipe and Neumann conditions for radial boundaries, because Equation 5.9 indicates that radial fields are given by derivatives of $E_z$. After meshing the geometry and defining the boundary conditions, we specify the coefficients $m, d, c, a$, and $f$ and call the eigenvalue solver `solvepdeeig()`, which returns the eigenvalues $k^2 = \omega_{mnp}^2/c^2$ for $m = 0$ in the range $1 < k^2 < 800$ and the fields $E_z$

```
cfun=@(location,state)-location.y;
dfun=@(location,state)-location.y;
specifyCoefficients(model,'m',0,'d',dfun,'c',cfun,'a',0,'f',0);
result=solvepdeeig(model,[1,800]);
k2=result.Eigenvalues; Ez=result.Eigenvectors;
freq_GHz=sqrt(k2')*3e8/(2*pi*1e9)
```

In the MATLAB script, which is reprinted in Appendix B.5, we generate the figure shown on the left-hand side in Figure 6.8 with contour lines of $E_z$ that show the field to be distorted near to the beam pipe and even slightly penetrates into the pipe. On the right-hand side

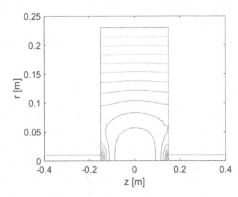
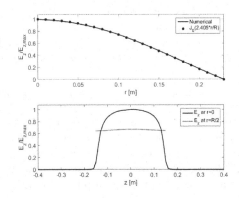

**Figure 6.8** The presence of the beam pipe distorts the electric field in a pill-box cavity (left) and on the axis (bottom right, solid line), but the influence is moderate at larger radius (dotted line). Very little distortion is visible along the vertical radial line at $z = 0$ (top right).

the upper graph shows the radial dependence of $E_z$ along a radial line in the center of the cavity as a solid line. The asterisks show the analytical result from Equation 5.21. Here we see that the pipe has very little influence. The solid line in the lower graph shows $E_z$ along the center line and we find that the field penetrates into the beam pipe and is significantly reduced near the ends of the cavity. The dotted line shows $E_z$ at $r = R/2$ and, since it is further removed from the beam pipe, is much more constant. When varying the beam pipe radius, we found that the eigenvalues, and the frequency of the fundamental mode changes. It is now a simple exercise to vary the radius $R$ of the cavity to recover the desired frequency, here 500 MHz.

Since all fields in the cavity are available in numerical from, it is a straightforward exercise to calculate the loss factor and shunt impedance from integrals of the fields over the surfaces. In most cases specific design codes for RF applications calculate these quantities automatically. Instead of discussing the details of cavity design further, we turn to accelerating structures that are normally used in linear accelerators.

### 6.4.3 Disk-loaded waveguides

It is possible to accelerate beams with multiple and separate cavities, but in that case each cavity has to be synchronized, or "phased," with the arrival time of the particles. This requires a significant amount of hardware, such as phase shifters. This complication can be avoided by connecting many cavities in series, only separated by an iris through which the particles and the RF power propagate from one cavity, in this case often called a *cell*, to the next. See Figure 6.9 for an illustration. By suitably choosing the radius $R$ and length $d$ of the cells, as well as the aperture $a$ and thickness $w$ of the irises, it is possible to adjust the coupling from one cell to the next and control the phase of the fields in the cells to match the arrival time of the particles. The structure with multiple cells separated by the irises in between resembles a waveguide with corrugations and is aptly called *disk-loaded waveguide*. Instead of considering the sequence of cells that are coupled through apertures, the irises, we may also consider it as a waveguide that is loaded with washer disks, spaced

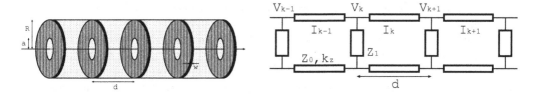

**Figure 6.9** Physical realization of a disk-loaded waveguide (left) and the electrical model as an equivalent transmission line (right). See the text for explanations of the symbols.

by a distance $d$. In this view the purpose of the disks is to reduce the phase velocity of waves propagating in the waveguide to match the speed of the particles. According to Equation 6.5, in an unloaded waveguide, it is always larger than the speed of light. This ensures that the particles experience the same phase of the field, typically close to the maximum, in each of the cells.

In the following we consider $TM_{01}$ waves that accelerate particles with their longitudinal electric field $E_z$. The disks reflect the waves and cause a longitudinal modulation of the field. Owing to the periodicity with the longitudinal spacing of the disks $d$, and with the help of Floquet's theorem, we can express $E_z$ as a propagating wave modulated with some periodic function $g(r, z) = \sum_n f_n(r) e^{-2\pi i n z/d} = g(r, z + d)$. Here we express the periodic function as a Fourier series in $z$ with coefficients $f_n(r)$ that depend on the radial coordinate $r$. Thus we arrive at

$$E_z(r, z, t) = e^{i(\omega t - k_z z)} g(r, z) = e^{i(\omega t - k_z z)} \left( \sum_{n=-\infty}^{\infty} f_n(r) e^{-2\pi i n z/d} \right) \tag{6.36}$$

instead of the simple harmonic Ansatz for the spatial dimension $z$ from Equation 5.4. The functions $f_n(r)$ with the radial dependence is determined by inserting Equation 6.36 into Equation 5.10, which leads to

$$r^2 \frac{d^2 f_n}{dr^2} + r \frac{df_n}{dr} + r^2 \left[ \frac{\omega^2}{c^2} - \left( k_z + \frac{2\pi n}{d} \right)^2 \right] f_n = 0 . \tag{6.37}$$

We recognize this equation as Bessel's differential equation for index zero [23], such that we can write the solution as an ordinary Bessel function of integer order $f_n(r) = J_0(\hat{K}_n r)$ with the expression in the square brackets abbreviated as $\hat{K}_n^2 = \omega^2/c^2 - (k_z + 2\pi n/d)^2$, provided $\hat{K}_n^2 > 0$. If, on the other hand $K_n^2 < 0$, the solution is a modified Bessel function with $f_n(r) = I_0(\hat{K}_n r)$. Inserting the solution for $K_n^2 > 0$ into Equation 6.36 we obtain

$$E_z(r, z, t) = \sum_{n=-\infty}^{\infty} E_{n,0} J_0(\hat{K}_n r) e^{i\omega t - i(k_z + 2\pi n/d)z} \tag{6.38}$$

which describes waves for different values of $n$, called *space harmonics*. Comparing the phase velocity of a space harmonic $v_{p,n} = \omega/(k_z + 2\pi n/d)$ with that of the fundamental $v_{p,0} = \omega/k_z$, we find for their ratio

$$\frac{v_{p,n}}{v_{p,0}} = \frac{1}{1 + 2\pi n/k_z d} . \tag{6.39}$$

By choosing $n$ suitably large we can make the phase velocity as small as we like and, in particular, make it smaller than the speed of light and match it to the speed of the particles.

To visualize the interaction of a particle that we assume to move very close to the speed of light and the longitudinal electric field component $E_z$ we select $\omega = 2\pi f$ and the distance $d$ such that $2df = c$. In that case, the fields in adjacent cells point in opposite directions, but once the particle arrives in the next cell, the polarity is reversed and it is accelerated once again. This mode of operation is called $\pi$–mode, because the fields in adjacent cells are 180 degrees out of phase. Equivalently, considering a snapshot in time, one period (or wavelength) of the field repeats after two cell. Other modes, where the fields repeat after three cells, so-called $\pi/3$ or $2\pi/3$–modes or after four cells, so-called $\pi/2$–mode are used. We refer to the specialized literature [33] for further discussions.

Instead of the electro-magnetic description in terms of fields we can describe the disk-loaded waveguide as an electric circuit consisting of a transmission line with impedance $Z_0$, and propagation constant $k_z$, and impedances $Z_1$, typically capacitive, that represent the washer disks and repeatedly shunt part of the current with period $d$. We use the model shown on the right-hand side in Figure 6.9 to relate the currents $I_k$ and voltages $V_k$ in the circuit with the help of the ABCD matrices from Section 6.2

$$\begin{pmatrix} V_k \\ I_k \end{pmatrix} = \begin{pmatrix} \cos k_z d & iZ_0 \sin k_z d \\ (i/Z_0)\sin k_z d & \cos k_z d \end{pmatrix} \begin{pmatrix} 1 & 0 \\ 1/Z_1 & 1 \end{pmatrix} \begin{pmatrix} V_{k+1} \\ I_{k+1} \end{pmatrix} . \tag{6.40}$$

This description closely resembles that to describe the effect of an additional quadrupole on the tune in a storage ring, a system we will analyze in Section 8.3.2. Here $k_z d$ takes the role of the unperturbed tune and $Z_1$ the focal length of the perturbing quadrupole. Therefore, we can also calculate the perturbed propagation constant $k_z'$ for one cell from the trace of the matrix and find

$$\cos k_z' d = \cos k_z d + (iZ_0/2Z_1) \sin k_z d . \tag{6.41}$$

The impedance $Z_1 = 1/i\omega C_1$ of the perturbing washer disk is usually capacitive with capacitance $C_1$, and the impedance $Z_0$ and propagation constant $k_z$ of a loss-less transmission line are given by $Z_0 = \sqrt{L'/C'}$ and $k_z = \omega\sqrt{L'C'}$, respectively. Here we used the notation introduced in Figure 6.3. Inserting in Equation 6.41, we obtain the dispersion relation that relates the propagation constant $k_z'$ to the frequency $\omega$ of the wave

$$\cos k_z' d = \cos(\omega/\omega_{LC}) - a\frac{\omega}{\omega_{LC}}\sin(\omega/\omega_{LC}) \tag{6.42}$$

with the abbreviation $\omega_{LC} = 1/d\sqrt{L'C'}$ and $a = C_1/2C'd$ characterizes the perturbation.

Figure 6.10 shows a graphical representation of the dispersion relation for $a = 0.2$ (solid lines) and $a = 0.5$ (dashed lines), where we calculate $k_z'd$ as a function of $\omega/\omega_{LC}$ and then swap the axes. We see that the periodicity of the cosine function causes all branches of the dispersion relation to appear in the range between 0 and $\pi$. Note that there are forbidden frequencies at which waves cannot propagate in the perturbed waveguide; so-called *stopbands*. For $a = 0.5$ they are indicated by the dotted lines in Figure 6.10. Waves can thus only propagate for frequencies that are in a *passband,* defined as the range that is not in a stopband. The existence of passbands and stopbands is fundamental to perturbed periodic systems. Since Equation 6.42 is formally the same as Equation 8.32 that describes the changed tune in a storage ring due to a perturbing quadrupole, we can identify the stopbands in Figure 6.10 with the integer and half-integer stopbands in the tune diagram of a storage ring. Moreover, the perturbed crystal potentials described by the Kronig-Penney model [18] cause the formation of energy bands and forbidden band-gaps—corresponding to the stopbands—in solids.

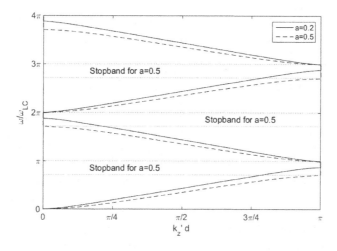

**Figure 6.10** Dispersion diagram with stopbands for a disk-loaded waveguide.

Disk-loaded waveguides are operated either as *traveling-wave structures* or as *standing-wave structures.* The latter are equipped with a coupler on one end and terminated by an electrical short on the other end, such that a standing wave pattern inside the structure ensues. In most cases the accelerating field in adjacent cells is pointing in opposite directions; it differs by $\pi$ in phase and is called $\pi$–mode. Electrically, this mode of operation resembles that of a transmission line feeding power into a resonator. If, on the other hand,'the disk-loaded waveguide has couplers on both ends, the RF-power passes through the structure, which then is called traveling-wave structure. In this case the fraction of the RF-power, not used to accelerate the beam, is directed to a load, where the remaining power is dissipated. Traveling-wave structures can thus be thought of a different section of transmission line with matching section—the couplers—on either end. The phase-advance between adjacent cells in this case can be anything admissible from the dispersion diagram in Figure 6.10, but often $2\pi/3$ is chosen, such that the field pattern repeats every three cells.

We now turn to a number of practical aspects when operating radio-frequency components.

## 6.5  TECHNOLOGICAL ASPECTS

The theoretical considerations discussed in the previous sections are indispensable when first designing a radio-frequency system, but there are a number of practical aspects that appear when operating them.

### 6.5.1  Normal-conducting

A pertinent feature of normal-conducting acceleration structures, often made of copper, are the ohmic losses on the inner surfaces of the structure. This limits the loss factor $Q_0$ to be on the order of $10^5$, which causes the fields to decay rapidly and limits the pulse length for high-gradient linacs to the order of micro-seconds and the repetition rate of the linac to a few 100 pulses per second, each of which contains one or a few closely separated bunches. Moreover, the dissipated energy must be removed from the structure and requires *cooling* water, usually de-ionized. The varying temperatures cause varying thermal expansions of the material and that causes detuning of the structures from their design frequency. This is compensated with a regulation system that adjusts the cooling water temperature to maintain a constant temperature of the structure.

After their installation and powering them for the first time, new accelerating structures or cavities often exhibit internal discharges, even at field strengths well below their design fields. These discharges are typically detectable by increased emission of X-rays, increased vacuum pressure, and reflected RF-power. Often they are related to field-emitted electrons, so-called *dark currents*, that may resonantly bounce between surfaces within the structure; a process called *multipacting*. Or contaminations, such as gas or dust are desorbed from the inner surfaces and are subsequently ignited by dark currents to form a plasma. These effects are typically intermittent and are cured during a process called *conditioning* in which the RF-power and the fields are gradually increased until some discharge happens, whence the power is reduced for a little while, before increasing it again. In this way the fields are inched upwards to and beyond their design's level.

In order to make linear accelerators as short as possible the longitudinal accelerating electric fields $E_z$ are made a large as possible. With linac structures the fields of the inner surfaces are two to three times larger than the accelerating field on the axis which lies between a few MV/m up to 100 MV/m for the structures intended for CLIC. These high fields cause spontaneous discharges, even after conditioning, and pose a limit to the reliability of future high-energy linear accelerators.

The structures for linear accelerators are often based on radio-frequency systems operating between 3 as 12 GHz with correspondingly short wavelengths. This also causes the accelerating structures to be rather small with irises in the disk-loaded waveguides that measure a few mm in diameter. This poses extreme requirements on the steering of the particle beams, in particular, because transversely displaced beams excite transverse modes in the structures and those can disturb following bunches. These fields are commonly called *wake fields* and we will discuss them further in Chapter 12 but next we will briefly address a few topics relevant for the operation of super-conducting cavities.

### 6.5.2 Super-conducting

The reason to choose super-conducting cavities for an accelerator is to minimize the dissipation of energy on the inner surfaces. Despite a vanishing resistance to direct currents show super-conductors a finite resistance to alternating currents caused by the radio-frequency fields. This finite resistance stems from the finite inertia of the Cooper-pairs that are the charge carriers in the superconducting state; once they flow, they flow without losses, but turning them around dissipates energy and is the main cause for the large, but finite $Q_0$ in the range of $10^9$ to $10^{10}$ for super-conducting cavities. These losses need to be removed and are a determining factor for the specification of the cryogenic system that cools the cavities below the critical temperature $T_c$ of the super-conducting material, usually Niobium or one of its alloys. The only agent able to cool to the required temperatures is Helium that turns liquid below 4.2 K. Improved performance of the cavities can be achieved by cooling with super-fluid Helium at temperatures below 1.9 K. We will return to cryogenics in Section 13.6. A further consequence of the low losses is the ability to accelerate much longer, typically several ms-long pulse-trains of bunches.

Only the inner surfaces that define the fields in the cavities are made of Niobium. In order to allow efficient cooling from the Helium that flows in a vessel surrounding the cavities the Niobium is very thin and therefore not very rigid. Pressure fluctuations in the surrounding

Helium bath or other mechanical perturbations can cause the cavities to deform and thereby detune the resonance frequency. This mechanism is called *microphonics*. Considering the very small bandwidth $\Delta f \approx f_0/Q_0$, which is on the order of Hz for a bare (without power-coupler installed) cavity, even minute deformations will change the resonance frequency unacceptably. A second mechanism that deforms the cavity are the large electric fields that pull on the inner surfaces proportional to the square of the electric field. The origin of this force is the same by which the plates of a capacitor attract each other. The capacitance $C \propto A/x$ depends on area $A$ and distance $x$ between plates, such that we obtain for the force $F \propto d(CU^2/2))/dx \propto (U/x)^2$. And $U/x$ equals the electric field between the capacitor plates. When operating cavities in pulsed mode this mechanism will periodically detune the cavity and needs to be corrected. Tuners with piezo-electric actuators are normally used, because they are fast enough to compensate deformations during the ms-long, pulse trains of bunches.

The inner surfaces of the cavities need to be extremely smooth to avoid any protrusions that enhance the electric field locally and cause increased field-emissions, because field-emitted electrons are accelerated by the same fields and will hit the super-conducting surfaces uncontrollably. They deposit their kinetic energy at the impact-site and locally heat the material above the critical temperature, such that it loses super-conductivity. At the now normal-conducting impact-site the fields will dissipate more energy such that the site expands and eventually the entire cavity will turn normal-conducting; a process called *quenching*. Thus reliable operation requires extreme smoothness of the inner surfaces and this is ensured during manufacturing and preparation of the cavities, which includes several steps, including chemically etching, electro-polishing, and high-pressure water rinsing. Once the cavities are installed, one has to avoid to deposit new protrusions, such as dust particles, on the inside of the cavities. This explains the need for clean-rooms in the manufacturing process and extreme care after installation in the accelerator, especially with venting the vacuum system.

With many aspects of the RF system covered, we turn to its interaction with the beam.

## 6.6  INTERACTION WITH THE BEAM

When a beam passes a resonating structure, such as a cavity, the image currents that accompany the beam excite electro-magnet fields in the structure. The energy stored in these fields was provided by the beam and in the next section we address the question of how much energy is actually lost.

### 6.6.1  Beam loading

This question can be answered [33, 34] by considering two short bunches, each having charge $q$ that arrive in the cavity half an oscillation period apart. We assume that the first particle induces a voltage $\hat{V}$ in the cavity and loses some fraction $\eta$ of this voltage as energy such that its energy changes by $\Delta E_1 = -\eta q \hat{V}$. Half a period later, when a second bunch arrives, the voltage induced from the first bunch has reversed its polarity, and increases the energy of the second bunch by $q\hat{V}$. At the same time the second bunch creates a voltage pulse that cancels the one from the first bunch and thereby loses energy $-\eta q \hat{V}$. The total change of energy of the second bunch is $\Delta E_2 = q\hat{V} - \eta q \hat{V}$. Since the fields in the cavity are canceled, the stored energy is zero and therefore the sum of the energy changes of the two bunches must cancel, such that $0 = \Delta E_1 + \Delta E_2 = -\eta q \hat{V} + (1 - \eta)q\hat{V}$, which implies $\eta = 1/2$. Thus we find, based on very few assumptions, the *fundamental theorem of beam loading* [34]: a bunch that excites a voltage $\hat{V}$ in a cavity, loses energy by "seeing" half that voltage itself. It also implies that the bunch-induced voltage opposes the current flow; it is always retarding. At the same time, the beam extracts energy from the cavity and is thereby accelerated, but in order to balance the energy, it must leave behind a field that cancels part of the accelerating fields. This mechanism is commonly referred to as *beam loading.* The additional current that drives the cavity is already shown in Figure 6.7 on page 154 as the current source $I_c$.

We now need to relate this current source $I_c$ to the average circulating beam current $I_b$. Since the beams are bunched and we assume that the bunches are much shorter than the wavelength of the RF, we can represent them as a sequence of delta-functions that appear with a period $T_0$, the revolution time in the ring. For bunches made of $N$ charges we have

$$I(t) = Ne \sum_{k=-\infty}^{\infty} \delta(t - kT_0) = \frac{Ne}{T_0} \sum_{m=-\infty}^{\infty} e^{im\omega_p t} = \frac{Ne}{T_0} + \frac{2Ne}{T_0} \sum_{m=1}^{\infty} \cos(m\omega_0 t) \qquad (6.43)$$

with $\omega_0 = 2\pi/T_0$. Here we find that the driving current $I_m = 2Ne/T_0$ of harmonic $m$ is twice that of the average current $I_b = Ne/T_0$. Moreover, we already know that the voltage induced in a cavity is opposed to the driving current, such that for the current source $I_c$ in Figure 6.7 we obtain $I_c = -2I_b$. In the following paragraphs we will first investigate the steady-state, also called *continuous wave,* or CW-operation of beam with current $I_b$ interacting with a cavity. Later we consider the pulsed operation, normally encountered in linear accelerators.

## 6.6.2 Steady-state operation

From Figure 6.7 we see that both the generator current $I_g$ and the beam current $I_c = -2I_b$ drive a load that consists of the RLC circuit that represents the cavity and the load resistor $Z_0$ "seen" through the transformer with winding ratio $n$. Their combined impedance $Z_t$ is then given by $1/Z_t = 1/n^2 Z_0 + (1 + iQ\delta)/R$ at a reference point inside the cavity can then be expressed as

$$Z_t = \frac{R}{1 + \beta + iQ_0\delta} , \qquad (6.44)$$

where the detuning $\delta$ is defined in Equation 6.20 and we use the definition of the coupling factor $\beta = R/n^2 Z_0$ to simplify the equation. The voltage $V_c$ in the cavity that is seen by the beam must be $V_c = \hat{V}e^{i\phi_s}$ with the peak cavity voltage $\hat{V}$ and synchronous phase angle $\phi_s$. Since the generator current $I_g$ is stepped down with the winding ratio $n$, we obtain

$$\hat{V}e^{i\phi_s} = Z_t \left( \frac{I_g}{n} e^{i\psi_g} - 2I_b \right) = \frac{R\left( (I_g/n)e^{i\psi_g} - 2I_b \right)}{1 + \beta + iQ_0\delta} , \qquad (6.45)$$

which determines the generator current $I_g$ and phase $\phi_g$. Note that for negligible beam current, and operating on the resonance frequency with $\delta = 0$, we have $\hat{V}e^{i\phi_s} = R(I_g/n)e^{i\psi_g}/(1+\beta)$ such that the shunt impedance $R$ of the cavity determines the required current to be $I_g/n = (1+\beta)\hat{V}/R$ and the generator phase must be equal to the synchronous phase: $\psi_g = \phi_s$. However, with non-negligible beam current, these values will change, and we will have to adjust other parameters, such as the coupling $\beta$ and the detuning $\delta$ in order to minimize the reflections to the generator.

To determine the reflections, we need to know the impedance $Z_L$, shown in Figure 6.7 that the generator "sees," but with the beam current taken into account. To find $Z_L$, we again use Equation 6.17 with non-zero $I_c$, and find the impedance $Z_L$ to be

$$Z_L = \frac{V_L}{I_L} = \frac{(R/n^2)\hat{V}e^{i\phi_s}}{(1 + iQ_0\delta)\hat{V}e^{i\phi_s} - 2I_bR} = \frac{\beta Z_0\hat{V}e^{i\phi_s}}{(1 + iQ_0\delta)\hat{V}e^{i\phi_s} - 2I_bR} \quad (6.46)$$

instead of the zero-current value given by Equation 6.19. This impedance is perfectly matched to the load impedance $Z_0$ of the generator, provided $Z_0 = Z_L$ and this requirement leads to a condition for the coupling $\beta = 1 + iQ_0\delta - (2I_bR/\hat{V})e^{-i\phi_s}$. Since the coupling must be real-valued, we obtain the following conditions for its real and imaginary part

$$\beta = 1 + \frac{2I_bR}{\hat{V}}\cos\phi_s \quad \text{and} \quad Q_0\delta = -\frac{2I_bR}{\hat{V}}\sin\phi_s \; , \quad (6.47)$$

which indicates that a non-zero beam current $I_b$ causes both the matched coupling $\beta$ and the detuning $\delta$ of the cavity from resonance differ from the values found in Section 6.4.3, which were valid in the limit of negligible beam current. Realizing that the power transferred to the beam $P_b$ is given by $P_b = I_b\hat{V}\cos\phi_s$ and the power dissipated in the cavity $P_c$ can be written as $P_c = \hat{V}^2/2R$, we can express the optimum coupling parameter as $\beta = 1 + P_b/P_c$. Moreover, in the literature the detuning $\delta$ is usually expressed in terms of the *detuning angle* $\psi$ by $\tan\psi = Q_L\delta = Q_0\delta/(1 + \beta)$. Inserting into Equation 6.47 then leads to

$$\tan\psi = -\frac{\beta - 1}{\beta + 1}\tan\phi_s \quad \text{with} \quad \beta = 1 + \frac{P_b}{P_c} \quad (6.48)$$

for the values of the detuning angle $\psi$ and the coupling $\beta$ that match the generator to the cavity with beam loading. Here we operate with a current $I_b$ and synchronous phase angle $\phi_s$.

Now that we know the values of $\beta$ and $\delta$, or equivalently $\psi$ that minimize reflections, we can insert them in Equation 6.45 and solve for the generator current $I_g$ and phase $\psi_g$ that provide the required operating conditions

$$\frac{I_g}{n}e^{\psi_g} = (1 + i\tan\psi)\frac{(1 + \beta)\hat{V}e^{i\phi_s}}{R} + 2I_b \quad (6.49)$$

with $\beta$ and $\tan\psi$ from Equation 6.48. These last two equations specify the parameters of the RF system, such as coupling $\beta$ and the generator current $I_g$ in terms of parameters that come from beam physics requirements, the beam current $I_b$, the peak voltage $\hat{V}$, and the synchronous angle $\phi_s$. As a corollary these equations also show that varying beam currents spoil the matching and lead to reflections.

This change of beam current is maximum at the time of injecting the beam into an accelerator and leads to what is called *transient beam loading*, the topic of the next section.

## 6.6.3  Pulsed operation and transient beam loading

This abrupt change of beam current typically occurs when injecting a beam into a storage ring or when operating a high-intensity linear accelerator in pulsed mode; the European Spallation Source ESS may serve as an example. In such cases, the RF-system starts to fill the cavities up to a certain level before the beam is injected and is accelerated by extracting power from the cavities.

Since we intend to understand the transient behavior of the system, we need to consider the time-dependence of the currents flowing through each of the circuit elements from Figure 6.7

$$\frac{I_g}{n} - 2I_b = \frac{V_c}{n^2Z_0} + \frac{V_c}{R} + \frac{1}{L}\int V_c dt + C\frac{dV_c}{dt} = \frac{1 + \beta}{R}\left[V_c + \hat{\omega}Q_L\int V_c dt + \frac{Q_L}{\hat{\omega}}\frac{dV_c}{dt}\right] \quad (6.50)$$

where we use the relations $\hat{\omega}^2 = 1/LC$, $Q_0 = R\sqrt{L/C}$, and $Q_L = Q_0/(1+\beta)$ that we already encountered in Section 6.2 to simplify the notation. In deriving Equation 6.50 we moved the line impedance $Z_0$ and the generator current $I_g$, shown in Figure 6.7, to a reference point inside the cavity, which lies on the "other side" of the transformer. This accounts for the factor $n^2$ for the impedance $Z_0$ and the factor $n$ for the current $I_g$.

For given time-dependent drive-currents for generator and $I_g$ beam $I_b$ Equation 6.50 describes the response of the cavity, expressed by the voltage $V_c(t)$ and an arbitrarily fast time-scale where every oscillation is visible. It is, however, more instructive to factor out the fast oscillations $e^{i\omega t}$ and only consider the time-dependence of a slowly varying amplitude functions $I_g = I_{gs}e^{i\omega t}$ and $I_c = I_{cs}e^{i\omega t}$ for the currents and likewise for the cavity voltage $V_s(t)$ with $V_c(t) = V_s(t)e^{i\omega t}$. Using the latter expression to evaluate the derivative and integral over $V_c$ we find

$$\frac{dV_c}{dt} = e^{i\omega t}\left[i\omega V_s + \frac{dV_s}{dt}\right] \quad \text{and} \quad \int V_c dt \approx \frac{e^{i\omega t}}{\omega^2}\left[-i\omega V_s + \frac{dV_s}{dt}\right] \tag{6.51}$$

where we use partial integration twice and then neglect a term with $d^2V_s/dt^2$ to obtain the second expression. Inserting into Equation 6.50 and after canceling the common exponential factor, we arrive at

$$\frac{I_{gs}}{n} - 2I_{bs} = \frac{1+\beta}{R}\left[\left(1 + iQ_L\left(\frac{\omega}{\hat{\omega}} - \frac{\hat{\omega}}{\omega}\right)\right)V_s + \left(\frac{Q_L\hat{\omega}}{\omega^2} + \frac{Q_L}{\hat{\omega}}\right)\frac{dV_s}{dt}\right]$$

$$\approx \frac{1+\beta}{R}\left[(1 + iQ_L\delta)V_s + \tau_f\frac{dV_s}{dt}\right] \tag{6.52}$$

with $\delta$ from Equation 6.20 and the cavity-filling time $\tau_f \approx 2Q_L/\hat{\omega}$ where we assume that the drive frequency $\omega$ is close to the resonance frequency $\omega \approx \hat{\omega}$.

Finding the response $V_s$ to the current stimuli from the ordinary linear differential equation in Equation 6.52 is easily accomplished by introducing the Laplace-transforms of the involved function, for example

$$\tilde{V}_s(s) = \int_0^\infty e^{-st}V_s(t)dt \tag{6.53}$$

and likewise for the generator and beam currents. Since the derivative with respect to time $t$ is replaced by a multiplication with $s$, we obtain, after solving for $\tilde{V}(s)$

$$\tilde{V}_s(s) = \frac{R/(1+\beta)}{1 + iQ_L\delta + \tau_f s}\left(\frac{\tilde{I}_{gs}(s)}{n} - 2\tilde{I}_{bs}(s)\right). \tag{6.54}$$

Instead of finding and inserting Laplace transforms of the time-dependent currents in Equation 6.54 and then, after finding the inverse Laplace-transform of the resulting expression, we use MATLAB, because it provides powerful functions to analyze systems, similar to the one in Equation 6.54.

In MATLAB, we represent the transfer function of the system, that is often given by the quotient of two polynomials in $s$, by providing the two polynomials as input to the function `tf()`. It returns an object `sys` that represents our system and that can be used as input to powerful simulation functions. Note that polynomials in MATLAB are represented as arrays of the coefficients with the highest power as the first element. The transfer function $T(s) = [R/(1+\beta)]/(1 + iQ_L\delta + \tau_f s)$ from Equation 6.54 is thus specified by

```
sys=tf(R/(1+beta),[tauf,1+i*QL*delta]);
```

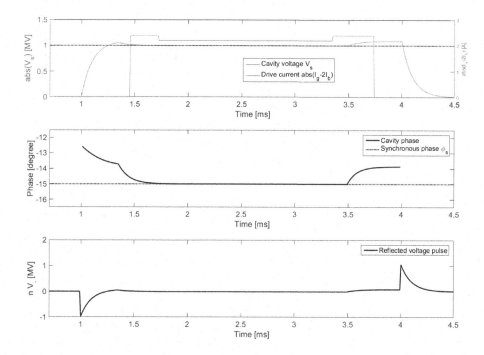

**Figure 6.11** The upper plot shows the sum of generator and beam current $|I_g - 2I_b|$ (dashed) and the cavity voltage $|V_s|$. We start filling the cavity at $t = 1$ ms and the cavity voltage increases determined by the filling time $\tau_f$. The beam is injected between $t = 1.35$ ms and 3.5 ms. The middle graph shows the phase of the cavity voltage (solid) and the desired synchronous phase $\phi_s$ (dotted). The lower graph shows the voltage pulse that travels back to the generator.

where we assume that the constants are defined and set to the appropriate values previously. The current stimuli we can provide as two arrays t and u which contain the times t and corresponding values u at the time steps. We fill the latter with the sum of the two currents $u = I_{gs}(t)/n - 2I_{bs}(t)$. Note that we specify the transfer function as the Laplace transform with independent variable $s$, whereas the stimulus $u$ is given as a function of time. MATLAB conveniently handles all translations between the different representations internally. With both system and stimulus defined, we simulate the system with the command

```
Vs=lsim(sys,u,t);
```

which returns the cavity voltage $V_s$ at time steps $t$.

Figure 6.11 shows the results of a simulation with cavity peak voltage $\hat{V} = 1$ MV, shunt resistance $R = 1$ M$\Omega$, synchronous phase $\phi_s = -15$ degrees, cavity filling time $\tau_f = 0.1$ ms, and beam current $I_b = 100$ mA. Beforehand, we adjust the coupling $\beta$ and tuning angle $\tan \psi = Q_L \delta$ to the values given by Equation 6.48 and use these values to determine the generator current and phase from Equation 6.49. In the simulation shown in Figure 6.11, the generator current is turned on after 1 ms, which causes the cavity voltage $V_s$ to increase with a time constant given by the cavity filling time $\tau_f$. The cavity phase, shown in the middle graph differs from the desired value, because the generator phase $\psi_g$ was optimized for $I_b = 100$ mA, but no beam is present between 1 and 1.35 ms. After 1.35 ms, however,

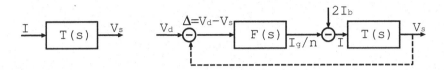

**Figure 6.12** On the left a transfer function, represented by its Laplace transform, that maps a current to the cavity voltage is shown on the left. On the right-hand side a closed-loop feedback controller $F(s)$ is added.

the beam is injected a little too late, and, at this time the peak field in the cavity already exceeds the desired value of 1 MV, but is reduced to the correct value that is indicated by the dotted line, again on a time scale given by $\tau_f$. Simultaneously, the cavity phase approaches the required value of $-15$ degrees, as indicated by the dotted line in the lower graph. After 3.5 ms the beam current stops, but the generator keeps providing current, which then causes cavity voltage and phase to deviate from their desired values.

The detuning angle $\tan\psi$ and coupling $\beta$, used in Figure 6.11, were determined by matching the impedance of the transmission line $Z_0$ to that of the cavity $Z_L$, which was given by the ratio $Z_L = V_L/I_L$ in Figure 6.7. But Figure 6.11 shows that the voltage $V_L = V_s/n$ varies with time, and, in particular, during the start and end of the pulse, when the impedance is different from the matched value. And this causes a voltage pulse $V_-$ to travel back towards the generator. The current $I_L$ is given by $I_g$ without the part that is dissipated in $Z_0$, such that $I_L = I_g - V_L/Z_0 = I_g - n\beta V_s/R$ and the voltage is $V_L = V_s/n$. For the reflected voltage pulse we obtain $nV_- = n(V_L - Z_0 I_L)/2 = V_s - (R/2\beta)I_g/n$, which is shown in the lowest graph in Figure 6.11. Clearly visible are the reflections at the start and end of the pulse. Circulators are often installed in the transmission line to direct the reflections to a load in order to protect the generator.

In the simulation the deviation of cavity voltage and phase from their desired values can be reduced by measuring the actual values with a second antenna, connected to the cavity, comparing this signal to the desired values, and using the difference to adjust the generator current and phase in order to minimize the difference. This regulation or control system is often called the *low-level RF* system, or LLRF, and is the topic of the next section.

### 6.6.4 Low-level RF system

In the previous section we pre-calculated the generator current and phase beforehand and adjusted the coupling $\beta$ and detuning angle $\tan\psi = Q_L\delta$ in Equation 6.54 accordingly. Then we just operated the generator to provide the current that will result in the desired cavity voltage and angle after the beam is injected. This situation corresponds to the left-hand image in Figure 6.12. Some current $I$ excites a dynamical system represented by a transfer function $T(s)$, defined on page 168, and that produces a voltage $V_s$ according to Equation 6.54. The situation, where the measured cavity voltage is used in a *feedback process* [35], is shown on the right-hand side in Figure 6.12. The dashed line symbolizes the measured cavity voltage $V_s$ picked up in the cavity with a small antenna. It is then used to calculate the difference signal $\Delta = V_d - V_s$ between the desired cavity voltage $V_d$ and $V_s$. This difference is passed to a, so far unspecified, transfer function $F(s)$, which represents the *feedback controller* that produces the generator current $I_g/n$ as output. After subtracting the effect of the beam current, we obtain a current $I = I_g/n - 2I_b$, which is passed to the transfer function $T(s)$ that represents the cavity. Its output is, of course, the cavity voltage $V_s$.

Instead of the transfer function $T(s)$ that maps input current to output voltage, we now want to find the transfer function $H(s)$ that maps the desired voltage $V_d$ to the cavity voltage $V_s$. If we tacitly assume that both the controller $F(s)$ and the cavity $T(s)$ are given by their Laplace-transforms, we have

$$V_s = T(s)\left[F(s)(V_d - V_s) - 2I_b\right] \quad \text{or} \quad V_s = \frac{T(s)}{1 + F(s)T(s)}\left[F(s)V_d - 2I_b\right] . \quad (6.55)$$

If $F(s)$ and $T(s)$ are given by quotients of polynomials in the Laplace-variable $s$, finding $H(s) = T(s)/(1 + T(s)F(s))$ only requires the multiplication of polynomials, provided that we know the polynomial for $F(s)$. Since the output cavity voltage $V_s$ is fed back to the input and compared with the desired voltage $V_d$, the transfer function $H(s)$ is called a *closed-loop* controller. One of the commonly used types of feedback processes is the PID-controller [35], defined by

$$F(s) = K_p + K_i/s + K_d s \quad (6.56)$$

where $K_p$ is a *proportionality* term, $K_i$ describes *integration* in Laplace-space, and $K_d$ *differentiation*. Finding suitable feedback coefficients $K_p, K_i$, and $K_d$ is covered in the literature discussing the speed, robustness, and stability of such feedback systems. Here we investigate a simple proportional regulator, where only $K_p$ is non-zero. This leads to the following expression for the closed-loop transfer function $H(s)$

$$K_p H(s) = \frac{g/(1+g)}{1 + i\tan(\psi)/(1+g) + \tau_f s/(1+g)} \quad \text{with} \quad g = \frac{K_p R}{1 + \beta}, \quad (6.57)$$

which has the same structure as the transfer function $T(s)$, except that the detuning is reduced by the factor $(1+g)$ and, more importantly the time constant for filling the cavity $\tau_f$ is reduced by the same factor. Increasing $g$ by increasing the feedback constant $K_p$ will thus result in a quicker response of the cavity voltage $V_s$ to required changes, albeit at the expense of large generator currents $I_g/n$, which appears as an intermediate variable in the feedback diagram on the right-hand side in Figure 6.12. This may in practice be impossible and will therefore limit the performance of the feedback system. A further limitation of the simple proportional controller is its inability to regulate $V_s$ exactly to $V_d$. This is easy to see by considering the steady state that is reached after a long time. It is characterized by vanishing temporal derivatives, or equivalently, by setting $s$ in the transfer function $H(s)$ to zero [35]. Apart from a small imaginary part, that we can ignore for the moment, we have in equilibrium $K_p H(0) = g/(1 + g)$ and thus $V_s = gV_d/(1 + g)$. Only for large feedback gain $g$ the cavity voltage $V_s$ will properly track the desired value $V_d$. A remedy to achieve proper tracking is to add an integral controller with $K_i$ to the feedback, which, on the other hand, may lead to oscillations and those can be remedied by adding a suitable differential contribution $K_d$. Implementing different controllers and coding the respective transfer functions $H(s)$ in MATLAB with the tf() function to investigate their performance is left as an exercise.

We point out that we discussed a linear controller that operates on complex voltages, rather than real values, that are more accessible experimentally. In practice, the slowly varying signals, such as $V_s$, are obtained by down-mixing the fast-oscillating signals, picked up by the antenna in the cavity, with two signals from a local oscillator to lower frequencies. Using two signals from the oscillator, with a relative phase-shift of 90 degrees, one obtains a cosine-like *in-phase* and a sine-like *quadrature* signals that are equivalent to the complex representation used earlier in this section. Moreover, we only discussed controlling the cavity amplitude and phase, but other variables, such as the detuning angle $\tan\psi$, which depends on the cavity resonance frequency $\hat{\omega}$ may change dynamically due to the Lorentz force detuning mentioned in Section 6.5.2. The feedback to compensate this, mentioned there, is part of the LLRF system. Some of the changes that are known can be compensated by pre-programming the feedback set-points, such as $V_d$, which is called *feed-forward* control.

After discussing the hardware to accelerate the beams we turn to figuring out how the beam behaves in the accelerator and that is the task of the *instrumentation and diagnostics* system.

## QUESTIONS AND EXERCISES

1. What is the cutoff frequency for a rectangular waveguide with $a = 2b = 10\,\text{cm}$ (a) for a $TE_{10}$ mode; (b) for a $TE_{01}$ mode?

2. Argue, why square waveguides are very unusual.

3. Display the longitudinal magnetic field $H_z$ of the following modes using MATLAB's `surf()` function: (a) $TE_{00}$; (b) $TE_{10}$; (c) $TE_{20}$.

4. Calculate the transverse fields $H_x, H_y, E_x$, and $E_y$: (a) for $TE_{10}$ mode; (b) for a $TE_{20}$ mode and display them using MATLAB's `surf()` and `quiver()` functions.

5. How does the cutoff frequency of the fundamental TE–mode in a WR-340 waveguide change, if the top right corner is dented. You can assume that the dent cuts off an equilateral triangle at half the height, as shown here ◣. Simulate this geometry with the PDE toolbox and show the transverse fields $H_x, H_y, E_x$, and $E_y$ in the way shown in Figure 6.1.

6. Use the PDE toolbox to calculate the TM modes in a triangular waveguide with equal sides having a length of 10 cm. What are the cutoff frequencies? Use `pdesurf()` to display $E_z$ for the two modes with the lowest cutoff frequencies.

7. Calculate the impedance of a network with a $Z_0$ series impedance, and two blocks of $1/n$ transformers with output shunted by $Z_0$ in series.

8. Network analyzers can measure the reflection coefficient $\Gamma = S_{11}$ and display the absolute value of $S_{11}$ in decibels. To better understand these measurements, prepare plots showing $20\log_{10}(|\Gamma|)$ as a function of frequency $f = \omega/2\pi$ (a) for a normal-conducting cavity with $Q_0 = 50000$ for values of the coupling $\beta = 0.05, 0.5, 1, 2$. (b) Repeat for a super-conducting cavity with $Q_0 = 10^9$.

9. You are worried that the barometric pressure buckles the vertical faces of the cavity in Figure 6.8 and detunes the cavity. Assume that the points at $R/2$ are displaced by 1 to 5 mm and determine the change of frequency with indentation.

10. For shunt impedance $R = 1\,\text{M}\Omega$ and cavity voltage $\hat{V} = 1\,\text{MV}$ calculate the required generator current $I_g$ and phase $\psi_g$ as a function of the beam current $I_b$ when operating the system in steady-state at the synchronous phase $\phi_s = -15°$. (a) What are the values $I_{g0}/n$ and $\psi_{g0}$ in the zero-current limit? (b) Plot the normalized generator current $(I_g/n)/(I_{g0}/n)$ and the generator phase $\phi_g$ as a function of the normalized beam current $I_b/(I_{g0}/n)$. (c) Plot the corresponding values for the coupling $\beta$ and detuning angle $\psi$.

11. Consider the `transient_beamloading.m` MATLAB script that was used to prepare Figure 6.11 and implement the P-controller, discussed in Section 6.6.4, in the simulation.

# CHAPTER 7

# Instrumentation and Diagnostics

In the previous chapters we discussed how to guide and accelerate particles and in this chapter will discuss devices and methods to measure the intensity of the beams, their position, and their sizes. In Chapter 2 we found that the distribution of particles in beams is usually described by their moments, and we often use their zeroth, first, and second moments to approximate the distribution by a Gaussian. Therefore, let us now turn to instruments to measure the moments and start with devices to measure the total charge, equivalently, the beam current.

## 7.1 ZEROTH MOMENT: CURRENT

The simplest method to measure the beam current, provided the beam power is sufficiently low, is to just dump the beam into a metal block that is connected via an Ampere-meter to ground. If positively charged ions are diagnosed electrons from the absorber material will rush towards the ions and try to compensate their charge, thus constituting an electronic current that is measured. Just using a metallic block is not very accurate and the more refined version, called *Faraday cup*, shields secondary electrons by magnetic fields or biasing the absorber with positive voltages. A sketch is shown on the top of the left-hand side in Figure 7.1. If very short particle bunches or other rapidly changing currents need to be analyzed, careful attention must be paid to stray capacitances and inductances. At higher beam currents and beam powers, the absorber must be cooled and one must ensure that the beam is really stopped inside the absorber. This type of beam current measurement device is invasive, because it stops the beam. Therefore, Faraday-cups are often retractable and only inserted into the beam path if the current needs to be measured.

Another device to measure the current is a *wall-gap monitor*, shown on the bottom of the left-hand side in Figure 7.1. It is based on measuring the image currents in the beam pipe. The vacuum chamber shields the magnetic fields that the beam generates, such that no magnetic fields are detectable outside the pipe. Inside the vacuum chamber the magnetic field is given by Ampere's law $\oint \vec{B}\vec{dl} = \mu_0 I_{beam}$, which implies that in the vacuum chamber a current propagates, equal in magnitude, but with the opposite polarity. This wall current can be measured by inserting a ceramic in the beam pipe and bypassing the ceramic with resistors such that the wall currents are forced to travel through the resistors. Across these resistors a voltage drop develops, which is measured with a sensitive voltmeter. Provided that careful attention is paid to shielding and electrical design, very fast signals into the

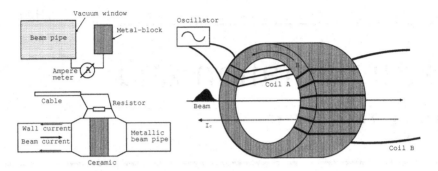

Figure 7.1   A Faraday cup is based on stopping the beam in a target and measuring the current flowing back to ground (top left). A wall-gap monitor (bottom left) blocks the image currents with a ceramic and measures the voltage drop across a bypass-resistor. The operation of current transformers (right) is explained in the text.

GHz range can be resolved. They correspond to a time resolution even below the nano-second time scale. Note that wall-gap monitors do not intercept the beam and are therefore non-invasive current measurement systems.

In the wall-gap monitor we picked up the image currents with a resistor, but we can also directly detect the magnetic field of the beam by placing a circular ferrite core encircling beam inside the beam pipe. A bunched beam induces a temporally varying flux inside the ferrite core that can be detected by winding several turns of wire around it, which act as secondary windings of a transformer, hence the name *current transformer*. This is illustrated on the right-hand side in Figure 7.1, where a bunch, passing from left to right through the toroidal ferrite core, induces a voltage in coil B, that is subsequently amplified and made visible on, for example, an oscilloscope.

Note that current transformers are sensitive to the magnetic field from temporally changing beam currents, which works well for bunched beams, but does not work for un-bunched, so-called coasting beams. They are common in nuclear physics rings that are equipped with an electron cooler. A device to measure the average or DC component of the current is the *direct current charge transformer* or *DCCT*. The operation method is also illustrated on the right-hand side in Figure 7.1. Similar to the normal beam transformer discussed in the previous paragraph, a ferrite toroid encircles the beam. In a DCCT, two coils are wound onto the ring. Coil A is coupled to an oscillator which excites the toroid with a sinusoidal current at a frequency of typically 10 kHz, such that the magnetic field inside the ferrite saturates. The voltage that is induced in coil B then contains not only the fundamental harmonic from the oscillator, but also harmonics at higher frequencies. Since the hysteresis curve is anti-symmetric, only odd harmonics are present. If, on the other hand, a beam current is passing through the toroid, the hysteresis curve will be shifted upwards or downwards and will not be asymmetric anymore. Therefore, even harmonics appear in coil B. If a second wire, carrying a current $I_c$, is passed through the toroid, and adjusted to cancel the second harmonic, we can very accurately measure the compensating current. The measurement is thus based on a null-measurement, which is typically very accurate. Note, however, that the bandwidth of the DCCT is limited by the exciting frequency of the driving oscillator.

Now that we know the number of particles in the beam, let us find out where they are and when they arrive.

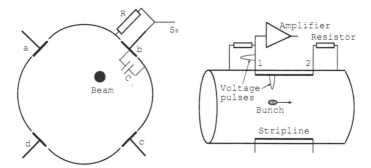

Figure 7.2   A button beam-position monitor (left) intercepts the image currents flowing in the beam pipe and a strip-line monitor (right) detects the voltages induced by the magnetic field, which accompanies the beam, in the small area between the strip-line and the beam-pipe walls.

## 7.2   FIRST MOMENT: BEAM POSITION AND ARRIVAL TIME

Most position monitors are based on the detection of the electro-magnetic fields that the beam produces. In the beam's frame of reference, the charges are at rest and the electric field a particle generates is purely radial and given by $\vec{E}^b = (E^b_{x'}, E^b_{y'}, E^b_{z'}) = e\vec{r}'/4\pi\varepsilon_0|\vec{r}'|^3$, where $\vec{r}'$ is the vector pointing from the particle that creates the field to the observation point of the field. Since the particle is at rest in its own rest frame, it does not create a magnetic field. In the laboratory frame, which moves with relative speed $\beta = v/c$ with respect to the beam's rest frame, we need to Lorentz-transform the fields and find for their components in the lab frame

$$E_x = \gamma E^b_{x'}, \quad E_y = \gamma E^b_{y'}, \quad E_z = E^b_{z'}, \quad B_x = -\beta\gamma E^b_{y'}/c, \quad B_y = \beta\gamma E^b_{x'}/c, \quad B_z = 0,$$
(7.1)

where we assume that the beam moves in the $z$–direction. We see that the field components in the direction of motion remain unaffected, but the transverse components increase by the relativistic factor $\gamma = 1/\sqrt{1-\beta^2}$, such that in the ultra-relativistic limit the fields are predominantly transverse. Moreover, the magnitude of the electric and magnetic components are approximately equal with $|E_x| \approx |cB_y|$ and $|E_y| \approx |cB_x|$, such that the fields in the laboratory frame closely resemble a TEM-wave.

Since the fields are purely transverse in the ultra-relativistic limit, we can determine them from two-dimensional calculations of quantities that obey Maxwell's equations and by solving an equivalent planar electro-static problems. First, we consider so-called button beam-position monitors, shown on the left-hand side in Figure 7.2. They consist of four small isolated button electrodes, mounted in the vacuum chamber, that pick up part of the wall currents. The buttons act similar to capacitors that are charged by the wall currents. The BPM therefore only measures the AC component of the beam current, which makes it especially suitable for short bunches. Since the buttons are small and intercept only a small portion of the fields, they are especially suitable for intense bunches. The positions can be deduced from the signals in the four buttons in the following way

$$x = k_x \frac{(S_a + S_d) - (S_b + S_c)}{S_a + S_b + S_c + S_d} \quad , \quad y = k_y \frac{(S_a + S_b) - (S_c + S_d)}{S_a + S_b + S_c + S_d}$$
(7.2)

where $k_x$ and $k_y$ depend on the geometry of the beam pipe where the BPM is located.

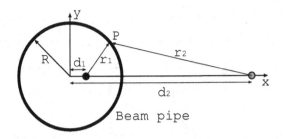

**Figure 7.3** The geometry to calculate the fields in the presence of a beam pipe.

For a round beam pipe the proportionality constants $k_x$ and $k_y$ can be analytically calculated from the normal components of the electric field on the surface of the beam pipe. This normal component draws charges from the conducting material and accumulates them on the wall nearest to the beam. These charges co-propagate with the beam and constitute the wall currents. The normal component of the electric field, or, equivalently, the wall currents at the locations of the buttons are proportional to the signals $S_a, S_b, S_c$, and $S_d$ in Equation 7.2. We calculate them using the method of image charges by considering the geometry in Figure 7.3. We note that the potential of a point charge in two dimensions $z = x + iy$ is given by $V(x, y) = \ln(z)/2\pi i\varepsilon_0$, in analogy to the potential of filament currents, we found in Section 4.2, such that the fields are $E_y + iE_x = 1/2\pi\varepsilon_0 z$. The contribution of a charge, displaced by $d_1$ from the center of the pipe, to the electric potential at point $P$ is $\log(r_1)/2\pi\varepsilon_0$ and a similar expression for the contribution from the image charge with opposite polarity. The sum of these expressions determines the total potential $V(x, y)$ at any point $x + iy = Re^{i\theta}$ on the beam pipe

$$V(x, y) = \frac{1}{4\pi\varepsilon_0} \ln\left[\frac{(x - d_1)^2 + y^2}{(x - d_2)^2 + y^2}\right] = \frac{1}{4\pi\varepsilon_0} \ln\left[\frac{d_1^2 \left(1 + (R/d_1)^2 - 2(R/d_1)\cos\theta\right)}{R^2 \left(1 + (d_2/R)^2 - 2(d_2/R)\cos\theta\right)}\right] \quad (7.3)$$

and we observe that the expressions with the angle $\theta$ cancel, provided that we have $R/d_1 = d_2/R$. Thus, if we place the image charge at $d_2 = R^2/d_1$ the beam pipe is an equipotential surface because the potential does not depend on $\theta$ anymore.

The combined electric field $\hat{E} = \hat{E}_y + i\hat{E}_x$ at a point on the beam pipe we calculate by adding the contributions from the two charges $E_y + iE_x = \lambda/2\pi\varepsilon_0 i(z - d_j)$, where $\lambda$ is the longitudinal charge density that is related to the beam current by $I = \lambda\beta c$. For $\hat{E}$ we find

$$\hat{E}_y + i\hat{E}_x = \frac{\lambda}{2\pi\varepsilon_0 i} \left[\frac{1}{(x - d_1) + iy} - \frac{1}{(x - d_2) + iy}\right] = \frac{\lambda}{2\pi\varepsilon_0 R} \left[\frac{-ie^{-i\theta}(R^2 - d_1^2)}{R^2 + d_1^2 - 2Rd_1\cos\theta}\right] \quad (7.4)$$

where we used $d_2 = R^2/d_1$. We see that $\hat{E}$ is horizontal at $\theta = 0$ and points in the vertical direction at $\theta = \pi/2$. Moreover, the field is normal to the surface at every point on the beam pipe, as it should. Note that placing the charges on the horizontal axis does not restrict the generality, because the geometry is rotationally symmetric. The field of a charge, located at $d_1(\cos\phi + i\sin\phi)$ is given by the expression in Equation 7.4 with $\theta$ replaced by $\theta - \phi$.

Since the signals $S$ on the four buttons in the BPM shown in Figure 7.2 are proportional to the field at the location of the button, we use Equation 7.4 to calculate them. For simplicity, we assume that the beam is only displaced horizontally by $d_1 \ll R$ and the buttons are located at $\theta = \pi/4, 3\pi/4, 5\pi/5$ and $7\pi/4$. Keeping only linear terms in Equation 7.4 we

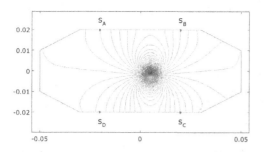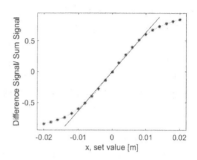

Figure 7.4 The field pattern in an octagonal BPM with four electrodes indicated by asterisks in the left-hand side. The figure on the right-hand side shows the derived signal $(S_b + S_c - S_a - S_d)/(S_a + S_d + S_b + S_c)$ while moving the test charge along a horizontal line in the mid-plane of the BPM.

find that the signals are approximately given by $S = 1 + 2(d_1/R)\cos\theta$, such that we obtain

$$\frac{S_a + S_d - S_b - S_c}{S_a + S_d + S_b + S_c} = \frac{-(8/\sqrt{2})d_1/R}{4} = -\frac{d_1}{R/\sqrt{2}} \,. \tag{7.5}$$

Comparing with Equation 7.2 we see that we have $|k_x| = R/\sqrt{2}$. In a similar fashion, we can determine $|k_y| = R/\sqrt{2}$. The response of the BPM is only linear for small deviations $d_1$ from the center of the beam pipe. For large deviations and general positions $d_1(\cos\phi + i\sin\phi)$, we can still use Equation 7.4 with $\theta$ replaced by $\theta - \phi$ to calculate the fields on the buttons and generate a linearity map of the BPM response, but we leave this as an exercise.

For round beam pipes it was possible to analytically calculate the fields on the beam pipe using the method of image charges. If, on the other hand, the geometry is more complicated, we have to resort to numerical methods. The MATLAB PDE toolbox allows us to calculate the two-dimensional fields from a point charge in a beam pipe that is defined by a polygon. Such a MATLAB script closely resembles those to calculate magnetic fields. We first have to define the geometric objects, here the octagonal shape of the vacuum chamber and the round beam, and assemble them to a single geometric object. In the next step we define the boundary conditions, here Dirichlet conditions on the vacuum chamber, and then mesh the geometry. Next, we define the coefficients of the partial differential equation, m, d, c, a, and f in MATLAB's generic representation of PDEs, Equation 4.22, and solve the equation for the scalar potential. We find the electric field from calculating the gradient and then determine its value at the location of the electrodes, indicated by asterisks on the left-hand side in Figure 7.4, which also shows the field lines. The figure of the right-hand side in Figure 7.4 shows the signal that is proportional to the horizontal position $(S_b + S_c - S_a - S_d)/(S_a + S_d + S_b + S_c)$ as function of the real position while moving the test charge along a line in the mid-plane of the BPM. We see that the response of the BPM is only linear for positions below 1 cm and saturates for larger positions. This nonlinearity can, however, be taken into account numerically. For small deviations, we can determine the value of $k_x$ from a linear least squares fit to the points in the range $|x| < 1$ cm shown in Figure 7.4 as the solid straight line. The resulting $k_x$ in this case is $k_x = 1/63.2 = 0.0158$ m. The MATLAB file for the simulation can be found in Appendix B.5.

In button BPMs only a small portion of the fields, or wall currents, contribute to the signal. This implies that the signal induced in the buttons is rather small and makes these BPM mainly suitable for high-intensity beams. Button BPMs normally are the preferred choice for synchrotron radiation sources with their short high-intensity bunches. On the other hand, in accelerators with very low-intensity beams the sensing electrodes are made much larger and one often uses a diagonally sliced box inside the beam pipe. It has the advantage that the sense-electrodes are particularly large, making these BPM suitable for, for example, heavy-ion storage rings with low-intensity beams of rare isotopes. Furthermore, their signal-response from the transverse position is highly linear, even at large displacements, because the signal intensity is directly given by the fraction of the electrode exposed to the wall current, and that is linear due to the diagonally and linearly cut shoe-box. As in other position-sensitive devices the position is then deduced from ratio of the difference to the sum of the signals from the two electrodes. A sketch of such a shoe-box position monitor is shown on the left-hand side in Figure 7.5. Normally one box, which can either have rectangular or round cross-section is used for the horizontal and another one for the vertical direction.

Where button BPMs predominantly couple to the electric field that accompanies the bunches, couple *strip-line BPMs*, shown schematically on the right-hand side in Figure 7.2, to the accompanying magnetic fields. A conducting strip-line is mounted isolated on the inside of the vacuum chamber. Its width and height above the vacuum chamber determine its impedance and it must be properly matched to the terminating resistors, also shown in Figure 7.2, in order to avoid reflections. When the beam passes under the strip-line, part of the accompanying magnetic field enters the region between strip-line and beam pipe. The changing flux induces a voltage pulse, one of them traveling upstream towards port 1. A second pulse travels downstream along the strip-line, and parallel to the bunch, towards port 2. When the bunch arrives at port 2, it will induce a second voltage pulse with opposite polarity, because the enclosed flux now decreases. One half of this voltage pulse travels upstream towards port 1, where it will be recorded at a time, determined by the length of the strip-line. In the downstream port 2 the pulses from the beam entering and exiting the region with the strip-line cancel, provided the signal speed on the strip-line and the bunch are equal. Thus, we can record a signal on the upstream port 1, but the signals on the downstream port 2 cancel, which makes strip-line monitors directional; they can discriminate beams going one way from those going in the opposite direction. This capability makes them particularly useful near interaction points of colliding beams facilities with counter-propagating beams. The signals from four strip-lines arranged in much the same way as in button BPMs allows us to extract the beam position with the help of Equation 7.2.

Yet another position-sensitive device is a *cavity BPM* which is based on a resonating structure, a pillbox cavity is shown on the right-hand side in Figure 7.5 that supports a spatially anti-symmetric mode that has a zero on the beam axis, such as a $\text{TM}_{110}$-mode. We refer to Section 5.1 regarding the labeling of the modes. The geometric size of these BPMs is typically several cm, such that the frequencies are in the GHz range. If the beam is centered, it does not couple to this mode and does not excite it, but if it is off-center, the beam couples to the electric field and thereby excites the mode. The excited mode can then, in turn, excite the electrons in the antenna shown on the bottom, providing a means to detect the signal. The information whether the beam is too high or too low is obtained from comparing the phase of the signal from the antenna with a reference signal. Cavity BPMs can achieve very good position resolutions in the $\mu$m range and are, for example, used in the European XFEL, where the bunches in a bunch train are only separated by

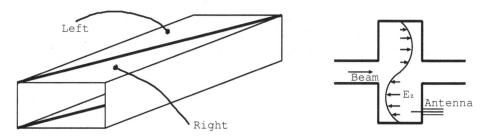

Figure 7.5 The large electrodes of a shoe-box BPM (left) allow us to detect the position of low-intensity beams, whereas cavity BPMs (right) allow us to detect the positions of very short and intense bunches.

about 220 ns. In this case, the modes in the cavity BPM are strongly damped in order to be able to independently determine the positions of successive bunches.

All of the above position monitors generate short electric pulses on their output ports. These pulses depend on the length of the bunches, but are also affected by the monitor itself. This can most easily be understood by realizing that, for example, a button of a button BPM has a capacitance $C$, shown as dashed symbol in Figure 7.2, against the surrounding beam pipe, which is charged by the passing beam, and discharged through the terminating resistor $R$. This setup constitutes a high-pass filter with cutoff frequency $\omega_c = 1/RC$ through which the signal from the beam is passed. In the next stage, often a *hybrid* is used to generate the sum and difference of the signals from the electrodes. A hybrid is a passive high-frequency device in which one of the signals from two input ports is inverted, and added to the input from the other. One output ports carries the sum and a second the difference of the input signals. After amplification and low-pass filtering the signals are digitized and all further processing is done digitally.

The averaged beam position of a long sequence of repetitively arriving bunches, as they occur in storage rings, generate signals at harmonics $\omega_n$ of the repetition frequency. A slowly varying position shows up as amplitude modulation. Extracting the slowly varying position information uses the *super-heterodyne* method, which is also used in old-fashioned radios. Here the signal from the BPM is mixed (multiplied) with an externally generated frequency $\tilde{\omega}$ from a *local oscillator* and this leads to the appearance of sum and difference frequencies $\omega_n + \tilde{\omega}$ and $\hat{\omega} = \omega_n - \tilde{\omega}$. After low- or band-pass filtering, only the *intermediate frequency* $\hat{\omega}$ is left. It still carries the information about the slowly varying position as modulation. Since the frequency $\hat{\omega}$ is much more slowly varying than the original, we can use a slower and less expensive digitizer. Moreover, by reducing the bandwidth, or equivalently, averaging over longer times, the position resolution is much improved. Systems to determine the closed orbit in rings often use these narrow-band methods.

Beam position monitors are arguably the most important diagnostic devices in any accelerator. They are used to verify that the beam passes through the center of the beam pipe, which minimizes beam losses. Moreover, connecting the raw difference signal from a position monitor to a spectrum analyzer, often in conjunction with exciting the beam with a short-pulsed kicker magnet. The analyzer directly shows the transverse betatron oscillation frequency around the harmonics of the revolution frequency, which can be used to determine the tune. This is only the crudest, but very direct method, to determine an equally crude estimate of the tune, and more elaborate methods are available. But more on that in later sections.

After the instrumentation for position measurements let us briefly discuss diagnostics to determine the beam energy and the arrival time. The average energy of the beam can be inferred from its position on a position monitor or a screen at a location with a large dispersion function $D_s$. If, in the latter case, the beta function $\beta_s$ at the screen is deliberately made small, the energy sensitivity is maximized provided the figure of merit, given by $D_s/\beta_s$, is maximized.

The arrival timing of bunches in the range up to a little better than a nano-second can be done by observing the sum signal from a position monitor on an oscilloscope with an analog bandwidth of several GHz. The timing requirements for modern free-electron lasers are, however, four to six orders more stringent. They lie in the femto-second range and we need to employ optical methods, inspired by developments in the laser community. This is additionally attractive, because the ultra-fast synchronization is needed to facilitate pump-probe experiments of the FEL pulse in conjunction with an external laser. One method to synchronize an electron beam with an external laser is based on lengthening a very short (fs range) laser pulse with an optical grating, which leads to a long (ps range) light pulse with a longitudinal correlation of wavelength inside the laser pulse, called *chirp*. On a grating spectrometer the laser pulse then shows a very broad spectrum, with short wavelength on one side and long on the other. Coupling to the electron bunch is achieved by placing an bi-refringent crystal between a crossed pair of a polarizer and an analyzer, such that no laser pulse actually reaches the spectrometer. The electric field of a passing bunch affects the crystal in such a way that it rotates the polarization of the simultaneously passing laser pulse, such that only a short fraction of the lengthened laser pulse now passes the second polarizer. The relative arrival time between electrons and lasers is encoded in the part of the spectrum visible on the spectrometer on a shot-by-shot basis.

After this brief overview of diagnostics of the first moment we turn to beam-size measurements.

## 7.3   SECOND MOMENT: BEAM SIZE

We now turn to the discussion of devices to measure the transverse and longitudinal size of the beam. Of course, the most straightforward way of measuring the transverse beam size is by placing a luminescent screen into the beam and observing the emitted light with a video camera and digitizing the picture on a computer as is shown on the left-hand side in Figure 7.6. This is obviously an invasive measurement, especially at low energies, where the beam is absorbed in the screen. There are often problems with these type of screens due to blind spots where the screen material is not as effective as on the surrounding area. Limited dynamic range and saturation of the intensity are other problems as well as cameras failing due to high radiation doses.

A variant of the luminescent screen are screens based on *optical transition radiation* (OTR) where the screen is made of a very thin foil of, for example $Al_2O_3$, which disturbs some beams, such as high-energy electron beams, only modestly. The beam pushes the electrons in the screen out of the way and after the beam has passed the electrons relax back to their original positions. These electrons experience an acceleration and therefore emit electro-magnetic radiation, which is then picked up by a camera, possibly after being transported in an optical beam line made of lenses and mirrors.

Another monitor is based on the ionization of very thin, typically a few $\mu$m, thick wires, as illustrated on the right-hand side in Figure 7.6, which is called a *secondary emission monitor* (SEM) grid. The electrons are knocked out of the wire by the intercepted beam and cause a small current to flow back to the wires that is first amplified and then measured in an

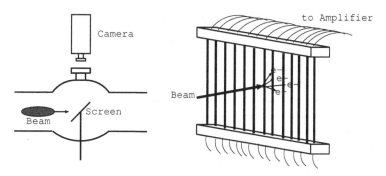

Figure 7.6 The beam position and size are often determined from observing the image on a screen (left) or by measuring the currents caused to replenish the knocked out electrons from a secondary-emission grid (right).

ampere-meter. Reading all currents simultaneously makes SEM-grid capable of measuring the profile of individual pulses. Of course all wires must be read out in parallel, which requires one amplifier per wire and they must be well-balanced. Moreover, a large number of wires must be passed from the inside of the beam pipe, where the SEM-grid is located to the outside, where the amplifiers are located, which requires vacuum feedthroughs with many connectors. All of this make SEM grids rather expensive. The recorded profiles are normally displayed as a histogram of the current in the wires as a function of the wire position from which the center and the width of the distribution then are easily determined.

Instead of placing a large number of wires permanently in the beam, we can use a *wire-scanner,* which consists of a fork that quickly moves a thin wire through the beam. In this case only a single amplifier for the wire is needed, but some mechanical devices such a stepper-motors or a pneumatic piston are required in order to move the wire. Of course, the position of the wire must be known while it traverses the beam, which is usually done by resistive or optical position encoders. Often it is possible to avoid electronic readout of the secondary emission current from the wire but observe nearby fast ion chambers for radiation protection. They will generate a signal related to the number of beam particles hitting the wire. It should also be noted that the wires are normally only several micrometers thick but nevertheless represent a thick target for the beam and perturb it, which can be critical especially in circular accelerators.

A variant of the wire scanner is the *Magnesium Jet profile monitor,* developed in Novosibirsk for VEPP-3 with a copy installed in CELSIUS in Uppsala. In the Magnesium Jet profile monitor, the wire is replaced by a thin stream of Magnesium vapor that is transversely swept across the beam while the position of the nozzle is recorded. The Magnesium is ionized by the beam and the electrons are attracted by a transverse electric field towards a photomultiplier, which produces a signal proportional to the number of Magnesium atoms hit by the beam. A variant of the same theme is the *residual gas monitor* where the Magnesium Jet is replaced by the residual gas that is normally present in the beam pipe to some extent. The gas is ionized and the electrons are guided by magnetic fields onto a position-sensitive multi-channel plate, thus providing information about the transverse beam size.

Instead of moving a wire or gas into the path of the beam, in *laser-wire scanners* a tightly focused laser beam that is transversely scanned across the beam. The fundamental interaction is Compton scattering of the laser photons and electrons. The photons, scattered in the forward direction of the beam, are detected by a scintillating crystal with a photo-multiplier tube and displayed as a function of the position of the laser-wire. In an installation in the PETRA-ring at DESY [36] the laser is a pulsed and frequency-doubled Nd:YAG laser with about 60 mJ in a 5 ns long laser pulse that is focused to an rms spot size of about 4.5 μm. A piezo-controlled rotating mirror is used to scan the laser across the electron beam once every 50 ms. Since integrating the scarce Compton-scattered photons for some time is necessary, a full scan of the electron beam profile takes about a minute.

Electrons and positrons at high energies emit synchrotron light in bending dipole magnets, which can be imaged onto a camera with, for example, a pinhole or a lens and thereby provide a direct image of the transverse electron beam distribution. In modern synchrotron-light sources the beam size is often too small to be directly observed. As we shall see in Chapter 10, the emitted radiation, observed in the horizontal bending plane, is horizontally polarized. Observing from a vertical angle $\alpha_y$, however, a non-zero vertical polarization component is observable. It is zero at $\alpha_y = 0$, but finite vertical size of the emitting beam increases this and this effect can be used to determine the vertical beam size [37]. Calibrating the dilution factor with numerical simulations of the optical system allows us to measure beam sizes on the order of a few μm.

If even smaller beam sizes, on the order of 100 nm or below, need to be determined, we can use a so-called *Shintake-monitor* [38]. It is based on splitting a laser beam and guide the two parts to counter-propagate. The resulting standing-wave pattern acts as a "quasi-photon-grid." Transversely moving the high-energy electron beam across it, results in a varying number of scattered Compton-photons. The modulation-depth of the counted photons in turn depends on the electron beam-size and is used to determine it.

Now we have covered a number of monitors to determine the transverse beam size, even below the μm-range and need to turn our attention to the determination of the longitudinal beam size. If it is on the order of ns, observing the sum-signal from a position monitor on a *oscilloscope* gives an indication of the bunch length. We have, however, to keep in mind that the signal is modified by the frequency response of the monitor and processing electronics. For button pickups it is that of a high-pass filter and the signal on an oscilloscope is a mixture of the bunch profile and its derivative. It is possible to compensate this effect numerically by deconvolution. Much shorter electron bunches down to the pico-second range (bunch length in the mm range) can be measured with a *streak camera,* which is based on the conversion of synchrotron-light pulses to electrons on a photo cathode. The electrons are subsequently accelerated and transversely deflected in a rapidly varying electric field. This "smears" or "streaks" the electrons transversely onto a multi-channel plate to increase the intensity before an image is created on a phosphorous screen that is observed with a CCD camera. The longitudinal intensity distribution of the light pulse is thus converted to a transverse distribution on the screen. Synchronicity is guaranteed by locking the transversely deflecting voltage to the radio-frequency system and thus the arrival time of the bunches. Orthogonal to the fast deflecting plates, some systems have a second pair of slower-deflecting plates installed. They permit to simultaneously observe the variation of the bunch length on a longer, typically μs, time scale.

Instead of streaking secondary electrons emitted by a photo-cathode, we can use a *transversely deflecting structure* (TDS) to "streak" the primary electron beam directly, albeit at the expense of using a much larger RF-structure and an additional klystron [39], both of which incur significant additional costs. The benefit, on the other hand, is that the range to which longitudinal profiles can be determined is extended down to the fs-range, where the bunches are only μm long. This is important to verify the stringent requirements on the bunch length and peak current for free-electron lasers, discussed further in Section 10.5. A TDS is based on a cavity with a transversely deflecting mode, such as $TM_{110}$. Timing

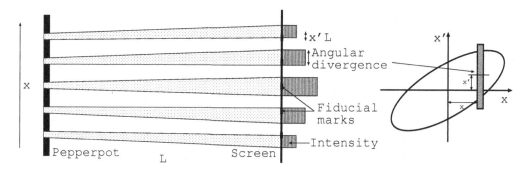

**Figure 7.7** The emittance can be determined by passing the beam through a plate with small holes, the pepper-pot, and observing spots on a screen. Their position reveals information about the beam size on the plate and the widths about the angular divergence.

the TDS in such a way that the electron bunches arrive near the zero-crossing of the field, deflects the head of the bunch to one side and the tail to the other. On a downstream screen, observed by a camera, the longitudinal profile appears converted to the transverse plane. Longer bunches produce a wider spot.

The momentum spread of a beam can be determined from the transverse beam size on a screen or wire scanner at a location with dispersion, provided the ratio $D/\beta$ of dispersion $D$ and beta function $\beta$ is large, as already discussed near the end of the previous section. In a synchrotron or a storage ring, we found near the end of Section 5.3 that the maximum phase excursion $\hat{\phi}$ and momentum offset $\hat{\delta}$ of a particle are related by $\hat{\phi} = (\eta \omega_{\rm rf}/\Omega_s)\hat{\delta}$. For the bunch length $\sigma_s$ and the momentum spread $\sigma_\delta$ it therefore follows that $\sigma_s = (\eta \omega_{\rm rf}/\Omega_s)\sigma_\delta$, such that an observation of the bunch length directly infers the momentum spread.

Let us now return to transverse beam size measurements and use them to determine the emittance and the beta functions.

## 7.4   EMITTANCE AND BETA FUNCTIONS

In low-energy accelerators the beam matrix as well as the emittances and Twiss parameters are determined with a *pepper-pot*, which uses an absorber with a number of small holes, very similar to the lid of a pepper-dispenser to spice one's dinner, and, after a distance $L$, a luminescent screen. Most of the beam incident on the pepper-pot is absorbed, but a small fraction, proportional to the local beam intensity at location $x$, passes through the holes and impinges on the downstream screen, where it forms an image of the holes on the pepper-pot. These image spots on the screen can be characterized by their intensity, their displacement $x'L$ from a fiducial mark, and their width, which is approximately proportional to the angular divergence of the incident beam at position $x$. Figure 7.7 illustrates the geometry in one dimension. The position $x$ of the hole and the displacement $x'L$ determine the position in the phase space diagram, shown on the right, and the width, together with the intensity of the spot determines the angular distribution at one slice. Repeating this analysis for each hole results in a reconstructed phase space of the beam incident on the pepper-pot. A real device has a two-dimensional array of holes and is commonly used to measure the beam matrix in both transverse directions. Since most of the beam is absorbed on the pepper-pot, they are only used in low-energy machines.

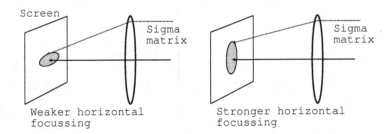

**Figure 7.8** In a *quadrupole scan*, the beam size, measured on a screen or with a wire scanner, is changed by varying the excitation of a quadrupole. The beam matrix $\sigma_0$ before the quadrupole can be determined by correlating the beam-size changes with the excitations.

A common way to experimentally determine the emittance and beta function is by using a number of beam size measurements, either by varying a quadrupole, or by observing the beam size on a number of screens or wire scanners. We will restrict ourselves to the one-dimensional case where the beam sigma-matrix is a $2 \times 2$ matrix that has three independent parameters, either $\sigma_{11}, \sigma_{12}$, and $\sigma_{22}$ or $\varepsilon, \beta$, and $\alpha$. All we *can* directly measure is the beam size but we still want to be able to determine the other parameters such as emittance $\varepsilon$ or $\langle xx' \rangle$ as well. We obviously need at least three measurements to determine the three independent parameters. We start by determining the moments first and later calculate the emittance and Twiss parameters, such as $\alpha$ and $\beta$.

Figure 7.8 shows the spot on a luminescent screen for two different settings of an upstream quadrupole. On the left-hand side, the quadrupole is focusing weakly and the beam is horizontally wide, but vertically small. If we increase the quadrupole excitation, the horizontal beam size will shrink, but the vertical beam size will increase, because quadrupoles focus in one plane but defocus in the other. We will now see how the horizontal beam size will change as a function of the quadrupole excitation for an input beam specified by its moments. For this we need the transfer matrix from the entrance of the quadrupole to the screen. It is given by that of a thin quadrupole and a drift space

$$R = \begin{pmatrix} 1 & l \\ 0 & 1 \end{pmatrix} \begin{pmatrix} 1 & 0 \\ -1/f & 1 \end{pmatrix} = \begin{pmatrix} 1 - l/f & l \\ -1/f & 1 \end{pmatrix}, \tag{7.6}$$

where $l$ is the distance from the quadrupole to the screen and $f$ is the focal length of the quadrupole. We now transport the input beam $\sigma$ with the transfer-matrix $R$ from Equation 7.6 to the screen by $\bar{\sigma} = R\sigma R^t$ and obtain for $\bar{\sigma}_{11} = \bar{\sigma}_x^2$, which equals the beam size squared on the screen

$$\begin{aligned} \bar{\sigma}_{11} &= R_{11}^2 \sigma_{11} + 2R_{11}R_{12}\sigma_{12} + R_{12}^2 \sigma_{22} \\ &= (1 - l/f)^2 \sigma_{11} + 2l(1 - l/f)\sigma_{12} + l^2 \sigma_{22} \\ &= \left(\frac{l}{f}\right)^2 \sigma_{11} - \left(\frac{l}{f}\right)(2\sigma_{11} + 2l\sigma_{12}) + (\sigma_{11} + 2l\sigma_{12} + l^2 \sigma_{22}), \end{aligned} \tag{7.7}$$

which shows that the squared beam size on the screen is a quadratic function, a parabola, of the relative quadrupole excitation $l/f$. The coefficients of the parabola are related to the beam parameters of the incoming beam. By setting the quadrupole to at least three different values and recording the corresponding beam sizes on the screen we can therefore determine the three parameters of the sigma matrix $\sigma_{ij}$ of the input beam. And from those three moments we can in turn determine the emittance and the Twiss parameters $\alpha$ and $\beta$ from

$$\varepsilon = \sqrt{\det \sigma} = \sqrt{\sigma_{11}\sigma_{22} - \sigma_{12}^2} , \qquad \beta = \frac{\sigma_{11}}{\varepsilon} , \qquad \text{and} \quad \alpha = -\frac{\sigma_{12}}{\varepsilon} , \qquad (7.8)$$

which follows directly from Equation 3.78.

Instead of measuring the emittance in a quadrupole scan we can measure the beam size in at least three different locations, where the intermediate transfer matrices must be known. The transfer matrices from the start location to the screens or wire scanners labeled $n$ are denoted by a superscript $n$ such as $R^n$. The measured beam size squared at screen $n$ we denote by $\bar{\sigma}_n^2$. The dependence of the measurements on the incoming beam matrix $\sigma_{ij}$ at an upstream location, is then determined by the following set of equations

$$\begin{aligned}
\bar{\sigma}_1^2 &= (R^1)_{11}^2\sigma_{11} + 2R_{11}^1 R_{12}^1\sigma_{12} + (R^1)_{12}^2\sigma_{22} \\
\bar{\sigma}_2^2 &= (R^2)_{11}^2\sigma_{11} + 2R_{11}^2 R_{12}^2\sigma_{12} + (R^2)_{12}^2\sigma_{22} \\
\bar{\sigma}_3^2 &= (R^3)_{11}^2\sigma_{11} + 2R_{11}^3 R_{12}^3\sigma_{12} + (R^3)_{12}^2\sigma_{22} ,
\end{aligned} \qquad (7.9)$$

which can be written in matrix form as

$$\begin{pmatrix} \bar{\sigma}_1^2 \\ \bar{\sigma}_2^2 \\ \bar{\sigma}_3^2 \end{pmatrix} = \begin{pmatrix} (R^1)_{11}^2 & 2R_{11}^1 R_{12}^1 & (R^1)_{12}^2 \\ (R^2)_{11}^2 & 2R_{11}^2 R_{12}^2 & (R^2)_{12}^2 \\ (R^3)_{11}^2 & 2R_{11}^3 R_{12}^3 & (R^3)_{12}^2 \end{pmatrix} \begin{pmatrix} \sigma_{11} \\ \sigma_{12} \\ \sigma_{22} \end{pmatrix} . \qquad (7.10)$$

We can solve for the unknown input beam $\sigma_{ij}$ by simple matrix inversion of the matrix that appears in Equation 7.10 and that we denote by $A$. If there are more measurements on one or more additional screens, we just add one or more equations to Equation 7.9, which leads to more rows in the corresponding matrix Equation 7.10 which then is over-determined and we can solve it in the least-square sense using the methods discussed in Appendix B.1.

We can easily take measurement errors for the beam size measurements into account and use them to determine error bars for the derived input-beam matrix elements $\sigma_{ij}$. We observe that each individual beam-size measurement $\bar{\sigma}_k$ is afflicted by a measurement error $\Delta\bar{\sigma}_k$, such that the measurement errors $E_k$ of $\bar{\sigma}_k^2$ is $E_k = 2\bar{\sigma}_k \Delta\bar{\sigma}_k$. Left-multiplying Equation 7.10 with the diagonal matrix that contains the inverse measurement errors $1/E_k$ on the diagonal gives the proper weight, dependent in the measurement error $\Delta\bar{\sigma}_k$, to each measurement. Both the measurements $\bar{\sigma}_k^2$ and the matrix elements of $A_{kj}$ are divided by the corresponding $E_k$ leading to $\bar{\sigma}_k^2/E_k$ and $B_{kj} = A_{k,j}/E_k$. Solving for the input matrix $\sigma_{ij}$ proceeds in the same way by using the pseudo-inverse, but now using the matrix $B$. Following the discussion in Appendix B.1, the error bars of the fit parameters $\sigma_{ij}$ are given by the square-root of the diagonal elements of the covariance matrix $C$, given by $C = (B^t B)^{-1}$ .

These measurements can be extended to determine the coupled $4 \times 4$ beam matrix if there are either skew quadrupoles available to change the coupling or ways to measure the correlation $\sigma_{xy}$. The latter we can determine from a spot on a screen of from a wire scanner wires oriented diagonally in the $xy$–plane.

## 7.5   SPECIALTY DIAGNOSTICS

BPMs and screens constitute the work-horse diagnostics of most accelerators, but there are a number of special diagnostic methods that were developed for special purposes. We start with the ability to record beam positions on a turn-by-turn basis.

### 7.5.1 Turn-by-turn position monitor data analysis

The ability to record transverse beam positions on many successive turns in a storage ring, often for several bunches independently, enables a number of very attractive diagnostic opportunities. The most obvious is to simply pass the positions x from a successive number of turns to a Fast-Fourier transform, just as we did in Section 3.3.2. In MATLAB, the command

```
plot((0:N-1)/N,2*abs(fft(x))/N)
```

achieves this and produces a figure with the fractional part of the tune on the horizontal axis and oscillation amplitudes on the vertical axis. The latter is properly normalized and has the same scale and units as the positions x. Note that the spectrum above $Q = 0.5$ is the mirror image, or *alias*, of the spectrum in the range $0 < Q < 0.5$, which is a consequence of sampling only once per turn and the Nyquist theorem, such that all frequencies appear aliased into the frequency range between 0 and 0.5, called the *baseband*. Therefore, usually only that range is shown when displaying the spectra of sampled data. The fractional tunes $Q$ and $1 - Q$ are therefore indistinguishable, but can be disentangled by changing a focusing quadrupole by a small amount. If the tune spectrum moves to higher values, the original tune is between 0 and 0.5; if it moves to smaller value, it lies between 0.5 and 1, where it actually increases. Only its alias in the baseband moves in the other direction.

Instead of passing positions from many turns to an FFT routine it is possible to obtain an estimate for the tune from only three consecutive turns $x_n, x_{n+1}$, and $x_{n+2}$. If we denote the transfer matrix for one turn and starting at the BPM by $R$, we use the inverse of $R$ to express the angle at turn $n + 1$ by $R_{12}x'_{n+1} = R_{22}x_{n+1} - x_n$. Moreover, the position $x_{n+2}$ is given by position and angle from the turn before $x_{n+2} = R_{11}x_{n+1} + R_{12}x'_{n+1}$. Inserting $R_{12}x'_{n+1}$ from the previous equation yields a linear relation among the positions from three consecutive turns

$$x_{n+2} + x_n = (R_{11} + R_{22})x_{n+1} = 2\cos(2\pi Q)x_{n+1} \, , \tag{7.11}$$

where we use Equation 3.60 to express the trace of the transfer matrix by an estimate for the fractional tune $Q$. A note of caution on the use of this equation is in order: if the dynamics is purely linear, the equation is correct, but using it to estimate an instantaneous tune from beam positions whose dynamics is governed by other, and possibly non-linear forces, violate its validity, and using it may be afflicted by systematic errors. This caveat notwithstanding we may use Equation 7.11 repeatedly for a number of consecutive turns and obtain a linear system of equations we can use to determine the trace of the transfer matrix $R$

$$\begin{pmatrix} \vdots \\ x_{n+1} + x_{n-1} \\ x_{n+2} + x_n \\ \vdots \end{pmatrix} = \begin{pmatrix} \vdots \\ x_n \\ x_{n+1} \\ \vdots \end{pmatrix} (R_{11} + R_{22}) \tag{7.12}$$

and using the pseudo-inverse $(A^t A)^{-1} A^t$ from Appendix B.1 to solve it for $R_{11} + R_{22} = 2\cos 2\pi Q$. The estimate of the error bars for the derived tune is left as an exercise.

In order to obtain sufficient signal amplitude for either the FFT or the three-turn equation the beam is often excited with a pulsed magnet to initiate the betatron oscillations. But it turns out that the oscillation amplitude decreases over a moderate number of turns. The reason is that the kicked bunch is an ensemble of many particles, each performing free betatron oscillations, albeit a slightly different frequencies, either due to the finite chromaticity and momentum spread or due to the amplitude-dependent tune shift caused by non-linear magnets, a topic we return to in Section 11.7. Here we only consider the effect of the chromaticity $\mu' = 2\pi Q'$ and note that the linear motion of particles in normalized phase space $(\tilde{x}, \tilde{y})$ can be succinctly written as

$$\tilde{x}_{n+1} + i\tilde{y}_{n+1} = e^{in(\mu + \mu'\delta)}(\tilde{x}_0 + i\tilde{y}_0) , \tag{7.13}$$

where the particle is assumed to have the momentum offset $\delta$ and the distribution $\psi(\delta)$ of the momentum offset can be described by a Gaussian $\psi(\delta) = e^{-\delta^2/2\sigma_\delta}/\sqrt{2\pi}\sigma_\delta$ with rms width $\sigma_\delta$. Since BPMs only observe average position of the entire ensemble of particles, we need to average Equation 7.13 over the distribution of momenta and find the average position $\langle \tilde{x}_n + i\tilde{y}_n \rangle$ to be

$$\langle \tilde{x}_n + i\tilde{y}_n \rangle = e^{-(\mu'\sigma_\delta)^2 n^2/2} e^{in\mu}(\tilde{x}_0 + i\tilde{y}_0) . \tag{7.14}$$

Here we observe that all particles, after having received an initial amplitude $\tilde{x}_0 + i\tilde{y}_0$ perform free oscillations with phase advance $\mu$, but the amplitude is reduced by the factor $e^{-(\mu'\sigma_\delta)^2 n^2/2}$. This factor has a quadratic dependence on the number of turns $n$ and thus does not describe exponential damping, but the loss of coherence due to the differing oscillation frequencies. The amplitude of each individual particle remains constant, but the average motion of the whole ensemble, and this is what BPMs observe, de-coheres; the particles get out of step, similar to what we found for the longitudinal motion, illustrated in Figure 5.5. The characteristic number of turns $n_c$ over which this de-coherence becomes important is given by $n_c = 1/\mu'\sigma_\delta$. We see that small chromaticities $\mu' = 2\pi Q'$ and small momentum spreads $\sigma_\delta$ are beneficial for experiments based on observing turn-by-turn position data of kicked beams.

So far, we only used the recorded positions from a single BPM, but adding a second one, with known transfer matrix $\tilde{R}$ between the two BPM, allows us to reconstruct the phase-space coordinates at one, here the first, BPM $x_1$ and $x_1' = (x_2 - \tilde{R}_{11}x_1)/\tilde{R}_{12}$ and experimentally determine the Poincaré plot of the dynamics in the ring by plotting the reconstructed phase space coordinates turn-by-turn. Note that a large value of $\tilde{R}_{12}$ is advantageous to accurately determine $x_1'$. This formalizes the common-sense approach to use two BPMs "with a 90° phase-advance between them." The pioneering experiment to do so was E778 [40] at Fermilab, where the dynamics in the presence of strong sextupoles was experimentally investigated and compared to simulations. In particular, the amplitude-dependent tune shift was measured and trapping of particles in resonance islands was observed.

We can determine the betatron phase-advance between two BPM [41] by first determining the betatron tune $Q$ with a FFT or any other method and then "mixing" the positions $x^a$ at BPM $a$ with both phases (sine and cosine) of the oscillation frequency

$$C_a = \sum_{k=1}^{n} x_k^a \cos(2\pi kQ) \quad \text{and} \quad S_a = \sum_{k=1}^{n} x_k^a \sin(2\pi kQ) \tag{7.15}$$

to obtain a phase value $\phi_a = \arctan(S_a/C_a) + \phi_0$ with an unknown phase offset $\phi_0$. Repeating the analysis with position from another BPM $b$, taken concurrently with BPM $a$ in order to have the same $\phi_0$, we obtain the phase at the second BPM as $\phi_b = \arctan(S_b/C_b) + \phi_0$ such that the phase difference between the two BPM is given by $\Delta\phi_{ba} = \phi_b - \phi_a$. Comparing the measured phase-advance differences $\Delta\phi_{ba}$ to those derived from a computer model, reveals locations with potential errors in the linear optics, such as quadrupole-gradient errors. Correlating measured phase advances from many or all BPMs in a ring allows us to determine [42] beta functions with good accuracy.

The next topic is based on utilizing the beam-beam interaction of colliding bunches at the interaction point for diagnostic purposes, which is used both in circular and linear colliders.

## 7.5.2 Beam-beam diagnostics

In order to maximize the luminosity, the beams in colliders are squeezed to very small transverse dimensions at the interaction point. This causes very high charge densities and corresponding electro-magnetic fields that affect the counter-propagating beam, and, since the beams move in opposite directions, the effects of electric and magnetic field components deflect in the same direction and need to be added. The transverse radial electric field $E_r$ of a round beam with Gaussian charge density $e^{-r^2/2\sigma^2}/2\pi\sigma^2$ is given by Gauss's law

$$2\pi r E_r = \frac{N}{2\pi\sigma^2\varepsilon_0} \int_0^r e^{-r'^2/2\sigma^2} r' dr' d\phi = \frac{N}{\varepsilon_0}\left(1 - e^{-r^2/2\sigma^2}\right) \tag{7.16}$$

where $N$ is the number of charges in the beam and $\sigma$ is the radial rms size. For electrons with energy $\gamma mc^2$ this leads to the deflection angle $\Delta\theta = -(2Nr_e/\gamma)\left(1 - e^{-r^2/2\sigma^2}\right)/r$ for a single particle. In Section 9.7 we will see that averaging over all particles of the counter-propagating beam with beam size $\tilde{\sigma}$ gives the (round-beam) beam-beam deflection angle for the whole beam, which is observable on position monitors in the vicinity of the interaction point

$$\langle\Delta\theta\rangle = -\frac{2Nr_e}{\gamma}\frac{1 - e^{-\Delta r^2/2(\sigma^2+\tilde{\sigma}^2)}}{\Delta r} . \tag{7.17}$$

This expression strongly resembles the deflection angle of a single particle, only the beam size is replaced by sum of the squares of the beam size, both for the field-producing and the deflected beam. Furthermore, the radius vector $r$ is replaced by the displacement $\Delta r$ of one beam with respect to the other. By transversely scanning one beam across the other and thereby changing $\Delta r$, while simultaneously recording the deflection angle $\langle\Delta\theta\rangle$ allows us to determine the initial displacement and $\sigma^2 + \tilde{\sigma}^2$ from a least-squares fit. This method was very successfully used in the SLC [43] to center the beams and use the beam sizes as estimate for the luminosity. Note that only the sum of squares of the beam sizes can be determined in this way, which is generally valid, even for elliptic or non-Gaussian beams [44].

Apart from deflecting the opposing beam as a whole, the deflection of the individual electrons or positrons causes them to emit synchrotron radiation, so-called *beamstrahlung* [45]. The number of photons is rather low, but they are highly energetic, because the deflecting fields are very strong. A monitor system [46] was installed adjacent to the interaction region in the SLC. It was based on first converting the beamstrahlung photons into electron-positron pairs that pass through a low-density gas where they produce Cerenkov photons, which are subsequently recorded by photo-multiplier tubes. The Cerenkov effect discriminates the few, but high-energy, beamstrahlung photons from the abundant low-energy photons. We will return to the discussion of beamstrahlung in Chapter 9 and now turn to the use of beam-beam diagnostics in circular colliders.

Expanding the beam-beam deflection angle in Equation 7.17 for small displacement $\Delta r$ shows that the deflection angle has a linear dependence and thus behaves like a quadrupole; and that affects the tunes of the beams in a storage ring. Since the slope of the deflection curve is maximum near the origin, maximizing the beam-beam tune shift will center the

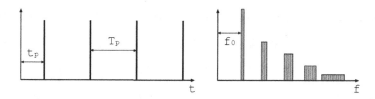

Figure 7.9  A pulse train of a periodically arriving single particle (left) and the spectral power density of an ensemble of many particles (right).

beams on top of each other and will maximize the luminosity. In colliders the tunes of the two rings are often equal and the beam-beam interaction couples their betatron oscillations, such that the coupled system will cause the oscillation frequencies to split into two modes and their separation directly gives the tune shift, as we will discuss in detail in Section 9.7. Scanning one beam across the other is often called a *van der Meer scan* and extending the scan range while recording the tune shift and, if possible, a sub-nuclear reaction, gives information about the beam size directly at the interaction point. Displacing one beam versus the other one causes the "quadrupole" to be offset and that will affect the closed orbit [47], which is visible on BPMs throughout the ring and can be used to determine the beam-beam deflection angle in the same way it is done in a linear collider.

### 7.5.3  Schottky diagnostics

The beams in storage rings are made of a large number of discrete particles and the finite momentum spread causes the particles to have slightly different revolution times, according to Equation 5.32, given by $\Delta f/f = \eta \sigma_\delta$. Each of the particles therefore produces a very weak "ping" on a detector such as a BPM. Sometimes, for intense beams, conventional BPMs are sufficient, but normally special monitors, often cooled to cryogenic temperatures in order to minimize thermal noise, are used to pick up the distribution of "pings." These are called Schottky signals because they resemble the noise generated by the individually arriving electron on the anode of vacuum-tubes, originally analyzed by W. Schottky in 1918. Using Schottky signals for diagnostic purposes to determine momentum spread, tune, and chromaticity is essentially non-invasive, which is often advantageous.

In order to quantitatively analyze the signals, we consider a single particle $p$, which takes the time $T_p$ for one revolution [48] and has a starting time $t_p$. An illustration is shown on the left-hand side in Figure 7.9. The current $i_p(t)$ that is detected by a monitor is then an infinite sequence of delta-function pulses with charge $e$, spaced by $T_p$

$$
i_p(t) = e \sum_{k=-\infty}^{\infty} \delta(t - t_p - kT_p) = \frac{e}{T_p} \sum_{m=-\infty}^{\infty} e^{im\omega_p(t-t_p)}
$$

$$
= \frac{e}{T_p} + \frac{2e}{T_p} \sum_{m=1}^{\infty} \cos(m\omega_p(t - t_p)) \, . \tag{7.18}
$$

Here the periodicity of the first equation implies that the second one can be written as a Fourier series with harmonics of the fundamental frequency $\omega_p = 2\pi/T_p$ with all Fourier coefficients evaluating to $1/T_p$. The third equality follows because the coefficients of negative and positive $m$ are equal. Using the second equation to Fourier transform $i_p(t)$ we find $\tilde{i}_p(\omega) = (2\pi e/T_p) \sum_{m=-\infty}^{\infty} \delta(\omega - m\omega_p)$ where we use the tilde to denote Fourier transforms.

We find that a sequence of equally spaced delta-pulses generates a frequency-comb that contains all harmonics $m\omega_p$ of the fundamental frequency $\omega_p$.

To observe longitudinal Schottky spectra we connect, for example, the sum signal of a BPM to a spectrum analyzer. An un-bunched, often called "coasting" beam with $N$ randomly distributed particles along the circumference, the starting times $t_p$ are randomly distributed over the revolution period and averaging causes only the constant term to survive. It represents the average current $I_0 = Ne/T_0$, where we assume that the spread in revolution times is small and $T_p \approx T_0$. On the other hand, the power deposited by the current that fluctuates due to its granularity and is proportional to square of the current variation $i_r^2 = \langle (i_p - e/T_p)^2 \rangle$, where the angle brackets denote averaging over the starting times $t_p$ that appear in the definition of $i_p(t)$ in Equation 7.18. It is given by $i_r \approx (2e/T_0)\sqrt{N/2}$, which increases with the square root of the number of particles, indicating the incoherent addition of the phase-averaged current pulses. It is independent of the harmonic $m$ and therefore the same for all harmonics. The spectral power density per frequency band of width $df$ is proportional to $i_r^2/df$, which is the quantity displayed on a spectrum analyzer. This is sketched on the right-hand side in Figure 7.9. The total power per band is equal and since the bandwidth increases the peak value decreases. The width of the peak $(\Delta f/f)_m$ of harmonic $m$ is related to the momentum spread $\sigma_\delta$ of the beam by $(\Delta f/f)_m = m\eta\sigma_\delta$.

Transverse Schottky spectra are observed by connecting the BPM difference-signal or dedicated Schottky monitor to a spectrum analyzer. This signal is proportional to the current $i_p$ of the periodically arriving particle, that is modulated by the transverse position $x_p(t) = a_p \cos(Q_p\omega_p t + \phi_p)$, where $Q_p$ denotes the fractional part of the tune, $a_p$ is the amplitude of the transverse oscillation, and $\phi_p$ is the random starting phase of the transverse oscillation. The frequency spectrum, we find from the dipole moment $d_p(t)$

$$d_p(t) = i_p(t)x_p(t) = \frac{ea_p}{T_0} \mathrm{Re}\left( \sum_{m=-\infty}^{\infty} e^{i(m+Q_p)\omega_p t+\phi_p} \right), \tag{7.19}$$

which shows that there are now spectral lines at $\omega = (m \pm q_p)\omega_p$. Similar to the previous paragraph, the spectral power density is proportional to $\langle d^2 \rangle = \langle a_p^2 \rangle (e/T_0)^2 N/2$, which is, again, independent of the harmonic $m$ and therefore equal for all harmonics. Note that it is proportional to the square of the oscillation amplitudes $\langle a_p^2 \rangle$, which is proportional to the emittance of the beam. The width of the spectral lines is now determined by the spread in revolution frequencies $\omega_p$, but also the spread in the tunes $Q_p$ among the different particles, for example, due to a finite chromaticity $Q'$. For their width, we find $\Delta f/f = [\eta(m \pm Q) \pm Q']\sigma_\delta$ to first order in the momentum spread. Here we see that the tune $Q$ can be determined by the difference of the central frequencies of the betatron sidebands.

If the beams are bunched by an RF-system, the arrival times of the particles are no longer evenly distributed but they will all appear approximately simultaneously in time, determined by the RF-frequency, rather than the energy of the particles. Moreover, the revolution time $T_0 = 2\pi/\omega_0$ of all particles is the same, and this gives rise to a much stronger coherent signal at the revolution harmonics that swamps all other, often minute signals. Despite a difficult signal acquisition and processing such systems were successfully used at Fermilab [49] and LHC [50]. Furthermore, with RF present, the particles may perform synchrotron oscillations that modulate their arrival time, represented by the additional term $\tau_p = \hat{\tau}_p \sin(\Omega_s t + \psi_p)$. Here $\Omega_s$ is the synchrotron frequency, $\hat{\tau}_p$ is the amplitude of the particle $p$, and $\psi_p$ its starting phase. The current $i_p$ for particle $p$ is then given by

$$i_p(t) = e \sum_{k=-\infty}^{\infty} \delta(t + \hat{\tau}_p \sin(\Omega_s t + \psi_p) - kT_0) = \frac{e}{T_0} \sum_{m=-\infty}^{\infty} e^{im\omega_0(t+\hat{\tau}_p \sin(\Omega_s t+\psi_p))} \tag{7.20}$$

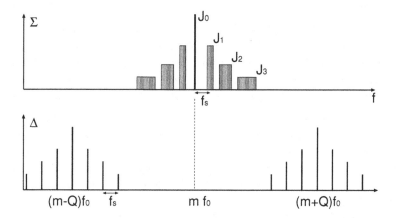

Figure 7.10 The longitudinal Schottky spectrum of a bunched beam (top) and the transverse Schottky spectrum (bottom) of the $m$–th revolution harmonic.

and with the relation $e^{iz\sin\phi} = \sum_{n=-\infty}^{\infty} J_n(z)e^{in\phi}$ from [23] we find

$$i_p(t) = \frac{e}{T_0} + \frac{2e}{T_0} \sum_{m=1}^{\infty} \mathrm{Re}\left[\sum_{n=-\infty}^{\infty} J_n(m\omega_0\hat{\tau}_p)e^{i(m\omega_0+n\Omega_s)t+in\psi_p}\right], \qquad (7.21)$$

where the sum over $m$ extends over the revolution harmonics $m\omega_0$ and the index $n$ describes the so-called Bessel-sidebands, characteristic of frequency-modulated signals. Note that for the sideband with $n = 0$ the dependence on the random starting phase $\psi_p$ drops out and all particles contribute in phase such that the power spectrum has a $N^2$–dependence, whereas for all sidebands with $n \geq 1$ the contribution to the power spectrum is proportional to $N/2$ instead, because averaging over the phases gives a vanishing contribution, but averaging the square of the current contributes to the spectral density. Furthermore, owing to the fact that $J_n(x)$ decreases with increasing sideband order $n$, the amplitude decreases as well.

The transverse spectrum of bunched beams, observed by connecting the difference signal of a BPM to a spectrum analyzer is given by

$$d_p(t) = ea_p \cos(Q\omega_0 t + \phi_p) \sum_{k=-\infty}^{\infty} \delta(t + \hat{\tau}_p \sin(\Omega_s t + \psi_p) - kT_0), \qquad (7.22)$$

where $\phi_p$ and $\psi_p$ are random starting phases of the transverse and longitudinal oscillations, respectively. After expressing the infinite sequence of delta functions by its Fourier-series and by using trigonometric and the Bessel-function identities, we obtain for particle $p$

$$d_p(t) = \frac{2ea_p}{T_0} \sum_{m=1}^{\infty} \mathrm{Re}\left[\sum_{n=-\infty}^{\infty} J_n\left((m \pm Q)\omega_0\hat{\tau}_p\right) e^{i((m\pm Q)\omega_0 t + n\Omega_s t + n\psi_p + \phi_p)}\right]. \qquad (7.23)$$

This expression shows that we can expect a comb of synchrotron sidebands separated by $\pm n\Omega_s$ around the upper and lower sidebands at $\pm\omega_0 Q$ of the revolution harmonics $m\omega_0$, which is illustrated in the lower plot in Figure 7.10. As is the case with un-bunched beams, the tune is encoded in the separation of the tune sidebands and if we can identify two sidebands belonging to the same revolution harmonic $m$, the tune is given by $Q = m(\omega_+ -$

$\omega_-)/(\omega_+ + \omega_-)$. In the following, we assume that the synchrotron-sidebands around the tune-sidebands are small, but the beam has, on the other hand, a significant momentum spread and, moreover, the chromaticity $Q'$ is non-zero. With these assumptions and realizing that both the tune $Q$ and the revolution frequency $\omega_0$ then depend on the momentum $\delta$, we can estimate the width of upper and lower tune-sidebands to be $\Delta\omega_+ = \omega_0[(m+Q)\eta + Q']\sigma_\delta$ and $\Delta\omega_- = \omega_0[(m-Q)\eta - Q']\sigma_\delta$. Solving for the chromaticity $Q'$ by adding and subtracting $\omega_+$ and $\omega_-$ we obtain

$$Q' = \eta \left[ m \frac{\Delta\omega_+ - \Delta\omega_-}{\Delta\omega_+ + \Delta\omega_-} - Q \right] . \tag{7.24}$$

We thus have to determine the tune $Q$ first and then can determine the chromaticity from the difference of the widths of the tune-sidebands. Recently [50], this method was successfully tested in the LHC.

After this discussion of devices and methods to diagnose the beam quality we turn to using them in order to analyze and the correct various imperfections that cause a real accelerator to behave differently from the prototype designed on a computer.

## QUESTIONS AND EXERCISES

1. In a quadrupole-scan emittance measurement the following rms beam sizes $\sigma_x$ were measured with a wire scanner 5 m downstream of a (thin) quadrupole, when the inverse focal length had the values indicated in the following table

| $1/f$ | $[1/m]$ | 0.1 | 0.2 | 0.3 | 0.4 | 0.5 |
|-------|---------|-----|-----|-----|-----|-----|
| $\sigma_x$ | $[mm]$ | 6.0 | 3.5 | 2.5 | 4.3 | 7.0 |

(a) Determine the horizontal beam matrix of the beam immediately before the quadrupole. (b) Deduce the emittance, beta function and $\alpha$ from that.

2. You are responsible for a transfer line that consists of FODO cells where the (thin-lens) quadrupoles have a focal length 4 m and the distance between quadrupoles is 5 m. In the line there is a wire scanner installed very close after each of three consecutive quadrupoles with polarity QF,QD,QF. (a) The three wire scanners measure horizontal beam sizes $\sigma_i$, with $i = 1, 2, 3$ with values 4.5, 2.5, 4.6 mm. Please design a measurement system, that relates the measured beam sizes to the sigma matrix at the first wire scanner and derive the emittance, beta function $\beta$, and $\alpha$ of the beam at the first wire scanner. (b) If the beam size measurements have an error bar of 0.1 mm, what are the error bars of the derived sigma matrix? (c) What are the error bars of the derived emittance, $\beta$, and $\alpha$?

3. Expand the algorithm to include a fourth wire scanner downstream of the next quadrupole in the FODO beam line and repeat the analysis.

4. Assume that three wire scanners are placed at locations, separated by phase advance $\phi$, all with $\beta = 1$ m and $\alpha = 0$, such that all transfer matrices are rotation matrices with angles that are multiples of $\phi$. Calculate the determinant of the matrix in Equation 7.10 to determine which phase advance leads to the most robust measurement system.

5. Consider the horizontal motion in a ring with tune $Q = 0.61$ and a position monitor at a location with $\beta = 5$ m and $\alpha = 0$. Iterate the motion of a particle for 1024 turns and determine the tunes by (a) Fourier transformation; (b) by using Equation 7.12.

# CHAPTER 8

# Imperfections and Their Correction

*Imperfections* are the discrepancies of a real, as-built, accelerator from the ideal computer model, we discussed in Chapter 3. The model typically describes a lattice in which all magnets are perfectly placed at their reference location, and all magnets are powered with their design currents. Moreover, typically only beam line elements for the accelerator are taken into account and magnetic fields that are not part of the bare lattice are omitted; examples for the latter are the solenoid magnets that are part of the detectors in high-energy accelerators or in electron coolers, but also the additional fields from undulator magnets in synchrotron light sources. In short, all influences that make a real accelerator deviate from its ideal behavior, we call "imperfection." In the section on corrections, we also address the compensation of chromatic effects, even though they are inherently caused by the finite momentum spread of the beam, rather than imperfections.

The main topic of this chapter, however, is the effect of undesired *magnetic fields*, those that are in the wrong place, on the reference trajectory of the beam. Figure 8.1 shows the types of fields we will consider. On the left-hand side, the vertical component of the magnetic flux density $B_y$ is constant across the beam and all particles in the beam receive the same kick $\Delta x'$. This effect is similar to that of a dipole corrector magnet, which also deflects all particles equally, irrespective of their position. If, on the other hand, $B_y$ varies linearly as we move from one side of the beam to the other, as shown in the middle of Figure 8.1, the

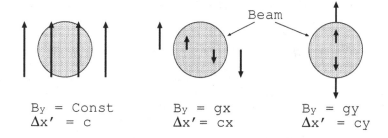

Figure 8.1 Types of fields that cause imperfections, which can be constant across the beam (left), show a gradient (middle) or a skew-gradient (right).

field behaves like an additional quadrupole and it will change the focusing of the beam and thus will change the beam sizes. In particular, a horizontally focusing force is accompanied by a vertically defocusing force, just as in conventional quadrupole magnets. Likewise, a vertically varying $B_y$, as shown on the right-hand side in Figure 8.1, characterizes skew-quadrupolar field, which couples the horizontal and vertical motion. In addition to the fields sketched in Figure 8.1, have solenoids in detectors or coolers a longitudinal component $B_s$ of the magnetic flux. Particles entering the magnet with a horizontal angle receive a vertical kick, which also couples the two transverse degrees of freedom.

In the following section we will briefly name a few of the most common sources that create these fields and how to include them in computer models.

## 8.1 SOURCES OF IMPERFECTIONS

Often imperfections have trivial causes, such as magnets with swapped power leads such that the magnetic field has opposite polarity, or power supplies that have small calibration or read-back errors, such that the magnets are incorrectly excited. Exciting magnets close or beyond the specified maximum may lead to saturation of the iron and corresponding deviations from the expected fields. Two magnets in close proximity may influence one another, because the fringe field of one magnet affects the saturation of the iron in an adjacent magnet. Incorrectly chosen shims are another source of potentially incorrect fields. In low-energy beam lines the magnetic field of the Earth, having an order of magnitude of $50\,\mu$T, can affect the orbit and may need to be taken into account. In LEP the solder used when brazing the vacuum chamber contained an alloy with nickel, which is ferromagnetic and caused unwanted magnetic field components that had to be corrected [51].

But by far the most common imperfections are due to misaligned magnets and we consider how to understand and simulate their effect next.

### 8.1.1 Misalignment and feed down

Longitudinally misaligning a magnet or other component by $d_s$ in an accelerator is easily accomplished by adding a drift space of length $d_s$ before the component and subtracting one with the same length immediately following the component. Misaligning transversely is equally easy to implement. For example, an element horizontally displaced by $d_x$, is treated by adding $d_x$ before entering the element and subtracting $d_x$ after exiting it, as should be evident from the left-hand side of Figure 8.2. First, we consider a beam line component, represented by a transfer matrix $\tilde{R}$, and horizontally displace it by $d_x$. In one transverse dimension, its overall effect on the beam, entering with initial phase-space coordinates $(x_i, x_i')$, is given by

$$\begin{pmatrix} x_f \\ x_f' \end{pmatrix} = \begin{pmatrix} -d_x \\ 0 \end{pmatrix} + \tilde{R}\left[\begin{pmatrix} d_x \\ 0 \end{pmatrix} + \begin{pmatrix} x_i \\ x_i' \end{pmatrix}\right] = \tilde{R}\begin{pmatrix} x_i \\ x_i' \end{pmatrix} + \left[\tilde{R} - 1\right]\begin{pmatrix} d_x \\ 0 \end{pmatrix}, \quad (8.1)$$

where the subscript $f$ denotes the final positions. We find that the component still behaves in the same way as the un-displaced one, as represented by the operation of $\tilde{R}$ on the incoming beam. There is, however, the additional perturbation that is added to the phase-space coordinates of the beam $\vec{q} = [\tilde{R} - 1]\vec{m}$ with misalignment vector $\vec{m} = (d_x, 0)^t$. It is instructive to consider the perturbation from a thin-lens quadrupole with transfer matrix $\tilde{R}$ given in Equation 3.5 for which we obtain

$$\vec{q} = \left[\tilde{R} - 1\right]\begin{pmatrix} d_x \\ 0 \end{pmatrix} = \begin{pmatrix} 0 & 0 \\ -\frac{1}{f} & 0 \end{pmatrix}\begin{pmatrix} d_x \\ 0 \end{pmatrix} = \begin{pmatrix} 0 \\ -\frac{d_x}{f} \end{pmatrix}, \quad (8.2)$$

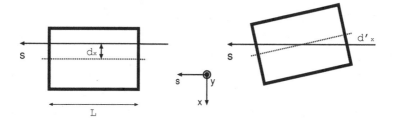

Figure 8.2 A component with a parallel displacement $d_x$ (left) and a tilt angle $d'_x$ (right) with respect to the direction of propagation of the beam.

which represents a change in the angle—a dipole kick. We thus find that a displaced thin quadrupole causes an additional dipole kick. The quadrupole does, however, focus equally well as one placed on the beam axis. This appearance of lower multipole kicks—the dipole kick—due to a displaced higher multipole—the quadrupole—is called *feed down* and is one of the most common sources of imperfections in accelerators.

Higher multipoles are used for several purposes in accelerators. For example, sextupoles are used to correct the dispersion. These magnets are often short and of moderate strength, such that the transverse position of the beam does not change within the multipole. We therefore can approximate the effect of these multipoles on the phase-space coordinates of the beam by a thin-element kick approximation, which is given by selecting a single multipole $n$ in Equation 3.37

$$\Delta x' - i\Delta y' = -\frac{k_n L}{n!}(x + iy)^n \, , \tag{8.3}$$

where $L$ is the length of the multipole and $k_n L$ is the integrated magnetic gradient, normalized to the beam energy. For $n = 1$ we recover the map of a thin-lens quadrupole $\Delta x' = -k_1 L x$ and $\Delta y' = k_1 L y$. The kick from a horizontally displaced multipole is then given by changing $x$ to $x + d_x$ in Equation 8.3 with the result

$$\Delta x' - i\Delta y' = -\frac{k_n L}{n!}(x + d_x + iy)^n = -\frac{k_n L}{n!}(x + iy)^n - \frac{k_n L}{n!}\sum_{k=0}^{n-1}\binom{n}{k}d_x^{n-k}(x + iy)^k \, . \tag{8.4}$$

We find the same behavior as before: the displaced multipole acts in the same way as the undisplaced one, but multipoles of lower orders $k < n$ are produced with their strength given by $(k_n L/n!)\binom{n}{k}d_x^{n-k}$. Displacements in the vertical direction or simultaneous displacement in both directions are handled in the same way.

As an example, we consider a sextupole with $n = 2$, horizontally displaced by $d_x$. Its kick is given by

$$\Delta x' - i\Delta y' = -\frac{k_2 L}{2}\left[(x + iy)^2 + 2d_x(x + iy) + d_x^2\right] \tag{8.5}$$

and we find an additional term describing an upright quadrupole with magnitude $k_2 L d_x$ and a dipole kick with magnitude $k_2 L d_x^2/2$. Figure 8.3 illustrates these contributions by showing the kick $\Delta x'$ from sextupole displaced to $-1\,\mathrm{mm}$. A beam, sketched as a the solid Gaussian, experiences an average kick with magnitude, indicated by the vertical dot-dashed line at $x = 0$. Furthermore, the beam experiences a linearly varying kick, shown as the dashed line, that resembles a quadrupolar kick.

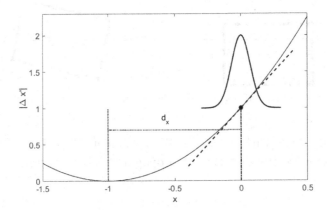

**Figure 8.3** The kick $\Delta x'$ from a sextupole, horizontally displaced by $d_x = 1\,\text{mm}$ (solid). A transversely small beam, indicated by the Gaussian, experiences an average deflection (dot-dashed) and a gradient (dashed), which illustrates feed-down of the sextupolar field to the lower multipoles.

Displacing the sextupole vertically, we find

$$\Delta x' - i\Delta y' = -\frac{k_2 L}{2}(x + iy + id_y)^2 = -\frac{k_2 L}{2}\left[(x + iy)^2 + 2id_y(x + iy) - d_y^2\right] \ . \quad (8.6)$$

We see that the term linear in $(x + iy)$ describes a skew-quadrupole, because, owing to the factor $2id_y$, a purely horizontal position $x$ produces a vertical kick $\Delta y'$. Generating a skew quadrupolar field will cause cross-plane coupling. When operating with flat beams this might cause the smaller beam size to increase significantly. If very flat beams are important, as they often are in synchrotron light sources, special care should be taken to vertically center the orbit in the sextupoles.

Displacing components parallel to the direction of the beam is not the only misalignment. If the component is extended it can also be rolled or tilted. An example is a girder, installed with an incorrect horizontal angle with respect to the direction of the beam.

## 8.1.2 Tilted components

The right-hand side in Figure 8.2 shows a component with length $L$ that is horizontally tilted by an angle $d'_x$. If its effect on the beam is characterized by a transfer matrix $\hat{R}$, we can describe the beam moving with an angle $d'_x$ through the tilted section by adding an angle $d'_x$ to the phase-space coordinate $x'$, but we also need to translate it by a distance $-d'_x L/2$ transversely as shown on the right-hand side in Figure 8.2. After propagating through the component we have to subtract the tilt angle $d'_x$ but also need to move in the same direction as before. For the combined effect we obtain

$$\begin{pmatrix} x_f \\ x'_f \end{pmatrix} = \begin{pmatrix} -d'_x L/2 \\ -d'_x \end{pmatrix} + \hat{R}\left[\begin{pmatrix} -d'_x L/2 \\ d'_x \end{pmatrix} + \begin{pmatrix} x_i \\ x'_i \end{pmatrix}\right] \quad (8.7)$$

$$= \hat{R}\begin{pmatrix} x_i \\ x'_i \end{pmatrix} + \left[\hat{R} + \begin{pmatrix} 1 & 0 \\ 0 & -1 \end{pmatrix}\right]\begin{pmatrix} -d'_x L/2 \\ d'_x \end{pmatrix} \ .$$

Again, we find the displaced component to behave in the same way as the properly placed one, but with an additional dipolar orbit perturbation

$$\vec{q} = \left[ \hat{R} + \begin{pmatrix} 1 & 0 \\ 0 & -1 \end{pmatrix} \right] \begin{pmatrix} -d'_x L/2 \\ d'_x \end{pmatrix} \tag{8.8}$$

added to the particle's phase-space coordinates. Generalizing to misalignments in two transverse directions is straightforward.

### 8.1.3 Rolled elements and solenoids

Elements, rolled with respect to the beam's direction of propagation, can be easily described by sandwiching the element between coordinate rotations, already discussed in Section 3.1.6.

The source of longitudinal fields $B_s$ are often solenoids and their transfer matrix is given in Equation 3.36. In order to take their effect on the beam into account, we simply replace the matrix for the drift space at the location with the solenoid by the matrix for a solenoid with the appropriate field.

### 8.1.4 Chromatic effects

Any unwanted effects that are due to the finite spread in particle momenta are called *chromatic*. The dispersion $D$ causes a finite momentum spread $\sigma_\delta$ to increase the beam size to $\sigma^2 = \varepsilon\beta + D^2\sigma_\delta^2$. We therefore often seek to minimize the dispersion at strategic locations, such as the interaction point of a collider.

The momentum-dependence of the focusing from quadrupoles causes the longitudinal position of the focal point, the waist, to depend on the particle momentum and this dilutes the minimum achievable beam size. In linear accelerators, this effect is called *chromaticity*. In a ring, *chromaticity* has a different meaning, namely the dependence of the tune on momentum, we denoted by $Q'_x$ and $Q'_y$ in Section 3.4.1. Of course, the origin is the momentum-dependence of the quadrupole focusing, as well. In rings, a finite chromaticity therefore causes the tunes to spread out in the tune diagram and may approach resonance lines, which perturbs the beams and may lead to increased emittances and even beam loss. Therefore, we need to correct the chromaticities $Q'_x$ and $Q'_y$. Some intensity-dependent instabilities depend on the chromaticity and require their adjustment to values close to zero.

### 8.1.5 Consequences

The consequence of all these unwanted additional fields due to misaligned components is a trajectory of the beam centroid that is bouncing back and forth due to the additional dipolar kicks and a beam envelope—the sigma matrix—that differs from the design, usually in an unfavorable way. In particular, a non-centered trajectory causes colliding beams to miss one another and thus reduces the luminosity. In a linear collider, non-zero trajectories can cause so-called wake-field kicks, discussed in Chapter 12, lead to increased emittance and a reduced luminosity. In synchrotron light sources, the emitted light may miss the experimental station with monochromators at the end of very long synchrotron light beam lines.

The quadrupolar fields cause the beam sizes to differ from the design and this will normally degrade the luminosity of a collider. Injecting mis-matched beams into a hadron accelerator leads to an increase in the emittance due to filamentation. Large beam sizes in a synchrotron light sources reduce the transverse coherence of the emitted light and in a free-electron laser this may prevent lasing at all. These are just a few consequences, but the effects are rather generic and we need to understand their generation in more detail, which is the topic for the next two sections. First, we look at beam lines.

## 8.2  IMPERFECTIONS IN BEAM LINES

In this section we consider single-pass, or non-periodic, beam lines in which the beam enters at one end and exits at the other. Examples are transfer lines between accelerators, but also linear accelerators. We start by looking at the results of a number of dipolar kicks in a beam line.

### 8.2.1  Dipole kicks and orbit errors

In Section 8.1, we found that dipolar kicks due to transverse misalignment (Equation 8.1) and tilt (Equation 8.7) are described by adding a constant displacement vector $\vec{q} = [1 - R]\vec{m}$ to all particles of the beam and therefore also to the centroid of the beam $\vec{X}$. Here $\vec{m}$ is the misalignment vector introduced in the previous section. Now we introduce the operator $\vec{q} + R$, which acts on a phase space vector $\vec{x}$ by first multiplying $\vec{x}$ by $R$ and then adding $\vec{q}$ to the result. In this way we obtain for the cumulative effect of all misalignments $\vec{q}_j$ upstream of an observation point labeled $n$

$$
\begin{aligned}
\vec{x}_n &= R_n \cdots (\vec{q}_{k+1} + R_{k+1})(\vec{q}_k + R_k) \cdots (\vec{q}_1 + R_1)\vec{x}_0 \\
&= R_n \cdots R_1 \vec{x}_0 + \sum_{j=1}^{n-1} (R_n \cdots R_{j+1})\vec{q}_j
\end{aligned}
\tag{8.9}
$$

where the interpretation is very simple. A particle with initial phase-space coordinates $\vec{x}_0$ has the final coordinates $\vec{x}_n$ that are given by the transfer matrix $R_n \cdots R_1$ from start to point $n$. Additionally, each misalignment vector $\vec{q}_j$ is propagated to the observation point by multiplying it with the transfer matrix $R^{n,j+1} = R_n \cdots R_{j+1}$ from the misaligned component to the observation point. Finally the contributions of all misalignment vectors are added. Conversely, in order to calculate the effect of a misalignment of component $j$, we first need to calculate the misalignment vector $\vec{q}_j$ and the transfer matrix $R^{n,j+1}$ from the exit of component $j$ to the observation point $n$. Then $R^{n,j+1}\vec{q}_j$ gives the contribution of misalignment $j$ to the observation point $n$. If we have a BPM at the observation point, the first and third components of $\vec{x}_n$ will show the position change due to misalignment $j$.

After being able to predict the result of dipolar errors and their misalignment vectors $\vec{q}$ let us look at the consequences of quadrupolar errors.

### 8.2.2  Quadrupolar errors and beam size

A quadrupolar error will cause the sigma matrix to deviate from its design value $\sigma$ and the changed sigma matrix $\hat{\sigma}$ is given by $\hat{\sigma} = Q\sigma Q^t$, where $Q$ is the transfer matrix describing the added quadrupolar field. Often it can be approximated by a thin-lens quadrupole with a transfer matrix given by Equation 3.7. The beam matrix $\bar{\sigma}$ at an arbitrary downstream location is then given by $\bar{\sigma} = R\hat{\sigma}R^t$ where $R$ is the transfer matrix from the location with the perturbing quadrupolar field to the screen. Now we confine ourselves to the 1D-motion in the horizontal plane in order to keep the algebra manageable, and instead of writing $2 \times 2$–matrix equations we treat the three independent components of the sigma matrix as a column vector. Multiplying the $2 \times 2$–matrices of $\bar{\sigma} = R\hat{\sigma}\hat{R}^t$ for the horizontal plane and comparing terms, we find

$$
\begin{pmatrix} \bar{\sigma}_{11} \\ \bar{\sigma}_{12} \\ \bar{\sigma}_{22} \end{pmatrix} = \begin{pmatrix} R_{11}^2 & 2R_{11}R_{12} & R_{12}^2 \\ R_{11}R_{21} & R_{11}R_{22} + R_{12}R_{21} & R_{12}R_{22} \\ R_{21}^2 & 2R_{21}R_{22} & R_{22}^2 \end{pmatrix} \begin{pmatrix} \hat{\sigma}_{11} \\ \hat{\sigma}_{12} \\ \hat{\sigma}_{22} \end{pmatrix}
\tag{8.10}
$$

and will denote the $3 \times 3$ matrix by $(RR)$ henceforth.

It is instructive to express both the transfer matrix and both sigma matrices $\bar{\sigma}$ and through Twiss parameters, where we employ Equation 3.77 to the transfer matrix. First, we rewrite $R = \bar{A}^{-1}\mathcal{O}A$ in terms of $3 \times 3$ matrices

$$
(RR) = \begin{pmatrix} \bar{\beta} & 0 & 0 \\ -\bar{\alpha} & 1 & 0 \\ \frac{\bar{\alpha}^2}{\bar{\beta}} & -2\frac{\bar{\alpha}}{\bar{\beta}} & \frac{1}{\bar{\beta}} \end{pmatrix}
\tag{8.11}
$$

$$
\begin{pmatrix} \cos^2\mu & 2\cos\mu\sin\mu & \sin^2\mu \\ -\cos\mu\sin\mu & \cos^2\mu - \sin^2\mu & \sin\mu\cos\mu \\ \sin^2\mu & -2\sin\mu\cos\mu & \cos^2\mu \end{pmatrix} \begin{pmatrix} \frac{1}{\beta} & 0 & 0 \\ \frac{\alpha}{\beta} & 1 & 0 \\ \frac{\alpha^2}{\beta} & 2\alpha & \beta \end{pmatrix},
$$

where the Twiss parameters without a bar are the design beta functions at the location of the perturbing quadrupole and those with a bar refer to downstream location, while $\mu$ is the phase advance between these locations. We introduce the following shorthand notation $(RR) = (\bar{A}\bar{A})^{-1}(\mathcal{O}\mathcal{O})(AA)$ for the $3 \times 3$ matrices appearing in Equation 8.11. In particular the matrix with the phase advance $\mu$ we call $(\mathcal{O}\mathcal{O})$ and rewrite it as

$$
(\mathcal{O}\mathcal{O}) = \frac{1}{2}\begin{pmatrix} 1 & 0 & 1 \\ 0 & 0 & 0 \\ 1 & 0 & 1 \end{pmatrix} + \frac{1}{2}\begin{pmatrix} \cos 2\mu & 2\sin 2\mu & -\cos 2\mu \\ -\sin 2\mu & 2\cos 2\mu & \sin 2\mu \\ -\cos 2\mu & -2\sin 2\mu & \cos 2\mu \end{pmatrix}.
\tag{8.12}
$$

We see that it has a constant contribution and a contribution that oscillates with twice the betatron phase advance. We thus expect the beam envelope, as given by the sigma matrix, to oscillate with twice the betatron phase advance around a baseline value that is determined by the constant term. Let us investigate this constant term further and apply $(AA)$ to the mismatched beam matrix $\hat{\sigma}$, expressed through the mismatched Twiss parameters

$$
(AA)\begin{pmatrix} \varepsilon\hat{\beta} \\ -\varepsilon\hat{\alpha} \\ \varepsilon\hat{\gamma} \end{pmatrix} = \varepsilon\begin{pmatrix} \hat{\beta}/\beta \\ \alpha\hat{\beta}/\beta - \hat{\alpha} \\ \alpha^2\hat{\beta}/\beta - 2\alpha\hat{\alpha} + \beta\hat{\gamma}/\hat{\beta} \end{pmatrix} = \varepsilon\begin{pmatrix} a \\ b \\ c \end{pmatrix}.
\tag{8.13}
$$

Here we introduced the abbreviations $a, b,$ and $c$ for the components of the vector with the Twiss parameters. Multiplying this expression from the left with $\mathcal{O}\mathcal{O}$ we obtain

$$
\varepsilon\left[\begin{pmatrix} \frac{1}{2} & 0 & \frac{1}{2} \\ 0 & 0 & 0 \\ \frac{1}{2} & 0 & \frac{1}{2} \end{pmatrix} + \begin{pmatrix} \frac{1}{2} & 0 & -\frac{1}{2} \\ 0 & 1 & 0 \\ -\frac{1}{2} & 0 & \frac{1}{2} \end{pmatrix}\cos 2\mu + \begin{pmatrix} 0 & 1 & 0 \\ -\frac{1}{2} & 0 & \frac{1}{2} \\ 0 & -1 & 0 \end{pmatrix}\sin 2\mu\right]\begin{pmatrix} a \\ b \\ c \end{pmatrix}
$$

$$
= \varepsilon\left[\begin{pmatrix} \frac{a+c}{2} \\ 0 \\ \frac{a+c}{2} \end{pmatrix} + \begin{pmatrix} \frac{a-c}{2} \\ b \\ -\frac{a-c}{2} \end{pmatrix}\cos 2\mu + \begin{pmatrix} b \\ -\frac{a-c}{2} \\ -b \end{pmatrix}\sin 2\mu\right]
\tag{8.14}
$$

$$
= \varepsilon\left[\begin{pmatrix} B_{mag} \\ 0 \\ B_{mag} \end{pmatrix} + \sqrt{B_{mag}^2 - 1}\begin{pmatrix} \cos(2\mu - \varphi) \\ \sin(2\mu - \varphi) \\ -\cos(2\mu - \varphi) \end{pmatrix}\right] \quad \text{with} \quad \tan\varphi = \frac{2b}{a - c}.
$$

Here we introduce the beta-beat mismatch factor $B_{mag} = (a + c)/2$, given by

$$B_{mag} = \frac{1}{2}\left[\frac{\hat{\beta}}{\beta} + \frac{\alpha^2\hat{\beta}}{\beta} - 2\alpha\hat{\alpha} + \frac{\beta\hat{\gamma}}{\hat{\beta}}\right] = \frac{1}{2}\left[\left(\frac{\hat{\beta}}{\beta} + \frac{\beta}{\hat{\beta}}\right) + \beta\hat{\beta}\left(\frac{\alpha}{\beta} - \frac{\hat{\alpha}}{\hat{\beta}}\right)^2\right]. \tag{8.15}$$

The beam size $\bar{\sigma}_x^2$ at locations downstream of the perturbation are then given by multiplying the first component by the beta function $\bar{\beta}$ at the downstream location

$$\bar{\sigma}_x^2 = \varepsilon\bar{\beta}\left[B_{mag} + \sqrt{B_{mag}^2 - 1}\cos(2\mu - \varphi)\right], \tag{8.16}$$

which shows that the beam size oscillates with an amplitude given by $\sqrt{B_{mag}^2 - 1}$ at twice the betatron phase advance $\mu$ around an average value that is increased by $B_{mag}$. This oscillation is the so-called *beta beat*, because it shows up as a modulation of the beta functions at places where they should be equal.

It is remarkable that both increase and oscillation amplitude only depend on the single parameter $B_{mag}$ and therefore we investigate its magnitude for a quadrupole error that is characterized by its focal length $f$. We calculate the beam matrix after the quadrupole error $\bar{\sigma}$ and compare it to the beam matrix $\sigma$ that had not experienced the quadrupole error. For $\bar{\sigma}$ we have

$$\varepsilon\left(\begin{array}{cc} \hat{\beta} & -\hat{\alpha} \\ -\hat{\alpha} & \hat{\gamma} \end{array}\right) = \left(\begin{array}{cc} 1 & 0 \\ -1/f & 1 \end{array}\right)\varepsilon\left(\begin{array}{cc} \beta & -\alpha \\ -\alpha & \gamma \end{array}\right)\left(\begin{array}{cc} 1 & -1/f \\ 0 & 1 \end{array}\right). \tag{8.17}$$

Evaluating the matrix multiplications we find $\hat{\beta} = \beta$ and $\hat{\alpha} = \alpha + \beta/f$. Inserting $\hat{\beta}$ and $\hat{\alpha}$ into the expression for $B_{mag}$ in Equation 8.15 we find that a localized quadrupole error causes a phase shift of $\tan\varphi = 2f/\beta$ and a beta mismatch of

$$B_{mag} = 1 + \frac{\beta^2}{2f^2}. \tag{8.18}$$

This allows us to rapidly evaluate the effect of a quadrupolar perturbation on a the beam sizes at downstream locations.

After having analyzed the effect of quadrupolar perturbations we turn our attention to that of skew-quadrupolar perturbations.

## 8.2.3 Skew-quadrupolar perturbations

Skew quadrupolar errors in a beam line are most severe for very flat beams with a large emittance ratio $\varepsilon_x/\varepsilon_y$ where they couple the large amplitude oscillations in one plane into the plane with the smaller emittance and severely increase the smaller emittance. We therefore consider an initially uncoupled $4 \times 4$ beam matrix $\sigma$ and a thin skew quadrupole with focal length $f$ whose transfer matrix is given by evaluating Equation 3.35, which leads to

$$S = \left(\begin{array}{cccc} 1 & 0 & 0 & 0 \\ 0 & 1 & 1/f & 0 \\ 0 & 0 & 1 & 0 \\ 1/f & 0 & 0 & 1 \end{array}\right). \tag{8.19}$$

After passing through the skew quadrupole, the $2 \times 2$ vertical beam matrix $\hat{\sigma}$ is

$$\left(\begin{array}{cc} \hat{\sigma}_{33} & \hat{\sigma}_{34} \\ \hat{\sigma}_{34} & \hat{\sigma}_{44} \end{array}\right) = \left(\begin{array}{cc} \sigma_{33} & \sigma_{34} \\ \sigma_{34} & \sigma_{44} + \sigma_{11}/f^2 \end{array}\right). \tag{8.20}$$

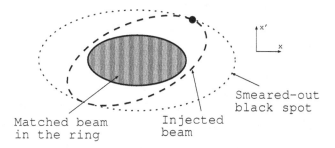

**Figure 8.4** Horizontal phase space at the injection point with the matched beam in the ring shown as the shaded ellipse. The dashed ellipse shows the phase-space of the injected beam with a small fraction denoted by the dark spot.

Calculating the projected vertical emittance $\hat{\varepsilon}_y$ of the perturbed beam by the determinant, we find

$$\hat{\varepsilon}_y^2 = \varepsilon_y^2 + \frac{\sigma_{11}\sigma_{33}}{f^2} = \varepsilon_y^2 \left(1 + \frac{\varepsilon_x}{\varepsilon_y}\frac{\beta_x\beta_y}{f^2}\right) . \tag{8.21}$$

We see that the vertical projected emittance $\hat{\varepsilon}_y$ increases significantly if the initial emittance ratio $\varepsilon_x/\varepsilon_y$ is large. Moreover, the coupling is proportional to the ratio of both beta functions, normalized to the focal length of the skew quadrupole. Thus we need to pay special attention to skew quadrupolar fields, often arising from quadrupoles misaligned with a roll angle, if we intend to transport very flat beams.

Often, at the end of a transfer beam line waits a circular accelerator and next we consider what happens if we inject a misaligned or mismatched beam into a ring.

### 8.2.4 Filamentation

The unavoidable momentum spread and the momentum dependence of the tune in a ring, due to a finite chromaticity, will cause any misaligned or mismatched beam to smear out in phase space. This causes its emittances to increase, which is particularly undesirable in hadron rings that do not have an inherent damping mechanism, as electron rings do. The smearing out is called *filamentation* and illustrated in Figure 8.4 where the shaded ellipse denotes a matched beam in the ring. The dashed line shows the phase space ellipse of the injected beam with a spot highlighted to the top right. Since the spot contains particles with many different momenta and therefore different tunes due to the chromaticity, the particles of the spot will smear out along the dotted ellipse, which has the same orientation as the shaded ellipse, but a larger area. Qualitatively, this leads to an increased emittance of the injected beam after many turns when the filamentation is complete.

To investigate this process quantitatively let us consider the horizontal plane at the injection point and assume that the beam coming from the transfer line has emittance $\varepsilon$ and Twiss parameters $\hat{\beta}$ and $\hat{\alpha}$. Also in this case, we can use Equation 8.16 to describe the beam size after $n$ revolutions $\sigma_n^2$ in the ring

$$\sigma_n^2 = \varepsilon\bar{\beta}\left[B_{mag} + \sqrt{B_{mag}^2 - 1} \ \cos(4\pi n(Q + Q'\delta) - \varphi)\right] \tag{8.22}$$

where $B_{mag}$ is given by Equation 8.15 and must be calculated from the Twiss parameters

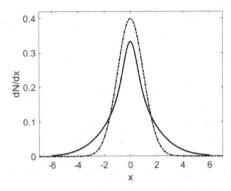
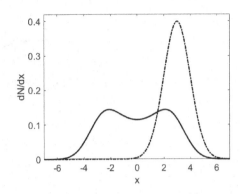

Figure 8.5 The left plot shows a matched beam (dashed) and the distribution after filamentation (solid) of a mismatched beam with $B_{mag} = 3$. The right plot shows the injected beam (dashed) with an rms beam size of 1 mm and injected displaced by 3 mm. After filamentation it becomes the double-humped distribution (solid).

of the injected beam and the periodic Twiss parameters of the ring at the injection point $\bar{\beta}$ and $\bar{\alpha}$. We replaced the phase advance $\mu$ in Equation 8.16 by the betatron tune $Q$ of the ring and the chromaticity $Q'$ and insert $\mu = 2\pi(Q + Q'\delta)$. If the incoming beam has a normalized Gaussian momentum distribution

$$\psi(\delta) = \frac{1}{\sqrt{2\pi}\sigma_\delta} e^{-\delta^2/2\sigma_\delta^2} , \tag{8.23}$$

we easily obtain the beam size after $n$ turns by averaging over the momentum distribution and obtain

$$\sigma_n^2 = \varepsilon\bar{\beta}\left[B_{mag} + e^{-2(2\pi Q'\sigma_\delta)^2 n^2}\sqrt{B_{mag}^2 - 1}\,\cos(4\pi n Q - \varphi)\right] . \tag{8.24}$$

The beam size oscillates with a frequency given by twice the tune. Moreover, the oscillation is damped by a factor $e^{-2(2\pi Q'\sigma_\delta)^2 n^2}$, which has a quadratic dependence on the turn number $n$ in the exponent. Thus the temporal evolution of the beam size shows a Gaussian dependence on the turn number $n$, rather than an exponential dependence. We can estimate the time scale of the damping to be $\hat{n} = 1/(9Q'\sigma_\delta)$. After $\hat{n}$ turns the amplitude is halved from its initial value and after a few times $\hat{n}$ turns the transient oscillations at twice the betatron tune will be damped and the beam will settle to an beam size whose increase is determined by $B_{mag}$. Here we only discussed the temporal evolution of the beam size after injection, but the other parameters in the beam matrix show the same increase, such that the net effect of a mismatched injected beam is an increase of the emittance of the stored beam, and the increase is described by $B_{mag}$. We need to point out that the injected beam after filamentation is not Gaussian [52] if $B_{mag}$ is large. The left-hand plot in Figure 8.5 shows that for $B_{mag} = 3$ the tails are more heavily populated with particles (solid) compared to the matched Gaussian (dashed).

Not only will a mismatch of the Twiss parameters of the injected beam cause the stored beam to filament due to its momentum spread. A "wrong" transverse position or angle also causes filamentation and increases the emittance. Moreover, the transverse distribution, shown in Figure 8.5, is very different from a Gaussian. If the beam has a large emittance ratio $\varepsilon_x/\varepsilon_y$ injecting a coupled beam will lead to an increased emittance. The increase can be described by Equation 8.21 with $f^2$ replaced by $2f^2$ as a consequence of the averaging over the momentum spread.

In this section we discussed how the imperfections affect the beam in beam line itself and after injecting into a ring. Next we turn to imperfections in the ring itself.

## 8.3  IMPERFECTIONS IN A RING

In contrast to a beam line, where imperfections only affect the beam downstream of the perturbation, the imperfections in a ring affect the beam everywhere. In equilibrium, after injection transients such as filamentation, have died away, the trajectory and the sigma matrix must be periodic and that poses additional constraints on the analysis.

The periodic trajectory in a ring is called the *closed orbit* and we first investigate how misalignment of elements affect it.

### 8.3.1  Misalignment and dipole kicks

We consider a single component that produces a misalignment vector $\vec{q}$, which we discussed in Sections 8.1.1 and 8.1.2. For the time being, we assume $\vec{q}$ to be general and adds a constant value to the four transverse phase-space coordinates $\vec{x} = (x, x', y, y')$. For example, a horizontal steering magnet only adds a constant value—its kick angle—to $x'$. We label the position at the location of the perturbation by the index $j$ and the requirement for the closed orbit to be periodic leads to

$$\vec{x}_j = R^{jj}\vec{x}_j + \vec{q}_j ,  \tag{8.25}$$

where $R^{jj}$ denotes the $4 \times 4$ transfer matrix for one turn in the ring that starts and ends immediately after location $j$ of the perturbation, and $x_j$ is the closed orbit immediately after the perturbation. The first superscript labels the location of the end point and the second that of the starting point for the transfer matrix. Solving for $\vec{x}_j$, we obtain

$$\vec{x}_j = (1 - R^{jj})^{-1}\vec{q}_j ,  \tag{8.26}$$

where 1 denotes the $4 \times 4$ unit matrix. The closed orbit at any other location denoted by the label $i$ can be calculated by propagating $\vec{x}_j$ by the $4 \times 4$ transfer matrix $R^{ij}$ from point $j$ to point $i$ and get

$$\vec{x}_i = R^{ij}\vec{x}_j = R^{ij}(1 - R^{jj})^{-1}\vec{q}_j = C^{ij}\vec{q}_j ,  \tag{8.27}$$

where we introduced the $4 \times 4$ response matrix $C^{ij}$ between location $j$ and $i$ defined by

$$C^{ij} = R^{ij}(1 - R^{jj})^{-1} .  \tag{8.28}$$

This expression allows us to calculate the closed orbit at any location $i$ in the ring for any perturbation at location $j$, provided we have access to the transfer matrices. The expression is valid for 1D or 2D systems with and without coupling. Note also, that the response matrix between points $j$ and $i$ has the same functionality as the transfer matrix $R^{ij}$ in a beam line; it relates changes in phase-space coordinate at position $j$ to those observable at position $i$. Only, the $C^{ij}$ have the periodicity requirement quasi "built-in" with the help of the term $(1 - R^{jj})^{-1}$.

If we only consider one dimension, say horizontal, and find out the effect of a steering magnet with a horizontal kick angle $\theta$, we use Equation 3.77 to write $R^{jj} = \mathcal{A}_x^{-1}\mathcal{O}^{jj}\mathcal{A}_x$, where $\mathcal{O}^{jj}$ is the rotation matrix with phase advance $\mu = 2\pi Q$ and $\mathcal{A}_x$ contains the Twiss parameters at the location of the steering magnet. Furthermore, we have $R^{ij} = \bar{A}_x^{-1} \mathcal{O}^{ij} \mathcal{A}_x$, where $\mathcal{O}^{ij}$ is the rotation matrix with phase advance $\mu_{ij}$ from the steering magnet to the position monitor and $\bar{A}_x$ contains the Twiss parameters at the location of the BPM. Inserting these representations Equation 8.28 and after some algebra, we find the change of the closed orbit $x_i$ at the location of the BPM

$$x_i = \left[ \frac{\sqrt{\beta_i \beta_j}}{2 \sin(\pi Q)} \cos(\mu_{ij} - \pi Q) \right] \theta . \qquad (8.29)$$

We identify the expression in the square brackets as the response coefficient $C_{12}^{ij}$ of the horizontal position at position $i$ (first super and sub-script) on the angle at position $j$ (second super and sub-script). Apart from the obvious divergence at integer tunes $Q$, we find the response coefficient to be proportional to the square root of the beta functions at the location of the steering magnet and the observation point.

In the previous paragraph we neglected that a ring, which operates with a radio-frequency system, forces the beam to maintain a constant revolution time. On the other hand, a corrector, which is located at a point $j$ with non-zero horizontal dispersion $D_j$, causes the circumference to change by $\Delta C = D_j \theta$. Since the RF system constrains the revolution frequency, according to Equation 5.32, the relative momentum of the particle must change by $\delta = -D_j \theta / \eta C$ and the particle moves on a trajectory that depends on the dispersion. At a BPM, located at position $i$ with horizontal dispersion $D_i$, the beam position additionally moves by $D_i \delta$. For the horizontal response coefficient we therefore obtain

$$C_{12}^{ij} = \left[ \frac{\sqrt{\beta_i \beta_j}}{2 \sin(\pi Q)} \cos(\mu_{ij} - \pi Q) - \frac{D_i D_j}{\eta C} \right] . \qquad (8.30)$$

For a generalization to coupled lattices and general misalignment vectors $\vec{q}$ see [53].

After this discussion of the consequences of imperfections that give rise to dipolar errors, we will look at the consequences of quadrupolar, or gradient errors in general.

## 8.3.2  Gradient imperfections

Gradient errors can be analyzed numerically by adding the perturbing quadrupoles to a lattice file and calculating the tunes and beta functions. If the ring is uncoupled, we can use the function R2beta, or, if the ring is coupled, because it contains skew quadrupoles or solenoids, we use edteng(). We then inspect the results graphically, as we did in Figure 3.17 in Section 3.5.

A deeper understanding, however, can be achieved by considering a simple one-dimensional model of a ring with tune $Q = \mu / 2\pi$ and adding a single, thin quadrupole at a location with Twiss parameters $\alpha, \beta$, and $\gamma$. Using the parameterization from Equation 3.51 for the unperturbed one-turn transfer matrix $\hat{R}$, and the transfer matrix from Equation 3.5, we call it $R_Q$, to describe the quadrupole, we obtain for the perturbed one-turn matrix

$$R_Q \hat{R} = \begin{pmatrix} \cos \mu + \alpha \sin \mu & \beta \sin \mu \\ -(\cos \mu + \alpha \sin \mu)/f + \gamma \sin \mu & \cos \mu - \alpha \sin \mu - (\beta/f) \sin \mu \end{pmatrix} . \qquad (8.31)$$

Using Equation 3.60, we find the perturbed tune $\bar{Q} = Q + \Delta Q$ from the trace of a transfer matrix. This leads to

$$2 \cos(2\pi(Q + \Delta Q)) = 2 \cos(2\pi Q) - \frac{\beta}{f} \sin(2\pi Q) . \qquad (8.32)$$

For weak perturbations we have $\beta/f \ll 1$ and $\Delta Q \ll 1$, such that the tune shift $\Delta Q$ is then given by

$$\Delta Q \approx \frac{\beta}{4\pi f} = \frac{1}{4\pi} \oint \beta(s)\Delta k_1(s)ds \ . \tag{8.33}$$

In the second equality we assumed that distributed gradient errors are described by a focusing function $\Delta k_1(s)$ with a azimuthally varying focal length $1/f \sim \delta k_1(s)ds$. Integrating around the ring then yields the tune shift $\Delta Q$, which is proportional to the unperturbed beta function $\beta$ at the location of the perturbation. This suggests to pay extra attention to the tolerances of components located near points where the beta function is large, such as the final focus quadrupoles.

A gradient perturbation not only generates a tune shift, but it also changes the beta functions. We therefore use Equation 3.60 to calculate the perturbed beta functions $\bar{\beta}$ from the elements of the matrix in Equation 8.31 and find

$$\bar{\beta} = \frac{\beta \sin(2\pi Q)}{\sin(2\pi(Q + \Delta Q))} \approx \beta \left[ 1 + 2\pi\Delta Q \cot(2\pi Q) \right] \ . \tag{8.34}$$

For the relative change of the beta function $\Delta\beta/\beta = (\bar{\beta} - \beta)/\beta$

$$\frac{\Delta\beta}{\beta} = 2\pi\Delta Q \cot(2\pi Q) \approx \frac{\beta}{2f} \cot(2\pi Q) \ , \tag{8.35}$$

which makes it obvious that the beta function diverges at half-integer values of the tune, because the cotangent has singularities at integer multiples of $\pi$. Note that here the argument of the cotangent is $2\pi Q$, whereas it was $\pi Q$ in Equation 8.29 for the effect of dipole imperfections on the closed orbit. This is a first indication that the multipolarity of the imperfection causes divergences at certain values of the fractional tune, here, for example, at integer and half-integer values.

To answer the question of how close to a half-integer tune we can operate the ring for a given gradient imperfection, we note that the cosine on the left-hand side of Equation 8.32 must stay in the range $\pm 1$. This implies for the right-hand side that

$$\left| \cos(2\pi Q) - \frac{\beta}{2f} \sin(2\pi Q) \right| \leq 1 \ . \tag{8.36}$$

Equation 8.36 describes the limit of stable operation and values of $Q$ outside this range lead to an unstable ring. These forbidden values for the tune are called a *stop bands* and are often shown as the width of resonance lines in the *tune diagram*. In a tune diagram, we display the tunes as a point $(Q_x, Q_y)$ in a two-dimensional plane with axes extending from the integer just below a tune value to the next integer. The right-hand plot in Figure 8.6 shows an example. Values outside the range of Equation 8.36 appear as the horizontal or vertical lines. Note that here we only calculated the effect of a single gradient errors in order to show qualitatively that gradient imperfections in a ring shrink the accessible range of tunes.

### 8.3.3 Skew-gradient imperfections

To analyze the consequences of skew gradients we consider a single skew quadrupole and calculate the one-turn transfer matrix for a ring that is shown on the left-hand side in Figure 8.6. A start and end-point of the ring, having tunes $Q_x$ and $Q_y$, are chosen as reference point and we assume that phase space variable of the reference point are expressed in

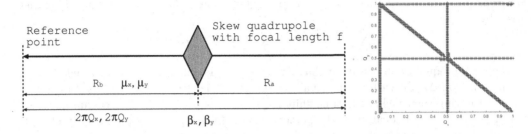

**Figure 8.6** A beam line with a single skew quadrupole (left) and a tune-diagram with stop bands caused by quadrupoles and a single skew-quadrupole (right).

normalized phase space coordinates with Twiss parameters factored out. Then the transfer matrices $R_a$ and $R_b$ are rotation matrices with the phase advances only. The phase advance from the skew quadrupole to the end is given by $\mu_x$ and $\mu_y$ and the beta functions at the skew quadrupole are $\beta_x$ and $\beta_y$, respectively. If we use the matrices $\mathcal{A}$ from Equation 3.51 to map from normalized to real phase space, the corresponding matrix $\tilde{S}$ for the skew quadrupole from Equation 8.19 that acts on normalized phase space variables is then given by

$$\tilde{S} = \begin{pmatrix} 1_2 & \tilde{Q} \\ \tilde{Q} & 1_2 \end{pmatrix} \quad \text{with} \quad \tilde{Q} = \begin{pmatrix} 0 & 0 \\ \sqrt{\beta_x \beta_y}/f & 0 \end{pmatrix} \tag{8.37}$$

where we introduce the 2 unit matrix $1_2$. For the one-turn transfer matrix $\hat{R}$ we write

$$\hat{R} = R_b \tilde{S} R_a = \left( R_b \tilde{S} R_b^{-1} \right) (R_b R_a) , \tag{8.38}$$

which shows that the left bracket contains a similarity transformation $R_b \tilde{S} R_b^{-1}$ that moves the effect of the skew quadrupole to the reference point and the right bracket contains the unperturbed transfer matrix for the ring. For the similarity transformation, we obtain

$$R_b \tilde{S} R_b^{-1} = \begin{pmatrix} O_x & 0 \\ 0 & O_y \end{pmatrix} \begin{pmatrix} 1_2 & \tilde{Q} \\ \tilde{Q} & 1_2 \end{pmatrix} \begin{pmatrix} O_x^t & 0 \\ 0 & O_y^t \end{pmatrix} = \begin{pmatrix} 1_2 & O_x \tilde{Q} O_y^{-1} \\ O_y \tilde{Q} O_x^{-1} & 1_2 \end{pmatrix} , \tag{8.39}$$

where we introduce the rotation matrices for the horizontal phase advance $O_x = O_x(\mu_x)$ and the corresponding one for the vertical plane. We note that $R_b \tilde{S} R_b^{-1} = 1 + \tilde{P}$ can be written as a small perturbation $\tilde{P}$ to a $4 \times 4$ unit matrix. Moreover, $\tilde{P}$ has only non-zero entries in the coupling part of the transfer-matrix that contain $O_x \tilde{Q} O_y^{-1}$ for which we then find

$$\begin{aligned} O_x \tilde{Q} O_y^{-1} &= \frac{\sqrt{\beta_x \beta_y}}{f} \begin{pmatrix} \sin\mu_x \cos\mu_y & -\sin\mu_x \sin\mu_y \\ \cos\mu_x \cos\mu_y & -\cos\mu_x \sin\mu_y \end{pmatrix} \\ &= \frac{\sqrt{\beta_x \beta_y}}{2f} \begin{pmatrix} \sin(\mu_x - \mu_y) + \sin(\mu_x + \mu_y) & -\cos(\mu_x - \mu_y) + \sin(\mu_x + \mu_y) \\ \cos(\mu_x - \mu_y) + \cos(\mu_x + \mu_y) & \sin(\mu_x - \mu_y) - \sin(\mu_x + \mu_y) \end{pmatrix} \end{aligned} \tag{8.40}$$

and for $O_y \tilde{Q} O_x^{-1}$ we only need to exchange $\mu_x$ and $\mu_y$. Here we observe that there are four independent coefficients: sine and cosine for $\mu_x - \mu_y$ and $\mu_x + \mu_y$, respectively.

If the ring contains several skew quadrupoles, they can be treated in the same fashion; we first move the one closest to the reference point, then the next one until the last skew quadrupole, which was originally closest to the right-hand side of the ring, is moved to the reference point. We then represent the transfer matrix for the entire ring with multiple skew quadrupoles by

$$\hat{R} = (1 + \tilde{P}_1)(1 + \tilde{P}_2) \cdots R_0 \approx (1 + \tilde{P}_1 + \tilde{P}_2 + \cdots)R_0 \tag{8.41}$$

with the unperturbed one-turn transfer matrix $R_0$. The approximation violates the symplecticity of the transfer matrix but shows that the four coefficients in $\tilde{P}$ add up to first order in the strength of the skew quadrupole excitation $\sqrt{\beta_x \beta_y}/f$. Assembling each of the two phases into an exponential, we see that the cumulative "strength" of all skew quadrupoles is proportional to

$$F_\pm = \sum_j \frac{\beta_{x,j}\beta_{y,j}}{2f_j} e^{i(\mu_{x,j} \pm \mu_{y,j})} \,, \tag{8.42}$$

where the sum extends over all skew quadrupoles in the ring. The beta functions are those at the skew quadrupoles and the phases $\mu_x$ and $\mu_y$ are between the skew quadrupole and the reference point. If the $F_\pm$ are zero, the ring is globally uncoupled at the reference point. Conversely, the coupling parameters $F_\pm$ can become large in rings that consist of a large number of equal cells in a constant relative phase relation, such that systematic imperfections have a fixed phase relation $\mu_x \pm \mu_y$ and coherently add up.

So, why is coupling bad for a ring? In planar rings the emittance is predominantly generated in the horizontal plane and ideally the vertical beam size is negligibly small. In the presence of coupling, however, the horizontal oscillations are coupled into the vertical plane and cause the vertical emittance to become non-zero. In synchrotron light sources this effect spoils the coherence properties of the emitted synchrotron radiation. But the most detrimental property is that it introduces stop-bands into the tune diagram in much the same way quadrupoles introduce half-integer stop bands, as described by Equation 8.36. To understand this, we model the ring by rotation matrices with tunes $Q_x$ and $Q_y$ with equal beta functions $\beta_x = \beta_y = \beta$ at the location with the quadrupoles, where we add an upright and a skew quadrupole with equal normalized strength $\beta/f = 0.1$. Scanning the tunes and plotting a plus sign, whenever one of the eigenvalues of the one-turn map exceeds unity, we obtain the stability diagram shown on the right-hand side in Figure 8.6. We see vertical and horizontal lines indicating the stopbands caused by the upright quadrupole, but also along the diagonal line $Q_x + Q_y = p$ where the motion becomes unstable around the sum resonance. This shows that skew quadrupoles indeed shrink the available space in the tune diagram. The situation is aggravated if non-linear elements such as sextupoles are present, with upright sextupoles exciting one type of resonances and skew-sextupoles exciting others. If however, skew quadrupoles couple the planes both types of resonances are driven and this will generate additional forbidden lines in the tune diagram. We will return to these matters in Chapter 11.

Now that we have see in what way imperfections deteriorate the performance of accelerators we will turn to correction methods that allow us to alleviate their impact.

## 8.4 CORRECTION IN BEAM LINES

We start by looking at correction algorithms for beam lines. They are conceptually simpler than rings, because correction elements only affect the beam downstream of the correction element. Often one only needs to correct the trajectory at one particular point of interest.

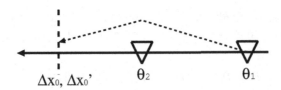

**Figure 8.7** Two steering magnets change the position by $\Delta x_0$ and the angle by $\Delta x_0'$.

### 8.4.1 Trajectory knobs and bumps

We use steering magnets that control the angles of the beam trajectory. For example, we might change the horizontal position at a reference point by $\Delta x_0$ or the horizontal angle by $\Delta x_0'$. A simple example is shown in Figure 8.7, where we use two steering magnets to change position and angle at a downstream point. Such a configuration is used to adjust the trajectory at the injection point into a ring. If there are only drift spaces of equal length $L$ between the correctors and the reference point, the effect of the steering magnets on $\Delta x_0$ and $\Delta x_0'$ are: $\Delta x_0 = 2L\theta_1 + L\theta_2$ and $\Delta x_0' = \theta_1 + \theta_2$. Requiring the position to change by $\Delta x_0$ but the angle to stay the same ($\Delta x_0' = 0$) implies that we need to use following linear combination of steering magnet excitations $(\theta_1, \theta_2) = (1/L, -1/L)\Delta x_0$. Such a linear combination of actuators—here steering magnets—that fulfill certain constraints (change $\Delta x_0$ but not $\Delta x_0'$) is often called a *knob*, because changing a single parameter causes the actuators to proportionally follow the set value of the parameter. Using the two steerers, we can also require to change the angle, but leave the position unaffected. This leads to a second knob: $(\theta_1, \theta_2) = (-1, 2)\Delta x_0'$. Since the knob that changes the position leaves the angle unaffected and vice versa, the two knobs are called *orthogonal knobs*.

If the beam line between the first steerer and the reference point is more complex than just drift spaces, we need to take the respective transfer matrices into account. We proceed by determining one equation for each constraint: one for $\Delta x_0$ and another for $\Delta x_0'$ and describe how each of the actuators affects the constraint. Assembling the equations in matrix form results in

$$\begin{pmatrix} \Delta x_0 \\ \Delta x_0' \end{pmatrix} = \begin{pmatrix} R_{12}^{01} & R_{12}^{02} \\ R_{22}^{01} & R_{22}^{02} \end{pmatrix} \begin{pmatrix} \theta_1 \\ \theta_2 \end{pmatrix}. \tag{8.43}$$

The superscript of the transfer matrices denote the end and start points, such that $R^{01}$ denotes the transfer matrix from steering magnet labeled by 1 to the reference point, labeled by 0. The subscripts have the usual meaning, such that $R_{12}^{01}$ describes the influence of the angle from steering magnet 1 on the position at the reference position and $R_{22}^{01}$ describes the influence of the angle change from steering magnet 1 on the angle at the reference point. One may call the matrix the response matrix of the actuators on the constraints. Inverting the equations yields

$$\begin{pmatrix} \theta_1 \\ \theta_2 \end{pmatrix} = \begin{pmatrix} R_{12}^{01} & R_{12}^{02} \\ R_{22}^{01} & R_{22}^{02} \end{pmatrix}^{-1} \begin{pmatrix} \Delta x_0 \\ \Delta x_0' \end{pmatrix} \tag{8.44}$$

and the columns of *the inverse of the response matrix are the knob coefficients* that will allow us to change one constraint without affecting the other.

In the previous example, the trajectory fulfills the constraint at the reference point, but it will oscillate further downstream in an uncontrolled way. To avoid this and only affect the trajectory locally, we add two additional steerers downstream of the reference point to steer the trajectory back onto the reference orbit. Figure 8.8 illustrates this configuration. The

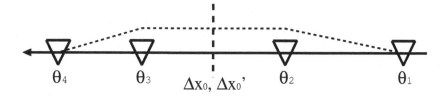

Figure 8.8 A closed bump with four steering magnets.

first two steerers, labeled 1 and 2, at the right adjust the orbit to change the position and angle by $\Delta x_0$ and $\Delta x_0'$, at the reference position. The following two correctors, labeled 3 and 4 later undo the changes to the trajectory, such that the changes to $\Delta x_0$ and $\Delta x_0'$, are no longer visible downstream of the last steerer. Such a configuration is called *closed bump*, or just *bump*.

To calculate the required excitations of the steerers for such a bump, we follow the same procedure as in the previous example. We only have two additional steerers and two additional constraints, namely that the combined effect of all four correctors causes the trajectory to be back on the reference trajectory. We denote the kick angles of the steerers by $\theta_j$ and the transfer matrix between corrector $j$ and the reference point by $R^{0j}$ and to the final point immediately after the last steerer by $R^{fj}$. Then we assemble the equations that describe how every corrector affects the constraints—the position and angle at the respective points. The four equation are assembled in the following equation:

$$
\begin{pmatrix} x_0 \\ x_0' \\ x_f = 0 \\ x_f' = 0 \end{pmatrix} = \begin{pmatrix} R_{12}^{01} & R_{12}^{02} & 0 & 0 \\ R_{22}^{01} & R_{22}^{02} & 0 & 0 \\ R_{12}^{f1} & R_{12}^{f2} & R_{12}^{f3} & R_{12}^{f4} \\ R_{22}^{f1} & R_{22}^{f2} & R_{22}^{f3} & R_{22}^{f4} \end{pmatrix} \begin{pmatrix} \theta_1 \\ \theta_2 \\ \theta_3 \\ \theta_3 \end{pmatrix} .
\tag{8.45}
$$

The interpretation of this equation is very intuitive. Consider the first equation, that defines the upper row of the matrix appearing in Equation 8.45. We want to change the position by $\Delta x_0$ at the reference point and only the upstream correctors 1 and 2 can affect it. The next line does the same for the change of angle $\Delta x_0'$. The third and fourth rows come from the requirement that the position and angle at the end of the bump must be zero, colloquially referred to as "closing the bump." As before, the knob coefficients are given by columns of the inverted matrix from Equation 8.45. In this case, only the first two columns that describe the dependence of the $\theta$ are meaningful. They give us the knobs for position and angle. The constraints to close the bump are fulfilled automatically.

The procedure to generate trajectory knobs can be easily extended and used to calculate knobs to change the position at several places simultaneously and synchronously, or to only change the position, without caring about the angle. If the beam line contains coupling elements, such as skew quadrupoles, we have to use off-diagonal transfer-matrix elements, for example $R_{14}$, that describe the dependence of the horizontal position on a vertical change of the angle at an upstream location. One of the most common requirements is to steer the trajectory to the center of all beam position monitors, and that is called *orbit correction*.

## 8.4.2  Orbit correction

Let us assume that we have $n$ position monitors and they report positions $x_i$ for $i = 1, \ldots, n$. Our task is to adjust the kick angles $\theta_j$ for $j = 1, \ldots, m$ of a number $m$ of steering magnets

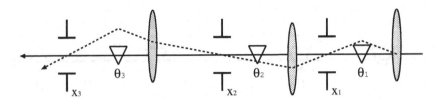

Figure 8.9 One-to-one orbit correction.

such that the monitors ideally all report zero. To be specific, here we consider the horizontal plane, but generalizations to the vertical plane and for coupled lattices follow the same pattern. In the spirit of the previous section, we interpret the required change of positions on the monitors as constraints to change the trajectory by $-x_i$ and use the $m$ steering magnets to achieve this objective. In order to calculate the changes in kick angles $\theta_j$ we construct the BPM-corrector response matrix in which each row corresponds to one constraint, each associated with one monitor. It is given in the following equation:

$$
\begin{pmatrix} -x_1 \\ -x_2 \\ \vdots \\ -x_n \end{pmatrix} = \begin{pmatrix} R_{12}^{11} & R_{12}^{12} & \cdots & R_{12}^{1m} \\ R_{12}^{21} & R_{12}^{22} & \cdots & R_{12}^{2m} \\ \vdots & \vdots & \ddots & \vdots \\ R_{12}^{n1} & R_{12}^{n2} & \cdots & R_{12}^{nm} \end{pmatrix} \begin{pmatrix} \theta_1 \\ \theta_2 \\ \vdots \\ \theta_m \end{pmatrix}. \tag{8.46}
$$

Obviously a steering magnet can only affect the reading of monitors downstream and all transfer matrices relating a steerer to an upstream monitor are zero. Once this response matrix is assembled, determining the $\theta_j$ from Equation 8.46 involves solving a set of linear equations.

Occasionally, some position monitors are more reliable than others and we can quantify the level of reliability by assigning BPM error bars $\sigma_i$ to a BPM labeled $i$. If the error bar is small, the BPM is reliable and if it is very large, we do not trust the BPM very much. It is now straightforward to assign $1/\sigma_i$ as a weight to each of the measurements, represented by a row in Equation 8.46. In practice, one constructs a matrix $W = \text{diag}(1/\sigma_i)$ with the inverses of the BPM error bars on its diagonal and multiplies Equation 8.46 from the left by $W$. This changes all the entries on the left-hand side from $-x_i$ to $-x_i/\sigma_i$ and the transfer matrix elements from $R_{12}^{ij}$ to $R_{12}^{ij}/\sigma_i$. The resulting equation has the same structure as before, but each measurement is weighted by the appropriate BPM error bar.

Before discussing different methods to solve Equation 8.46, we consider a short beam line with three interleaved position monitors and steering magnets. Figure 8.9 illustrates the geometry and the reasoning. The beam enters the beam-line section from the right and a misaligned quadrupole causes the trajectory, shown as a dashed line, to deviate from the reference orbit, shown by the solid horizontal line. The first steering magnet applies a deflection angle $\theta_1$ to steer the beam through the center of monitor 1 and makes $x_1$ zero. In the same way, steerer 2 adjusts the trajectory to make $x_2$ zero and steerer 3 zeroes $x_3$. The corresponding set of equations then has the following form

$$
\begin{pmatrix} -x_1 \\ -x_2 \\ -x_3 \end{pmatrix} = \begin{pmatrix} R_{12}^{11} & 0 & 0 \\ R_{12}^{21} & R_{12}^{21} & 0 \\ R_{12}^{31} & R_{12}^{32} & R_{12}^{33} \end{pmatrix} \begin{pmatrix} \theta_1 \\ \theta_2 \\ \theta_3 \end{pmatrix}, \tag{8.47}
$$

which, given initial monitor readings $x_i$ will force them to zero after applying the calculated angles $\theta_j$. In this beam line, steerer 2 and 3 cannot affect the position on monitor 1. This explains the zeros in the first row. In a similar fashion the zero on the second row is explained by steerer 3 being unable to affect the upstream monitor 2. The trigonal shape of the matrix allows successively solving on equation at a time, starting from the first. And this corresponds to using steerer 1 to fix monitor 1, then, given that monitor 1 is already corrected, adjust steerer 2 to fix monitor 2, and so on. The method of successively correcting the trajectory one monitor at a time is called *one-to-one* orbit correction. Note, however, that despite all BPMs showing zero, the orbit, shown by the dashed line, is not zero everywhere, but only at the BPMs. In cases, as our example, where the number of steerers and monitors is equal, the linear equations can be easily solved by inverting the response matrix.

In situations with more monitors than steerers we have $n > m$ and the set of system of linear equations is over-determined. The standard way to solve it is in the least-square sense by minimizing $\chi^2 = |-\vec{x} - A\vec{\theta}|^2$, where we denote the matrix in Equation 8.46 by $A$, the column vectors with $x_i$ by $\vec{x}$, and the column vector with $\theta_j$ by $\vec{\theta}$. Minimizing $\chi^2$ with respect to the components of $\vec{\theta}$ leads to $2A^t(-\vec{x} - A\vec{\theta}) = 0$ and solving for $\vec{\theta}$ yields the so-called *pseudo-inverse*

$$\vec{\theta} = -(A^t A)^{-1} A^t \vec{x} \ . \tag{8.48}$$

This allows us to correct the orbit in the least-squares sense, if there a fewer steerers than monitors available. In the special case that $A$ is square and non-degenerate $(A^t A)^{-1} A^t$ reverts to the conventional matrix inverse.

In situations where we have more steerers available than monitors $m < n$ there are more columns than rows in the matrix $A$ and we have too few constraints to uniquely determine the steerer excitations $\vec{\theta}$. Imagine a single BPM and many correctors; in that case the matrix $A$ is one long row vector, which cannot be "inverted" in the normal sense, because it is highly degenerate. Under such circumstances *singular value decomposition* (SVD) comes to the rescue.

SVD is a linear-algebra algorithm, which explicitly constructs three matrices $O, \Lambda$, and $U$ to write any matrix $A$, suitably augmented by zeros in order to make it square, as the product of these matrices in the following way

$$A = O\Lambda U^t \ , \tag{8.49}$$

where $\Lambda$ is a diagonal matrix with "eigenvalues" $\lambda$ on the diagonal and $O$ and $U$ are orthogonal matrices that have the property that their inverse equals their transpose.

The SVD decomposition has an intuitive interpretation. The effect of a matrix $A$ on any vector is to apply the orthogonal matrix $U^t$ first, which is just a rotation into a different coordinate system. Then the diagonal matrix $\Lambda$ stretches the respective axes in the new coordinate system by the eigenvalue and finally the matrix $O$ rotates the stretched vectors into some other direction. This intuitive view aids us in calculating the "inverse" of $A$. The "inverse" of $A$ in terms of the matrices $O, U$, and $\Lambda$ is simply given by

$$"A^{-1}" = U"\Lambda^{-1}"O^t \ . \tag{8.50}$$

Here "$\Lambda^{-1}$" is a special inverse of the diagonal matrix $\Lambda$, which is normally done by inverting the entries on the diagonal. But if one eigenvalue, say $\lambda_k$, is actually zero, we have to calculate $1/0$, which would lead to a meaningless result. On the other hand, the intuitive interpretation guides us to the realization that the problem is located in the subspace spanned by the eigenvectors (columns) of $O$ and $U$. The rule we apply is thus to invert the matrix where we can (where the eigenvalues are different from zero) and remove any

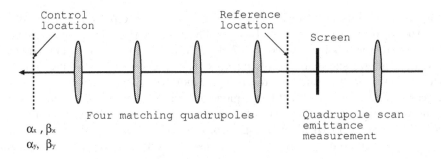

**Figure 8.10** Four quadrupoles are needed to independently adjust $\alpha_x, \beta_x, \alpha_y,$ and $\beta_y$ in an uncoupled beam line.

projection onto the subspace where the eigenvalues are zero. This leads to the strange recipe [54] to set $1/0$ to zero when calculating $\Lambda^{-1}$. As a by-product, this recipe minimizes the norm of the solution vector, such that we achieve our objective, to minimize the $x_i$, with the least possible rms change of corrector excitations.

Now that the trajectory is corrected, we need to adjust the beam- or sigma-matrix.

### 8.4.3 Beta matching

In Sections 8.2.2 and 8.2.4 we found that gradient errors lead to the undesirable beta beating and filamentation with increased emittance. In order to adjust the sigma matrix elements, or equivalently the Twiss parameters $\alpha$ and $\beta$ in both transverse planes, we need four quadrupoles to independently adjust $\alpha_x, \beta_x, \alpha_y,$ and $\beta_y$. Since a quadrupole always acts in both planes we need to consider both planes simultaneously. An additional difficulty is that the Twiss parameters or the sigma-matrix elements depend nonlinearly on the quadrupole excitations and additionally on the sigma matrix of the beam that arrives from upstream.

As a specific example, let us consider the beam line shown in Figure 8.10, where the control location on the left-hand side can be envisioned as the injection point into a ring at which we need to match the phase space ellipse of the injected beam to that of the stored beam. Furthermore, we have four quadrupoles at our disposal to adjust $\alpha_x, \beta_x, \alpha_y,$ and $\beta_y$ at the control location. If we have an emittance measurement system available, based, for example, on quadrupole scans from Chapter 7, we can experimentally determine the Twiss parameters at the reference location. Knowing the four Twiss parameters at this location we then use the matching procedure from Section 3.6.2 to adjust $\alpha_x, \beta_x, \alpha_y,$ and $\beta_y$ at the control location.

If we have no diagnostic equipment to determine the Twiss parameters upstream of the matching section with the four quadrupoles, we can still calculate orthogonal knobs that change one of the four constraints $\alpha_x, \beta_x, \alpha_y,$ and $\beta_y$ at a time without affecting the others. We base their calculation on the assumption that the beam at the reference location is the design beam from an optics code. This will cause the knobs to be only approximately orthogonal, and we may have to iterate their application. The procedure to empirically optimize some observable, say the stored current, is based on optimizing with respect to one knob at a time, and once the maximum is reached, start to optimize with respect to the next knob. After a few rounds of applying the four knobs the stored current should be maximized.

A change of $\alpha$ in the vicinity of a beam waist has an intuitive interpretation, because

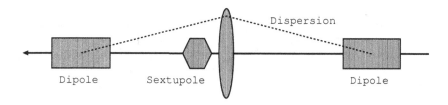

**Figure 8.11** Correcting the chromaticity by placing sextupoles at locations with non-zero dispersion.

$\alpha$ changes linearly with the distance from the waist with $\beta = \beta_0$ such that changing $\alpha$ is equivalent to changing the longitudinal position $\Delta s$ of the beam waist by $\Delta s = -\alpha \beta_0$.

### 8.4.4 Dispersion and chromaticity

In beam transport lines, the dispersion is usually measured by slightly changing the energy of the incoming beam and observing the trajectory change on position monitors. Often quadrupoles in a section of the transport line, in which the dispersion is non-zero by design, are used to adjust the dispersion to become zero in the final dipole of that section.

The momentum-dependent focusing plays an important role in transport lines with very large beta functions in very strong quadrupoles as is common in final focus systems. To compensate the chromaticity from the final-focus quadrupoles we need other, easily controllable, and momentum-dependent quadrupolar fields. For this purpose we place sextupoles at a location with horizontal dispersion $D_x$, as illustrated in Figure 8.11. There, the particles are sorted by their momentum offset $\delta$ and experience a different slope—the gradient varies with momentum—of the sextupolar field. Here we assume that the dispersion is predominantly in the horizontal plane. We can therefore use Equation 8.5 and replace the transverse displacement $d_x$ by $D_x \delta$

$$\Delta x' - i\Delta y' = -\frac{k_2 L}{2} \left[ (x + iy)^2 + 2D_x \delta(x + iy) + D_x^2 \delta^2) \right] , \qquad (8.51)$$

where $k_2 L$ is the integrated strength of the sextupole. The second term describes a quadrupolar field with strength, or inverse focal length, that is proportional to the relative momentum offset $\delta$

$$\frac{1}{f_\delta} = k_2 L D_x \delta . \qquad (8.52)$$

We see that changing the sextupole strength $k_2 L$ changes the strength of these momentum-dependent quadrupoles. Final focus systems therefore contain a section with dispersion deliberately made large. Sextupoles are placed in this section in order to compensate the large chromatic effects from the final focus quadrupoles.

After having corrected the beam lines we now turn to rings.

## 8.5 CORRECTION IN RINGS

Here we will correct the problems that imperfections in a ring cause and start with the closed orbit.

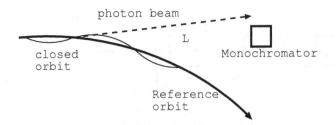

photon beam

L

closed
orbit

Monochromator

Reference
orbit

Figure 8.12 The angle of the source point of the emission of synchrotron radiation must be adjusted such that the photon beam (dashed) the monochromator at the end of an often long synchrotron radiation beam line.

### 8.5.1 Orbit correction

The closed and local trajectory bumps from Section 8.4.1 do not disturb the orbit outside the range of the bump, possibly except for a small change in the revolution frequency as discussed near the end of Section 8.3.1, and can equally well be used in a ring. Bumps, for example, are used to center the beams in collimators without perturbing the trajectory outside the collimation section. A second application is steering the beam orbit in a way that the synchrotron radiation hits the monochromator of an experiment as shown in Figure 8.12. The angles $x'$ and $y'$ at the source point of the synchrotron radiation inside a dipole magnet will, owing to the often large distance $L$, significantly move the photon beam on the monochromator. Including the location of the monochromator is trivially accomplished by adding two drift spaces with length $L$ and $-L$ at the source point and placing a "virtual position monitor" in between these two drift spaces. This addition will not affect the particle optics, but will allow us to include constraints and control points for the monochromator.

The global correction of the closed orbit in a ring follows the same general idea as trajectory correction in a beam line. We use the position monitor readings $-x_i$ as constraints to calculate the steering magnet excitations $\theta_j$. To set up the systems of equations that relate how each steerer affects each monitor we use the response coefficients, either calculated from Equation 8.28 or 8.30, and obtain

$$
\begin{pmatrix} -x_1 \\ -x_2 \\ \vdots \\ -x_n \end{pmatrix} = \begin{pmatrix} C_{12}^{11} & C_{12}^{12} & \cdots & C_{12}^{1m} \\ C_{12}^{21} & C_{12}^{22} & \cdots & C_{12}^{2m} \\ \vdots & \vdots & \ddots & \vdots \\ C_{12}^{n1} & C_{12}^{n2} & \cdots & C_{12}^{nm} \end{pmatrix} \begin{pmatrix} \theta_1 \\ \theta_2 \\ \vdots \\ \theta_m \end{pmatrix}, \tag{8.53}
$$

which is entirely analogous to the construction of Equation 8.46. The only difference is that here we need to use the response coefficients $C^{ij}$ that fulfill the periodicity constraint. Moreover, all entries in the matrix can be non-zero, because in a periodic system, any steerer can affect the position reading of every monitor. Adding constraints to leave the photon beam on the monochromators unaffected can be implemented by adding a line with the constraint not to change the position on a "virtual monitor" at the monochromator. The large matrix containing the response coefficients $C^{ij}$ is usually referred to as *orbit response matrix*.

Weighing the BPMs by their respective error bars is handled in the same way we discussed in Section 8.4.2. Likewise, solving Equation 8.53 to find the steering magnet excitations that null the monitor readings involves the same linear algebra operations we discussed in Section 8.4.2, depending on whether we have more monitors than steerers or vice versa. Also here, SVD is commonly used for degenerate or near-degenerate orbit response matrices. Another often used algorithm to solve Equation 8.53 is the *MICADO* algorithm [55], which is especially suitable for rings with many monitors and correctors, such as LHC with several hundreds of each. MICADO operates according to the following principle: first pick the corrector that improves the orbit the most and apply the correction. Then repeat this procedure with the remaining orbit and correctors. This appears like a lot of searching, but the algorithm can be very efficiently coded using Householder-transformations of the response matrix [55].

### 8.5.2 Dispersion-free steering

The steering magnets we use to correct the trajectory are small dipole magnets and as such also change the dispersion in the same way as large dipole magnets, discussed in Section 3.1.3, do. From the sixth column of the transfer matrix from Equation 3.24, we find that to first order in $l/\rho \ll 1$, only $R_{26}$ assumes the value $\phi = l/\rho$, which is the deflection angle. Transferring this observation to steering magnets, which change the direction of the beam by the angle $\theta$, we conclude that also steering magnet changes the angle of the dispersion $D'$ by $\theta = \phi$ at the location of the steering magnet.

The dispersion can be measured by changing the energy of the beam and observing the position change on position monitors. In a linac the injection energy can be changed and in a ring changing the frequency $\omega_{\rm rf}$ of the RF-system causes the beam to adjust its momentum offset by $\delta = -(\Delta\omega_{\rm rf}/\omega_{\rm rf})/\eta$ in order to maintain synchronicity with the RF.

The purpose of *dispersion-free steering* is to simultaneously adjust the position and the dispersion at the BPMs. The method is based on numerically calculating the dispersion response matrix $S^{ij} = dD_i/d\theta_j$ and correcting the position $x_i$ and the dispersion $D_i$ at the same time, by solving

$$
\begin{pmatrix} \vdots \\ -x_i \\ \vdots \\ -D_i \\ \vdots \end{pmatrix} = \begin{pmatrix} \vdots & \vdots & & \vdots \\ C_{12}^{i1} & C_{12}^{i2} & \cdots & C_{12}^{im} \\ \vdots & \vdots & & \vdots \\ S_{12}^{i1} & S_{12}^{i2} & \cdots & S_{12}^{im} \\ \vdots & \vdots & & \vdots \end{pmatrix} \begin{pmatrix} \theta_1 \\ \theta_2 \\ \vdots \\ \theta_m \end{pmatrix} . \tag{8.54}
$$

The upper $n$ rows are the same as those in Equation 8.53 and the second set of $n$ rows contain the dispersion response coefficients. Again, handling the BPM error bars is done in the same way we discussed in Section 8.4.2 and solving Equation 8.54 is done using standard methods, such as pseudo-inverse, SVD, or MICADO, we discussed previously.

Apart from correcting the orbit in a ring, the next most frequently corrected parameter is the tune.

### 8.5.3 Tune correction

We now assume that we have a system that measures the fractional tunes $Q_x$ and $Q_y$ and two independently powered quadrupoles, for simplicity assumed to be thin quadrupoles characterized by their focal lengths $f_1$ and $f_2$. If the quadrupoles are not thin, we use the corresponding integrated and normalized gradient $k_1 l$ instead of $1/f$. In any case, changing the gradient of the quadrupole adds just another gradient perturbation, as described by

Equation 8.33, and the effect of quadrupole 1 on the tunes is given by

$$\Delta Q_x = \frac{\beta_{1x}}{4\pi f_1} \quad \text{and} \quad \Delta Q_y = -\frac{\beta_{1y}}{4\pi f_1} \tag{8.55}$$

where $\beta_{1x}$ and $\beta_{1y}$ are the horizontal and vertical beta functions at the location of quadrupole 1. Corresponding equations are valid for quadrupole 2. Note the minus sign in the second equation, because the quadrupole is focusing in one plane and defocusing in the other. If the changes are small, the tune changes from the two quadrupoles can be added with the result

$$\Delta Q_x = \frac{\beta_{1x}}{4\pi f_1} + \frac{\beta_{2x}}{4\pi f_2} \quad \text{and} \quad \Delta Q_y = -\frac{\beta_{1y}}{4\pi f_1} - \frac{\beta_{2y}}{4\pi f_2}, \tag{8.56}$$

which can be written in matrix form

$$\begin{pmatrix} \Delta Q_x \\ \Delta Q_y \end{pmatrix} = \frac{1}{4\pi} \begin{pmatrix} \beta_{1x} & \beta_{2x} \\ -\beta_{1y} & -\beta_{2y} \end{pmatrix} \begin{pmatrix} 1/f_1 \\ 1/f_2 \end{pmatrix}. \tag{8.57}$$

The desired correction strengths $1/f_i$ can be obtained by inverting the $2 \times 2$ matrix. The correction is well-behaved, if the determinant of the matrix is large, which requires the beta functions at the two quadrupoles to differ significantly. Normally this is guaranteed by choosing $\beta_x$ to be larger than $\beta_y$ in one quadrupole and vice versa in the other. Note that sometimes several quadrupoles are powered in series by a single power supply. In such cases we need to use the sum of the respective horizontal and vertical beta functions in these quadrupoles. Inverting the $2 \times 2$ matrix in Equation 8.57 is trivial and gives the changes of the quadrupole excitations as a function of the desired tunes changes $\Delta Q_x$ and $\Delta Q_y$.

This correction only fixes the average tune, but the momentum spread causes a spread in tunes, because the focusing of the quadrupoles is momentum dependent and, as we discussed in Section 3.4.1, leads to a finite chromaticity.

### 8.5.4 Chromaticity correction

Before correcting the chromaticities $Q'_x$ and $Q'_y$ in a ring we measure it through changing the momentum offset $\delta = -(\Delta\omega_{rf}/\omega_{rf})/\eta$ by changing the RF-frequency $\omega_{rf}$ and measuring the tunes at different set points. The thus observed momentum-dependence of the tune is caused by the same momentum-dependence of the quadrupoles that causes the dilution of the spot size in final focus systems we discussed in Section 8.4.4. Thus, even in rings, we need controllable, and momentum-dependent quadrupolar fields and we also use sextupoles placed at locations where the dispersion is non-zero. They provide momentum-dependent focusing as described by Equation 8.52 and therefore change the tunes by

$$\Delta Q_x = \frac{k_2 L D_x \beta_x}{4\pi}\delta \quad \text{and} \quad \Delta Q_y = -\frac{k_2 L D_x \beta_y}{4\pi}\delta. \tag{8.58}$$

With two sextupoles, or independent families of sextupoles powered in series, we can adjust the natural chromaticities $Q'_x$ and $Q'_y$ discussed in Section 3.4.1. The response matrix that relates the excitation of the chromaticity-correction sextupoles to the change in chromaticities $\Delta Q'_x = \Delta Q_x/\delta$ and $\Delta Q'_y = \Delta Q_y/\delta$ is very similar to the one for the tune correction in Equation 8.57 and reads

$$\begin{pmatrix} \Delta Q'_x \\ \Delta Q'_y \end{pmatrix} = \frac{1}{4\pi} \begin{pmatrix} D_{1x}\beta_{1x} & D_{2x}\beta_{2x} \\ -D_{1x}\beta_{1y} & -D_{2x}\beta_{2y} \end{pmatrix} \begin{pmatrix} (k_2 L)_1 \\ (k_2 L)_2 \end{pmatrix}, \tag{8.59}$$

where $D_{ix}$ is the horizontal dispersion at sextupole $i$. Inverting the matrix gives us the required excitations $(k_2 L)_i$ for the two sextupoles to change the chromaticities by desired values $\Delta Q'_x$ and $\Delta Q'_y$.

In this and the previous section we considered corrections of gradient imperfections. In the next section we will deal with skew-gradients that couple the transverse planes.

### 8.5.5 Coupling correction

In Section 8.3.3 we found that there are four different parameters that characterize the strength of the difference and sum resonance, each having a sine and a cosine-like phase. We therefore need four independently powered skew quadrupoles to correct these terms independently. Sometimes, if tunes are chosen close to the diagonal of the tune diagram, mostly the difference resonance plays a role and needs to be compensated. In that case, two skew quadrupoles suffice. In the following, we consider, however, the general situation where all four resonance-driving terms need to be compensated. As with most correction algorithms, we construct a response matrix of the parameter that needs to be corrected to the device that affects it—here the correction skew quadrupoles.

$$\begin{pmatrix} \mathrm{Re}(F_-) \\ \mathrm{Im}(F_-) \\ \mathrm{Re}(F_+) \\ \mathrm{Im}(F_+) \end{pmatrix} = \begin{pmatrix} \cos(\mu_{x1} - \mu_{y1}) & \cdots & \cos(\mu_{x4} - \mu_{y4}) \\ \sin(\mu_{x1} - \mu_{y1}) & \cdots & \sin(\mu_{x4} - \mu_{y4}) \\ \cos(\mu_{x1} + \mu_{y1}) & \cdots & \cos(\mu_{x4} + \mu_{y4}) \\ \sin(\mu_{x1} + \mu_{y1}) & \cdots & \sin(\mu_{x4} + \mu_{y4}) \end{pmatrix} \begin{pmatrix} \kappa_1 \\ \kappa_2 \\ \kappa_3 \\ \kappa_4 \end{pmatrix} \tag{8.60}$$

with the normalized skew quadrupole strengths $\kappa_i = \sqrt{\beta_{xi}\beta_{yi}}/2f_i$. The phase advances $\mu$ are those from the skew quadrupoles to a reference location.

If we can experimentally determine coefficients $F_\pm$, for example, with turn-by-turn measurements of the lattice with skew-quadrupolar correction magnets turned off and then insert $-F_\pm$ on the left-hand side in Equation 8.60 and invert it to give us the required skew quadrupole excitations $\kappa_i$ that compensate the measured coupling. Alternatively, if we only determine the coupling by a closest tune measurement, as described in Section 3.5, we construct knobs to adjust one coefficient of the $F_\pm$ at a time and empirically tune the skew quadrupoles to minimize the measured tune separation.

Inverting matrices can fail or leads to magnet excitations, where two magnets "fight each other" and mostly even out their corrections. In such cases, the matrix is degenerate or near-degenerate. Conversely, the matrix is well-behaved, if its condition number, the ratio between the maximum and minimum eigenvalue, is close to unity, and, in the optimum case, all eigenvalues are the same [56]. This also implies that the four independent parameters of $F_\pm$ can be controlled equally well. Experimenting with the phase advances, we find that an optimum set of phase advances, one that gives unit condition number, is $\mu_x = 360, 270, 90, 0$ and $\mu_y = 270, 180, 90, 0$ degrees. Such a placement of correction magnets results in a robust correction scheme with, additionally, the smallest magnet currents.

All the correction schemes so far address one particular parameter at a time. In the next section we will discuss a method to globally debug the linear optics of a storage ring.

### 8.5.6 Orbit response-matrix based methods

The basic idea is to compare the orbit response coefficients—all the elements $C^{ij}$ in the matrix in Equation 8.53—expected from a computer model to measured values $\hat{C}^{ij}$. The latter are easily determined by observing the orbit to move on $n$ position monitors when changing one of $m$ steering magnets at a time. This is often referred to as *recording difference orbits*. We do not write subscripts to avoid cluttering the equations; normally one considers all horizontal response coefficients with subscript 12 and vertical with subscript 34 simultaneously. Given the measured values, the method then tries to explain the discrepancies between predicted and measured values by gradient errors $\Delta g_k$ of all quadrupoles in the ring. In the simplest incarnation of the algorithm, we consider

$$\hat{C}^{ij} = C^{ij} + \sum_k \frac{\partial C^{ij}}{\partial g_k} \Delta g_k \ , \tag{8.61}$$

where the change of response coefficients with the quadrupole gradients $\partial C^{ij}/\partial g_k$ are calculated from two simulation runs with a small difference in gradient excitation, and repeating this one quadrupole at a time. The method is very powerful, because the number of response coefficients $C^{ij}$ is very large, typically $2nm$, half of them horizontal and the other half vertical, and we only try to determine $n_q$ gradients $g_k$, a much smaller number. The linear system of equations is thus vastly over-determined and can be solved by the usual least square methods, even giving an estimate of the error bars for the determined gradients.

Equation 8.61 only describes the very basic idea. First proof-of-principle tests on SPEAR [57] also determined scale errors of position monitors $1 + \Delta x^i$ and of steering magnets $1 + \Delta y^j$ at the same time the gradients were determined. Later, the algorithm was refined at NSLS [58], and a modern implementation, called *LOCO*, is used at many synchrotron light sources. It is based on the following equation

$$\hat{C}^{ij} = C^{ij} + \sum_k \frac{\partial C^{ij}}{\partial g_k} \Delta g_k + C^{ij} \Delta x^i - C^{ij} \Delta y^j + \sum_l \frac{\partial C^{ij}}{\partial p_l} \Delta p_l \tag{8.62}$$

where $\Delta x^i$ is the amount that the reading of monitor $i$ differs from unity and $\Delta y^j$ the scale error of steering magnet $j$. Additional parameters $p_l$ that can be accounted for by numerically determining the derivative $\partial C^{ij}/\partial p_l$ and including parameter variations $\Delta p_l$ in the fitting procedure. Examples for these parameters are systematic orbit displacements in sextupoles, which cause additional quadrupole fields by feed down, as discussed in Section 8.1.1. Furthermore, including cross-plane response coefficients, such as $C_{14}^{ij}$ or $C_{32}^{ij}$, in the algorithm, permits to determine spurious sources of skew gradients or solenoidal fields.

Once all sources of discrepancy between the measured and modeled response coefficients are identified, the opposite of the found quadrupole gradient errors can be added to the set values of the quadrupoles. This normally leads to a ring whose lattice is much closer to the model than before the correction. Since Equation 8.62 only uses first-order terms in the Taylor-expansion of the response coefficients in the sought parameters, several iterations of the correction are required, until the algorithm converges, leading to equal model and measured response coefficients within the error bars.

All correction methods we discussed so far are slow or quasi-static and the time-behavior of monitors and correction elements are not taken into account. If very fast and automatic corrections are required, we need to use *feedback systems*.

## 8.5.7  Feedback systems

Feedback systems are normally used to match the correction speed to rapid changes of the sources of perturbations. Examples are ground vibrations due to environmental noise such as heavy trucks or subways passing nearby. Time-dependent eddy currents in superconducting magnets are another example. In such cases fast monitors of the orbit or the tune are needed and equally fast actuators, steering and quadrupole magnets, are required. The correction algorithms are the same as we discussed in earlier sections of this chapter. In feedback systems, they just operate at a higher update rate.

All modern synchrotron light sources with their extremely small beam sizes must stabilize the beam positions and their orbit feedback systems typically run with a bandwidth of several tens of Hz. Considering the large number of position monitors and steerers implementing the high rates normally requires dedicated processing hardware, such as digital signal processors (DSP) or custom-programmed field programmable gate arrays (FPGA).

In order to guarantee the stability of the beams in LHC, having huge stored energies, the tune, chromaticity, and coupling are continuously observed and immediately corrected.

## QUESTIONS AND EXERCISES

1. Consider the part towards the right from the IP of the beam line in the figure below. The total length of the section from corrector C1 to the IP has a length $2L$ and halfway there is a (thin-lens) horizontally focusing quadrupole with focal length $f$. Immediately downstream of the thin quadrupole a second corrector C2 is located. You can assume that the distance from C2 to the IP is equal to the distance from the quadrupole to the IP. (a) Calculate the multi-knob for horizontal position at the IP that leaves the angle at the IP unaffected. (b) Calculate the multi-knob for horizontal angle at the IP that leaves the position at the IP unaffected. (c) Calculate the corresponding vertical knobs.

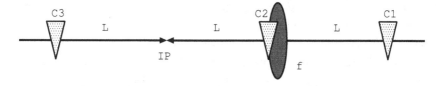

2. Start from the same geometry as used in the previous exercise, but assume that there is another corrector C3 a distance $L$ downstream of the IP with no extra magnets in between. In this exercise, we ignore the angle at the IP. For the calculation you can assume that the focal length $f$ is equal to the distance of the quadrupole to the IP. (a) Calculate the multi-knob for horizontal position at the IP, but make sure that the bump is "closed" and the orbit is unperturbed after the last corrector C3. (b) Calculate the corresponding vertical knob.

3. Correct the orbit in a straight beam line with five of the double cells, described in the file `doublet.bl`, available from this book's web page. Place steering magnets in the QF and position monitors in the QD. (a) First generate a "bad" orbit by sampling uniformly-distributed random displacements for quadrupoles with $-1 < d_x < 1\,\mathrm{mm}$ and determine the positions on the monitors. (b) Devise an orbit correction system and use it to correct the positions on the monitors. (c) Instead of displacing the quadrupoles independently, now move the two adjacent quadrupoles of a doublet by the same amount, because they sit on the same girder. Is the orbit distortion smaller or larger? Discuss your observation!

4. Use the ring from Exercise 10 in Chapter 3 and assume that there are position monitors and steering magnets installed immediately downstream of the focusing quadrupoles, only. (a) Prepare a function that receives the positions of monitor and perturbing element as input andreturns the $4 \times 4$ response matrix. You can ignore the effect of the radio-frequency system in this exercise. (b) One of the defocusing quadrupoles is displaced by $d_x = d_y = 1\,\mathrm{mm}$ in both planes. What orbit do the BPMs show? (c) If the displacements of all quadrupoles are uncorrelated and random with rms displacement $\langle d_x^2 \rangle = (0.5\,\mathrm{mm})^2$, what rms orbit do you expect to observe on the BPM. (d) Repeat

the previous exercise in the vertical plane with $\langle d_y^2 \rangle = (0.5\,\text{mm})^2$. (e) Calculate all orbit response coefficients $C^{ij}$ and design an orbit correction system to correct (f) the horizontal orbit and then (g) for the vertical orbit. Use uniformly distributed displacements and explore how well do the respective orbit correction systems reduce the orbit.

5. In the ring from Exercise 4, add a single horizontal steering magnet just upstream of the first focusing quadrupole in (a) and re-design the orbit correction system. Since you have one additional steering magnet to correct the orbit at the same number of position monitors, the correction should work better, right? (b) Does it? (c) Analyze, (d) fix, and (e) discuss the problem!

6. How do you have to change the code to account for displacements of long quadrupoles?

7. Assume that all quadrupoles have random errors in their strength of 1% (rms) in the ring from Exercise 4. (a) Write a Monte-Carlo simulation to randomly assign focusing errors to them and generate a histogram of tunes for 1000 seeds. (b) Make an analytic estimate of the rms width of the tunes. Do they agree with the width of the histograms? (c) Single out two quadrupoles and use them to correct the tunes of one random seed of quadrupole-excitation errors.

8. (a) Calculate the chromaticity of the ring from Exercise 4 unless you have already done so in Exercise 10 in Chapter 3. (b) Correct the chromaticity to $Q'_x = Q'_y = 0$ with two sextupoles. Find their excitations, if one is located adjacent to a QF and the other adjacent to a QD.

9. Draw a tune diagram with all resonance lines, defined by $nQ_x \pm mQ_y = 1$ with $n + m \leq 3$. Then add a dot for the tunes $Q_x = 0.28$ and $Q_y = 0.31$ to the diagram.

10. You are in charge of a small (unrealistic) ring with only three FODO cells, each having the same geometry as the cells used for Exercise 4, but omit the dipole magnets (I told you it is unrealistic!). (a) Adjust the phase advance of the cells to a little under 90° per cell, such that the design values of the tunes are $Q_x = Q_y = 0.72$. When operating the accelerator, you observe that the horizontal tune is a little off and you suspect that a quadrupole might have shorted coil-windings. You therefore plan to use Equation 8.61 and compare the ideal to the measured horizontal response coefficients one quadrupole at a time. (b) First, calculate the unperturbed horizontal response coefficients $C_{12}^{ij}$. (c) Then simulate a bad ring by changing the focal length of the first quadrupoles by 0.2 m and calculate the response coefficients $\hat{C}_{12}^{ij}$ for the perturbed ring. (d) Calculate the derivative of the response coefficients $\partial C^{ij}/\partial g_k$, where $g_k = f_k$ is the focal length of the quadrupoles. (e) Test one quadrupole at a time. Which one best explains the difference $\hat{C}^{ij} - C^{ij}$ between the measured and design response coefficients? Can you find the 0.2 m change in the focal length? (f) Stimulated by your success, you next try to fit for all quadrupole errors $\Delta g_k$ simultaneously by expressing Equation 8.61 in matrix form and solving the over-determined system using standard linear algebra techniques. (g) Inspect the covariance matrix in order to understand the error bars of your fit parameters.

11. In CELSIUS, with a circumference of 82 m, and momentum compaction factor $\alpha = 0.123$, the typical magnitude of the response coefficients was $C^{ij} \approx 10$ m and about $D_x \approx 4.5$ m for the horizontal dispersion. The operating range of the kinetic energies for protons was from 50 to 1360 MeV. At what energies do you expect the change in $\eta$ to change the response coefficients by 10%?

# CHAPTER 9

# Targets and Luminosity

In this chapter, we discuss the interaction of the beam with a target, which is the prime reason for the existence of many accelerator. We will discuss the relation between the count rate in experiments and the microscopic cross section, which is given by the luminosity. We will see how the reactions lead to reduction of the beam current that will reveal itself as a finite lifetime. Apart from the desired losses due to nuclear reaction, there are detrimental effects due to electro-magnetic interactions with the target material that will perturb the beam. In the first part, we will deal with fixed targets, where these detrimental effects are energy loss due to ionization of the target atoms and transverse Rutherford scattering from the target nuclei. In the second part, we consider colliding beams, where the beams are often squeezed to small beam sizes with extreme electromagnetic fields that perturb the counter-propagating beam.

We start by considering the count rate and its relation to the cross sections.

## 9.1  EVENT RATE AND LUMINOSITY

In a nuclear or high-energy physics experiment, one often tries to determine count rates of events with a particular signature. Examples are the creation of two oppositely charged muons, leaving the interaction region in a collider in opposite directions, or the occurrence of a number of jets, or three pions. Since these events are compared with other, more frequent events, a high count rate $R$ is desirable in order to achieve high statistical significance of the findings. Note that a count rate $R$ has units $1/s$.

Of course, the count-rate is also determined by the nuclear cross-section $\sigma$, which has units of an area [m$^2$] and can be visualized as a small area that causes an event to happen, if it is hit by a particle from the beam. The cross-section is commonly expressed in units of [cm$^2$] rather than [m$^2$] and another frequently used unit is the *barn*, which equals $10^{-24}$ cm$^2$ or $10^{-28}$ m$^2$ or 100 fm$^2$, where one femto-meter is roughly the diameter of a nucleon.

The constant of proportionality that relates the microscopic cross-section $\sigma$ to the experimental count rate $R$ is the *luminosity* $\mathcal{L}$, which thus fulfills

$$R = \sigma \mathcal{L} \qquad \text{with} \qquad \mathcal{L} = N_b f_0 \frac{N_t}{A} \tag{9.1}$$

for a fixed-target experiment. It quantifies the performance of the accelerator and the target system. It depends on the number of beam particles $N_b$ incident on the target with frequency $f_0$. Moreover, it depends on the area density $N_t/A$ of the number of target atoms $N_t$ per unit area $A$. Note that the luminosity is conventionally expressed in units of $1/\text{cm}^2\text{s}$.

For illustrative purposes let us consider a *solid target* of Lithium (Li) that is used for the production of neutrons in the $p + {}^7Li \rightarrow {}^7Be + n$ reaction. The target thickness $N_t/A$ can be calculated from the mass density $\rho_{Li} = 0.534\,\mathrm{g/cm^3}$, the geometric thickness $dx$ of the target material, and the mass of a Lithium atom $m_{Li} = 7m_u$ with the mass of a nucleon given by $m_u = 1.66 \times 10^{-24}\,\mathrm{g}$. For $dx = 0.1\,\mathrm{cm}$ we find $N_t/A = 4.6 \times 10^{21}\,\mathrm{/cm^2}$. Beryllium (Be) or carbon (C) are used as target material for experiments to produce neutrinos. The European Spallation Source will produce neutrons by impinging protons onto a fast-rotating wheel made of tungsten (W). The solid targets will either stop the beam entirely or significantly perturb the beam such that it is discarded immediately after the target.

In some storage rings, moderately dense targets, so-called *internal targets,* are placed in the path of the circulating beam. The target can be either a jet of gas, a stream of clusters of the target material, or a stream of microscopic pellets. The latter are created by injecting hydrogen, prepared to have temperature and pressure close to the triple point, through an oscillating vacuum injection capillary, which freezes the stream of hydrogen into micron-size spheres of frozen hydrogen. Pellets or clusters provide thicker targets compared to gas jets and make it easier to remove the gas after the collisions with the beam. Typical target densities $N_t/A$ encountered in nuclear storage rings are about $N_t/A \approx 10^{14}\,\mathrm{/cm^2}$ for gas jet targets and a few times $10^{15}\,\mathrm{/cm^2}$ for pellet targets, which also makes it an option for the target in the HESR anti-proton ring that is part of the FAIR facility. The first pellet target was operated in CELSIUS, the now defunct storage ring for nuclear physics experiments in Uppsala. Typically $N_b = 10^{10}$ protons were stored at a revolution frequency of about $f_0 = 3\,\mathrm{MHz}$, resulting in a current of $I = N_b e f_0 \approx 5\,\mathrm{mA}$. The number of particles impinging on the target per second is therefore given by $N_b f_0$. Achievable target density was about $N_t/A \approx 3 \times 10^{15}\,\mathrm{/cm^2}$, resulting in a luminosity of $\mathcal{L} \approx 10^{32}\,\mathrm{/cm^2 s}$.

In a storage ring the beam particles that participate in the wanted nuclear reactions with cross-section $\sigma$ are of course lost from the beam. The loss rate is given by Equation 9.1 as well as

$$\frac{dN_b}{dt} = -\sigma\mathcal{L} = -\sigma N_b f_0 \frac{N_t}{A} = -\frac{1}{\tau_L}N_b \,, \tag{9.2}$$

which is obviously proportional to the number of beam particle $N_b$ and leads to an exponential reduction of the number of beam particles with a time constant $\tau_L$ given by $1/\tau_L = \sigma f_0 N_t/A$, which is called the *luminosity lifetime.* If multiple reaction channels with different cross sections are involved, the cross sections of the respective reactions must be added.

These losses from the nuclear reactions are normally small, because the nuclear cross sections are very small. On the other hand, the cross sections for the electro-magnetic interactions with the electrons and target material are orders of magnitude larger and perturb the circulating beam much more. One of the main effects is the ionization of the target material, which leads to a random energy-loss of the beam. And that is the topic of the following section.

## 9.2 ENERGY LOSS AND STRAGGLING

The energy of the beam (MeV to TeV) is normally much higher than the binding energy of the electrons (eV to keV) and we can assume that a beam particle passes through a sea of free electrons. The electro-magnetic field created by the beam particle transfers momentum from the beam to the electrons, which acquire kinetic energy that the beam particle has lost. We describe this process quantitatively for a beam of protons by calculating the momentum transfer from a single proton traveling along the $z-$axis with velocity $v$ to an electron that has a minimum distance, or impact parameter, $b$ with respect to the

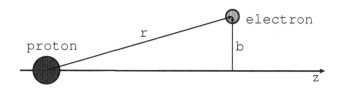

**Figure 9.1** The proton moves along the $z$−axis and gives a kick to the electron, which carries away some of the energy provided by the proton.

proton. Figure 9.1 illustrates the geometry. Only the transverse momentum transfer of the electron $\Delta p_\perp = \int_{-\infty}^{\infty} F_\perp dt$ is non-zero, because the electron moves very little during the passage of the proton and the longitudinal transfer in the $z$−direction cancels. The force $F_\perp = F_c b/r = e^2 b/4\pi\varepsilon_0 r^3$ is simply the transverse component of the Coulomb force $F_c = e^2/4\pi\varepsilon_0 r^2$ between proton and electron. With $F_\perp$ given and with $dt = dz/v$ and $r^2 = z^2 + b^2$ we calculate $\Delta p_\perp$ and find

$$\Delta p_\perp = \frac{e^2 b}{4\pi\varepsilon_0 v} \int_{-\infty}^{\infty} \frac{dz}{(z^2 + b^2)^{3/2}} = \frac{e^2}{2\pi\varepsilon_0 b v} , \tag{9.3}$$

where we make the approximation that the speed $v$ of the proton does not change appreciably. The impact from the beam particle, on the other hand, transfers the momentum $\Delta p_\perp$ to the electron, which then carries the kinetic energy $T_e = \Delta p_\perp^2/2m_e$. And this is energy lost by the beam particle. In order to calculate the energy loss of the beam due to *all* electrons we have to sum over all of them. There are $N_e(b) = \rho_e 2\pi b db dx$ electrons with impact parameters between $b$ and $b+db$, where $\rho_e$ is the density of the electrons. Integrating over $b$ thus yields the energy loss per unit length $dE/dx$ of the proton beam

$$\frac{dE}{dx} = 2\pi\rho_e \int \frac{\Delta p_\perp^2}{2m_e} b db = \frac{e^4 \rho_e}{4\pi\varepsilon_0^2 m_e v^2} \int \frac{db}{b} = \frac{e^4 \rho_e}{4\pi\varepsilon_0^2 m_e v^2} \log\left(\frac{b_{max}}{b_{min}}\right) . \tag{9.4}$$

Here we encounter the *Coulomb-logarithm* that depends on two artificially introduced cutoff parameters $b_{min}$ and $b_{max}$. The former is necessary, because very small impact parameters $b$ lead to divergent momentum transfers and the latter is necessary because the number of electron grows more rapidly than the electro-magnetic force drops off at large impact parameters. One therefore needs to heuristically determine the range of impact parameters relevant to the problem. Here we use the properties of the target material to find reasonable values for $b_{min}$ and $b_{max}$. Following [59], we realize that for very large impact parameters the proton fly-by takes a time $\Delta t$ and only weakly pushes the far-away electrons. In order to knock them out from the target atom, the impact must be short and intense. If the electron is in a bound state with an equivalent orbital frequency $\nu$ we expect ionization to occur, provided that the duration of the impact has the order of magnitude $\Delta t = b_{max}/\gamma v < 1/\nu$, where $\gamma$ takes care of the relativistic contraction of time. Solving for $b_{max}$, we obtain $b_{max} = \gamma v/\nu$. The lower value of impact parameters $b_{min}$, we obtain from Heisenberg's uncertainty principle, because the electron can only be localized to $\Delta x \approx h/\gamma m_e v$ which we take as a reasonable estimate for the lower bound for the impact parameter and get $b_{min} \approx h/\gamma m_e v$. Inserting these heuristic values for $b_{min}$ and $b_{max}$ in Equation 9.4, we get

$$\frac{dE}{dx} = \frac{e^4 \rho_e}{4\pi\varepsilon_0^2 m_e v^2} \log\left(\frac{\gamma^2 m_e v^2}{h\nu}\right) = \frac{e^4 \rho_e}{4\pi\varepsilon_0^2 m_e v^2} \log\left(\frac{2\gamma^2 m_e v^2}{I}\right) , \tag{9.5}$$

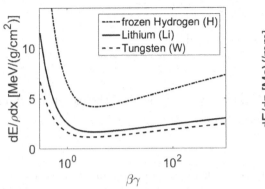
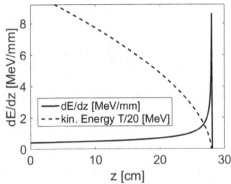

Figure 9.2 The left graph shows the energy loss $dE/\rho dx$ of protons as a function of the normalized beam momentum $\beta\gamma$ for targets made of frozen hydrogen, lithium, and tungsten. The right graph shows the deposited energy (solid) and remaining beam energy (dashes) of a proton beam in water as a function of depth $z$ in the target material.

where we estimate the characteristic frequency $\nu$ from the ionization potential $I = 2h\nu$ of the target material.

Equation 9.5 is a much simplified version of the *Bethe-Bloch equation* for the energy loss of charged particles in matter. The average energy loss per meter of a beam of particles with charge $Z_b$ and mass $M$ with momentum $\beta\gamma Mc$ is given by

$$\frac{dE}{\rho dx} = -KZ_b^2 \frac{Z_t}{A_t} \frac{1}{\beta^2} \left[ \frac{1}{2} \ln \left( \frac{2m_ec^2\beta^2\gamma^2 T_{max}}{I^2} \right) - \beta^2 - \frac{\delta}{2} \right] \tag{9.6}$$

with $K/A = 4\pi N_A r_e^2 m_e c^2/A = 0.307 \, \text{cm}^2/\text{g}$ for A=1 g/mol and

$$T_{max} = \frac{2m_ec^2\beta^2\gamma^2}{1 + 2\gamma m_e/M + (m_e/M)^2} \tag{9.7}$$

is the maximum kinematically achievable kinetic energy of the ionized electron with mass $m_e$. The target atoms are described by their Charge $Z_t$, their atomic mass number $A_t$, and their ionization potential. The small factor $\delta$ in Equation 9.6 describes volume effects. Details can be found in Section 33 of [60] entitled "Passage of particles through matter."

On the left-hand side in Figure 9.2, we show the energy loss scaled by the density $dE/\rho dx$ of the target material $\rho$ as a function of the normalized momentum $\beta\gamma$ for protons impinging on targets of frozen hydrogen, lithium, and tungsten. Despite widely different densities of the target material, the normalized energy loss for the different materials turns out to be close to $2 \, \text{MeV}/(\text{g}/\text{cm}^2)$ for values of $\beta\gamma > 5$, only hydrogen shows twice that value. We also note that for small momenta, the energy loss grows rapidly due to the inverse dependence on the speed $v$ or $\beta = v/c$ of the protons in Equation 9.6. This implies that once a particle has lost most of its kinetic energy, the remainder is lost very rapidly. This is illustrated on the right-hand side in Figure 9.2, where the solid line shows the deposited energy per unit length (not normalized to the density $\rho$) for a proton beam with initial kinetic energy $T = 200 \, \text{MeV}$ impinging on a target made of water. The dashes show the

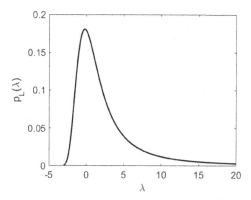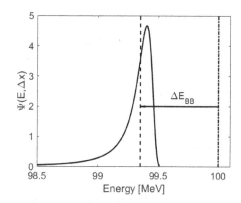

**Figure 9.3** The Landau probability distribution $p_L(\lambda)$ (left) and the energy distribution of an initially mono-energetic proton beam with kinetic energy $T = 100\,\text{MeV}$ after passing a 2 mm thick lithium target (right).

remaining kinetic energy $T$ (divided by 20) that the beam has at depth $z$ in the target material. Near $z = 28\,\text{cm}$ it drops to zero and the protons are stopped. Initially the energy losses near $z = 0$ are on the level of 0.5 MeV/mm, but grow to almost 10 MeV/mm after 25 cm. This final peak of the deposited energy is called the *Bragg peak*. The fact that most of the energy is deposited close to the end of the path in the target material is used in cancer proton-therapy. In particular, a proton beam with a kinetic energy of $T = 200\,\text{MeV}$ deposits most of its energy at a depth of around 28 cm in water or in human tissue while the tissue before the Bragg peak is exposed to much less ionization losses.

The Bragg peak on the right-hand plot in Figure 9.2 is very narrow, because in the simulation we used a mono-energetic proton beam and neglected the statistic nature of the ionization processes, where both effects broaden the peak. Broadening due to the latter effect for moderately thick targets is described by the Landau theory [61], which gives the distribution $\Psi(E, \Delta x)$ of the energy loss $E$ after traversing a target of thickness $\Delta x$ as

$$\Psi(E, \Delta x) = \frac{1}{\xi} p_L(\lambda) \qquad \text{with} \qquad p_L(\lambda) = \frac{1}{\pi} \int_0^\infty e^{-t\log(t)-\lambda t} \sin(\pi t)dt . \qquad (9.8)$$

Here $\lambda$ is given by $\lambda = (E - \Delta E_{BB})/\xi - 1 + \gamma_E - \beta^2 - \log(\xi/T_{max})$ with Euler's constant $\gamma_E \approx 0.577215$ and $\xi = 0.1534(Z^2/\beta^2)(Z_t/A_t)\rho\Delta x$ describes the thickness of the target. Here $\xi$ is given in units of MeV and $\rho\Delta x$ in units of g/cm$^2$. Moreover, $\Delta E_{BB}$ is the average energy loss in the target as described by Equation 9.6. The universal Landau distribution $p_L(\lambda)$ is shown on the left-hand side in Figure 9.3. We observe that the distribution has a distinct tail towards large values, which causes many beam particles to experience an energy loss much larger than the one described by the Bethe-Bloch equation. The right-hand side in Figure 9.3 shows the energy distribution of a proton beam with initial kinetic energy $T = 100\,\text{MeV}$ after passing through a 2 mm thick lithium target. The energy loss from the Bethe-Bloch equation $\Delta E_{BB}$ is indicated by the horizontal arrow that starts at the initial energy $T = 100\,\text{MeV}$. The shape of the distribution is given by the Landau distribution function $p_L(\lambda)$ and shows again the extended low-energy tail. We point out that Landau's theory is valid only for a thin target, a limit that we violate to some extent in the above example, and a more elaborate theory, due to Vavilov [61], covers thicker targets.

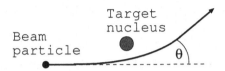

Figure 9.4 A beam particle, closely passing the nucleus of a target, is transversely deflected by its electric potential, where the underlying process is Rutherford scattering.

The discussion in the previous paragraphs described the energy loss of protons or other heavy beam particles. For the *energy loss of electrons*, a few additional aspects, especially at high energies, are important. Low-energy electrons also lose most of their energy by ionizing the target material. In thin targets the dominant effect is ionization and a slightly modified version [60] of Equation 9.6 can be used to calculate the loss. At higher energies, the electrons lose most of their energy by the emission of *bremsstrahlung*. It depends on the square of the nuclear charge $Z_t^2$ of the target material and explains the use of heavy, high–$Z_t$ materials, such as tungsten or tantalum, as targets for the production of x-rays. Electrons with ultra-relativistic energies exhibit an increased production of $e^+e^-$–pairs. At these energies, both electrons or photons cause electromagnetic cascades in the target material, which are characterized by the radiation length $X_0$, the exponential attenuation length of the incident particle's initial energy. Values of radiation lengths for many materials are tabulated in [60] and a few selected values are given below in Table 9.1.

After discussing the effect of the targets on the beam energy, we will consider transverse scattering of beam particles for the nuclei of the target atoms.

## 9.3 TRANSVERSE SCATTERING, EMITTANCE GROWTH, AND LIFETIME

Apart from the energy loss due to ionizations, the charged beam particles will be elastically deflected by the nuclei of the target material, which is given by the Rutherford-scattering cross section

$$\frac{d\sigma}{d\Omega} = \left( \frac{Z_b Z_t e^2}{8\pi\varepsilon_0 \beta pc} \right)^2 \frac{1}{\sin^4(\theta/2)} . \tag{9.9}$$

It describes the probability to deflect a particle into an angular range $d\Omega = \sin\theta d\theta d\phi$ and has the well-known $1/\theta^4$ dependence for small angles, which implies that most particles experience small deflections. The small deflections are due to large impact parameters of the beam particle with respect to the scattering nucleus. Averaging over all impact parameters leads to the rms scattering angle, which can be estimated from a slightly simplified equation from [60]

$$\theta_{rms} \approx Z_b \frac{13.6\,\mathrm{MeV}}{\beta pc} \sqrt{\frac{\Delta x}{X_0}} , \tag{9.10}$$

where $Z_b$ is the charge of a beam particle with momentum $p$ and $\Delta x$ is the thickness of the target material with radiation length $X_0$. It is tabulated for various materials in [60] and we quote a few common values in Table 9.1. Considering, for example, a proton with kinetic energy $T = 100\,\mathrm{MeV}$ that passes through a 2 mm lithium target already discussed in the previous section. We find its rms scattering angle $\theta_{rms} \approx 2.6\,\mathrm{mrad}$. On the other hand, a proton with momentum $p = 1\,\mathrm{GeV/c}$ traversing a frozen-hydrogen pellet with a diameter of $d = 30\,\mu\mathrm{m}$ only experiences an rms deflection angle of $\theta_{rms} \approx 3.4\,\mu\mathrm{rad}$.

Table 9.1 Radiation length of several materials [60]. The values $X_0$ in the upper row are normalized to the material density $\rho$ and the values $x_0$ in the lower row are given as a distance.

| Material | $H_2O$ | $H_2$(liq.) | C | Fe | Cu | Al | W | Li |
|---|---|---|---|---|---|---|---|---|
| $X_0$ [g/cm$^2$] | 36.1 | 63.1 | 42.7 | 13.8 | 12.86 | 24.1 | 6.76 | 82.8 |
| $x_0$ [cm] | 36.1 | 891 | 18.8 | 1.76 | 1.44 | 8.9 | 0.35 | 155 |

If an internal target is installed in a storage ring, the beam particles repeatedly receive random deflection angles and since all particles will be kicked independently this will cause the *emittance to grow*. We analyze this effect with a simple model, in which we assume that the target is located at a waist with $\hat\alpha = 0$. The beta function at the target is $\hat\beta$ and the tune of the ring is $Q = \mu/2\pi$, such that we construct the one-turn transfer matrix $\hat R$ from Equations 3.51 and 3.52. At the target a particle receives the transverse kick $\theta_k$ on turn number $k$, where the angles at different turns have average zero, such that we have $\langle\theta_k\rangle = 0$. Moreover, they are statistically independent, which implies $\langle\theta_k\theta_m\rangle = \theta_{rms}^2\delta_{km}$ with the Kronecker symbol $\delta_{km}$ that is unity for $k = m$ and zero otherwise. This allows us to calculate the particle's phase space coordinates in normalized phase space $\tilde x_n$ and $\tilde x'_n$ after $n$ turns by summing up all the kicks at turns $m$ and propagating them through the remaining $n - m$ turns

$$\begin{pmatrix} \tilde x_n \\ \tilde x'_n \end{pmatrix} = A\sum_{m=1}^{n} \hat R^{n-m}\begin{pmatrix} 0 \\ \theta_m \end{pmatrix} = \sum_{m=1}^{n} \mathcal{O}^{n-m}\begin{pmatrix} 0 \\ \sqrt{\hat\beta}\theta_m \end{pmatrix}, \qquad (9.11)$$

where we used the definitions of $A$ and $\mathcal{O}$ from Equation 3.52. After $n$ turns, the Courant-Snyder action variable $2J_n = \tilde x_n^2 + \tilde x_n'^2$ of the particle is then given by

$$2J_n = \sum_{k=1}^{n}\sum_{m=1}^{n}(0, \sqrt{\hat\beta}\theta_k)\left(\mathcal{O}^{n-k}\right)^T \mathcal{O}^{n-m}\begin{pmatrix} 0 \\ \sqrt{\hat\beta}\theta_m \end{pmatrix} = \hat\beta\sum_{k=1}^{n}\sum_{m=1}^{n}\cos((k-m)\mu)\theta_k\theta_m .$$

$$(9.12)$$

Averaging over many realizations of kick angles $\theta_k$ and all particles of the beam we obtain

$$\langle 2J_n\rangle = \hat\beta\sum_{k=1}^{n}\sum_{m=1}^{n}\cos((k-m)\mu)\langle\theta_k\theta_m\rangle = \hat\beta\sum_{k=1}^{n}\cos(0)\theta_{rms}^2 = n\hat\beta\theta_{rms}^2 , \qquad (9.13)$$

where we used the statistical properties of the kick angles from above. Since the emittance $\varepsilon = \langle J\rangle$ of the ensemble of many particles is defined as the Courant-Snyder invariants $J$ of the individual particles averaged over the beam distribution, we find that $\varepsilon = n\hat\beta\theta_{rms}^2/2$ grows linearly with the number of turns. We therefore obtain

$$\frac{d\varepsilon}{dn} = \frac{\hat\beta}{2}\theta_{rms}^2 \quad \text{or} \quad \frac{d\varepsilon}{dt} = \frac{\hat\beta}{2T_0}\theta_{rms}^2 \qquad (9.14)$$

with the revolution time $T_0$ for the growth rate of the emittance $d\varepsilon/dt$. We observe that it is proportional to the beta function $\hat\beta$ at the target, where it is normally made small in order to maximize the luminosity. This has the additional benefit that the emittance growth is small.

The emittance growth described by Equation 9.14 is rather slow and can sometimes be compensated by some beam-cooling or damping mechanism. This can either be damping due to synchrotron radiation, a topic we will discuss in detail in Chapter 10, by stochastic cooling or by electron cooling, topics we will briefly mention in Section 13.4. Typical damping times $\tau_d$ are in the range of ms to s. We add this damping heuristically to the right equation in Equation 9.14 by adding a term $-\varepsilon/\tau_d$ to arrive at

$$\frac{d\varepsilon}{dt} = -\frac{\varepsilon}{\tau_d} + \frac{\hat{\beta}}{2T_0}\theta_{rms}^2 \, , \tag{9.15}$$

such that the scattering angle $\theta_{rms}$ will cause the emittance to assume an equilibrium value $\varepsilon_0 = \tau_d\hat{\beta}\theta_{rms}^2/2T_0$ after a few damping times $\tau_d$, which is a slow process on the time scale of the revolution time $T_0$.

The previous paragraph discussed distant target-nucleus fly-bys of beam particles, which causes the emittance to grow. If, on the other hand, the fly-by is very close, the beam particle is deflected to such an extent that it is very quickly lost during the next turn, because its amplitude exceeds the beam-pipe aperture $r_a$ at some point in the ring and the above damping mechanisms act on too long time scales to be able to counteract this. The maximum angle $\theta_a$ is given by $r_a = \sqrt{\beta_a\hat{\beta}}\theta_a$, where $\sqrt{\beta_a\hat{\beta}}$ is the maximum transfer matrix element $R_{12}$ that relates the scattering angle to the position $r_a$ with beta function $\beta_a$ at the aperture limit. The cross section $\sigma(\theta > \theta_a)$ to exceed a scattering angle $\theta_a = r_a/\sqrt{\beta_a\hat{\beta}}$ then follows from integrating the differential cross section from Equation 9.9 for all angles $\theta > \theta_a$ with the result

$$\sigma(\theta > \theta_a) = \int_0^{2\pi} d\phi \int_{\theta_a}^{\pi} \frac{\hat{a}^2\sin(\theta)d\theta}{\sin^4(\theta/2)} = 8\pi\hat{a}^2 \int_{\theta_a}^{\pi} \frac{\cos(\theta/2)d(\theta/2)}{\sin^3(\theta/2)} = \frac{4\pi\hat{a}^2}{\tan^2(\theta_a/2)} \, , \tag{9.16}$$

where we introduced the abbreviation $\hat{a} = Z_b Z_t e^2/8\pi\varepsilon_0\beta pc$ to describe the factor in brackets in Equation 9.9 and used the substitution $z = \sin(\theta/2)$ when evaluating the integral.

The cross section $\sigma(\theta > \theta_a)$ describes almost immediate beam loss due and the rate is given by Equation 9.2

$$\frac{dN_b}{dt} = -\sigma(\theta > \theta_a)\mathcal{L} = -\frac{4\pi\hat{a}^2 N_b f_0}{\tan^2(\theta_a/2)}\frac{N_t}{A} = -\frac{1}{\tau_s}N_b \, , \tag{9.17}$$

which defines the lifetime due to large-angle scattering off of the target atoms as $1/\tau_s = (4\pi\hat{a}^2 N_b f_0/\tan^2(\theta_a/2))(N_t/A)$ with $\theta_a = r_a/\sqrt{\beta_a\hat{\beta}}$. Note that any other atoms or molecules in the path of the beam particles, such as the residual gas in the beam pipe, causes additional losses and reduces the lifetime of the beam, a topic we will return to when discussing vacuum-related issues in Section 13.5.

So far, we have discussed the consequences of a fixed target on the dynamics of the beam. In the next sections, we will consider what happens when the target is a counter-propagating beam.

## 9.4  COLLIDING BEAMS

The reason to use counter-propagating colliding beams is the much higher energy $\sqrt{s} = E_{cm}$ available in the center-of-mass system compared to fixed-target experiments. We calculate $E_{cm}$ from the relativistic four-momenta $\mathcal{P}_1 = (E_1, \vec{p}_1 c)$ and $\mathcal{P}_2 = (E_2, \vec{p}_2 c)$ of two colliding particles as the invariant

$$s = E_{cm}^2 = (\mathcal{P}_1 + \mathcal{P}_2)^2 = (E_1 + E_2)^2 - (\vec{p}_1 c + \vec{p}_2 c)^2 = m_1^2 c^4 + m_2^2 c^4 + 2E_1 E_2 - 2c^2\vec{p}_1\vec{p}_2 \, , \tag{9.18}$$

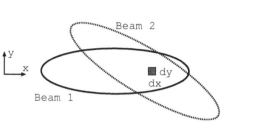
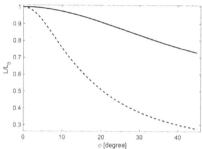

Figure 9.5   Sketch to illustrate the calculation of the luminosity in beam-beam colli-sions (left) and the luminosity loss as a function of the tilt angle $\phi$ for beams with aspect ratio $r = 3$ (solid) and $r = 10$ (dashes).

where we used $E_1^2 = m_1^2 c^4 + c^2 \vec{p}_1 \vec{p}_1$ and $E_2^2 = m_2^2 c^4 + c^2 \vec{p}_2 \vec{p}_2$. In a fixed-target experiment, one of the particles is at rest, say particle 2, with $\vec{p}_2 = 0$ and the energy available to create new particles is $E_{cm,f}^2 = m_1^2 c^4 + m_2^2 c^4 + 2m_2 c^2 E_1$. In a colliding-beams experiment with counter-propagating beam particles of equal mass $m_1 = m_2$ the momenta are equal, but have opposite sign $\vec{p}_1 = -\vec{p}_2$, such that the available energy in the center-of-mass system is $E_{cm,c}^2 = (E_1 + E_2)^2 = 4E_1^2$ which is always larger than $E_{cm,f}^2$, because momentum conservation dictates that part of the energy is needed to maintain the center-of-mass motion.

This higher available energy, however, comes at the price of higher technical complexity and price, because two beams need to be produced and stored. Moreover, the luminosity is significantly smaller, because there are fewer collision targets with the counter-propagating beam than there are in a solid target. In colliders with particles and their anti-particles, it is possible to store both beams in the same magnetic structure, because the sign of the charge and the velocity in the Lorentz-force equation cancel and the beams follow the same trajectory. One has, however, to pay special attention to prevent multiple bunches to collide outside the detector. Moreover, the production of anti-particles is often difficult and expensive. Therefore, in LHC two proton beams are brought into collisions. Since the two beams have the same charge, two independent magnet systems to guide the two beams around the ring are required.

But if the highest possible beam energy is the objective of an accelerator, a collider is the machine of choice, and we therefore need to determine its luminosity.

## 9.5   BEAM-BEAM LUMINOSITY

In contrast to the calculation of the luminosity in fixed target experiments, where we as-sumed that the target material had a uniform density of scattering centers $N_t/A$, the two beams in colliding-beam experiments have a transverse distribution, often assumed to be Gaussian, $N_1\psi_1(x, y)$ and $N_2\psi_1(x - X, y - Y)$. Here $N_1$ and $N_2$ are the number of particles in beam 1 and 2, respectively and $X$ and $Y$ describe the relative displacement of the second beam's center with respect to the first. Figure 9.5 shows the one-sigma contours of the two beams as solid and dotted ellipses. In order to calculate the contribution of the shaded patch of width $dxdy$ to the total luminosity, we realize that the number of particles in beam 1 within the patch is $dN_1 = \psi(x, y)dxdy$. The target density $N_t/A$ of scattering centers in the second beam is $N_t/A = N_2\psi_2(x - X, y - Y)$. For the contribution to the luminosity of the shaded area $\mathcal{L}(x, y)$ we then obtain $\mathcal{L}(x, y) = f_c N_1 N_2 \psi_1(x, y)\psi(x - X, y - Y)$ and for the total luminosity we have to integrate over $x$ and $y$ with the result

$$\mathcal{L} = f_c N_1 N_2 \int_{-\infty}^{\infty} \int_{-\infty}^{\infty} \psi_1(x, y)\psi_2(x - X, y - Y)dxdy \ , \tag{9.19}$$

where $f_c$ is the repetition frequency of the collisions. Here we observe that the luminosity $\mathcal{L}$ is given as the convolution of the transverse distributions of the two beams. In particular, if the two distributions are Gaussian with beam-matrices, restricted to the two spatial transverse directions $\sigma^a$ and $\sigma^b$, the convolution is characterized by $\Sigma = \sigma^a + \sigma^b$ or

$$\begin{pmatrix} \Sigma_{xx} & \Sigma_{xy} \\ \Sigma_{xy} & \Sigma_{yy} \end{pmatrix} = \begin{pmatrix} \sigma_{xx}^a & \sigma_{xy}^a \\ \sigma_{xy}^a & \sigma_{yy}^a \end{pmatrix} + \begin{pmatrix} \sigma_{xx}^b & \sigma_{xy}^b \\ \sigma_{xy}^b & \sigma_{yy}^b \end{pmatrix} \tag{9.20}$$

and the luminosity $\mathcal{L}$ given by

$$\mathcal{L}(X, Y) = \frac{N_1 N_2 f_c}{2\pi\sqrt{\det \Sigma}} \exp\left[-\frac{1}{2}(X, Y)\Sigma^{-1}\begin{pmatrix} X \\ Y \end{pmatrix}\right] \ . \tag{9.21}$$

For round ($\sigma_{xx}^a = \sigma_{yy}^a = \sigma_{xx}^b = \sigma_{yy}^b = \sigma_r^2$) beams with equal rms beam size $\sigma_r$ Equation 9.21 reduces to the well-known expression $\mathcal{L}_r = (N_1 N_2 f_c/4\pi\sigma_r^2)e^{-(X^2+Y^2)/4\sigma_r^2}$. Obviously, it is proportional to the number of particles in each beam and the collision frequency $f_c$. Moreover, the luminosity is larger, the smaller the beam size $\sigma_r$ at the interaction point is. From the exponential dependence on the beam-center displacements $X, Y$ we see that the luminosity is reduced by 22 % if the beams miss each other by $\sqrt{X^2 + Y^2} = \sigma_r$, whereas for $0.2\sigma_r$ the reduction is only 1 %, which may be used to define tolerances for the alignment of the respective beams.

For beams with large aspect ratios $r = \sigma_x/\sigma_y$ the relative tilt angle $\phi$ between the major axes of the beams has a significant impact on the luminosity. We calculate it by considering diagonal matrices with $\sigma_{xx}^a = \sigma_{xx}^b = \sigma_x^2$ and $\sigma_{yy}^a = \sigma_{yy}^b = \sigma_y^2$ in Equation 9.20 and then rotate $\sigma^b$ by the tilt angle $\phi$. For $\det \Sigma$ we obtain $\det \Sigma = 4\sigma_x^2\sigma_y^2 + (\sigma_x^2 - \sigma_y^2)^2 \sin^2 \phi$. The luminosity for centered beams is, according to Equation 9.21, proportional to $1/\sqrt{\det \Sigma}$. The right-hand side in Figure 9.5 shows the luminosity, normalized to the value at $\phi = 0$, as a function of the tilt angle $\phi$. The solid line shows the luminosity loss for a beam with aspect ratio $r = 3$ and the dashed lines for $r = 10$. Apparently, beams with large aspect ratio are very sensitive to the tilt angle and require careful compensation of the cross-plane coupling, in particular, from the solenoidal fields that are often part of the detector.

In the previous paragraphs, we assumed that the transverse beam size during the collision does not change, which is a valid assumption, provided the bunch length $\sigma_s$ is small compared to the beta function $\beta^*$ at the interaction point. If, on the other hand, $\beta^*$ approaches $\sigma_s$, different longitudinal slices of the bunches have different transverse beam sizes during the interaction and the luminosity is reduced. This called the *hourglass effect* [62]. It is illustrated on the left-hand side in Figure 9.6 which shows the vertical beam size $\sigma_y/\sigma_{y0}$ in the vicinity of the interaction point at $s = 0$ as the solid line for $\beta_y^* = 1$ cm. The dashed and dot-dashed lines denote the longitudinal distribution of one beam moving towards the right and one moving towards the left, immediately before the centers of the beams meet at $s = 0$. The shaded areas denote longitudinal slices of beams that contribute to the luminosity, where the transverse beam size $\sigma_y$ of the dark area is only about half that of the lightly shaded slices. If the bunch length $\sigma_s$ were much smaller than the rate of growth of $\sigma_y$, which is $\beta_y^*$, all slices had transverse beam size $\sigma_{y0}$. In order to quantitatively evaluate the

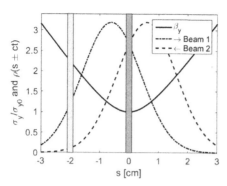 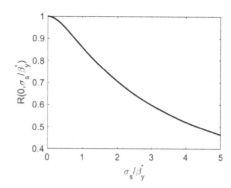

Figure 9.6 Two bunches (dashes and dot-dashes), with bunch length $\sigma_s$ comparable to the minimum $\beta_y^*$ (solid), colliding head-on at $s = 0$. The transverse beam sizes of the longitudinal bunch slices are larger for $s \neq 0$ (light grey) than they are for $s = 0$ (dark grey), which reduces the luminosity and depends on the ratio $\sigma_s/\beta_y^*$ (right).

reduction in luminosity, we consider the distributions of two counter-propagating beams, $\rho_+$ and $\rho_-$, given by

$$\rho_\pm(x, y, s, t) = \frac{N_\pm}{(2\pi)^{3/2}\sigma_x(s)\sigma_y(s)\sigma_s} \exp\left[-\frac{x^2}{2\sigma_x^2(s)} - \frac{y^2}{2\sigma_y^2(s)} - \frac{(s \pm ct)^2}{2\sigma_s^2}\right].$$ (9.22)

For simplicity, we assume that their speed in the laboratory frame can be approximated by the speed of light $c$, the beams collide head-on, and their beam sizes are equal, but vary with longitudinal position $s$ according to $\sigma_x^2(s) = \sigma_{x0}^2(1 + s^2/\beta_x^{*2})$ and $\sigma_y^2(s) = \sigma_{y0}^2(1 + s^2/\beta_y^{*2})$. With the bunch collision frequency $f_c$, we thus obtain for the luminosity

$$\mathcal{L} = 2cf_c \int_{-\infty}^{\infty} ds \int_{-\infty}^{\infty} dx \int_{-\infty}^{\infty} dy \int_{-\infty}^{\infty} dt\, \rho_+(x, y, s, t)\rho_i(x, y, s, t).$$ (9.23)

Inserting the distributions from Equation 9.22 we find that the integrals over $t, x$, and $y$ are Gaussian and can be evaluated trivially with the result

$$\mathcal{L} = \frac{N_+N_-f_c}{4\pi^{3/2}\sigma_s} \int_{-\infty}^{\infty} \frac{e^{-s^2/\sigma_s^2}ds}{\sigma_x(s)\sigma_y(s)} = \frac{N_+N_-f_c}{4\pi^{3/2}\sigma_s\sigma_{x0}\sigma_{y0}} \int_{-\infty}^{\infty} \frac{e^{-u^2}du}{\sqrt{1 + \frac{\sigma_s^2 u^2}{\beta_x^{*2}}}\sqrt{1 + \frac{\sigma_s^2 u^2}{\beta_y^{*2}}}}.$$ (9.24)

For vanishingly short bunch length $\sigma_s \to 0$ the integral evaluates to $\sqrt{\pi}$ and we find $\mathcal{L} = N_+N_if_c/4\pi\sigma_{x0}\sigma_{y0}$ which is the same as we find from Equation 9.21 for equal beams. The reduction from this value due to finite $\sigma_s/\beta_{x/y}^*$ is described by the reduction factor $R(\sigma_s/\beta_x^*, \sigma_s/\beta_y^*)$ that we define by

$$R(\sigma_s/\beta_x^*, \sigma_s/\beta_y^*) = \frac{1}{\sqrt{\pi}} \int_{-\infty}^{\infty} \frac{e^{-u^2}du}{\sqrt{1 + \frac{\sigma_s^2 u^2}{\beta_x^{*2}}}\sqrt{1 + \frac{\sigma_s^2 u^2}{\beta_y^{*2}}}}.$$ (9.25)

In many colliders, especially for $e^+e^-$, the vertical beta function and beam size are much

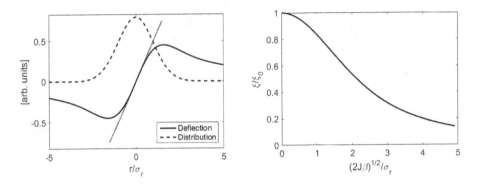

**Figure 9.7** Left: The distribution function $\rho(r)$ (dashes) and the transverse deflection function $d(r, \sigma_r) = (1 - e^{-r^2/2\sigma_r^2})/r$ (solid) for round Gaussian beams. The dotted line indicates the slope of the deflection function at the origin. Right: the amplitude-dependent tune shift for round beams given by Equation 9.27.

smaller than the horizontal, and often the luminosity is maximized by decreasing the vertical beta function $\beta_y^*$ at the interaction point, such that we can ignore the horizontal plane when evaluating the reduction factor and we have to evaluate $R(0, \sigma_s/\beta_y^*)$. On the right-hand side in Figure 9.6, we show it as a function of $\sigma_s/\beta_y^*$ calculated in a MATLAB script that numerically evaluates the integral. The reduction is about 14 % for $\sigma_s \approx \beta_y^*$. Note that decreasing the beta function still increases the luminosity, but simply not as much as one can expect from Equation 9.21.

In order to understand sub-nuclear reactions with very small reaction cross-sections, dedicated colliders, so-called *factories,* seek to reach the highest luminosities by colliding a large number of very intense and closely spaced bunches. They are used to investigate rare decays of $\Phi$ or B-mesons in order to reveal the secrets of CP-violation or to find Higgs-bosons in order to understand the origin of the mass of subatomic particles. For example, in LHC the bunches are spaced by about 8 m. These closely spaced bunches not only collide at the IP, but also at several longitudinal positions adjacent to the IP. In these *parasitic crossings* the beams will perturb each other, unless the transverse distance between the counter-propagating bunches is made large. And this requires the bunches to collide with a *crossing angle* at the IP at the expense of a reduced luminosity, because different longitudinal slices of the colliding bunches are transversely offset with respect to each other. This leads to a reduction of the luminosity, which can be recovered by turning the bunches in the plane where they cross, such that the slices are lined up properly. This scheme is called *crab crossing* and requires special radio-frequency deflectors, so-called *crab cavities.*

Decreasing the transverse beam sizes by decreasing the beta function at the interaction point we already discussed, but increasing the number of particles within each bunch is yet another method to increase the luminosity. At some point, however, the electro-magnetic fields that the many particles, compressed to small transverse sizes, create, are so strong that they perturb the other beam and limit the luminosity. We will consider the consequences of these fields in the next section.

## 9.6  INCOHERENT BEAM-BEAM TUNE SHIFT

The electric field $\vec{E}$ generated by an ultra-relativistic beam of particles only has transverse components when observed in the laboratory frame. Moreover, it is accompanied by transverse magnetic fields, given by $\vec{B} = (\vec{v}_0/c^2) \times \vec{E}$, where $\vec{v}_0 = \vec{\beta}_0 c$ is velocity of the particles. A particle with opposite charge and moving in the opposite direction $-\vec{v}_s$ experiences the Lorentz force $\vec{F} = -e[\vec{E} - \vec{v}_0 \times \vec{B}] = -e[\vec{E} - \vec{\beta}_0 \times (\vec{\beta}_0 \times \vec{E})] = -e[\vec{E} + \vec{\beta}_0^2 \vec{E}] \approx -2e\vec{E}$. Thus we only need to calculate the electric field $\vec{E}$ of a given transverse charge distribution. In the ultra-relativistic limit the magnetic field adds the same contribution.

We start by calculating the electric field of a round gaussian charge distribution moving towards the negative $s$−axis and having transverse rms beam size $\sigma_r$, which has the charge density $\rho(r, s, t) = (Ne/(2\pi)^{3/2}\sigma_r^2\sigma_s)e^{-r^2/2\sigma_r^2}e^{-(s+ct)^2/2\sigma_s^2}$. We derive the electric field $E(r, s)$ at radius $r$ from applying Gauss's law to a cylinder with radius $r$ at longitudinal position $s$ with length $ds$. Here $\varepsilon_0$ times the integral of the electric field on the surface is $2\pi\varepsilon_0 r E(r, s)ds$ and this is equal to the integral over the enclosed charges $Neds \int_0^{2\pi} d\phi \int_0^r dr'r'\rho(r', s, t) = (Neds/\sqrt{2\pi}\sigma_s)\left(1 - e^{-r^2/2\sigma_r^2}\right)e^{-(s+ct)^2/2\sigma_s^2}$. Solving for $E(r, s)$ we find $E(r, s) = (Ne/(2\pi)^{3/2}\sigma_s\varepsilon_0)e^{-(s+ct)^2/2\sigma_s^2}d(r, \sigma_r)$, where we introduce the deflection function $d(r, \sigma_r) = (1 - e^{-r^2/2\sigma_r^2})/r$. The force $\vec{F}$ is purely radial and its magnitude $F_r$ is given by $F_r = -2eE(r, s) = -(2Ne^2/(2\pi)^{3/2}\varepsilon_0\sigma_s)e^{-(s+ct)^2/2\sigma_s^2}d(r, \sigma_r)$. A particle, moving in the opposite direction has the longitudinal position $s = ct$, and experiences a radial momentum transfer $\Delta p_r = \int_{-\infty}^{\infty} F_r dt$ where we must evaluate $F_r$ at $s = ct$. After a little algebra we find $\Delta p_r = -(Ne^2/2\pi\varepsilon_0 c)d(r, \sigma_r)$. Dividing $\Delta p_r$ by the longitudinal momentum of the deflected particle, $p_0 \approx \gamma_0 mc$ in the ultra-relativistic limit, we find the radial angular deflection $\Delta r' = \Delta p_r/p_0 = -(2Ne^2/4\pi\varepsilon_0 mc^2\gamma_0)d(r, \sigma_r)$. After introducing the classical particle radius $r_p = e^2/4\pi\varepsilon_0 mc^2$, we obtain the relation for the beam-beam deflection angle for round beams

$$\Delta r' = -\frac{2Nr_p}{\gamma_0}\frac{1 - e^{-r^2/2\sigma_r^2}}{r} \qquad \text{and} \qquad \Delta x' = \frac{x}{r}\Delta r', \qquad \Delta y' = \frac{y}{r}\Delta r' . \qquad (9.26)$$

Here the minus sign applies to colliding beams of opposite charge and, for example, for electrons and positrons, in that case the classical particle radius $r_p$ is the classical electron radius $r_p = r_e = 2.818 \times 10^{-15}$ m.

Figure 9.7 shows the transverse Gaussian charge-distribution as a dashed line and the deflection function $d(r, \sigma_r)$ as a solid line. The latter is anti-symmetric and assumes its maximum near $r/\sigma_r \approx 1.6$. For large $r$, it has a $1/r$–dependence and close to the origin it has the linear dependence $r/2\sigma_r^2$, which is indicated by the dotted line in Figure 9.7. For the change of the horizontal deflection angle, we likewise obtain a linear dependence $\Delta x' = -(Nr_p/\sigma_r^2\gamma_0)x = -x/f_r$ and in the vertical plane $\Delta y' = -y/f_r$ with an effective focal length $f_r$. Particles crossing an oncoming bunch thus experience a focusing force in both planes that is characterized by the focal length $1/f_r = Nr_p/\sigma_r^2\gamma_0$. Following the reasoning from Section 8.3.2 an additional quadrupolar field in a ring causes the tunes $Q_x$ and $Q_y$ to differ from their unperturbed values by $\Delta Q_x = \beta_x/4\pi f_r$ and $\Delta Q_y = \beta_y/4\pi f_r$. They are commonly denoted *incoherent beam-beam tune shift* parameters for round beams $\xi_x = Nr_p\beta_x/4\pi\sigma_r^2\gamma_0$ and $\xi_y = Nr_p\beta_y/4\pi\sigma_r^2\gamma_0$. These are the tune shifts that individual particles with small betatron amplitudes experience, hence they are called "incoherent."

In Figure 9.7 we see that the slope is linear for small deviations near the origin, but particles with betatron amplitudes larger than $\sigma_r$ will experience the linear part of the beam-beam force only a fraction of the time. The tune shift will therefore depend on the amplitude of the particle and is derived in [63] with the result

$$\xi = \xi_0 \frac{1 - e^{-\alpha/2} I_0(\alpha/2)}{\alpha/2} \ . \tag{9.27}$$

$\xi_0$ is the tune shift close to the origin and $\alpha$ is the ratio of the square of the betatron amplitude $2J_{x,y}\beta_{x,y}$ and the beam size $\alpha = 2J_{x,y}\beta_{x,y}/2\sigma_r^2$ for the horizontal and vertical plane. On the right-hand side in Figure 9.7 we show $\xi/\xi_0$ as a function of the betatron amplitude normalized to the beam size $\sqrt{2J_x\beta_x}/\sigma_r$. Clearly the particles with amplitudes exceeding $\sigma_r$ experience a significantly reduced tune shift, which is intuitively clear, because the beam-beam force decreases with $1/r$ for large amplitudes, and large-amplitude particles spend the dominant fraction of their oscillation periods in the region with weak fields. Consequently their tune shift is reduced. This amplitude-dependence of the tune shift causes a tune spread among the particles in a beam, which extends all the way from the bare tune of the particles with very large amplitudes to the tune shift $\xi_0$ of the particles at the center. Large bunch populations $N_1$ and $N_2$ cause the tune spread to grow significantly and may cover destructive resonance lines in the tune diagram. This can cause the beam size to increase and even cause particle losses. Moreover, the non-linear dependence of the deflection angle on the radial position $r$ makes the beam-beam force highly non-linear. It can therefore excite many non-linear resonances [63]. The particle intensities or the beam currents at which the performance of the collider deteriorates, either by increased background in the detector, or increased beam size such that the luminosity no longer increases linearly with the number of stored particles, is referred to as *beam-beam limit* [64]. Colliders usually operate just below this limit and typical values for hadron colliders, summed over all interaction points, are $\xi \approx 0.003$ whereas electron-positron colliders achieve about ten times higher values $\xi \approx 0.03$.

In the previous paragraphs we focused on round beams, which mostly pertain to hadron colliders, because the, normally equal, emittances are determined at the particle source and then maintained during acceleration. In $e^-e^+$ colliders, the horizontal emittance is usually much larger than the vertical, and the beams are very flat at the interaction point. We will therefore briefly discuss the incoherent tune shifts and the fields generated by beams with a transversely elliptic cross section. The fields were derived for upright elliptic beams with a Gaussian distribution by Bassetti and Erskine in [65]. For arbitrarily oriented elliptic Gaussian beams the deflection angles $\Delta y' + i\Delta x' = (2r_p N/\gamma_0) F_0(x_1, x_3, \sigma)$ are derived in [44] with $F_0$ given by

$$F_0(x_1, x_3, \sigma) = \frac{\sqrt{\pi}}{\sqrt{2(\sigma_{11} - \sigma_{33} + 2i\sigma_{13})}} \left\{ w(z_1) - e^{-g} w(z_2) \right\} \tag{9.28}$$

with the complex error function $w(z) = e^{-z^2} (1 - \mathrm{erf}(-iz))$ from [23] and the abbreviations $g = \frac{1}{2} \sum_{j,k=1,3} \sigma_{jk}^{-1} x_j x_k$,

$$z_1 = \frac{x_1 + ix_3}{\sqrt{2(\sigma_{11} - \sigma_{33} + 2i\sigma_{13})}} \quad \text{and} \quad z_2 = \frac{(\sigma_{33} - i\sigma_{13})x_1 + i(\sigma_{11} + i\sigma_{13})x_3}{\sqrt{\sigma_{11}\sigma_{33} - \sigma_{13}^2}\sqrt{2(\sigma_{11} - \sigma_{33} + 2i\sigma_{13})}} \ .$$

Here we use the notation $x_1 = x$ and $x_3 = y$. In the limit $\sigma_{13} \to 0$ and $\sigma_{11} = \sigma_{33} \to \sigma_r^2$, the function $F_0$ reduces to the deflection function $d(r, \sigma_r)$ from the previous paragraph [44]. From a Taylor-expansion of $F_0$ around the origin, which results in $F_0 \approx -y/\sigma_y(\sigma_x + \sigma_y) - ix/\sigma_x(\sigma_x + \sigma_y)$, we determine beam-beam tune shift for elliptic beams. The focal lengths become $1/f_x = 2Nr_p/\sigma_x(\sigma_x + \sigma_y)\gamma_0$ and $1/f_y = 2Nr_p/\sigma_y(\sigma_x + \sigma_y)\gamma_0$. With these values the tune shift parameters for elliptic beams become $\xi_x = Nr_p\beta_x/2\pi\sigma_x(\sigma_x + \sigma_y)\gamma_0$ and $\xi_y = Nr_p\beta_y/2\pi\sigma_y(\sigma_x + \sigma_y)\gamma_0$.

Treating the beam-beam interaction as a linearly focusing element that causes a tune shift of a particle is only a first approximation. In order to take the non-linear character of the beam-beam force into account, one has to resort to *multi-particle simulations*. In *weak-strong simulations*, one of the beams, the "strong" one, is assumed to have a much higher intensity than the other one and remains unaffected by the beam-beam interaction. Therefore, its electric field is constant and is treated as a non-linear lens, by using either Equation 9.26 or Equation 9.28 to describe the deflections of particles of the other, the "weak," beam. In these models, only the stability of the weak beam can be analyzed. If, on the other hand, both beams carry a large charge, *strong-strong simulations* are employed. Here, both beams are simulated by a large number of sample particles and in this way can represent any distribution. In order to maintain self-consistency, we have to calculate the electro-magnetic fields responsible for the deflections from solving the Poisson equation, which, however, is computationally expensive and leads to long simulation times.

In this section we described the effect of one beam on the individual particles of the other beam. In the coming section, we analyze how the center of mass of the beam—the centroid—is affected, which is commonly referred to as *coherent beam-beam interactions*.

## 9.7   COHERENT BEAM-BEAM INTERACTIONS

Here we consider that beam 1 with transverse distribution $\psi_1$ creates the field that deflects beam 2, which has distribution $\psi_2$. We proceed to calculate the change of the centroid angles $\Delta X'$ and $\Delta Y'$ of beam 2 by averaging the deflections angles of the individual particles $\Delta x'$ and $\Delta y'$ over their distribution function $\psi_2$. Instead of explicitly averaging over the electric field, which is proportional to $F_0$ in Equation 9.28, we note that the deflection angles of individual particles can be written by the convolution of the field-generating distribution $\psi_1(\vec{x} - \vec{X}_1)$ with a center displaced by $\vec{X}_1$, and a Green's function $G(\vec{x}) = 2iNr_p/\gamma(x + iy)$. For a particle at position $\vec{y}$, this representation yields the following deflection angles

$$\Delta y' + i\Delta x' = \int d^2x\, G(\vec{y} - \vec{x})\psi_1(\vec{x} - \vec{X}_1). \tag{9.29}$$

All integrals extend over the two transverse—horizontal and vertical—directions and all quantities that are ornamented with a vector arrow are assumed to have a horizontal and a vertical component. Averaging $\Delta y' + i\Delta x'$ over the distribution $\psi_2(\vec{y} - \vec{Y}_2)$ gives us the centroid deflection angle of beam 2

$$\begin{aligned}\Delta Y_2' + i\Delta X_2' &= \int d^2y\, \psi_2(\vec{y} - \vec{Y}_2) \int d^2x\, G(\vec{y} - \vec{x})\psi(\vec{x} - \vec{X}_1) \\ &= \int d^2y\, G(\vec{X}_2 - \vec{X}_1 - \vec{y}) \int d^2x\, \psi_2(\vec{x} - \vec{y})\psi_1(\vec{x}), \end{aligned} \tag{9.30}$$

where we changed the order of integration in the second equality. We now recognize the second integral over $d^2x$ as the convolution of the distributions of the field-producing beam 1 and the deflected beam 2. Comparing with Equation 9.19 we note that this convolution is proportional to the luminosity. And finally, comparing the second line with Equation 9.29, we observe that the centroid-deflection angles $\Delta Y_2' + i\Delta X_2'$ are given as the "electric field" of the convolution of the two distributions. This result holds for any distribution, and especially for Gaussians, for which convolutions can be calculated by adding the covariance or sigma matrices, as shown in Equation 9.20.

In order to explicitly calculate the centroid-deflection angle from the collision of two Gaussian distributions, we can use the equations for the electric field, either for a round beam in Equation 9.26 or for elliptic beams in Equation 9.28 after replacing the individual transverse sigma matrices by their sum as shown in Equation 9.20. We immediately use this insight and calculate an estimate for the *coherent beam-beam tune shift* $\Xi_x$ and $\Xi_y$ for colliding beams with equal beam sizes in both planes, such that we find $\Sigma_x = \sqrt{2}\sigma_x$ and $\Sigma_y = \sqrt{2}\sigma_y$. Calculating the first-order Taylor-expansion of $F_0$ with $\Sigma_{x,y}$ instead of $\sigma_{x,y}$, we find $\Xi_x = Nr_p\beta_x/4\pi\Sigma_x(\Sigma_x + \Sigma_y)\gamma_0 = \xi_x/2$ and $\Xi_y = \xi_y/2$. The coherent tune shifts appear to be only half of the incoherent values within the limit of our theory that assumes rigid bunches. A more refined analysis [66] finds values between 1.21 and 1.33 instead of 2 for beams of different aspect ratios $\sigma_x/\sigma_y$. The smaller values indicate that only a fraction of particles near the core of the distribution oscillate, rather than the entire bunch oscillating rigidly.

In Section 7.2 we found that centroids of the beam distribution are detectable on beam position monitors and we can therefore use them to observe the consequences of the beam-beam deflections. Since the oscillations on the two colliding beams are coupled by the beam-beam interaction, this will affect the tunes of the two rings. Moreover, the additional deflection angles at the interaction point are visible as a change in the closed orbit, as we saw in Section 8.3.1. To probe the beam-beam deflections usually one beam is transversely scanned across the other beam with a closed bump, such as shown in Figure 8.8. Such a scan is commonly named after *Simon van der Meer* who first used it to maximize the luminosity in the ISR at CERN. In colliders with separate rings, such as the ISR or LHC, magnetic deflectors in regions where the beams are separated are used. On the other hand, in colliders where both beams are guided in a common magnetic structure, which is often the case in $e^-e^+$–colliders, electro-static deflectors are used. Using either type of deflector to implement the bump, one simultaneously observes the tunes with a system discussed in Section 7.5.1 and observes changes of the closed orbit. Using as many position monitors as possible is advantageous to increase the sensitivity to detect the often small beam-beam deflection angles. In the companion software for this book, a simple MATLAB simulation is available to illustrate the methods. The parameters are loosely based on an early design of the B-factory at SLAC and more details can be found in [67]. In the code, first the fixed point of the closed orbit with beam-beam deflections is found by iterating a variant of Equation 8.27. The iterations are required, because the beam-beam deflection angle from Equation 9.28 is non-linear, but a few iterations usually suffice for normal operating parameters. The upper plots in Figure 9.8 show the angle of the closed-orbit while scanning a horizontal (upper left) and vertical (upper right) bump. The amplitude of 50 to 100 $\mu$rad will lead to orbit changes on position monitors in the ring of a several 100 $\mu$m which is visible on modern position monitors. This information can then be used to find the bump amplitude at which the beams are centered, which also maximizes the luminosity. Moreover, it is possible to determine the convoluted beam sizes $\Sigma$, as defined in Equation 9.20, from the shape of the of deflection curves. In addition to analyzing the closed orbit itself, slightly perturbing the closed orbit allows us to find the tunes by Fourier-transforming recorded positions. The lower plots in Figure 9.8 show the horizontal (lower left) and the vertical (lower right) tunes as a function of the bump amplitude. For large bump amplitudes, the tunes of the two rings are not coupled and therefore equal. Decreasing the amplitude, we observe that the two tunes split and that one mode stays at the unperturbed tune and does not change. This mode is characterized by the two beams oscillating in phase and is called the $\sigma$−mode. The other mode is characterized by both beams oscillating with opposite phase and is commonly called $\pi$−mode. The maximum separation between $\sigma$ and $\pi$−mode coincides with centered

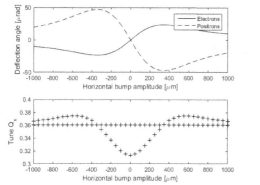
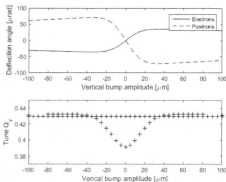

Figure 9.8 The centroid deflection angles (top) for electron and positron beams and the tunes of the coupled system of the two rings (bottom). The horizontal plane is shown on the left and the vertical on the right.

beams and the separation of the tunes in this simplified model with rigid bunches is given by $\Xi_x$ and $\Xi_y$. Maximizing the tune split between the two modes will also cause the beams to collide head-on and well-centered.

The beam-beam collisions cause the betatron oscillations of the two rings to couple, which showed up as the tune split into $\sigma$ and $\pi$−modes in Figure 9.8. If the beams are colliding head-on, the linear slope of the beam-beam deflections causes this coupling. It is qualitatively similar to the coupling between horizontal and vertical betatron oscillations discussed in Section. 8.3.3 and, likewise, can cause stop bands. If the two beams collide at several interaction points, for example at the ATLAS, CMS, LHCb, and ALICE detectors in LHC, many different bunches couple and the mode spectrum becomes very complex. In [68] the stability of these modes is investigated by constructing coupled matrix models of the interacting bunches and analyzing their eigenvalues as a function of the unperturbed tunes and identifying stop bands—regions in which the eigenvalues are complex-valued.

After the discussion of the beam-beam interaction in circular colliders, we will briefly touch upon the special features of linear colliders, caused by the extremely small beam sizes in the range of nm to $\mu$m at the interaction point.

## 9.8　LINEAR COLLIDERS

Already the first linear collider, the Stanford Linear Collider (SLC), which operated from 1988 until 2002, achieved beam sizes at the interaction point below $1\,\mu$m. The beam sizes in two proposed machines, the ILC [69] and CLIC [70], will be two to three orders of magnitude smaller. The electro-magnetic fields created by the large number of charges squeezed to nm sizes are enormous and cause the emission of extremely energetic photons and furthermore perturb the "other" beam by focusing it with a focal length comparable to the length of the bunch.

It is instructive to calculate the focal length due to the electro-magnetic field. Near the end of Section 9.6, we found that it is given in the vertical plane by $1/f_y = 2Nr_p/\sigma_y(\sigma_x + \sigma_y)\gamma_0$. For the SLC, which operated at energies of $45\,$GeV with round beams of about $\sigma_r = 1\,\mu$m and intensities around $N = 3 \times 10^{10}$ particles per bunch, we find $f \approx 1\,$mm which is approximately equal to the bunch length $\sigma_s$. We therefore expect the beam size to decrease during the collision, which causes the luminosity to slightly increase. The parameter that quantifies this processes is the *disruption parameter* $D_y = \sigma_s/f = 2Nr_p\sigma_s/\sigma_y(\sigma_x + \sigma_y)\gamma_0$, here written for the vertical dimension where the effect is often stronger due to the smaller vertical beam size at the interaction point. For CLIC at 1.5 TeV with $3.72 \times 10^9$ particles per bunch and beam sizes of $\sigma_x = 40$ nm, $\sigma_y = 1$ nm, and $\sigma_s = 44\,\mu$m we find $D_y = 3.8$.

In order to investigate the effect on the luminosity, we prepare a MATLAB script called disruption.m. It simulates the collision of two bunches that are subdivided into 100 longitudinal slices and we collide all pairs of slices. For the deflections we make the simplifying assumption that the force is purely linear and can be described by the horizontal and vertical focal lengths and is therefore modeled by matrices. In the simulation we first define parameters for CLIC with 1.5 TeV beam energy and initialize arrays to hold the beam matrices siga and sigb and the centroids Xa, Xb for all slices of the two beams. The main part of the simulation is the loop over time and the slices that actually meet. It is reproduced here:

```
for t=0:2*Nslices-2     % loop over time steps
  disp(['at time step t=' num2str(t)])
  for j=1:Nslices  % loop over the slices that meet at time t
    k=t+2-j;          % index in the other beam
    if ((k<1) || (k>Nslices)) continue; end      % slices do not meet
    disp(['   slice ' num2str(j) ' meets slice ' num2str(k)])
    [sa,xa,sb,xb,lumi]=collide_slices(siga(:,:,j),Xa(:,:,j),Na(j), ...
       sigb(:,:,k),Xb(:,:,k),Nb(k),dz,gamma);
    siga(:,:,j)=sa; Xa(:,:,j)=xa;  % copy back updated values
    sigb(:,:,k)=sb; Xb(:,:,k)=xb;
    lumi_total=lumi_total+lumi;    % sum the luminosity per slice
  end
  % display result for this time step
    :
end
```

In the function collide_slices() first the focal lengths and the length of the drift spaces between slices are constructed and then used to propagate the centroids and beam matrices. Please inspect the script that accompanies this book and is available at this book's web page.

On the left- and right-hand side in Figure 9.9 the bunch intensities (top), the horizontal (center) and vertical (bottom) beam size are displayed at 271 fs (left) and 582 fs after the simulation started. The solid lines pertain to the left-moving bunch and the dashed lines for the bunch moving towards the right. The left plots show the parameters after the heads of the two bunches start to overlap. The two lower plots show the beam size along the bunches and we find that the respective heads of the bunches are already starting to decrease due to the mutual focusing. The plot on the right shows the situation after the centers of the bunches have already passed each other and the horizontal beam size of the head of the bunches is reduced from its initial value of 40 nm to about 30 nm. The vertical beam size on the lower plot shows several intermediate waists within the bunch. This indicates that a number of slices are focused within the bunch and then overshoot before being refocused into the next waist. This is a clear indication of disruption. That the beam size is squeezed to smaller values is often referred to as the *pinch effect.* The luminosity calculated by summing over all colliding slices is about twice the value calculated with hourglass effect for $\sigma_s/\beta_y \approx 0.65$ taken into account.

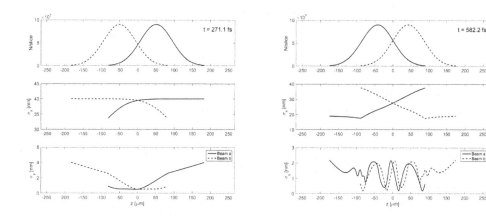

**Figure 9.9** The longitudinal distribution (top) of two colliding beams (solid is moving to the left, dashed to the right), the horizontal (middle) and vertical (bottom) beam size for two instances during the collision.

The second consequence of significantly deflecting the particles during the course of the collision is the emission of synchrotron radiation, in this case called *beamstrahlung*. It turns out that the recoil from the emission of photons with very high energies significantly affects the momentum distribution of the beams, which makes the interpretation of the events recorded in the detector difficult. Let us look into this topic further. In Chapter 10 we will see that the critical energy $\varepsilon_c = 3\gamma_0^3 \hbar c/2\hat{\rho}$ of the emitted photons depends on the energy of the emitting electrons $\gamma m c^2$ and the bending radius $\hat{\rho}$ of the deflection. A particle, transversely displaced by $y = \sigma_y$, is deflected by $\theta = \sigma_y/f_y$ over the distance of the bunch length, which we use to estimate the bending radius $\hat{\rho} \approx \sigma_s/\theta = \gamma\sigma_s(\sigma_x + \sigma_y)/2Nr_p \approx 0.25\,\mathrm{m}$. Here we used the parameters for CLIC from page 238. From this bending radius, we obtain for the critical energy $\varepsilon_c \approx 30\,\mathrm{TeV}$, which is significantly higher than the particle's energy $E_0 = 1.5\,\mathrm{TeV}$. Despite the large value of $\varepsilon_c$, energy conservation dictates that the spectrum is truncated at the beam energy. Nevertheless, a considerable number of photons are emitted with energies comparable to the beam energy. Therefore the initially almost mono-energetic beam develops a considerable low-energy tail during the collision and the luminosity comes from a very wide distribution of particle energies and complicates the interpretation of the events recorded in the detector.

It is common to characterize the emission of beamstrahlung, especially in extreme regimes, by the parameter $\Upsilon = 2\varepsilon_c/3E_0$. It is used to describe the interaction of beamstrahlung photons with the remainder of the counter-propagating bunch, where the photons create $e^+e^-$−pairs in the process. These additional particles have low energies and will contribute to the background in the detector. If the collision is characterized by $\Upsilon < 0.6$, beamstrahlung photons generated early on during the collision interact with individual particles in the tail of the bunches and create *incoherent* $e^+e^-$−pairs. One fundamental process, among several others, is the Bethe-Heitler process $e^\pm\gamma \to e^\pm e^+e^-$. On the other hand, for values of $\Upsilon$ larger than about 0.6 the beamstrahlung photons produce large numbers of so-called *coherent* $e^+e^-$−pairs by interacting with the collective field of the counter-propagating beam. In both cases, one half of a pair has the "wrong" polarity and will be deflected into the detector where it contributes to the background.

Beamstrahlung photons are extreme examples of synchrotron radiation, the topic of our next chapter.

## QUESTIONS AND EXERCISES

1. You want to investigate a nuclear reaction with Uranium that has a cross section of $\sigma = 1\,\mu$barn. The cyclotron produces protons with a kinetic energy of 100 MeV with a constant current of $100\,\mu$A. The data acquisition system of your experiment has an efficiency of $\eta = 1.7 \times 10^{-3}$ to detect the nuclear reaction and you need a count rate of at least 20 kHz in order to get enough data over the weekend to complete your thesis. At what rate (particles/second) do protons impact on the target? How thick must the target at least be?

2. Calculate the energy loss of a proton beam with kinetic energy of a) 100 MeV and b) 180 MeV in a 28 mm thick lithium target. By how much do the angular divergences of the respective proton beams increase when passing the same target?

3. Analyze the dependence of the luminosity on the aspect ratio $r = \sigma_x/\sigma_y$ for two colliding beams with otherwise equal beam parameters. To make the different configurations commensurable, ensure that the cross-sectional area $A = \pi\sigma_x\sigma_y$ stays the same for all configurations.

4. Find expressions for the beam-beam deflection angle for the center of mass $\Delta X'$ and $\Delta Y'$ of (a) round beams; (b) beams with arbitrary transverse aspect ratios. Implement them in a MATLAB script, such that you can specify the beam parameters (sizes, intensities, energies) and the script generates a plot of the respective deflection angles for the two beams in both planes.

5. Write a weak-strong beam-beam simulation by representing the ring by a single transfer matrix from IP to IP and a single non-linear lens that represents the kick from the beam-beam interaction (a) for a round kick-producing beam and (b) calculate the amplitude-dependent tune shift. Is it consistent with Equation 9.27?

6. Add treating the displacement Xa and Xb of the centroids to the collide_slices() function, used in the disruption.m script. (a) What is the limit of validity of using the linearized focussing? (b) Explore the reduction in luminosity as the beams collide with an offset of $\sigma_y/2$.

# CHAPTER 10

# Synchrotron Radiation and Free-Electron Lasers

While all charged particles emit electro-magnetic radiation, only electrons and positrons do so in appreciable quantities, because the emitted power is inversely proportional to the fourth power of the mass of the particles. Therefore it is strongly suppressed for heavier particles, although LHC is an exception we will discuss in Section 14.1. As a consequence, in this chapter, we will focus on electrons.

Already a few decades after Maxwell formulated his theory, Larmor derived expressions that describe the emission of radiation from non-relativistic electrons. Later, Liénard generalized them to electrons at relativistic velocities and found that the power $P_\gamma$ emitted from an electron with velocity $\vec{v} = \vec{\beta}c$ and acceleration $\dot{\vec{\beta}}$ is given by

$$P_\gamma = \frac{e^2\gamma^6}{6\pi\varepsilon_0 c}\left(\dot{\vec{\beta}}^2 - \left(\vec{\beta}\times\dot{\vec{\beta}}\right)^2\right) \quad \text{and} \quad \hat{P} = \frac{e^2\gamma^4}{6\pi\varepsilon_0\rho^2} = \frac{C_\gamma cE^4}{2\pi\rho^2} = \frac{C_\gamma e^2 c^3}{2\pi}B^2 E^2 \quad (10.1)$$

with $C_\gamma = e^2/3\varepsilon_0(mc^2)^4 = 8.846 \times 10^{-5}\,\mathrm{m/GeV^3}$ for electrons. We will assume that the motion is ultra-relativistic with $\beta \approx 1$ henceforth. Here, $\hat{P}$ is the power emitted by a particle with energy $E = \gamma mc^2$ that follows a circular path with radius $\rho$ in a constant magnetic field with flux density $B$, which, for example, is produced by a dipole magnet in a storage ring. The energy $U_0$, emitted in one turn by a particle with the reference energy $E_0$, is given by $U_0 = (C_\gamma E_0^4/2\pi)\oint ds/\rho^2 = (C_\gamma E_0^4/2\pi)I_2$, where we introduce the second radiation integral $I_2 = \oint ds/\rho^2$. For an iso-magnetic ring $\rho$ is constant and we obtain for the total emitted power $U_0 = C_\gamma E_0^4/\rho$ with $\int_0^C ds/\rho = 2\pi$. The emission of this *synchrotron radiation* has a profound influence on the properties of the electron beam, both through its energy dependence and the fact that the radiation is emitted as photons in a quantum-mechanical, and therefore an inherently random, process. We therefore briefly state the spectral properties of the radiation from dipole magnets, from which the relevant statistical properties of the radiation can be determined, before discussing how they determine the properties of the electron beams. In later sections, we will return to a more detailed discussion of the radiation itself, specialty magnets, such as undulators, and dedicated accelerators for free-electron lasers.

But let us now return to the radiation emitted by ultra-relativistic electrons in a dipole magnet, which was first analyzed by Schwinger in [71]. He found that the total emitted power $P_\gamma$ from Equation 10.1 has the spectral distribution

$$\frac{dP}{d\omega} = \frac{P_\gamma}{\omega_c} S\left(\frac{\omega}{\omega_c}\right) \quad \text{with} \quad S(y) = \frac{9\sqrt{3}}{8\pi} y \int_y^\infty K_{5/3}(x)dx \quad \text{and} \quad \omega_c = \frac{3\gamma_0^3 c}{2\rho}. \qquad (10.2)$$

Here $\omega_c$ is the *critical frequency* and $\varepsilon_c = \hbar\omega_c$ is the *critical energy* of the photon spectrum. Half the power is emitted at frequencies below $\omega_c$ and the other half above. We find the number spectrum of the photons $d\dot{\mathcal{N}}/d\omega$, emitted with a given frequency $\omega$, from dividing the power spectrum $dP/d\omega$ by the photon energy $\hbar\omega$. Subsequently integrating over all frequencies $\omega$ yields the number of photons $\dot{\mathcal{N}}$ emitted per unit time $\dot{\mathcal{N}} = (15\sqrt{3}/8)P_\gamma/\hbar\omega_c = 5\alpha c\gamma_0/2\sqrt{3}\rho$ with the fine-structure constant $\alpha = e^2/4\pi\varepsilon_0\hbar c$. Likewise, by averaging the photon energy $\varepsilon = \hbar\omega$ over the number spectrum $d\dot{\mathcal{N}}/d\omega$, we find the mean energy of the photons $\langle\varepsilon\rangle = (8/15\sqrt{3})\varepsilon_c$ and the second moment of photon energies $\langle\varepsilon^2\rangle = (11/27)\varepsilon_c^2$.

With the average energy loss due to synchrotron radiation described by Equation 10.1 and the statistics of the photon emission by $\dot{\mathcal{N}}$ and $\langle\varepsilon^2\rangle$ we are ready to determine their influence on the beam.

## 10.1  EFFECT ON THE BEAM

First we discuss how the longitudinal motion, previously discussed in Section 5.3, is affected, followed by the transverse motion and implications for the lifetime of the beams.

### 10.1.1  Longitudinally

Here, we assume that the energy loss $U_d(E)$ in Equation 5.27 is caused by the energy loss from synchrotron radiation, which, according to Equation 10.1, depends on the magnetic flux density field and on the energy as $B^2E^2$. We therefore expand $U_d(E_0+\Delta E) = U_d(E_0) + (\partial U_d/\partial E)\Delta E$ around the reference energy and use Equation 5.29 to express $\Delta E = \beta_0^2 E_0\delta$. Inserting into Equation 5.30 and using Equation 5.27, we obtain

$$\frac{d\delta}{dt} = \frac{e\hat{V}}{T_0 E_0}\sin\phi - \frac{1}{T_0 E_0}\left(U_d + \frac{\partial U_d}{\partial E}\Delta E\right) = \frac{e\hat{V}}{T_0 E_0}(\sin\phi - \sin\phi_s) - \frac{1}{T_0}\frac{\partial U_d}{\partial E}\delta \qquad (10.3)$$

where, compared to Equation 5.30, an additional term, proportional to $\delta$, appears. It is negative and describes damping, which should be intuitively clear, because higher-energy particles radiate more and thereby lose more energy than particles with the reference energy, whereas lower-energy particles radiate less. In both cases, the radio-frequency system seeks to keep the particle on-energy with the net effect of damping the longitudinal motion with time constant $1/\tau_E = (\partial U_d/\partial E)/2T_0$.

It remains to calculate $\partial U_d/\partial E$, where we have to keep in mind that, for constant magnetic flux density $B$, particles with higher energy not only radiate more, but also move on a trajectory with dispersion $D_x$. The latter effect causes the trajectory in dipole magnets to be longer by a factor $(1 + x_E/\rho)$ with $x_E = D_x\Delta E/E_0$. Moreover, if the dipoles have an embedded gradient $dB/dx$, the radiation emitted during one turn is

$$U_d(E) = \oint \hat{P}\frac{ds}{c} \approx \oint \hat{P}_0\left(1 + \frac{2}{B}\frac{\partial B}{\partial x}x_E + \frac{x_E}{\rho}\right)\frac{ds}{c}. \qquad (10.4)$$

where $\hat{P}_0(E)$ is power emitted by particles moving on the reference trajectory with energy $E$.

Differentiating $U_d(E)$ with respect to $E$ leads to

$$\frac{\partial U_d}{\partial E} = \oint \left( \frac{2\hat{P}_{00}}{E_0} + \frac{2\hat{P}_{00}D_x}{BE_0}\frac{\partial B}{\partial x} + \frac{D_x\hat{P}_{00}}{\rho E_0} \right) \frac{ds}{c} , \tag{10.5}$$

where $\hat{P}_{00} = \hat{P}_0(E_0) = C_\gamma c E_0^4/2\pi\rho^2$ is the power radiated by particles having the reference energy $E_0$. Furthermore, the integral of $\hat{P}_{00}$ over one turn is the total energy radiated $U_0 = \oint \hat{P}_{00}ds/c$

$$\frac{\partial U_d}{\partial E} = \frac{U_0}{E_0}\left[ 2 + \frac{1}{U_0c}\oint \hat{P}_{00}D_x\left( \frac{1}{\rho} + \frac{2}{B}\frac{\partial B}{\partial x} \right) ds \right] . \tag{10.6}$$

For the damping time for energy oscillation $1/\tau_E$, we then obtain

$$\frac{1}{\tau_E} = \frac{U_0}{2T_0E_0}(2+\mathcal{D}) \quad \text{with} \quad \mathcal{D} = \frac{I_4}{I_2} \quad \text{and} \quad I_4 = \oint \frac{D_x}{\rho}\left( \frac{1}{\rho^2} + 2k_1 \right) ds , \tag{10.7}$$

where we introduce the normalized gradient $k_1 = (\partial B/\partial x)/B\rho$ from Equation 3.8. $\mathcal{D}$ is the *damping partition number* and $I_4$ is the fourth synchrotron radiation integral, which is often small. We therefore find that the ratio of the longitudinal damping time $\tau_E$ and the revolution time $T_0$ is mainly determined by the ratio of beam energy $E_0$ and the radiated energy $U_0$. This ratio is often on the order of $10^3$, such that the damping time is on the order of a few thousand turns in the ring.

Radiation damping alone would cause the momentum spread to shrink to zero, but this is prevented by the stochastic emission of the radiation which acts like a source of noise and "heats" the distribution of particles. The balance of damping and excitation then determines the momentum spread. Typically the damping time $\tau_E$ is much slower than the synchrotron oscillation period and since the time scales differ significantly, we can treat the effect of damping and excitation as small perturbations on the dynamics of the scaled momentum $\delta$. From turn number $n$ to $n+1$ the momentum $\delta$ changes according to

$$\delta_{n+1} = \left( 1 - \frac{2T_0}{\tau_E} \right)\delta_n + \frac{\hat{u}}{E_0} = \xi\delta_n + \frac{\hat{u}}{E_0} , \tag{10.8}$$

where $\hat{u}$ is a random process that describes the fluctuations of the energy loss from emitting synchrotron radiation and $\xi = 1 - 2T_0/\tau_E$. If we denote the average of the momentum width $\delta^2$ over the beam by $\langle\delta^2\rangle$ we find

$$\langle\delta_{n+1}^2\rangle = \xi^2\langle\delta_n^2\rangle + 2\xi\langle\delta_n\hat{u}/E_0\rangle + \langle\hat{u}^2/E_0^2\rangle = \xi^2\langle\delta_n^2\rangle + \langle\hat{u}^2\rangle/E_0^2 . \tag{10.9}$$

Here $\langle\delta_n\hat{u}\rangle$ averages to zero, because the momenta $\delta_n$ and $\hat{u}$ are uncorrelated. We can now calculate $\langle\hat{u}^2\rangle$ as the integral of the photon emission rate $\dot{\mathcal{N}}$ multiplied by the second moment of the photon energy distribution $\langle\varepsilon^2\rangle = (11/27)\varepsilon_c^2$, which yields $\langle\hat{u}^2\rangle = \oint \dot{\mathcal{N}}\langle\varepsilon^2\rangle ds/c = (55\alpha\hbar^2c^2\gamma_0^7/24\sqrt{3})\oint ds/|\rho|^3$. This allows us to determine the equilibrium momentum spread $\sigma_\delta^2$ from requiring $\sigma_\delta^2 = \langle\delta_{n+1}^2\rangle = \langle\delta_n^2\rangle$ and find

$$\sigma_\delta^2 = \frac{\tau_E}{4CE_0^2}\oint \dot{\mathcal{N}}\langle\varepsilon^2\rangle ds = C_q\gamma_0^2\frac{I_3}{I_2(2+\mathcal{D})} \tag{10.10}$$

with $C_q = (55/32\sqrt{3})(\hbar/mc) \approx 3.83 \times 10^{-13}$ m and the third synchrotron radiation integral $I_3 = \oint ds/|\rho|^3$. For reference, we mention that for most electron or positron rings the momentum spread $\sigma_\delta$ turns out to be on the order of $10^{-3}$. From the momentum spread and the assumption that the beam is matched in the longitudinal phase space, we find the longitudinal equilibrium size to be

$$\sigma_\phi = \frac{\omega_{\rm rf}|\eta|}{\Omega_s}\sigma_\delta = \frac{h|\eta|}{\nu_s}\sigma_\delta \tag{10.11}$$

with the harmonic number $h$ and the synchrotron tune $\nu_s$. The physical bunch length $\sigma_s$ is related to the length in phase $\sigma_\phi$ by $\sigma_s = \sigma_\phi\lambda_{rf}/2\pi$. Now that we know the longitudinal equilibrium beam size it is time to look at the transverse sizes.

## 10.1.2 Vertically

Synchrotron radiation is emitted in the particle's forward direction and decreases both the transverse and longitudinal momenta. Only the latter is restored in the radio-frequency cavities, while the transverse momenta remain unaffected. On average, the total momentum lost in one turn $U_0/c$ is added to the longitudinal momentum of the electron, only. Since the transverse angles are the ratio of transverse to the longitudinal momenta, we find the vertical angles $y'$ to be reduced by $\Delta y' \approx -yU_0/E_0$. Inserting $y' + \Delta y'$ into the definition of the Courant-Snyder invariant from Equation 3.44 yields $\Delta J_y = -(\alpha yy' + \beta y'^2)U_0/E_0$. Averaging over the particle distribution results in $\Delta\varepsilon_y = \langle \Delta J_y \rangle = -\varepsilon_y U_0/E_0$, which follows from Equation 3.78 with $\langle yy' \rangle = -\varepsilon_y\alpha_y$ and $\langle y'^2 \rangle = \varepsilon_y\gamma_y$. Since $U_0$ is radiated during one turn, we obtain the vertical damping time $\tau_y$ from

$$\frac{d\varepsilon_y}{dt} \approx \frac{\Delta\varepsilon_y}{T_0} = -\frac{U_0}{T_0 E_0}\varepsilon_y = -\frac{2}{\tau_y}\varepsilon_y \quad \text{with} \quad \frac{1}{\tau_y} = \frac{1}{2T_0}\frac{U_0}{E_0}, \tag{10.12}$$

where the damping time of the emittance is twice as fast as that of individual particles, hence the factor 2 in the definition of the damping time. Comparing with Equation 10.7, we note that the vertical damping time $\tau_y$ is approximately twice the longitudinal damping time $\tau_E$.

Most storage rings are designed to be planar and therefore the vertical dispersion is zero, which implies that no vertical oscillations are excited by the mechanism discussed in Section 3.4.3, and the vertical emittance is ideally zero. Apart from a very small contribution due to the small spread of emission angles of synchrotron radiation, vertical oscillations and consequently the vertical emittance $\varepsilon_y$ is caused by coupling of horizontal oscillations into the vertical plane and other imperfections. By carefully correcting these imperfections, $\varepsilon_y$ can usually be reduced to below 1% of the horizontal emittance $\varepsilon_x$; the topic of the next section.

## 10.1.3 Horizontally

Already in Section 3.4.3 we found that betatron amplitudes are excited when a particle changes its energy at a location where the dispersion is non-zero, because after the change, the equilibrium trajectory "jumps away" from the particle and the particle starts oscillating around the new trajectory. This process is illustrated in Figure 3.15. When the change of energy is caused by the emission of synchrotron radiation, there are two contributions to the change: as before, the average energy loss is responsible for damping and the fluctuations around the average stochastically excite oscillations.

We address the damping process first, and note that the change of the betatron motion $(dx, dx')$ due to the change of the particle momentum $d\delta$ as a consequence of the emission of radiation in a short section $ds$ of a combined function dipole magnet is

$$\begin{pmatrix} dx \\ dx' \end{pmatrix} = -\begin{pmatrix} D_x \\ D'_x \end{pmatrix} d\delta \quad \text{with} \quad d\delta \approx -\frac{\hat{P}_{00}}{cE_0}\left(1 + \frac{2}{B}\frac{\partial B}{\partial x}x + \frac{x}{\rho}\right)ds \quad (10.13)$$

with the horizontal dispersion $D_x$ and its derivative $D'_x$. The emitting particle loses energy, which accounts for the minus sign in the right equation. Furthermore, the particle's position does not change in the emission process; it is the equilibrium orbit that changes. This accounts for the minus sign in the left equation. Moreover, we point out that the expression in the bracket in the equation for $d\delta$ closely resembles the integrand of Equation 10.4, because in both cases, it describes the increased emission from particles that slightly deviate from their reference orbit. Just as in the previous section, we now need to calculate the change of the action variable $\Delta J_x$ due to the emission of radiation and find

$$dJ_x = \gamma_x x dx + \alpha_x(xdx' + x'dx) + \beta_x x'dx' \quad (10.14)$$

and after first inserting $dx$, $dx'$ and $d\delta$ from Equation 10.13. Subsequently averaging over the particle distribution, we obtain

$$d\varepsilon_x = \langle dJ_x \rangle = \varepsilon_x \frac{\hat{P}_{00} D_x}{cE_0}\left(\frac{2}{B}\frac{\partial B}{\partial x} + \frac{1}{\rho}\right)ds \quad (10.15)$$

for the horizontal emittance growth in the short section with length $ds$ of a dipole magnet. Furthermore, we used $\langle x \rangle = 0$, $\langle x' \rangle = 0$, $\langle xx' \rangle = -\varepsilon_x\alpha_x$, and $\langle x^2 \rangle = \varepsilon_x\beta_x$. After integrating over all dipoles in one turn, we arrive at the following expression for the variation of the emittance

$$\Delta\varepsilon_x = \varepsilon_x \oint \frac{\hat{P}_{00} D_x}{cE_0}\left(\frac{2}{B}\frac{\partial B}{\partial x} + \frac{1}{\rho}\right)ds = \mathcal{D}\frac{U_0}{E_0}\varepsilon_x . \quad (10.16)$$

The last equality follows from recognizing the integral to be the same that appeared in Equation 10.6 and following the same steps to express it through the synchrotron radiation integrals $I_2$ and $I_4$ as well as $\mathcal{D} = I_4/I_2$. Here we find that the sign of the right-hand side is positive, which indicates that the emittance grows. Luckily, it is also damped through the acceleration in the acceleration cavities in the same way as the vertical motion, described in the previous section. As a consequence the horizontal emittance experiences net damping according to

$$\left.\frac{d\varepsilon_x}{dt}\right|_d = -\frac{U_0}{T_0 E_0}\varepsilon_x + \mathcal{D}\frac{U_0}{T_0 E_0}\varepsilon_x = -\frac{2}{\tau_x}\varepsilon_x \quad \text{with} \quad \frac{1}{\tau_x} = \frac{1}{2T_0}\frac{U_0}{E_0}(1 - \mathcal{D}) . \quad (10.17)$$

We observe that the sum of the inverse damping times $1/\tau_E + 1/\tau_y + 1/\tau_x = 2U_0/E_0T_0$, which is *Robinson's damping criterion* [72]. It is valid under very general assumptions and one can exploit it, for example, to reduce the horizontal damping time $\tau_x$ at the expense of the longitudinal $\tau_E$ by varying the frequency of the RF-system. This forces the particles to circulate on a dispersion orbit and experience additional dipole fields in all quadrupoles with dispersion, which will affect $I_4$ and thereby $\mathcal{D}$ in Equations 10.7 and 10.17. In passing, we note that the quantities $\mathcal{J}_x = 1 - \mathcal{D}$, $\mathcal{J}_y = 1$, and $\mathcal{J}_z = 2 + \mathcal{D}$ are often referred to as *damping partition numbers*. Robinson's criterion then reads $\mathcal{J}_x + \mathcal{J}_y + \mathcal{J}_z = 4$.

We already discussed in Section 3.4.3 how the emission of photons as a stochastic process excites betatron oscillations. We recognize the magnitude of the excitation $\delta^2_{rms}$ in Equation 3.99 as that of a short section $ds$ of a magnet with $\delta^2_{rms} = \dot{\mathcal{N}}\langle\varepsilon^2\rangle ds/cE_0^2$ such that we obtain the emittance growth rate to be

$$\left.\frac{d\varepsilon_x}{dt}\right|_{ex} = \frac{2}{3}C_q r_e \gamma_0^5 I_5 \quad \text{with} \quad I_5 = \oint \frac{\gamma_x D_x^2 + 2\alpha_x D_x D'_x + \beta_x D'^2_x}{|\rho|^3}ds , \quad (10.18)$$

where we introduce the fifth synchrotron radiation integral $I_5$.

We found that the horizontal emittance is subject to damping (Equation 10.17) and to excitations (Equation 10.18). The balance of both effects determines the horizontal equilibrium emittance $\varepsilon_{x0}$ that is determined from $0 = d\varepsilon_x/dt = -2\varepsilon_x/\tau_x + d\varepsilon_x/dt|_{ex}$. Solving for $\varepsilon_x$ yields

$$\varepsilon_{x0} = \frac{1}{3}C_q r_e \gamma_0^5 \tau_x I_5 = C_q \gamma_0^2 \frac{I_5}{I_2 - I_4} \ , \tag{10.19}$$

which indicates that for a given magnetic structure the horizontal emittance grows with the square of the energy. All information about the details of the beam optics, such as the beta functions and the dispersion is encoded in the radiation integrals. It is also noteworthy, that the horizontal emittance in an electron or positron ring is completely determined by the magnets through the radiation integrals and the energy at which the ring operates.

In the previous sections we only calculated the second moments or rms quantities, but the central limit theorem actually guarantees that the limiting distribution after being subjected to a large number of random events turns out to be Gaussian. But Gaussian distributions extend to very large amplitudes and if the size of the beam pipe is finite, a few particles will be lost at aperture limits. This will have an influence of the stored current as a function of time, the *quantum lifetime* that we will determine in the next section.

## 10.1.4 Quantum lifetime

We saw that the equilibrium beam sizes in electron rings are determined by the balance of damping and excitation. In the presence of an aperture limit, the latter process will transport particles to larger amplitudes such that they are eventually lost. In order to calculate the rate of particle loss $dN/dt$ let us assume that we can describe the horizontal distribution of $N$ particles in the variables of normalized phase space by $dN/rdrd\phi = (N/2\pi\tilde\sigma^2)e^{-r^2/2\tilde\sigma^2}$ where we use the variables $r, \phi$ instead of $\tilde x$ and $\tilde x'$. The variable $\tilde\sigma$ denotes the equilibrium transverse rms beam size in normalized phase space, which is related to the real beam size $\sigma_x$ by $\sigma_x^2 = \beta_x\tilde\sigma^2$. Integrating over $\phi$ and introducing $\xi = r^2$ as a new variable, we find $dN/d\xi = (N/2\tilde\sigma^2)e^{-\xi/2\tilde\sigma^2}$. Now we assume that the aperture limit only intercepts very few particles and leaves the equilibrium distribution mostly unaffected. In equilibrium, there is a balance of the inward flow of particles due to damping and the outward flow due to the excitation. In the presence of an aperture limit only the outward flow is left. We can, however, estimate it by the inward flow, which allows us to use the damping rate $d\xi/dt = -2\xi/\tau_x$, and we obtain $dN/dt = -(N\xi/\tau_x\tilde\sigma^2)e^{-\xi/2\tilde\sigma^2}$. If we now specify $\xi$ at the aperture limit $\tilde a$ as $\xi = \tilde a^2$, we find for the loss rate

$$\frac{dN}{dt} = -\frac{N\tilde a^2}{\tau_x\tilde\sigma^2}e^{-\tilde a^2/2\tilde\sigma^2} = -\frac{N}{\tau_q} \quad \text{with} \quad \tau_q = \tau_x\frac{\tilde\sigma^2}{\tilde a^2}e^{\tilde a^2/2\tilde\sigma^2} \ , \tag{10.20}$$

where $\tau_q$ is the *quantum lifetime*. Note that the ratio of aperture limit and beam size in normalized phase space equals the ratio in normal space and we can replace $\tilde a/\tilde\sigma$ by $a_x/\sigma_x$, where $a_x$ is the physical size of the aperture limit.

Calculating the quantum lifetime $\tau_q$ for several values of $a_x/\sigma_x$, we find for $a_x/\sigma_x \approx 7$ the lifetime $\tau_q$ to be about $10^9$ longer than the damping time $\tau_x$, which is on the order of several hundred hours. This leads to the requirement to make the beam pipe radius or any other limiting apertures at least seven times as large as the rms beam size in order to prevent the quantum lifetime from becoming a significant limitation.

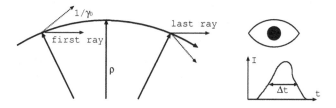

Figure 10.1 The duration between first and last rays, emitted by electrons with energy $\gamma_0 m_e c^2$ and traveling on a circular path with radius $\rho$, determines the duration of the observed light pulse and thus the typical frequency of the emitted radiation.

## 10.2 CHARACTERISTICS OF THE EMITTED RADIATION

We now turn to the discussion of the emitted radiation itself and will first discuss two key parameters qualitatively: the angular distribution, the critical frequency, and the emitted power. These quantities can be estimated from very basic assumptions [73]. Let us therefore consider an electron moving with velocity $v_0$ in the $z$–direction, that, in its rest frame, emits a photon towards the $x$–direction. This photon has the velocity $v_x = c$ and $v_z = 0$. We can now ask ourselves at which angle $\Theta'$ we observe this photon in the laboratory frame and calculate the velocities $v'_x$ and $v'_z$ with the Lorentz transformation [74]

$$v'_x = \frac{v_x}{\gamma_0(1 + v_z v_0/c^2)} = \frac{c}{\gamma_0} \quad \text{and} \quad v'_z = \frac{v_z + v_0}{1 + v_z v_0/c^2} = v_0 \qquad (10.21)$$

such that we obtain for the angle $\Theta' = v'_x/v'_z = c/\gamma v_0 \approx 1/\gamma_0$, where we assume that the speed of the electron is ultra-relativistic $v_0 \approx c$. We find that all radiation emitted into the forward hemisphere is compressed towards the forward direction with an opening angle $\pm 1/\gamma_0$.

From the width of the opening angle we can estimate the duration $\Delta t$ of a flash of light an observer detects by calculating the difference in time the first ray is emitted and the time the last ray is emitted in a dipole magnet with bending radius $\rho$. Figure 10.1 illustrates the geometry with an electron moving along the arc with energy $E_0 = \gamma_0 mc^2$. The first and last rays are denoted by solid arrows and they deviate from the direction of the electron by $\pm 1/\gamma_0$. The time between the emissions is the difference of the time $\Delta t_e$ the electron follows the circular path and the time the photon follows the sagitta. The electron's path length is $\Delta l = 2\rho/\gamma_0$ and the time to traverse the arc is $\Delta t_e = \Delta l_e/\beta_0 c$. The length of the sagitta is $\Delta l_p = 2\rho \sin(1/\gamma_0)$ and the time the photon needs to traverse it, is $\Delta t_p = \Delta l_p/c$. The time difference $\Delta t$ then follows from

$$\Delta t = \Delta t_e - \Delta t_p = \frac{2\rho}{\gamma_0 \beta_0 c} - \frac{2\rho \sin(1/\gamma_0)}{c} \approx \frac{\rho}{\gamma_0^3 c} \quad \text{and} \quad \omega_{typ} \approx \frac{1}{\Delta t} = \frac{\gamma_0^3 c}{\rho} , \qquad (10.22)$$

where we used $1 - \beta_0 \approx 1/2\gamma_0^2$. This reasoning leads to an estimate for the typical frequency $\omega_{typ}$ that is rather close to the critical frequency $\omega_c = 3\gamma_0^3 c/2\rho$ from Equation 10.2. For convenience, we also report the critical energy $\varepsilon_c = \hbar\omega_c$ given in engineering units $\varepsilon_c[\text{keV}] = 0.665 E_0^2[\text{GeV}]B[\text{T}]$.

The emitted power can be derived from the last equality in Equation 10.1 by integrating the instantaneously emitted power $P_0$ over the length $L$ of the magnet and multiplying with the number of emitting electrons $N_e$. Recasting the equation into engineering units we obtain

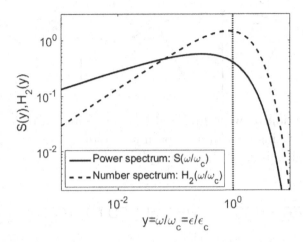

Figure 10.2 The spectral power (solid) and photon number (dashed) shows as a function of the scaled frequency $y = \omega/\omega_c$, normalized to the critical frequency $\omega_c$.

for the power $P_d$, emitted in a dipole magnet, $P_d[\text{kW}] = 1.266E_0^2[\text{GeV}]B^2[\text{T}]L[\text{m}]I[\text{A}]$. The latter equation is useful to estimate the heat load deposited in the beam pipe or delivered to the experimental stations, especially the monochromators. In both cases active cooling might be necessary.

These three parameters allow a rapid assessment of the radiation emitted from an electron or positron storage ring, but in order to plan an experiment, more information about the spectrum of the radiation is needed.

## 10.2.1 Dipole magnets

We already mentioned the power spectrum $S(\omega/\omega_c)$ emitted from electrons with energy $E_0$ in dipole magnets with bending radius $\rho$ in Equation 10.2. Figure 10.2 shows $S(\omega/\omega_c)$ as a function of $\omega/\omega_c$ as the solid line on a double-logarithmic scale. The critical frequency $\omega_c$ is indicated by a dotted line. Half of the power $P_d$, where $P_d$ is given at the end of the previous section, is radiated above $\omega_c$ and the other half below.

In most experiments, monochromators are used to select small ranges of frequencies $\Delta\omega/\omega$ before impinging the radiation onto experimental samples. Most of the radiation is emitted into a small angular region $\Delta\theta\Delta\psi$ around the forward direction with $\theta = 0$ and $\psi = 0$. Here $\theta$ is the angle with respect to the horizontal and $\psi$ with respect to the vertical direction. For an experiment it is important to know the number of photons $\dot{N}_p$ that are radiated per unit of time into the opening aperture of the monochromator $\Delta\theta\Delta\psi$ and into the spectral range $\Delta\omega/\omega$. In practical units, it is given by [75]

$$\dot{N}_p = 1.327 \times 10^{16} E^2[\text{GeV}]I[\text{A}]H_2(\omega/\omega_c)\Delta\theta[\text{mrad}]\Delta\psi[\text{mrad}]\left(\frac{\Delta\omega}{\omega}\right), \qquad (10.23)$$

where $\Delta\omega/\omega = 10^{-3}$ is often quoted in the literature as $0.1\%BW$, or $0.1\%$ of the bandwidth. The dependence on the frequency through the function $H_2(y) = y^2K_{2/3}^2(y/2)$, with the modified Bessel function of the second kind $K_{2/3}$ [23], is shown as the dashed line in

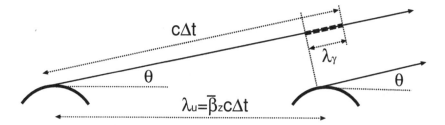

**Figure 10.3** Radiation emitted at an angle $\theta$ from two consecutive periods of an undulator interferes constructively provided the wavelength of the radiation $\lambda_\gamma$ satisfies Equation 10.25.

Figure 10.2. Like the power spectrum $S(y)$, it peaks near the critical frequency $\omega_c$. Note that $\Delta\theta\Delta\psi$ is the angular size of the monochromator slits. Equation 10.23 allows us to estimate the number of photons that pass a monochromator with a bandwidth of $\Delta\omega/\omega = 10^{-3}$ and impinge per second on a sample.

It is worth pointing out that the radiation, emitted near $\theta = \psi = 0$, is linearly polarized in the bending plane, which is, in most accelerators, horizontal. The radiation emitted above or below the bending plane shows elliptic polarization. Dipole magnets were the first sources of synchrotron radiation used for experiments, but soon special *insertion devices,* such as undulators and wigglers, were installed in storage rings. In the next section we will briefly describe the radiation they produce.

## 10.2.2 Undulators and wigglers

We already encountered the magnetic structure of undulators and wiggler in Section 4.5.3 and build small prototypes in Appendix A.3. These magnets are constructed as a sequence of short dipole magnets with alternating polarity, which force the electrons to follow a sinusoidally oscillating trajectory. The total deflection of these devices is zero and they can therefore be inserted in a straight section of an accelerator, hence they are often referred to as *insertion devices.* In many cases one transverse component, here the vertical $B_y$, of the magnetic flux density varies sinusoidally $B_y(z) = B_0 \cos(2\pi z/\lambda_u)$ with peak value $B_0$ and period $\lambda_u = 2\pi/k_u$ along the length of the device as described by Equation 4.57. From integrating the equations of motion once, we obtain for the velocities

$$v_x = -\beta_0 c \frac{K}{\gamma} \sin(k_u z) \quad , \quad v_z = \beta_0 c \left[1 - \frac{K^2}{2\gamma_0^2} \sin^2(k_u z)\right] \quad \text{with} \quad K = \frac{e B_0 \lambda_u}{2\pi m_e c} , \quad (10.24)$$

$k_u = 2\pi/\lambda_u$, and suitably chosen initial conditions. Here $\beta_0 c$ is the reference velocity of the electrons and this equals the longitudinal velocity component in the middle of a magnet pole. The electrons assume their maximum transverse velocity at $k_u z = \pi/2$ where the trajectory has an angle $v_x/v_z \approx K/\gamma_0$ for $K \ll \gamma$ with respect to the axis of the device. Since the natural opening angle of the radiation emitted in the forward directions is $\Theta \approx 1/\gamma_0$, we observe a weakly modulated intensity from devices if $K \approx 1$. In this case the magnets are referred to as *undulators.* In contrast, the radiation from devices with $K \gg 1$, called *wigglers,* appears in short bursts at the instances when the direction of emission coincides with the line of sight to the observer.

In order to determine the spectral characteristics of undulators we observe that the average longitudinal component of the velocity is given by the average of $v_z$ from Equation 10.24 over one period with the result $\bar{\beta}_z = \beta_0(1 - K^2/4\gamma^2)$. With this average speed, the time it takes for an electron to travel the distance $\lambda_u$, is given by $\Delta t = \lambda_u/\beta_0 c$. During this time, the radiation emitted one oscillation period earlier, has traveled by $s = c\Delta t$. The radiation with wavelength $\lambda_\gamma$ from the two periods interferes constructively, provided it is emitted in phase, which implies that

$$\lambda_\gamma = c\Delta t - \lambda_u \cos\theta \approx \frac{\lambda_u}{2\gamma_0^2}\left(1 + \frac{K^2}{2} + \gamma_0^2\theta^2\right) , \tag{10.25}$$

where we refer to Figure 10.3 to illustrate the geometry. Note that the wavelength $\lambda_\gamma$ is given by the undulator period divided by the square of $\gamma_0$. Moreover, it is increased by the magnetic field through the term proportional to $K^2$. The radiation emitted off-axis with $\theta \neq 0$ has a longer wavelength as well. Therefore, the spectral lines, integrated over a finite vertical angle, are widened towards longer wavelengths, or equivalently, towards lower frequencies. Note also that the spectrum predominantly contains odd harmonics of the wavelength given by Equation 10.25, due to mixing the transverse oscillations at the first harmonic with the even harmonic from the longitudinal oscillations described by the $\sin^2(k_u z)$–dependence of $v_z$ in Equation 10.24. The finite number of periods of the undulator $N_u$ causes the emitted radiation pulse to have the same number of periods. Fourier-transforming a wave train with $N_u$ periods shows that the width of the spectrum is approximately $\Delta\omega/\omega \approx 1/N_u$ for the fundamental harmonic, and $1/nN_u$ for the $n$–th harmonic. In order to plan experiments with radiation from undulators, the number of emitted photons $\dot{N}_p$ in a bandwidth of typically $\Delta\omega/\omega = 10^{-3}$ is a key quantity. In engineering units and at harmonic $n$, it is given in [75] by the expression

$$\dot{N}_p = 1.744 \times 10^{17} N_u^2 E^2[\text{GeV}]I[\text{A}]F_n(K)\Delta\theta[\text{mrad}]\Delta\psi[\text{mrad}]\left(\frac{\Delta\omega}{\omega}\right) , \tag{10.26}$$

where the $F_n(K)$ assume a maximum of approximately 0.4 for harmonics $n = 1, 3, 5, 9$ and for values of $K$ in the range $1 < K < 3$. We refer to [75] for the definition of $F_n(K)$ and further details.

The spectrum from a wiggler shows a significantly longer wave length of the first harmonic due to the $K^2$–dependence, but contains many short-wavelength harmonics due to the emission in short bursts along the line of sight to an observer. As a consequence, the spectrum in the limit of large $K$ closely resembles that of a dipole magnet shown in Figure 10.2 with the maximum critical energy of the emitted photons [75] given by $\varepsilon_{c,max}[\text{keV}] = 0.665E_0^2[\text{GeV}]B_0[\text{T}]$ with the peak field $B_0$.

The total power $P_u$, emitted from either undulator or wiggler, is given by the same expression as for a dipole magnet, but with rms value of the magnetic field $B_0/\sqrt{2}$ used instead of the dipole field. In engineering units, it therefore reads $P_u[\text{kW}] = 0.633 E_0^2[\text{GeV}]B_0^2[\text{T}]L_u[\text{m}]I[\text{A}]$, where $L_u = N_u\lambda_u$ is the length of the undulator or wiggler with $N_u$ periods and $I$ is the beam current.

Experiments to determine the structure of crystals are based on recording diffraction patterns from samples irradiated with synchrotron radiation, which requires a high degree of transverse coherence of the radiation. This, in turn, implies that the radiation must not only be emitted with a small angular divergence, but also from a source with small transverse size. In an accelerator the latter requirement implies that the emitting electrons must have small emittances. The figure of merit for the transverse coherence is the *brilliance* $\mathcal{B}$, which is given by the number of photons emitted per second and 0.1 %$BW$ and per area and angular divergence. It is thus given by $\dot{N}_p$ from Equation 10.23 or 10.26, divided by

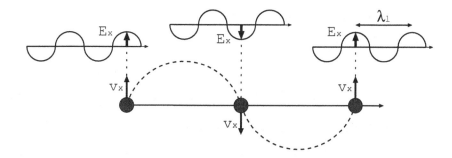

**Figure 10.4** The electron (dark dot) always gains energy proportional to $v_x E_x$, if it passes one period of the undulator (dashed), while a transversely polarized electromagnetic wave (solid) with electric field $E_x$ overtakes it by one period of its wave length $\lambda_l$. This defines the *resonance condition* from Equation 10.28.

$(2\pi)^2 \Sigma_x \Sigma_y \Sigma_{x'} \Sigma_{y'}$ where the $\Sigma$ denote the convolution of electron beam size and divergence with the corresponding quantities of the diffraction limited radiation [75]. Maximizing the brilliance is the major incentive to minimize the emittances in synchrotron light sources.

In this section we treated the spontaneous emission, where the electrons emit radiation independently of each other and hence with random phases. This causes the radiation to be incoherent and the intensity to be proportional to the number of electron. A dramatic increase of the intensity and consequently, the brilliance, can be achieved by causing all electrons to emit radiation with the same phase. In that case, the electric fields add in phase and the intensity will be proportional to the square of the number of electrons, which results in a huge gain. Free-electron lasers, the topic of the next sections, provide a mechanism to achieve this.

## 10.3 SMALL-GAIN FREE-ELECTRON LASER

Apart from the small recoil from the emission of synchrotron radiation that affects the emittance through damping and quantum excitation, as described in Section 10.1, we have neglected any reaction of the radiation of the motion of the electrons. On the other hand, in the presence of strong coherent electro-magnetic radiation, polarized in the plane of electron oscillation in the undulator, energy can be systematically exchanged between radiation and electrons. The radiation can be either emitted by the electrons themselves or provided externally.

### 10.3.1 Amplifier and oscillator

In order to understand the transfer of energy let us assume for the moment that the radiation is provided by an external laser, is horizontally polarized, and has wavelength $\lambda_l = 2\pi/k_l = 2\pi c/\omega_l$. The horizontal component of the electric field is then given by $E_x(z,t) = \hat{E}\cos(k_l z - \omega_l t + \psi_0)$ with peak value $\hat{E}$ and phase $\psi_0$. Since the electron motion has a transverse velocity component, given by Equation 10.24, it has to work against the electric field of the laser and the energy exchange is given by $dW = -eE_x(z,t)v_x(t)dt$. After inserting $E_x$ and $v_x$ we

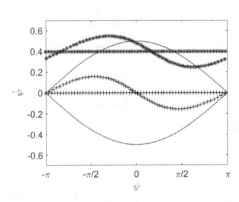 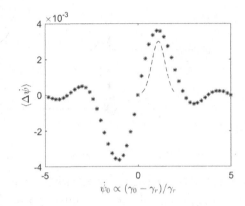

**Figure 10.5** The evolution of electrons, initially evenly distributed in phase $\psi_0$, for energy offsets proportional to $\dot\psi_0 = 0$ and 0.4 (left). In the latter case an asymmetry develops, which causes the average of $\langle\dot\psi\rangle$ to slightly increase. Plotting $\langle\dot\psi\rangle$ as a function of $\dot\psi_0$ results in the well-known FEL-gain curve (right).

find

$$\frac{dW}{dt} = -\frac{ecK\hat E}{2\gamma_0}(\sin\psi_+ - \sin\psi_-) \quad \text{with} \quad \psi_\pm = (k_l \pm k_u)\bar\beta_z ct - \omega_l t + \psi_0 \qquad (10.27)$$

and the average longitudinal position of the electron $z = \bar\beta_z ct$. The second term, proportional to $\sin\psi_-$, is rapidly oscillating and averages to zero. The first term with $\psi_+$, however, can be made constant and thus guarantees a constant sign of the energy exchange. The requirement for the constancy of $\psi_+$, the so-called *ponderomotive phase*, is $0 = d\psi_+/dt = (k_l+k_u)\bar\beta_z c-\omega_l$ and leads to the *resonance condition*

$$\lambda_l = \frac{\lambda_u}{2\gamma_0^2}\left(1 + \frac{K^2}{2}\right), \qquad (10.28)$$

where we used $\beta_0^2 = 1 - 1/\gamma_0^2$ and $\bar\beta_z = \beta_0(1 - K^2/4\gamma_0^2) \approx 1 - (1 + K^2/2)/2\gamma_0^2$, which follows from Equation 10.24. This equation describes a constraint between the electron energy $\gamma_0$, the laser wavelength $\lambda_l$ and the undulator parameters $\lambda_u$ and $K$ that ensures $dW/dt$ to have a constant value for a given electron during its passage through the undulator. Figure 10.4 illustrates this resonance condition. During the time that the electron performs one oscillation in the undulator the radiation must overtake the electrons by the distance of one laser wavelength. If this condition is fulfilled the relative sign of horizontal velocity $v_x$ and electric field $E_x$ is always the same. Note that different electrons have different starting phases $\psi_0$ and their $dW/dt$ may be different, though constant during the passage through the undulator.

Equation 10.28 describes a situation, where the energy exchange $dW/dt$ for each electron with $\gamma_0$ does not change with time. But since the electrons in a beam usually have a distribution of energies, it is important to understand the dynamics of electrons with energy deviation $\delta = (\gamma - \gamma_0)/\gamma_0$. Using $d\delta/dt = (1/\gamma_0 m_e c^2)dW/dt$ and only taking the term with $\psi_+$ in Equation 10.27 into account, we obtain

$$\frac{d\delta}{dt} = -\frac{e\hat E K}{2m_e c\gamma_0^2}\sin\psi_+ \quad \text{and} \quad \frac{d\psi_+}{dt} = \frac{k_l c}{2}\left(1 + \frac{K^2}{2}\right)\left(\frac{1}{\gamma_0^2} - \frac{1}{\gamma^2}\right) \approx 2k_u c\delta, \qquad (10.29)$$

where the second equation follows from the definition of $\psi_+$ in Equation 10.27 and expressing $\bar{\beta}_z$ through the deviation to its value on resonance. Differentiating the equation for $\psi_+$ with respect to time and inserting $d\delta/dt$, we find

$$\ddot{\psi}_+ + \hat{\Omega}^2 \sin\psi_+ = 0 \qquad \text{with} \qquad \hat{\Omega}^2 = \frac{e\hat{E}Kk_u}{\gamma_0^2 m_e} . \tag{10.30}$$

This is the same equation we found in Section 5.4 to describe the motion of particles in the potential of a radio-frequency system, with the phase space of the latter system shown in Figure 5.3 on page 128. Here the peak field $\hat{E}$ takes the place of the radio-frequency voltage $\hat{V}$ and $\hat{\Omega}$ the place of the synchrotron frequency $\Omega_s$ from Section 5.4. Therefore, we can also use the MATLAB function `pendulumtracker.m` to numerically analyze the dynamics of electrons in a free electron laser. In order to simplify the notation, we omit the plus sign in subscript of $\psi_+$ in the following. Since the ponderomotive phase $\psi$ changes over a distance of one laser wavelength, we can assume that the electrons are evenly distributed in phase. This is shown by the two distributions with constant initial energy offset $\delta = \dot{\psi}/2k_u c$ for $\dot{\psi}_0 = 0$ and $\dot{\psi}_0 = 0.4$ shown as horizontal lines with plus signs and asterisks on the left-hand side in Figure 10.5. When tracking electrons that start on the resonance energy $\dot{\psi}_0 = 0$ over 0.1 times the oscillation period $T = 2\pi/\hat{\Omega}$, we observe that of the electrons with $\psi < 0$ slightly move upwards, towards positive $\dot{\psi}$. Conversely those starting with $\psi > 0$ move downwards, towards negative $\dot{\psi}$. Since the curve is anti-symmetric, half the electrons lose energy and the other half gain energy, such that the beam as a whole does not change its average energy. The situation is slightly different for a electrons that start with $\dot{\psi}_0 = 0.4$ shown as the S-shaped curve displayed with asterisks which shows a small asymmetry. The energy gained by electrons moving above their initial value of $\dot{\psi}_0 = 0.4$ is slightly different to those below.

We now repeat this simulation for a number of starting values $\dot{\psi}_0$, and calculate the average of the difference of final $\dot{\psi}_f$ and starting values $\langle \Delta\dot{\psi} \rangle = \langle \dot{\psi}_f - \dot{\psi}_0 \rangle$. The right-hand side in Figure 10.5 displays $\langle \Delta\dot{\psi} \rangle$ as a function of $\dot{\psi}_0$, which is the typical *gain curve* of a small-gain free-electron laser. For a given undulator with $\lambda_u$ and $K$ and fixed laser wavelength $\lambda_l$ varying the energy of the electrons will cause them to lose or gain energy proportional to this gain curve. Since energy is conserved, the energy lost by the electrons increases the amplitude of the electro-magnetic field of the laser. In the simulation we used an initial distribution of electrons, all having the same energy $\delta \propto \dot{\psi}$. The finite energy spread of realistic beams can be taken into account by averaging over the gain curve. The small inset dashed curve on the right-hand side in Figure 10.5 illustrates this and we immediately see that an energy spread that is wider than the maximum of the gain function will significantly deteriorate the average gain of the beam.

In the previous paragraph, we numerically evaluated the energy exchange between electrons and laser. From a perturbative expansion of the first integral of Equation 10.30 $\dot{\psi} = \dot{\psi}_0\sqrt{1 + 2(\hat{\Omega}^2/\dot{\psi}_0^2)(\cos\psi - \cos\psi_0)}$ in the parameter $\varepsilon = \hat{\Omega}/\dot{\psi}_0$, which is small for moderate laser power, we can calculate $\langle \Delta\dot{\psi} \rangle$ up to second order and find

$$\langle \Delta\dot{\psi} \rangle = -\hat{\Omega}^4 \frac{1 - \cos(\dot{\psi}_0 t) - (\dot{\psi}_0 t/2)\sin(\dot{\psi}_0 t)}{\dot{\psi}_0^3} = \frac{\hat{\Omega}^4 t^3}{16}\frac{d}{d\xi}\left(\frac{\sin^2\xi}{\xi^2}\right) \tag{10.31}$$

with the abbreviation $\xi = \dot{\psi}_0 t/2$. Note that change of energy $\langle \Delta\dot{\psi} \rangle$ is proportional to the energy density of the laser, because we have $\hat{\Omega}^4 \propto \hat{E}^2$ from Equation 10.30. Moreover, it is proportional to the third power of the length $L_u = N_u\lambda_u$ of the undulator, because $t^3 \propto L_u^3$.

Being able to write the energy loss as the derivative of $\sin^2 \xi / \xi^2$ constitutes the essence of *Madey's theorem* [76], which relates the energy loss of electrons in a free-electron laser to the power spectrum of the spontaneously emitted synchrotron radiation in the undulator magnet. The latter can be expressed by the square of the Fourier transform of a sine with $N_u$ periods, which has the characteristic $\sin^2 \xi / \xi^2$–dependence.

The transfer of energy from the electrons to the laser is usually rather modest, on the order to a few percent, but enclosing the undulator in a collinear optical resonator allows us to repetitively amplify the same pulse and operate the system as an oscillator. Such an oscillator normally can seed itself, because the fundamental wavelength of the spontaneous radiation emitted on-axis, given by Equation 10.25, coincides with the resonance condition for the free-electron laser from Equation 10.28. In such systems the distance between the mirrors, which determines the length of the optical resonator, must be carefully adjusted to guarantee synchronicity between the arrival of electrons and the light pulses. The FEL process provides a mechanism that locks the longitudinal resonator modes in such a way that short light pulses emerge with lengths in the ps range and pulse energies of $\mu$J to mJ. Part of the light trapped in the resonator is usually coupled out from the resonator and used for experiments.

A number of free-electron lasers were installed in storage rings, but their output power is limited, because the FEL process significantly disturbs the electron beam and increases the momentum spread to exceed the width of the gain curve, such that light pulses are no longer amplified. With light pulses gone, the beam recovers due to damping described by Equation 10.7 until the momentum spread is small enough to be able to amplify light pulses again. The average power that can be extracted from a storage-ring FEL $\bar{P}$ is determined by the balance of increasing the momentum spread by the FEL and damping by synchrotron radiation. It is given by the *Renieri limit*, $\bar{P} = P_{sr}/2N_u$, where $P_{sr}$ is the power emitted by synchrotron radiation and $N_u$ is the number of undulator periods.

Operating free-electron laser oscillators at wavelengths significantly below the visible range is hampered by the limited availability of high-reflectivity mirrors. Moreover, the photon energies below this regime are a few eV–in the same range as the binding energies of chemical bonds—such that color centers, which reduce the reflectivity further, are easily created. In order to reach shorter wavelengths other operating regimes are needed and it turns out that electron beams with extremely high peak currents of kA passing through very long undulators exhibit a collective instability that allows the radiation to grow exponentially. This is the topic of the next section.

## 10.4  SELF-AMPLIFIED SPONTANEOUS EMISSION

A closer look at the FEL process reveals that it consists of five mechanisms. First, when operating the FEL as a oscillator, the spontaneous radiation provides the "seed" that is subsequently amplified. Second, the undulator couples the transverse motion of the electrons to the electric field of the radiation as described by Equation 10.27. Third, since the energy exchange depends on the relative phase of the electron motion and the radiation, the energy of the electrons is modulated on the length scale of the radiation wavelength. Fourth, each half-period of the undulator acts as a small bunch compressor chicane, as described in Section 3.7.9, and causes the energy modulation to give rise to a longitudinal density modulation. The period of this *micro-bunching* is the radiation wavelength as well. Fifth, the micro-bunched electrons radiate coherently and thus increase the intensity of the radiation.

In Section 10.3 we assumed that electron current was modest and the amplitude of the radiation was assumed to stay approximately constant during the electron's passage through the undulator. If, on the other hand, the peak current is very high and a large number of electrons are micro-bunched and start radiating coherently, the intensity of the radiation grows along the undulator. This, in turn, increases the energy modulation and the ensuing micro-bunching causes the radiation to grow even more. The process can become unstable and leads to an exponential growth of the radiation [77, 78]. Since the growth is initiated by the spontaneously emitted radiation in the undulator, it is called *self-amplified spontaneous emission*, or SASE.

The detailed analysis of the SASE process can be found in the original publications [77, 78] and textbooks [79, 80] but here we follow [81], where a very illustrative approach is presented that describes all relevant features. The dynamics of the coupled system of electrons and radiation is described by Equation 10.30 in rescaled variables with the time replaced by the fractional distance along the undulator $\tau = c\bar{\beta}_z t/L_u$, the phase $\zeta = \psi_+ + \pi/2$, and its derivative $\nu = d\zeta/d\tau$, which is proportional to the deviation of the electron energy to the resonant energy by virtue of the right equation in Equation 10.29. These transformations lead to the left of the following equations

$$\frac{d^2\zeta}{d\tau^2} = \frac{d\nu}{d\tau} = |a|\cos(\zeta + \phi) \qquad \text{and} \qquad \frac{da}{d\tau} = -j\langle e^{-i\zeta}\rangle \,, \qquad (10.32)$$

where $a$ is the normalized electric field $a = |a|e^{i\phi}$ with amplitude $|a| = \hat{\Omega}^2 = e\hat{E}Kk_u/\gamma_0^2 m_e$, already defined in Equation 10.30. The right equation in Equation 10.32 describes the change of both amplitude $|a|$ and phase $\phi$ of the radiation along the undulator. It is proportional to the scaled electron beam current $j = 2\pi N_u (eKN_u\lambda_u)^2 \rho_e/\varepsilon_0 \gamma_0^3 m_e c^2$ and to the average of the ponderomotive phases $\langle e^{-i\zeta}\rangle$. The average needs to be taken over the evenly distributed initial phases $\zeta_0$ given by $\langle g(\zeta_0)\rangle = (1/2\pi)\int_0^{2\pi} g(\zeta_0)d\zeta_0$. The average over the exponential with the phase $\zeta_0$ thus equals the Fourier-harmonic of the electron distribution at the resonant wavelength and describes the micro-bunching mentioned in the previous paragraph.

In Appendix B.5, we discuss scripts `sase_simulation.m` to numerically simulate Equation 10.32 for SASE and `oscillator.m` for oscillator configurations. Here, on the other hand, we analyze the startup of the FEL process under the assumptions that the beam current $j$ is large but the radiation amplitude $a$ is small. This allows us to write a perturbative expansion, in powers of $a$, of the electron beam variables. To first order in $a$, we write $\zeta = \zeta_0 + \nu_0\tau + \zeta_1$ where $\zeta_1$ is proportional to $a$. Inserting this in Equation 10.32, we find $d^2\zeta_1/d\tau^2 = |a|\cos(\zeta_0 + \nu_0\tau + \phi)$ and $da/d\tau = ij\langle e^{-i(\zeta_0+\nu_0\tau)}\zeta_1\rangle$ where we used $\langle e^{-i(\zeta_0+\nu_0\tau)}\rangle = 0$. For simplicity, we choose the initial energy of the electron beam to equal the resonance energy, such that we have $\nu_0 = 0$. Differentiating $da/d\tau$ two more times and inserting the $\ddot{\zeta}_1$ we obtain

$$\frac{d^3a}{d\tau^3} = ij|a|\langle e^{-i\zeta_0}\cos(\zeta_0 + \phi)\rangle = \frac{ij}{2}a \qquad \text{leading to} \qquad \alpha^3 = \frac{ij}{2}\,, \qquad (10.33)$$

where we use the ansatz $a = Ae^{\alpha\tau}$ for the growth rate $\alpha$. This cubic equation for $\alpha$ has three roots, one of which has a positive real part and is given by $\alpha = (j/2)^{1/3}(\sqrt{3}+i)/2$. Thus, the radiation exhibits exponential growth proportional to $e^{(j/2)^{1/3}(\sqrt{3}+i)\tau/2}$. Naturally, this exponential growth cannot continue indefinitely and once the electrons have completed approximately one oscillation, half the electrons donate energy to the radiation and the other half receive energy from the radiation. At this point the process *saturates*. Most SASE FELs are based on undulators long enough to allow the radiation to reach saturation.

Instead of parameterizing the growth rate by using the scaled beam current $j$, another commonly found parameter is the *Pierce parameter* $\hat{\rho} = (j/2)^{1/3}/4\pi N_u$, which typically has the order of magnitude of $10^{-3}$. It conveniently characterizes several phenomena. First, as $j$ before, it defines the exponential e-folding length, the *power gain length* $L_g = \lambda_u/4\pi\sqrt{3}\hat{\rho}$., of the radiation along the undulator. Second, the saturation of the FEL occurs after $N_{sat} \approx 1/\hat{\rho}$ undulator periods. Third, it describes the efficiency of the FEL process. The power that can be extracted from the FEL is given by $P_{FEL} \approx \hat{\rho}P_{beam}$. The properties of this radiation, which leaves the FEL after the end of the undulator, will be our next topic.

SASE FELs start from spontaneously emitted radiation in early sections of the undulator. Initially the electrons are uncorrelated and so are the photons. Consequently, they are emitted with random phases. For the micro-bunching process to start, on the other hand, a sizeable radiation intensity with a well-defined phase, must be present. So we calculate the probability for a large number of plane waves with equal amplitudes $a$ but with random phases $\phi_k$ to form a plane wave $Ae^{i\Phi} = \Sigma_k ae^{i\phi_k}$ with some phase $\Phi$ and amplitude $A$ [80] and output intensity proportional to $A^2$. Adding random complex numbers of equal magnitude is equivalent of a random walk in two dimensions with constant step size $a$ and random direction given by $\phi_k$. We note in passing that this describes a diffusion process. Writing $x_k = a\cos\phi_k$ we find that the mean squared deviation of the $x_k$ is given by $\langle x^2 \rangle = a^2\langle\cos^2\phi\rangle = a^2/2$ and likewise for $\langle y^2 \rangle = a^2/2$ and after adding $N$ random complex numbers the mean squared deviation is $\sigma^2 = Na^2/2$. Consequently the distribution in the $x$ and $y$ is given by a Gaussian distribution $\psi(x, y) = (1/2\pi\sigma^2)e^{-(x^2+y^2)/2\sigma^2}$. Since we are interested in the power level of the radiation, which is proportional to $A^2$, we find the average power to be $\langle A^2 \rangle = \langle x^2 + y^2 \rangle = \int(x^2 + y^2)\psi(x,y)dxdy = 2\sigma^2$. Experimentally, one typically measures the distribution of the radiation normalized to its average value $u = A^2/\langle A^2 \rangle = A^2/2\sigma^2$ and transforming the distribution function $\psi(x, y)$ to the new variable, we obtain the exponential distribution $\Psi(u)du = e^{-u}du$, which indicates that the most probable power is zero. This coincides with the expected deviation of a random walk from the starting point, which is also zero. We thus find that the power distribution of the radiation that seeds the exponential gain phase of the FEL is an exponential distribution. Since the radiation slips ahead of the electrons by one wavelength per undulator period (the resonance condition), only the radiation from nearby electrons can overlap. Therefore only nearby longitudinal slices of the electron bunch are responsible for the startup of one particular FEL pulse; conversely several slices of a bunch can radiate simultaneously and independently. The intensity of each of these radiation pulses, or *modes*, is sampled from an exponential distribution and the sum of energies $U = \sum_m^M u_m$ then can be shown [80] to have the probability distribution function $p_M(U)dU = U^{M-1}e^{-U}/\Gamma(M)$ with the generalized factorial, the Gamma function $\Gamma(M)$ [23]. It turns out that the experimentally accessible quantity is not the total energy emitted per pulse $U$, but $U$ normalized to its average $\langle U \rangle$, thus $v = U/\langle U \rangle = U/M$. Changing the variables in the probability distribution function then leads to

$$p_M(v)dv = \frac{M^M v^{M-1}}{\Gamma(M)}e^{-Mv}dv \qquad \text{with} \qquad v = \frac{U}{\langle U \rangle}. \tag{10.34}$$

Figure 10.6 shows the distribution of pulse energies, normalized to their average, for $M = 1, 3$, and $10$. We observe that for $M = 1$ the distribution is exponential, as discussed previously, and for larger $M$ the peak of the distribution shifts closer towards its average value at $v = 1$. This spread in intensities then seeds the FEL process and is exponentially amplified, but the stochastic nature of the seed intensities is reflected in statistical variations of the intensities of the FEL pulses delivered to user experiments. From a histogram of the measured pulse energies of a large number of FEL pulses, the number of simultaneously radiating modes $M$ can be determined by fitting Equation 10.34 to the histogram. Apart

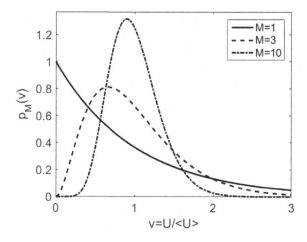

Figure 10.6 Distribution of pulse energies in a SASE pulse, normalized to its average, for $M = 1, 3$, and 10 independently radiating modes.

from the variation in pulse energy, the wavelength of the radiation varies stochastically as a consequence of being the offspring of spontaneously emitted radiation.

In order to overcome the startup of the SASE FEL from noise and the ensuing randomness of the output power and wavelength of the radiation, different schemes have been proposed to seed a FEL and thereby provide stable conditions for the laser to start the exponential gain phase. For very short wavelengths there are, however, no sufficiently strong coherent sources available and one has to resort to other methods. One method is based on several optical lasers to prepare a coherent micro-bunched structure [82] in the bunch that will radiate in later sections. An ingenious method [83, 84] is based on filtering the radiation emitted from an early part of the undulator. This stretches the length of the radiation pulse and allows the filtered wavelength to seed trailing electrons in the same bunch.

## 10.5   ACCELERATOR CHALLENGES

The radiation, spontaneously emitted and subsequently amplified by a short longitudinal slice of an electron bunch, is slightly faster than the electrons and can only affect other electrons over a distance $N_u \lambda$ ahead in the bunch. This implies that only electrons within a short slice cooperate to form the instability that leads to the exponential gain. Moreover, the exponential gain is proportional to the electron density and we therefore needs as many electrons as possible *in a short longitudinal slice* of the bunch and this explains the need for extremely large *peak* currents, rather than total number of electrons in a bunch. The currents need to be on the order of several kA which can, for example, be achieved with a bunch charge of 100 pC and a bunch length of 20 fs. Since the bunch length directly after the cathode is limited by space charge to a few ps, the bunches must be longitudinally compressed a few hundred times. For this purpose several stages of bunch compressors, as discussed in Section 3.7.9, are installed in the accelerator. The resulting high peak currents make the beam susceptible to a number of instabilities, discussed in Chapter 12, and mitigating methods are required, such as laser heaters, also mentioned there.

The transverse coherence of the radiation, especially at very short wavelengths, requires beams with extremely small emittances. These beams must be created with a very small emittance in the electron gun and then the growth of the emittance must be prevented along the accelerator. Immediately after the gun, space-charge compensation schemes and the focusing at low energies are of particular importance. The electron beams have a very small momentum spread $\sigma_\delta$ that makes the beams susceptible to so-called micro-bunching instabilities along the linear accelerator, and especially in the bunch compressors. Increasing $\sigma_\delta$ in a controlled way in a laser heater is one way to alleviate this problem.

In order to synchronize pump-probe experiments, where the FEL and an external lasers are used to excite, and subsequently probe samples, the timing of the entire accelerator complex must be controlled to a level of tens of fs. This is particularly demanding for the low-level radio-frequency system that adjusts the fields in the acceleration structures. These challenges were met with SASE FELs operating in the visible and ultraviolet spectral range at Argonne National Laboratory [85], later extended to shorter wavelengths at DESY in TTF-FEL, now FLASH [86], in Hamburg. Angström wavelengths were reached in 2009 at LCLS in the US [87] and later also at SACLA in Japan [88].

## QUESTIONS AND EXERCISES

1. Calculate the energy that every electron radiates in every revolution in (a) LEP with a circumference of 27 km and a bending radius of $\rho = 3$ km at beam energies of 50 and 100 GeV; (b) in a 3 GeV synchrotron light source with a circumference of 200 m and a bending radius of $\rho = 10$ m. (c) What are their approximate longitudinal and transverse damping times? (d) What is the critical energy $\varepsilon_c$ of the emitted radiation? (e) What is the angular width of the emitted radiation?

2. Numerically evaluate the five synchrotron radiation integrals $I_1 = \oint (D/\rho) ds, I_2 = \oint \left(1/\rho^2\right) ds, I_3 = \oint \left(1/|\rho|^3\right) ds, I_4 = \oint (D/\rho) \left(1/\rho^2 + 2k_1\right) ds$ and $I_5 = \oint \left(\mathcal{H}/|\rho|^3\right) ds$, for the ring consisting of eighteen FODO cells, discussed in Exercise 10 of Chapter 3. $D$ is the horizontal dispersion and $\mathcal{H}$ is defined in Equation 3.100. Derive the momentum compaction factor, the equilibrium momentum spread and the horizontal equilibrium emittance assuming the beam energy is 1 GeV.

3. Repeat the previous exercise for the doublet ring from Exercise 11 of Chapter 3.

4. For an experiment with a 160 MeV electron beam you need to design an undulator with $K \approx 2$ that needs to be resonant at a wavelength $\lambda = 880$ nm. What undulator period $\lambda_u$ do you need to specify? What is the peak field $B_0$? Does a beam pipe with outer diameter of 30 mm fit in the gap?

5. The long undulators in SASE FELs consist of about 5 m-long undulators with interleaved phase shifters to ensure the constancy of the phase $\phi$ in Equation 10.32. Based on the sase_simulation.m script with $j = 200$ and $a_0 = 10^{-2}$ from Appendix B.5, investigate the dependence of the amplitude $|a|$ at the end of the undulator on the tolerance of the phase-shifter settings. Assume that, at 50 evenly spaced times in the simulation, uniformly distributed random phases are added to $a$.

6. Qualitatively explore the influence of the momentum spread on the oscillator FEL, discussed in Appendix B.5, by increasing the initial momentum spread dx(2) in the range from 0.1 to 10. Use the scaled current $j = 1$ and initial amplitude $a_0 = 10^{-8}$ and display the final amplitude $|a|$ as a function of the initial momentum spread. Feel free to explore the dependence on other parameters as well.

# CHAPTER 11

# Non-linear Dynamics

In most of Chapter 3 we discussed accelerator components that were represented by matrices. There, we also found the focal length of quadrupoles to depend on the momentum offset and cause the chromaticity to become non-zero. This effect becomes more important, the stronger the quadrupoles are. Therefore, in most accelerators, the chromaticity is corrected by adjusting sextupoles that are installed at locations with non-zero dispersion. This was described in detail in Section 8.5.4. We found that compensating the chromaticity depends on the product of the integrated sextupole strength $k_2L$ and dispersion $D_x$ and correcting a large chromaticity requires strong sextupoles. This is particularly relevant for linear colliders with strong quadrupoles adjacent to the interaction point where the beta functions are extremely large. A second example are modern synchrotron light sources with their strongly focusing quadrupoles to keep the beta functions and dispersion small in order to minimize the emittance. Since the dispersion is small, the sextupoles must be very strong. But these very strong sextupoles do not only compensate the chromaticity, they also give non-linear kicks to particles with the correct momentum and thus distort their transverse motion. It is the consequence of these non-linear kicks that we will investigate in this chapter.

To do so, we will later use modern methods in a framework of Hamiltonians and Lie-maps, but we start by exploring the motion qualitatively with a very simple one-dimensional model of a ring with a single sextupole.

## 11.1 A ONE-DIMENSIONAL TOY MODEL

We assume the transfer matrix of the ring to be given by the parameterization from Equation 3.51 and the kick of the sextupole by $\Delta x' = -(k_2L/2)x^2$ from Equation 3.37, such that we can describe the map from one turn, labeled $n$, to the next by

$$
\begin{pmatrix} x_{n+1} \\ x'_{n+1} \end{pmatrix} = \mathcal{A}^{-1} \begin{pmatrix} \cos\mu & \sin\mu \\ -\sin\mu & \cos\mu \end{pmatrix} \mathcal{A} \begin{pmatrix} x_n \\ x'_n - \frac{k_2L}{2}x_n^2 \end{pmatrix} , \tag{11.1}
$$

where $\mathcal{A}$ is defined in Equation 3.52. Expressing this equation by variables in normalized phase space, denoted by a tilde, we find

$$
\begin{pmatrix} \tilde{x}_{n+1} \\ \tilde{x}'_{n+1} \end{pmatrix} = \begin{pmatrix} \cos\mu & \sin\mu \\ -\sin\mu & \cos\mu \end{pmatrix} \begin{pmatrix} \tilde{x}_n \\ \tilde{x}'_n - \frac{k_2L}{2}\beta^{3/2}\tilde{x}_n^2 \end{pmatrix} . \tag{11.2}
$$

The factor before the quadratic kick can be transformed to unity by using scaled variables $\hat{x} = (k_2L/2)\beta^{3/2}\tilde{x}$, resulting in an equation with a single free parameter, the phase advance

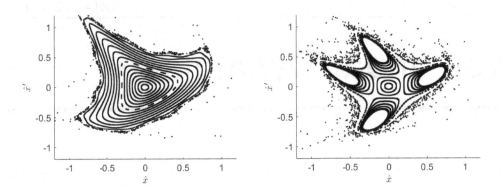

Figure 11.1 Phase space portrait for the map given by Equation 11.3 for phase advances $\mu = 2\pi Q$ for $Q = 0.31$ (left) and $Q = 0.2526$ (right).

for one turn $\mu$

$$\begin{pmatrix} \hat{x}_{n+1} \\ \hat{x}'_{n+1} \end{pmatrix} = \begin{pmatrix} \cos\mu & \sin\mu \\ -\sin\mu & \cos\mu \end{pmatrix} \begin{pmatrix} \hat{x}_n \\ \hat{x}'_n - \hat{x}_n^2 \end{pmatrix}, \tag{11.3}$$

where we can express $\mu = 2\pi Q$ by the tune $Q$. We now use this map to follow a particle for a large number of turns, a process called *tracking*, and display the phase-space coordinates, once per turn, in a so-called *Poincaré plot*.

Figure 11.1 shows such a phase space plot of $\hat{x}'$ versus $\hat{x}$ when iterating the map with starting coordinates $\hat{x}'_0 = 0$ and varying $\hat{x}_0$ from 0.05 to 1.5 in steps of 0.05. With these starting coordinates we iterate 1024 times and plot $(\hat{x}, \hat{x}')$ in every iteration. Each trajectory thus consists of 1024 points, unless the point lies outside the visible area of the plot in which case the iteration stops. On the left, the trajectories for $Q = 0.31$ are shown. Near the origin the trajectories are almost circular, but progressively deform towards a triangular shape for larger starting coordinate $\hat{x}_0$, because the tune is close to 1/3. The innermost trajectories appear to be contiguous, but at some amplitude break up and all trajectories further outside become chaotic. For starting coordinates larger than $\hat{x}_0 = 1.1$ the trajectories diverge, indicating the limit of stability, called the *dynamic aperture*. It is not an immediate physical boundary, but once a particle exceeds this limit, its amplitude increases without limit, and the particle eventually hits a physical aperture; the beam pipe or a collimator.

Roughly the same behavior is visible on the phase space plot for the tune $Q = 0.2526$, shown on the right-hand side in Figure 11.1. Here the tune is close to 1/4, which explains the four-fold periodicity. Again, almost circular trajectories near the origin are distorted as the starting coordinate for the trajectory increase. From the fourth trajectory to the tenth, the trajectories are confined to the *islands* and for larger starting coordinates the trajectories become chaotic and eventually unstable. The existence of islands with a four-fold periodicity indicates that the tune is even closer to 1/4 than the bare tune $Q = 0.2526$ and the tune appears to depend on the amplitude of the oscillations. We thus find *amplitude-dependent tune shift* that is entirely due to the dynamics defined by the sextupole added to the otherwise linear ring. It provides a non-linear force that affects the particles, similar to the sinusoidal restoring force that appeared in the description of the longitudinal motion in Equation 5.36, and resulted in the amplitude-dependence of the synchrotron tune shown on the right-hand side in Figure 5.3.

From an operational point of view, the dynamic aperture is the most serious limitation, because it constrains the space available for the transverse oscillations and, if it is too small, can cause beam loss.

## 11.2   TRACKING AND DYNAMIC APERTURE

Our task is thus to determine the dynamic aperture of a ring often consisting of many thousand elements, some of them represented by transfer matrices and many others by non-linear kicks. Since systems with non-linear forces, such as the sextupoles, are in general non-integrable, there are no generally applicable theoretical methods to calculate the dynamic aperture. We therefore have to resort to numerical methods and simply use tracking to follow particles with different starting conditions to see whether they are stable or whether they follow divergent trajectories. Here the question arises of how long to follow the particle? Normally, one takes a pragmatic approach and chooses a maximum number of turns to follow and chooses a maximum amplitude at which the particle is considered lost; typically on the order of the beam pipe radius. For the number of turns to follow, often a few 1000 turns is a good choice to determine the so-called *short-term dynamic aperture*. For electron accelerators, it is often the relevant quantity, because damping from synchrotron radiation operates on a similar time scale and counteracts slowly diverging particle trajectories. For accelerators without damping, the number of turns is often chosen to be on the order of $10^5$ to $10^6$ turns. Since many sample particles with different starting positions are followed for a large number of turns, these calculations are very time-consuming and rely on highly optimized computer codes.

We illustrate these methods by extending the toy model with a single sextupole to two transverse dimensions, such that the map from one turn to the next now reads

$$
\begin{pmatrix} \hat{x}_{n+1} \\ \hat{x}'_{n+1} \\ \hat{y}_{n+1} \\ \hat{y}'_{n+1} \end{pmatrix} = \begin{pmatrix} \cos\mu_x & \sin\mu_x & 0 & 0 \\ -\sin\mu_x & \cos\mu_x & 0 & 0 \\ 0 & 0 & \cos\mu_y & \sin\mu_y \\ 0 & 0 & -\sin\mu_y & \cos\mu_y \end{pmatrix} \begin{pmatrix} \hat{x}_n \\ \hat{x}'_n - (\hat{x}_n^2 - \hat{y}_n^2) \\ \hat{y}_n \\ \hat{y}'_n + 2\hat{x}_n\hat{y}_n \end{pmatrix}, \quad (11.4)
$$

where we assume that the beta functions in both planes are equal. Iterating this equation from given starting positions and testing in the routine whether a boundary is exceeded is easily encoded in a routine that we call survived_turns()

```
function out=survived_turns(N,R,x0,y0,dx,dy)
x=[x0;0;y0;0]; out=N;
for k=1:N
   thetax=(x(1)-dx)^2-(x(3)-dy)^2; thetay=-2*(x(1)-dx)*(x(3)-dy);
   x=R*[x(1);x(2)-thetax;x(3);x(4)-thetay];
   if ((abs(x(1))>3) || (abs(x(3))>3)) out=k; return; end
end
```

It receives the requested maximum number of iterations N, the linear transfer matrix R, the starting positions $\hat{x}_0, \hat{y}_0$ and the misalignment of the sextupole $\hat{d}_x, \hat{d}_y$ as input, and returns the survived number of turns in the variable out. We use this function to probe the dynamic aperture along rays extending with angle $\phi$ from the origin with a bisection method. The following code fragment implements this

```
for phi=0:pi/100:pi
   nmax=12; dr=10; r=dr;
   while (nmax)
```

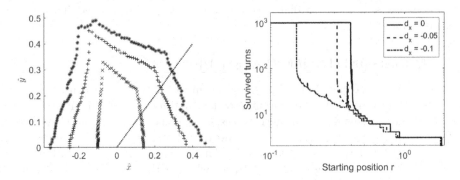

**Figure 11.2** On the left the dynamic aperture for the map given by Equation 11.4 is shown for tunes $Q_x = 0.31$ and $Q_y = 0.28$. The sextupole is horizontally misaligned by $\hat{d}_x = 0$ (asterisk), $\hat{d}_x = -0.05$ (plus sign), $\hat{d}_x = -0.1$ (cross). The plot on the right shows the survived number of turns as a function of the distance $r$ of the origin from the starting position $\hat{x}_0 = r\cos\phi$ and $\hat{y}_0 = r\sin\phi$ with $\phi = 45°$ for configurations with the same misalignments.

```
        nmax=nmax-1; x0=r*cos(phi); y0=r*sin(phi); dr=0.5*dr;
        if (survived_turns(NN,R,x0,y0,0,0)==NN) r=r+dr; else r=r-dr; end;
    end
    plot(x0,y0,'k*');
end
```

The outer loop iterates over the angles $\phi$ and in the loop we initialize the requested number of bisections nmax, the maximum amplitude to probe dr and the starting radius r. The bisection proceeds until nmax is zero and in each iteration the starting position x0 and y0 is initialized and the size of the search interval dr is halved. Then, depending on whether the particle survives the requested iterations, the starting radius is reduced or increased. After the requested number of bisections, the final positions, which define the dynamic aperture, are plotted.

On the left-hand side in Figure 11.2 we show the resulting dynamic aperture for a perfectly aligned sextupole with $\hat{d}_x = \hat{d}_y = 0$ by the curve with the asterisks. Under this curve the starting positions lead to stable motion and those outside lead to divergent motion. If the sextupole is misaligned by $\hat{d}_x = -0.05$, the dynamic aperture is reduced to the region under the curve defined by the plus signs and for $\hat{d}_x = -0.1$ to that under the crosses. Apparently, misaligned magnets have a significant influence on the dynamic aperture. The reason is, of course, feed-down, which causes multipoles of lower order to appear, as discussed in Section 8.1.1. And these affect both linear optics produce additional non-linearities.

On the right-hand side in Figure 11.2, we show the number of survived turns as a function of the starting position along the dotted line shown in the graph on the left-hand side and defined by $\hat{x}_0 = r\cos\phi$ and $\hat{y}_0 = r\sin\phi$ for $\phi = 45°$ for the three configurations with the misaligned sextupole. This type of presenting the dynamic aperture is called a *survival plot*. Here, the configuration with the aligned magnet is shown by the solid line and for large amplitudes the number of surviving turns is very small but with decreasing starting positions and starting radius r the surviving number of turns increases, until it rather abruptly reaches the maximum number of turns, here 1000. This abrupt change defines the dynamic aperture. Repeating the same simulation with a misaligned sextupole shows the abrupt transition to the maximum number of turns occurring at smaller starting radius, which is consistent with the observation on the graph on the left-hand side.

Preparing plots of the dynamic aperture and survival plots, similar to those shown in Figure 11.2 for other accelerators is straightforward and usually extensively done during the design phase for a new accelerator. Usually, the non-linear elements are represented by a sequence of thin-lens kicks and drifts spaces, despite having a finite length. This has the advantage that the map for the element is symplectic and does not lead to unphysical artefacts. Moreover, also the longitudinal motion is taken into account when tracking the particles and often many different random configurations of misalignments are analyzed in order to understand the robustness of the accelerator against perturbations. A major problem is the computing speed, especially for large accelerators with thousands of magnets and other components. Therefore methods were invented to concatenate the many elements to a compact representation to describe the map from one turn to the next.

One method is based on the realization that the concatenation of linear and non-linear elements involves the substitution of multi-variate polynomials of the phase-space coordinates into other polynomials. This process is based on manipulating mathematical structures, called *differential algebras*. It can be coded efficiently and is used to compute transfer maps to arbitrary order [89]. The order of the maps is only limited by the available computer memory, but still, for large accelerators the maps must be truncated, an approximation that is called *truncated power series algebra* (TPSA). Since the truncated maps are not necessarily symplectic, they can lead to unphysical results, such as spurious damping or growth of oscillations. This limits their use in determining the dynamic aperture and nowadays the standard method, facilitated by the ever-increasing speed of the computers, is still to track particles element-by-element.

On the other hand, the map-based methods allow us to efficiently extract physical quantities, such as high-order chromaticities or the dependence of the tune on the misalignment of octupoles. Both tracking and TPSA provide a global assessment of a given accelerator, even an extremely large one, but their use in finding out where to place correction magnets or how cancellations of perturbations occur, is limited. We therefore describe a second method in the following sections. It is based on associating a *Hamiltonian* or *Lie-generator* with each non-linear element and then concatenating and analyzing the Hamiltonians.

## 11.3  HAMILTONIANS AND LIE-MAPS

In mechanical systems, already discussed in Section 2.2, the Hamiltonian $H(q,p)$ depends on the phase space coordinates $q$ and $p$ and describes the forces that determine the temporal evolution of the system [8] via Hamilton's equations: $dq/dt = \partial H/\partial p$ and $dp/dt = -\partial H/\partial q$. They are straightforward to generalize to several degrees of freedom. Moreover, we assume that the Hamiltonian does not explicitly depend on time and therefore describes time-invariant dynamics. The temporal derivative of an arbitrary function $f(q,p)$ is then given by

$$\frac{df}{dt} = \frac{\partial f}{\partial q}\frac{dq}{dt} + \frac{\partial f}{\partial p}\frac{dp}{dt} = \frac{\partial f}{\partial q}\frac{\partial H}{\partial p} - \frac{\partial f}{\partial p}\frac{\partial H}{\partial q} = [f, H] = [-H, f] =: -H : f , \qquad (11.5)$$

where we used Hamilton's equation in the second equation to replace $dq/dt$ and $dp/dt$ by partial derivatives of the Hamiltonian. Furthermore, we use the definition of the *Poisson bracket* of two functions $f$ and $g$ as $[f, g] = (\partial f/\partial q)(\partial g/\partial p) - (\partial f/\partial p)(\partial g/\partial q) = -[g, f]$. In the last equality in Equation 11.5 we introduce a new notation to denote the Poisson bracket, which, however, clearly exhibits the character of the Hamiltonian as an operator to implement an infinitesimal propagation in time. This motivates the name *Lie-generator,* in resemblance to the infinitesimal generators of groups that depend on continuous parameters, named *Lie groups.* Pauli-matrices, as the infinitesimal generators of the rotation group, may serve as an example.

The infinitesimal step in time, described by Equation 11.5, can be generalized to finite time steps by first realizing that we can express multiple derivatives by iterated Poisson brackets, or, equivalently, as powers of Lie operators by $d^n f/dt^n = [-H, [-H, \ldots, [-H, f] \ldots]] =: -H :^n f$. In a second step we write the function $f(t + \Delta t)$ as a Taylor-series

$$f(t + \Delta t) = \sum_{n=0}^{\infty} \frac{\Delta t^n}{n!} \frac{d^n f}{dt^n} = \sum_{n=0}^{\infty} \frac{\Delta t^n}{n!} : -H :^n f = e^{:-H:\Delta t} f \ . \tag{11.6}$$

In the third equality, motivated by the fact that the infinite sum is just a representation of the exponential, we introduce the short-hand notation for the *Lie map* $e^{:-H:\Delta t}$ to avoid writing many infinite sums. When forced to explicitly calculate $e^{:-H:\Delta t} f$, we have to use, however, the representation by infinite sums and multiple-iterated Poisson brackets.

From Section 2.2, we know, that instead of the time $t$, we normally use the longitudinal position $s$ as the independent variable, but the equations that correspond to Equation 11.5 are (formally) the same with the substitution $t \to s$. Moreover, the Hamiltonian that describes thin-lens multipole magnets turn out to be related to $A_s$ in Equation 2.2 and to the complex potentials from Equation 4.12. For example, the Hamiltonian of an upright sextupole is

$$\tilde{H}_S = H_S(x, x', y, y')\Delta s = \text{Re}[-F_S(x + iy)L/B\rho] = (k_2 L/6)(x^3 - 3xy^2) \ , \tag{11.7}$$

where $F_S$ was defined in Section 4.2. Similarly, the imaginary part of $-F_S(x + iy)L/B\rho$ describes the Hamiltonian for the corresponding skew-magnet. Since the Hamiltonian describes the propagation of any function of the phase-space coordinates, it also describes the propagation of the coordinates themselves and, for example, the horizontal position $x_a$ after the sextupole is given by

$$x_a = e^{:-\tilde{H}_S:} x = \sum_{n=0}^{\infty} \frac{1}{n!} : -\tilde{H}_S :^n x = x + [-\tilde{H}_S, x] + \frac{1}{2}[-\tilde{H}_S, [-\tilde{H}_S, x]] + \cdots = x \ , \tag{11.8}$$

because explicitly calculating the Poisson bracket shows that $[-\tilde{H}_S, x] = 0$. We thus find the expected result that the transverse position does not change in the thin-lens sextupole. Likewise, the transverse angle $x'_a$ after the sextupole is given by

$$x'_a = e^{:-\tilde{H}_S:} x' = x' + [-\tilde{H}_S, x'] + \frac{1}{2}[-\tilde{H}_S, [-\tilde{H}_S, x']] + \ldots \tag{11.9}$$

and we have to evaluate the Poisson brackets, which leads to

$$
\begin{aligned}
[-\tilde{H}_S, x'] &= -\left[ \frac{\partial \tilde{H}_S}{\partial x} \frac{\partial x'}{\partial x'} - \frac{\partial \tilde{H}_S}{\partial x'} \frac{\partial x'}{\partial x} + \frac{\partial \tilde{H}_S}{\partial y} \frac{\partial x'}{\partial y'} - \frac{\partial \tilde{H}_S}{\partial y'} \frac{\partial x'}{\partial y} \right] \\
&= -\frac{\partial \tilde{H}_S}{\partial x} = -\frac{k_2 L}{2}(x^2 - y^2) \tag{11.10}
\end{aligned}
$$

because the only non-zero derivative of a coordinate is $\partial x'/\partial x' = 1$. Moreover, all higher Poisson brackets vanish, because the first Poisson bracket $-\partial H_S/\partial x$ only depends on the

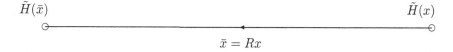

$$\bar{x} = Rx$$

Figure 11.3 Pushing Hamiltonians to the end of the beam line.

spatial phase-space coordinates $x$ and $y$, and the Poisson bracket of two functions that only depend on spatial coordinates always vanishes. Therefore, we have $[-\tilde{H}_S, [-\tilde{H}_S, x']] = 0$. For the change in angle, we thus recover the result we already used in Equation 11.4, $x'_a = x' - (k_2L/2)(x^2 - y^2)$. In a similar way we find the vertical position unchanged $y_a = y$, and the vertical kick to be $y'_a = y' - \partial\tilde{H}_S/\partial y = y' + (k_2L/2)2xy$.

It is straightforward to convince oneself that for any thin multipole defined through a complex potential $F(z)$ and Hamiltonian $\tilde{H} = \mathrm{Re}[-F(x + iy)L/B\rho]$ the phase-space coordinates after the magnet $x_a, x'_a, y_a$, and $y'_a$ are given by

$$x_a = x, \qquad x'_a = x' - \frac{\partial\tilde{H}}{\partial x}, \qquad y_a = y, \quad \text{and} \quad y'_a = y' - \frac{\partial\tilde{H}}{\partial y}. \tag{11.11}$$

The imaginary part of $F(z)$ describes the effect of the corresponding skew magnet. Using the Hamiltonian formalism is not restricted to thin-lens magnets. It is easy to see that $\tilde{H} = (L/2)(x'^2 + y'^2)$ results in the map for a drift space and after some lengthy algebra $\tilde{H} = (L/2)(x'^2 + k_1x^2 + y'^2 - ky^2)$ reproduces the map for the thick quadrupole from Equation 3.15, where the calculation of iterated Poisson brackets lead to in a Taylor-series expression for sine and cosine.

## 11.3.1 Moving Hamiltonians

At this point, we observe that a Hamiltonian depends on the phase-space variables at the location where the corresponding magnet is placed. Two magnets at different places depend on different variables, and, in order to make their effect on the beam commensurable, we have to transform the variables, or, somewhat sloppily expressed as "moving the Hamiltonian," such that both Hamiltonians depend on the same variables. To illustrate how this is done, we consider Figure 11.3 in which the beam propagates from right to left and first meets a magnet with Hamiltonian $\tilde{H}(x)$ that depends on the phase-space coordinates $x$ on the right-hand side in the figure. Then the beam passes through a linear beam line, represented by a transfer matrix $R$, which maps phase-space variables $x$ to those on the left-hand side $\bar{x}$. The combined action of the Lie-map $e^{:-\tilde{H}(x):}$ first and subsequent linear transport by $R$, we denote by the map $\mathcal{M}$, which is given by

$$\mathcal{M} = Re^{:-\tilde{H}(x):} = Re^{:-\tilde{H}(x):}R^{-1}R = e^{:-\tilde{H}(Rx):}R = e^{:-\tilde{H}(\bar{x}):}R, \tag{11.12}$$

where we insert $R$ and its inverse on the right-hand side and then use the non-trivial identity, proven in [90], $Re^{:-\tilde{H}(x):}R^{-1} = e^{:-\tilde{H}(Rx):}$, which describes the so-called *similarity transform*. It states that a Lie-map, sandwiched between a matrix $R$ and its inverse can be simplified by transforming the variables of the Hamiltonian using the same transfer matrix $R$. Comparing the first and last equality in Equation 11.12 shows that we can exchange the operation of the Lie-map and the linear transport, if we change the variables of the Hamiltonian to the variables at the location. Thus, all we have to do in order to move a Hamiltonian from one place to the other is to change its variables to those of the new location.

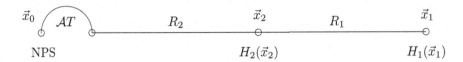

**Figure 11.4** Pushing multiple Hamiltonians to the end of the beam line that optionally lies in normalized phase space.

We need to point out that Equation 11.12 should be understood as a short-hand notation for the sequence of elements with each operator describing the respective element. We use Lie-maps for some elements and transfer matrices for others. The equation is a mnemonic representation of the beam line. We also use the convention that the operations are applied from right to left, much like matrix multiplications work. This differs from much of the literature on Lie methods, but visualizing beam lines becomes easier, even though we have to pay special attention when concatenating elements.

In order to understand the joint action of all elements in a beam line and the cancellations of aberrations, we move all Hamiltonians to the same reference point. Let us assume for the moment that it is the end point of a beam line. Optionally, we may left-multiply the last transfer matrix by a matrix, given by $\mathcal{AT}$ from Equation 3.104 and [13] to map the reference point to normalized phase space (NPS). This will prove particularly useful when describing rings. In the following sections, we tacitly absorb this transformation into $R_2$. To calculate the map $\mathcal{M}$ for the entire beam line, depicted in Figure 11.4, we first transport the Hamiltonian $H_2$, which is closest to the reference point, to the reference point, then the second, and so forth, such that we only have linear transport between the Hamiltonian to transport and the reference point. Finally, we have all Hamiltonians in the beam line located at the reference point. Moreover, all moved Hamiltonians are expressed in terms of the coordinates at the end of the beam line (and, optionally, in normalized phase space) such that all Hamiltonians are now commensurate.

Let us illustrate the method by closely analyzing the system shown in Figure 11.4. The beam propagates from right to left and first encounters Hamiltonian $H_1(\vec{x}_1)$, then it passes through linear transport with transfer matrix $R_1$, followed by a second Hamiltonian $H_2(\vec{x}_2)$ and matrix $R_2$. The matrix $R_1$ maps coordinates $\vec{x}_1$ to $\vec{x}_2 = R_1\vec{x}_1$. The map $\mathcal{M}$ for the entire beam line and the subsequent manipulations in order to transport the Hamiltonians to the end of the beam line is shown in the following equation

$$
\begin{aligned}
\mathcal{M} \;&=\; R_2 e^{:-H_2(\vec{x}_2):} R_1 e^{:-H_1(\vec{x}_1):} \\
&=\; R_2\, e^{:-H_2(\vec{x}_2):} \overbrace{R_2^{-1} R_2}^{=1} R_1\, e^{:-H_1(\vec{x}_1):} \overbrace{R_1^{-1} R_2^{-1} R_2 R_1}^{=1} \\
&=\; R_2\, e^{:-H_2(\vec{x}_2):} \underbrace{R_2^{-1}\, R_2 R_1\, e^{:-H_1(\vec{x}_1):} R_1^{-1} R_2^{-1}}\, R_2 R_1 \\
&\qquad\quad \underbrace{\phantom{e^{:-\tilde{H}_2(\vec{x}_0):}}}_{e^{:-\tilde{H}_2(\vec{x}_0):}} \qquad\qquad \underbrace{\phantom{e^{:-\tilde{H}_1(\vec{x}_0):}}}_{e^{:-\tilde{H}_1(\vec{x}_0):}} \\
&=\; e^{:-\tilde{H}_2(\vec{x}_0):} e^{:-\tilde{H}_1(\vec{x}_0):} R_2 R_1 \;.
\end{aligned}
\tag{11.13}
$$

In the first equality we insert unit matrices of the type $R^{-1}R$ in order to obtain similarity transformations of the type $R_2 e^{:-H_2(\vec{x}_2):} R_2^{-1}$. In the second equality, they are simplified to $e^{:-\tilde{H}_2(R_2\vec{x}_2):} = e^{:-\tilde{H}_2(\vec{x}_0):}$. Finally, we obtain a description of the beam line with all linear transfer matrices traversed first, followed by the sequence of Hamiltonians that are now expressed in the coordinates of normalized phase space at the end of the beam line.

## 11.3.2 Concatenating Hamiltonians

In order to understand cancellations of aberrations caused by the individual non-linearities we need to *concatenate the Hamiltonians* to form a single one that represents the cumulative effect of all non-linearities. Since each coefficient represents an independent aberration, we have a non-redundant representation of the effect of all non-linearities. The concatenation is accomplished by the *Campbell-Baker-Hausdorff* (CBH) [90] formula

$$e^{:H:}e^{:K:} = e^{:H:+:K:+(1/2)[:H:,:K:]+(1/12)[:H:-:K:,[:H:,:K:]]+...} , \tag{11.14}$$

where the order shown is sufficient for a consistent treatment of up to decapole order. In Equation 11.14 we follow the literature [90] and $H$ is traversed before $K$. We need to keep this in mind when later implementing the algorithm in software.

An immediate consequence of Equation 11.14 is that aberrations from two sextupoles, $H_1$ and $H_2$, located at positions with equal beta functions and separated by phase advances of $\phi_x = \phi_y = 180°$, cancel to all orders, because the transfer matrix between the two sextupoles reverses the sign of all phase-space coordinates, such that $H_2 = -H_1$. Choosing the reference point at the location of the second sextupole, we find that the first order in Equation 11.14 cancels, but also all Poisson brackets $[H_2, H_1] = [-H_1, H_1] = 0$ vanish, such that the joint effect of the two sextupoles cancels. Note that this cancellation pertains to the geometric aberration, not to the dispersion. In many accelerators, the SLC final focus system is an example, this method of placing chromaticity-correction sextupoles 180° phase advance apart, is used.

The main result of this section, however, is that transporting all Hamiltonians to a reference point and concatenating them to a single "Super-duper-pop-up-kick" $e^{:-\hat{H}:}$, preceded by the linear transport $R$, is the map $\mathcal{M}$, given by

$$\mathcal{M} = e^{:-\hat{H}:}R . \tag{11.15}$$

Here we use the convention of matrices, such that elements are traversed from right to left. The procedure described above thus cleans the beam line of non-linearities and collects them at the reference point. Adding the map into normalized phase space at the end allows us later to directly introduce action angle variables in a natural way. We also note that the transfer matrix $R$ is block-diagonal with two $2 \times 2$ rotation matrices on the diagonal, because it maps from normalized phase space to normalized phase space.

In the next section we illustrate the algorithm by developing MATLAB code to analyze beam lines with non-linearities up to octupole order. We confine ourselves to one dimension in order not to clutter the discussion with excessive indices.

## 11.4 IMPLEMENTATION IN MATLAB

There are 14 monomials in the one-dimensional phase-space variables $x$ and $x'$ up to fourth order in the Hamiltonian. Note that fourth order is sufficient to describe octupoles. We denote the two first-order monomials by $\vec{m}^{(1)} = (x, x')$, the three second-order monomials by $\vec{m}^{(2)} = (x^2, xx', x'^2)$, the four third-order monomials by $\vec{m}^{(3)} = (x^3, x^2x', xx'^2, x'^3)$, and the five fourth-order monomials by $\vec{m}^{(4)} = (x^4, x^3x', x^2x'^2, xx'^3, x'^4)$. Here the superscript in brackets denotes the order. Within a given order the ordering is defined by starting with the highest power of $x$ and then reducing it by one power at a time while increasing the power of $x'$ correspondingly. The coefficients before the monomials therefore define 14 independent aberrations up to fourth, or octupolar, order. They are given by the arrays $\vec{h}^{(1)}, \vec{h}^{(2)}, \vec{h}^{(3)}$, and $\vec{h}^{(4)}$ with sizes corresponding to the monomials of the same order. A

Hamiltonian $H(x, x')$ can thus be written as a scalar product of the aberrations $h^{(n)}$ and the monomials $m^{(n)}$

$$H(x, x') = \vec{h}^{(1)} \vec{m}^{(1)} + \vec{h}^{(2)} \vec{m}^{(2)} + \vec{h}^{(3)} \vec{m}^{(3)} + \vec{h}^{(4)} \vec{m}^{(4)} = \vec{H} \vec{M} , \qquad (11.16)$$

where we also introduce the array of all monomials $\vec{M}$ and all aberrations $\vec{H}$ to represent the sum of the scalar products in the respective orders. The dimension of $\vec{M}$ and $\vec{H}$ is thus 14 and $\vec{M} = (x, x', x^2, \dots, x'^4)$.

In the above sequence of monomials it is straightforward to construct Hamiltonians by simply placing the strength in the appropriate element of an array. The function thamlie() receives the strength and order of the multipole and returns the Hamiltonian H0

```
function H0=thamlie(strength,M) % Constructs Hamiltonian
H0=zeros(14,1);
switch (M)
  case 3    % sextupole
    H0(6)=strength/6;       % x^3
  case 4    % octupole
    H0(10)=strength/24;     % x^4
end
```

Note that $x^3$ appears as the sixth element and $x^4$ as the tenth element in the sequence of monomials $x, x', x^2, \dots, x'^4$. The numerical factor $1/6$ is taken from Equation 11.7.

Using the array $\vec{H}$ to represent the Hamiltonian coefficients we now need to implement similarity transformations, Poisson brackets, and the Campbell-Baker-Hausdorff equation in order to push the Hamiltonians around and then concatenate them. These manipulations require information about the location of the monomials in the array $\vec{H}$ and we therefore introduce two book-keeping arrays. The first one, MO, returns the power $n$ of variable $x^n$ in the Hamiltonian coefficient $H_j$, such that n=MO(j,1), while m=MO(j,2) returns the power $m$ of $x'^m$ in the same coefficient. The second array MM returns the position $j$ in which the coefficient for monomial $x^n x'^m$ is stored, such that j=MM(n+10*m). A negative returned value j indicates the absence of the monomial. This form of specifying n and m is determined by MATLAB's inability to handle zero-based arrays. The arrays MO and MM are computed prior to using any other function in a function called hamini. The following code snippet illustrates this for the second order.

```
for i1=1:N            % loop boundaries defines
  for j1=i1:N         % ordering of monomials
    ii=ii+1;                      % next slot in Hamiltonian
    MO(ii,:)=0;                   % init to zero
    MO(ii,i1)=MO(ii,i1)+1;   % increment if i1, j1 occurs
    MO(ii,j1)=MO(ii,j1)+1;
    MM(MO(ii,1)+10*MO(ii,2))=ii;  % store where?
  end
end
```

All other orders work in the same fashion, only the number of iterated loops differs. The arrays MO and MM play a central role in implementing the similarity transformations to move the Hamiltonians to the reference point.

First, we refer to Figure 11.3 to illustrate the transformation of the first-order Hamiltonian $H^{(1)} = \sum_{j=1}^{2} h_j^{(1)} x_j$ with coefficients that are linear in the coordinates $x_i$. The coordinates $\bar{x}_i$ at the reference point are given by $\bar{x}_i = \sum_{j=1}^{2} R_{ij} x_j$, such that we find

$$H^{(1)} = \sum_{j=1}^{2} h_j^{(1)} x_j = \sum_{j=1}^{2} h_j^{(1)} \sum_{i=1}^{2} R_{ji}^{-1} \bar{x}_i = \sum_{i=1}^{2} \tilde{h}_i^{(1)} \bar{x}_i \quad \text{with} \quad \tilde{h}_i^{(1)} = \sum_{j=1}^{2} h_j^{(1)} R_{ji}^{-1} . \quad (11.17)$$

Here we note that we can write this equation in matrix form as

$$\tilde{h}^{(1)} = \left(R^{-1}\right)^T h^{(1)} = S^{(1)} h^{(1)} \quad (11.18)$$

and we find that we change the coefficients of the Hamiltonian from being expressed in variables $x_j$ to $\bar{x}_i$ by multiplying the first-order coefficients $h^{(1)}$ by $S^{(1)} = \left(R^{-1}\right)^T$, the transpose of the inverse of the linear transfer matrix.

It can be easily shown that transforming the second-order monomials is accomplished in the same way as transforming the sigma-matrix elements in Equation 3.43, which can be rewritten as

$$\begin{pmatrix} \bar{x}_1^2 \\ \bar{x}_1 \bar{x}_2 \\ \bar{x}_2^2 \end{pmatrix} = \begin{pmatrix} R_{11}^2 & 2R_{11}R_{12} & R_{12}^2 \\ R_{11}R_{21} & R_{11}R_{22} + R_{12}R_{21} & R_{11}R_{21} \\ R_{21}^2 & 2R_{21}R_{22} & R_{22}^2 \end{pmatrix} \begin{pmatrix} x_1^2 \\ x_1 x_2 \\ x_2^2 \end{pmatrix} . \quad (11.19)$$

If we denote the matrix in the previous equation by $RR$ and vector of the second-order monomials by $(xx)_i$ and $(\bar{x}\bar{x})_j$, we write for the second-order Hamiltonian $H^{(2)}$

$$H^{(2)} = \sum_i h_i^{(2)} (xx)_i = \sum_i \sum_j h_i^{(2)} (RR^{-1})_{ij} (\bar{x}\bar{x})_j = \sum_j \tilde{h}_j^{(2)} (\bar{x}\bar{x})_j \quad (11.20)$$

and we find that the Hamiltonian coefficients transform according to

$$\tilde{h}^{(2)} = \left(RR^{-1}\right)^T h^{(2)} = S^{(2)} h^{(2)} . \quad (11.21)$$

Again, the coefficients of the Hamiltonian transform by multiplying it with the transpose of the inverse of $RR$, the matrix that maps the monomials, or $S^{(2)} = \left(RR^{-1}\right)^T$. The same reasoning applies for higher orders.

The matrices $S^{(2)}, S^{(3)}$, and $S^{(4)}$ can be easily generated using the book-keeping arrays MO and MM. Here we only illustrate the algorithm for the second order. The complete code is available online.

```
R=sinv(R);    %........the inverse and..(#)
S2=zeros(3); %........second order (x^2,x*x',x'^2)
ii=0;
for i1=1:N        % ordering of monomials
  for j1=i1:N
    ii=ii+1;      % column index
    for i2=1:N
      for j2=1:N
        IR(:)=0; IR(i2)=IR(i2)+1; IR(j2)=IR(j2)+1;
        jj=MM(IR(1)+10*IR(2))-N1;   % row index
        S2(ii,jj)=S2(ii,jj)+R(i1,i2)*R(j1,j2);
      end
    end
  end
end
S2=S2';   % (#)..the transpose
```

The function `sinv()` is a simplified matrix inversion for matrices with unit determinant. We use it invert the transfer matrix R and then find two double loops over the different monomials in the prescribed order. Note that in the loops we construct double-character indices `ii` and `jj`. They define the ordering of the monomials and are in the following used to address the components of a Hamiltonian, which follow the same ordering. The first double loop over `i1` and `j1` thus determines the column-index `ii` of the matrix S2 and the second double loop determines the powers of $x_1$ and $x_2$ and stores that in the array IR. Then MM finds the column of S2 in which to store the appropriate product of transfer matrices R. The matrices for higher orders are generated in the same way. Using these matrices the Hamiltonian is then transformed to the variables $\bar{x}_i$ in the function `propham()` that receives a Hamiltonian H0 and the inverse of a transfer matrix as input, calculates the matrices $S^{(n)}$, and multiplies the section of H0 that corresponds to a specific order, and finally returns the Hamiltonian H1 expressed in the new variables.

```
% Propagates a Hamiltonian through transfer matrix R
function H1=propham(R,H0)
N=2; N1=N; N2=N*(N+1)/2; N3=N2*(N+2)/3; N4=N3*(N+3)/4;
NM=length(H0); H1=zeros(NM,1);
[S1,S2,S3,S4]=adjoint2(R);
H1(1:N1)=S1*H0(1:N1);
H1(N1+1:N1+N2)=S2*H0(N1+1:N1+N2);
H1(N1+N2+1:N1+N2+N3)=S3*H0(N1+N2+1:N1+N2+N3);
H1(N1+N2+N3+1:N1+N2+N3+N4)=S4*H0(N1+N2+N3+1:N1+N2+N3+N4);
```

Here the variables N1, N2,... denote the number of monomials in the respective order. Applying this function to each Hamiltonian in the beam line results in all Hamiltonians being expressed by the variables of the reference point.

Having "pushed" all Hamiltonians to the reference point we now need to implement the Poisson bracket and first calculate the Poisson bracket for two Hamiltonians $f = h_{ii}x^{i_1}x'^{i_2}$ and $g = h_{jj}x^{j_1}x'^{j_2}$ each with a single coefficient. Here, for example, the exponents $i_1$ and $i_2$ are related to the double-character subscript $ii$ through the indexing array MO by `i1=MO(ii,1)` and `i2=MO(ii,2)`

$$
\begin{aligned}
[f, g] &= \frac{\partial f}{\partial x}\frac{\partial g}{\partial x'} - \frac{\partial f}{\partial x'}\frac{\partial g}{\partial x} \\
&= h_{ii}h_{jj}\left(i_1 x^{i_1-1}x'^{i_2}x^{j_1}j_2 x'^{j_2-1} - x^{i_1}i_2 x'^{i_2-1}j_1 x^{j_1-1}x'^{j_2}\right) \quad (11.22) \\
&= h_{ii}h_{jj}(i_1 j_2 - i_2 j_1)x^{i_1+j_1-1}x'^{i_2+j_2-1} .
\end{aligned}
$$

We find that both terms yield the same monomial $x^{i_1+j_1-1}x'^{j_1+j_2-1}$ with the coefficient $h_{ii}h_{jj}(i_1 j_2 - i_2 j_1)$. The MATLAB code to implement this result is the following:

```
function H3=PB(H1,H2)      % Poisson bracket
global MO MM
NM=length(H1); H3=zeros(NM,1);
for ii=1:NM       % index for H1
  if abs(H1(ii))<1e-10, continue; end
  i1=MO(ii,1); i2=MO(ii,2);
  for jj=1:NM     % index for H2
    if abs(H2(jj))<1e-10, continue; end
    j1=MO(jj,1); j2=MO(jj,2);
    x12=H1(ii)*H2(jj); l1=i1*j2-i2*j1;
```

```
      if (ll==0), continue; end
      k1=i1+j1-1; if (k1<0 || k1>4), continue; end
      k2=i2+j2-1; if (k2<0 || k2>4), continue; end
      if (k1+k2>4), continue; end  % limit to octupole order
      kk=MM(k1+k2*10); H3(kk)=H3(kk)+x12*ll;
    end
  end
```

In this function, we loop over all coefficients of H1 and over H2 and abort if the coefficient is very small. Otherwise, we use the array M0 to find the powers of $x$ and $x'$, calculate the new coefficient, and use the array MM to find where to place it in the result H3. Much of the code is needed to catch conditions that lie outside the chosen order.

Using the function for the Poisson bracket PB(), it is trivial to implement the Campbell-Baker-Hausdorff equation as

```
function H3=CBH(H1,H2) % Campbell-Baker-Hausdorff
H3=H1+H2+0.5*PB(H1,H2); % +PB(H1-H2,PB(H1,H2))/12;
```

which is sufficient up to octupolar order. These functions to move and concatenate Hamiltonians are sufficient to calculate the Hamiltonian H0 that represents the cumulative effect of all non-linearities in the beam line up to fourth order, which is accomplished in the function fulham()

```
function H0=fulham3(beamline)
nlines=size(beamline,1);  H0=zeros(14,1);
for k=nlines:-1:1
  if (beamline(k,1)==1003)        % it is a sextupole
    Htmp=thamlie(beamline(k,4),3); % strength, multipolarity
    R=TM(k,nlines);                % transfer matrix to the end
    Htmp=propham(R,Htmp);          % propagate hamiltonian
    H0=CBH(Htmp,H0);               % concatenate with what is already there
  else if (beamline(k,1)==1004)    % octupole
    :
  end
end
```

The function receives the beam-line description **beamline** and returns the Hamiltonian H0. In the code it loops backwards over all elements and if it encounters a sextupole, marked by code 1003, it constructs the Hamiltonian Htmp in the function **thamlie**, calculates the transfer matrix from the non-linearity to the reference point at the end of the beam line with the function TM and uses it to propagates the Hamiltonian to the reference point. Finally the current Hamiltonian Htmp is added to the previously accumulated H0. Note that here we need to obey the conventional ordering of Lie generators, because that is used in the function CBH().

## 11.5  TWO-DIMENSIONAL MODEL

In the one-dimensional toy model from the previous section there are only two phase-space variables $x$ and $x'$ and a Hamiltonian up to fourth order contains only 14 elements. In this section we extend the discussion to two dimensions and four phase-space variables $x, x', y$, and $y'$. The number of aberrations in first order is four, because there are only four monomials in that order. This corresponds to four degrees of freedom that can be globally corrected in a beam line or ring, namely the two transverse positions and two angles.

In second order, there are ten monomials $x^2, xx', xy, xy', x'^2, x'y, x'y', y^2, yy', y'^2$ and three of them, $x^2, xx', x'^2$ can be attributed to the horizontal phase space and are related to the horizontal values of $Q_x, \beta_x$, and $\alpha_x$. Likewise, $y^2, yy', y'^2$ are related to the corresponding vertical parameters. The four monomials $xy, xy', x'y$, and $x'y'$ are related to four degrees of freedom that describe coupling between horizontal and vertical plane. As a corollary, we find that four independently powered skew quadrupoles are sufficient to correct the coupling, corroborating our finding from Section 8.5.5.

In third order, there are 20 monomials, and, if the beam line in not coupled, only ten of them occur as a consequence of upright sextupoles. The other ten monomials are due to skew sextupoles. This separation into two groups of ten monomials is a consequence of deriving the Hamiltonians from the real or imaginary part of the complex potential of the magnetic field, given, for example, by Equation 11.7 for an upright sextupole. Here the power of $x$ is odd in each monomial, while the power of $y$ is even. This remains the case, even if we move the Hamiltonian to the reference point with an un-coupled transfer matrix; the powers of $x$ and $x'$ add up to an odd integer and for $y$ and $y'$ to an even integer. Since there are 10 monomials with this property, we need a maximum of ten sextupoles to compensate all aberrations, generated by other sextupoles, provided the beam line is un-coupled. Skew-sextupoles, on the other hand, generate ten aberrations that are described by monomials with even powers of $x$ and $x'$ and require up to 10 skew-sextupoles to correct. We will see in the next section that this splitting of the monomials causes only certain resonances to be excited, rather than all possible.

In fourth order, there are 35 monomials that split into two groups of 16 monomials, half of them excited by octupoles and the other half by skew-octupoles. Three monomials are special and in Section 11.7, we will find that they are closely related to *amplitude-dependent tune shift*. In fifth order, there are 56 monomials in $N = 4$ variables, and, similar to the third-order case, they fall in two groups of 28. We will not dwell on this further, but provide an expression for the number of different monomials (or aberrations) $M(m)$ in order $m$

$$M(m) = N \left( \frac{N+1}{2} \right) \left( \frac{N+2}{3} \right) \cdots \left( \frac{N+m-1}{m} \right) \qquad (11.23)$$

where $N$ is the number of phase-space variables. For $N = 4$ and order $m = 4$ and 5 we indeed find $M(4) = 35$ and $M(5) = 56$ in fifth order.

In the next section we will use the Hamiltonian description to find methods to correct the potentially detrimental effect of non-linearities.

## 11.6  KNOBS AND RESONANCE-DRIVING TERMS

In this section, we generalize the concept of linear knobs, previously introduced in Section 8.4.1, to the correction of non-linear aberrations. Instead of constructing knobs to change the trajectory, we now seek to change individual coefficients in a Hamiltonian, without perturbing others. Using Hamiltonians has the distinct advantage that the description is non-redundant and each coefficient describes an independent aberration. Moreover, the lowest order in the Campbell-Baker-Hausdorff formula is given by the sum of the contributing Hamiltonians, each of which contributes proportionally to its excitation $k_n L$.

To illustrate this method, we restrict ourselves to one dimension, in which case, there are four third-order monomials, and consequently four independent aberrations. We assume that there are four independently powered sextupoles located near the start of the beam line and several other sextupoles. Following the procedure from Equation 11.13 to move the sextupoles to the reference point, we find that the Hamiltonian $H_f$ for the full system can be written as

$$H_f = H_o^{(3)} + H_1^{(3)} + H_2^{(3)} + H_3^{(3)} + H_4^{(3)} + H^{(>)} , \qquad (11.24)$$

where $H_o^{(3)}$ is the contribution of the *other* sextupoles to the third order of the full Hamiltonian. $H_1^{(3)}, \ldots, H_4^{(3)}$ are the Hamiltonians of the correction sextupoles already moved to the reference point. $H^{(>)}$ describes the higher-order terms in the full Hamiltonian that arise as a consequence of the Campbell-Baker-Hausdorff formula. We ignore them for the moment and use our four correction sextupoles to compensate the four aberrations in third order, only. We can use the software from the previous section to numerically calculate the contribution of each sextupole, but in this case it is instructive to calculate contribution by hand. Sextupole $i$ is characterized by the Hamiltonian $(k_2 L_i/6)x^3$ and we need to express $x^3$ in terms of the coordinates at the reference point $\bar{x}$ and $\bar{x}'$. They are related through the transfer matrix $R^i$ from the sextupole to the reference point, such that we can write

$$
\begin{aligned}
(k_2 L_i/6)x^3 &= (k_2 L_i/6) \left( R_{22}^i \bar{x} - R_{12}^i \bar{x}' \right)^3 \qquad (11.25) \\
&= (k_2 L_i/6) \left( (R_{22}^i)^3 \bar{x}^3 - 3(R_{22}^i)^2 R_{12}^i \bar{x}^2 \bar{x}' + 3R_{22}^i (R_{12}^i)^2 \bar{x} \bar{x}'^2 - (R_{12}^i)^3 \bar{x}'^3 \right) ,
\end{aligned}
$$

which defines one column in the matrix in Equation 11.26. Adding the contributions from the four correction sextupoles to the third order in the full Hamiltonian at the reference point are given by the right-hand side of the following equation

$$
\begin{pmatrix} h_{\bar{x}^3} \\ h_{\bar{x}^2 \bar{x}'} \\ h_{\bar{x} \bar{x}'^2} \\ h_{\bar{x}'^3} \end{pmatrix} = -\frac{1}{6} \begin{pmatrix} (R_{22}^1)^3 & (R_{22}^2)^3 & (R_{22}^3)^3 & (R_{22}^4)^3 \\ -3(R_{22}^1)^2 R_{12}^1 & -3(R_{22}^2)^2 R_{12}^2 & -3(R_{22}^3)^2 R_{12}^3 & -3(R_{22}^4)^2 R_{12}^4 \\ 3R_{22}^1 (R_{12}^1)^2 & 3R_{22}^2 (R_{12}^2)^2 & 3R_{22}^3 (R_{12}^3)^2 & 3R_{22}^4 (R_{12}^4)^2 \\ (R_{12}^1)^3 & (R_{12}^2)^3 & (R_{12}^3)^3 & (R_{12}^4)^3 \end{pmatrix} \begin{pmatrix} k_2 L_1 \\ k_2 L_2 \\ k_2 L_3 \\ k_2 L_4 \end{pmatrix}
$$
$$(11.26)$$

and the coefficients $h$ on the left-hand side of the equation describe the contributions of all the other sextupoles that we try to compensate which accounts for the minus sign on the right-hand side. Finding excitations $k_2 L_i$ to do so is then a matter of a simple matrix inversion. Once the excitations of the four sextupoles are found, we can calculate the cumulative effect of all sextupoles, which results in a compensated third order but with fourth and higher order taken into account properly. At this point we could select five octupoles and repeat the exercise to compensate the forth order aberrations.

Note that correcting all aberrations in third order only requires four sextupoles in one dimension, but up to 20 in two dimensions. Adding such a large number of sextupoles is rarely done and one has to identify relevant aberrations and only correct those with a limited number of sextupoles. But which aberrations are relevant? This question can be answered for periodic systems, for example storage rings, that are sensitive to perturbations that are resonant with natural oscillation frequencies of the system—the tunes.

To illustrate how to find the *resonant-driving terms* in a given Hamiltonian we restrict ourselves again to the one-dimensional case and assume that the reference point was chosen to use variables of normalized phase space $\tilde{x}$ and $\tilde{x}'$. They are related to the physical coordinates by $\mathcal{A}$ from Equation 3.52. Then we can express $\tilde{x}$ and $\tilde{x}'$ through the action $J_x$ and angle $\psi_x$ by $\tilde{x} = \sqrt{2J_x} \cos \psi_x$ and $\tilde{x}' = \sqrt{2J_x} \sin \psi_x$, such that the powers of $\tilde{x}$ and $\tilde{x}'$ can be expressed by, for example, $\tilde{x}^3 = (2J_x)^{3/2} (3 \cos \psi_x + \cos 3\psi_x)/4$ and $\tilde{x}^2 \tilde{x}' = (2J_x)^{3/2} (\sin \psi_x + \sin 3\psi_x)/4$. In this way we change the basis to express the Hamiltonian.

We find

$$
\begin{pmatrix}
\mathcal{C}\left[(2J_x)^{3/2}\cos\psi_x\right] \\
\mathcal{C}\left[(2J_x)^{3/2}\sin\psi_x\right] \\
\mathcal{C}\left[(2J_x)^{3/2}\cos 3\psi_x\right] \\
\mathcal{C}\left[(2J_x)^{3/2}\sin 3\psi_x\right]
\end{pmatrix}
=
\begin{pmatrix}
3/4 & 0 & 1/4 & 0 \\
0 & 1/4 & 0 & 3/4 \\
1/4 & 0 & -1/4 & 0 \\
0 & 1/4 & 0 & -1/4
\end{pmatrix}
\begin{pmatrix}
\mathcal{C}[\tilde{x}^3] \\
\mathcal{C}[\tilde{x}^2\tilde{x}'] \\
\mathcal{C}[\tilde{x}\tilde{x}'^2] \\
\mathcal{C}[x'^3]
\end{pmatrix},
\tag{11.27}
$$

where $\mathcal{C}[\cdot]$ denotes the *coefficient of* the term in the brackets. The coefficients on the left-hand side are thus simply determined by the coefficients of the Hamiltonian in normalized phase-space by left-multiplying with a matrix, which depends on the numerical coefficients derived by expressing powers of trigonometric functions by those of multiple angles.

This way of expressing resonance driving terms can be illustrated by revisiting the example with two sextupoles discussed at the end of Section 11.3. After "pushing" the first Hamiltonian to the location of the second and concatenating them up to third order, we obtain $H^{(3)}$. After normalized by the excitation $k_2 L$, we find

$$
\begin{aligned}
H^{(3)}/k_2 L &= x_2^3 + x_1^3 = x_2^3 + (x_2\cos\phi - x_2'\sin\phi)^3 = (2J)^{3/2}\left[\cos^3(\psi) + \cos^3(\psi + \phi)\right] \\
&= (2J)^{3/2}\left[3\frac{\cos(\psi) + \cos(\psi + \phi)}{4} + \frac{\cos(3\psi) + \cos(3\psi + 3\phi)}{4}\right].
\end{aligned}
\tag{11.28}
$$

In the second equality, we expressed the coordinates $x_1$ and $x_1'$ at location of the first sextupoles by those of the second, before expressing it by action and angle variables $J, \psi$ in normalized phase space as $x_2 = \sqrt{2J}\cos\psi$ and $x_2' = \sqrt{2J}\sin\psi$. The last equality shows driving terms of the integer resonance, dependent on $\psi$, and of the third integer resonance, dependent on $3\psi$. It is trivial to see that choosing $\phi = 180°$ cancels both terms, consistent with earlier observations. On the other hand, choosing $3\phi = 180°$ or $\phi = 60°$ will cancel the second term, but not the first and this configuration will only drive the integer resonance, but not the one at a fractional tune of $1/3$. With only two sextupoles we are unable to independently control the integer resonance, because it requires $\phi = 180°$ and that also implies that the third order is canceled. Using four independently powered sextupoles with suitably chosen phase advance $\phi$ between them, it is possible to control all four aberrations shown on the left-hand side in Equation 11.27.

In these simple one-dimensional examples we derived the relations to relate the monomials in $x$ and $x'$ to the action and angle variables $J$ and $\psi$ by hand, but it is straightforward to generalize to higher orders and to more dimensions. For example, $\tilde{x}^2\tilde{y} = 2J_x\cos^2(\psi_x)\sqrt{2J_y}\cos(\psi_y) = 2J_x\sqrt{2J_y}\left[\cos(\psi_y)/2 + \cos(2\psi_x + \psi_y)/4 + \cos(2\psi_x - \psi_y)/4\right]$ and we find that this term contributes to the integer resonance $Q_y$ and those at $2Q_x + Q_y$ and $2Q_x - Q_y$.

In the discussion, so far, we found a way to express the cumulative effect of all non-linearities through Equation 11.15 as a "super-duper-pop-up kick" $e^{:-\hat{H}:}$ and transfer matrix $R$ and how to interpret the coefficients of the Hamiltonian $\hat{H}$ with the help of introducing action and angle variables in terms of resonance driving terms. Equation 11.15 does not, however, explicitly show the periodicity of the beam line and in the following section, we will discuss a method to make this periodicity obvious. A first-order analogy that illustrates the idea is based on writing all kicks as a super-kick $\vec{q}$ (in the sense of Section 8.3.1) that contains offset $d_x$ and angle offset $d_x'$ at the end of the beam line after passing through the beam line once. It does not, however, respect the periodicity of the beam line, which, on the other hand, is done by the closed orbit as given by Equation 8.26. In the next section we will introduce the equivalent mechanism for non-linearities, which is based on so-called *normal forms*.

## 11.7 NON-RESONANT NORMAL FORMS

To convert the map $\mathcal{M}$ from Equation 11.15 to a from that explicitly obeys the periodicity of the beam line, we try to find a map $e^{:-K:}$ that transforms it into the following form

$$\mathcal{M} = e^{:-\hat{H}:}R = e^{:-K:}e^{:-C:}Re^{:K:} , \tag{11.29}$$

where we require the map $e^{:-C:}$ to depend only on the action variables $J_x = (\tilde{x}^2 + \tilde{x}'^2)/2$ and $J_y = (\tilde{y}^2 + \tilde{y}'^2)/2$. This is similar to eigenvalue transformations. They use a map that brings a matrix into a particularly simple form, the diagonal form. Here, we require the map $e^{:-K:}$ to transform the map into a form that only depends on the action variables and a rotation, but that means that it turns the complicated phase space, visible on a Poincaré plot, into circles. Through $e^{:-C:}$, the oscillation frequency now depends on the action variable and therefore on the amplitude of the oscillation. It thus describes the amplitude-dependent tune shift.

To simplify the notation, we omit the colons in the following discussion. Starting from Equation 11.29 and right-multiplying by $e^{-K}R^{-1}$, we obtain

$$e^{-H}Re^{-K}R^{-1} = e^{-K}e^{-C} \tag{11.30}$$

but $Re^{-K}R^{-1}$ is a similarity transformation of the map Hamiltonian $K$ by the linear transport matrix $S$ that we use to propagate Hamiltonians. We therefore use $Re^{-K}R^{-1} = e^{-SK}$ and arrive at

$$e^{-H}e^{-SK} = e^{-K}e^{-C} , \tag{11.31}$$

which we now solve order by order. To this end, we write all involved entities as a series in the order and label the order by a superscript in brackets

$$H = H^{(3)} + H^{(4)} , \quad K = K^{(3)} + K^{(4)} , \quad C = C^{(4)} , \quad SK = S^{(3)}K^{(3)} + S^{(4)}K^{(4)} , \tag{11.32}$$

where $H^{(n)}$ and $K^{(n)}$ are the polynomial coefficients of order $n$ and all tune shift polynomials $C^{(2m+1)}$ are zero, because terms in the Hamiltonian, which only depend on the action variables, but not on the phases, can only appear in even orders.

In third order we keep only terms of that order and have

$$e^{-H^{(3)}}e^{-S^{(3)}K^{(3)}} = e^{-K^{(3)}} . \tag{11.33}$$

Application of the Campbell-Baker-Hausdorff formula results in

$$H^{(3)} + S^{(3)}K^{(3)} = K^{(3)} + \text{higher orders} \tag{11.34}$$

where the we neglected terms such as $[H^{(3)}, K^{(3)}]$. They are of order $3 + 3 - 2 = 4$, which is octupolar order, or even higher. Solving for $K^{(3)}$ results in

$$K^{(3)} = (1 - S^{(3)})^{-1}H^{(3)} , \tag{11.35}$$

where $H^{(3)}$ is the column vector of the third-order coefficients and $S^{(3)}$ is the matrix that transforms the third-order coefficients, such that the operation results in a column vector containing the coefficients of $K^{(3)}$. They define the map that, for example, removes most of the triangular shape of the phase space trajectories in a Poincaré plot close to a third-order resonance, visible on the left-hand side in Figure 11.1.

In order to calculate the map $K^{(4)}$ and tune shift polynomial $C^{(4)}$ up to fourth order, we keep terms to fourth order in Equation 11.31

$$e^{-H^{(3)}-H^{(4)}}e^{-S^{(3)}K^{(3)}-S^{(4)}K^{(4)}} = e^{-K^{(3)}-K^{(4)}}e^{-C^{(4)}} \tag{11.36}$$

and apply CBH to obtain a single expression in the exponent. We obtain

$$H^{(3)} + H^{(4)} + S^{(3)}K^{(3)} + S^{(4)}K^{(4)} - \frac{1}{2}[S^{(3)}K^{(3)} + S^{(4)}K^{(4)}, H^{(3)} + H^{(4)}] + \dots$$

$$= K^{(3)} + K^{(4)} + C^{(4)} - \frac{1}{2}[C^{(4)}, K^{(3)} + K^{(4)}] + \dots . \tag{11.37}$$

Here we need to remember to reverse the order of the CBH and Poisson brackets, because the elements are traversed from right to left. Collecting terms of fourth order only, we find

$$H^{(4)} + S^{(4)}K^{(4)} - \frac{1}{2}[S^{(3)}K^{(3)}, H^{(3)}] = K^{(4)} + C^{(4)} \tag{11.38}$$

and we assemble the terms containing the unknown $K^{(4)}$ and $C^{(4)}$ on the left-hand side

$$(1 - S^{(4)})K^{(4)} + C^{(4)} = H^{(4)} - \frac{1}{2}[S^{(3)}K^{(3)}, H^{(3)}] . \tag{11.39}$$

Here we realize that we cannot invert $(1 - S^{(4)})$, because it has zero eigenvalues; one eigenvalue in one dimension and three in two dimensions. Note that $S^{(4)}$ is based on a pure rotation matrix $R$ and inspecting the eigenvalues reveals that they correspond to the eigenvector monomials $(x^2 + x'^2)^2$, $(y^2 + y'^2)^2$, and $(x^2 + x'^2)(y^2 + y'^2)$. But these monomials are just proportional to $J_x^2, J_y^2$, and $J_x J_y$, respectively. This observation provides us with a method to separately determine the transformation $K^{(4)}$, and the tune shift polynomial $C^{(4)}$. We use singular value decomposition of the matrix $(1 - S^{(4)})$ to find the eigenvalues and eigenvectors

$$(1 - S^{(4)}) = U\Lambda V^T = \sum_i \lambda_i |u_i\rangle \langle v_i| \tag{11.40}$$

where $\Lambda$ is a diagonal matrix containing the eigenvalues where as $U$ and $V$ are orthogonal matrices that contain the respective eigenvectors $|u_i\rangle$ and $|v_i\rangle$. Here we borrow the notation with bra and ket vectors from quantum mechanics to visualize the construction of the projection operator $\hat{P}_j$ onto the subspace for $\lambda_j$. It is given by

$$\hat{P}_j = \frac{|v_j\rangle \langle u_j|}{\langle u_j|v_j\rangle} \quad \text{such that} \quad \hat{P}_j^2 = \hat{P}_j . \tag{11.41}$$

We then use this projector onto the null-space to project the right-hand side in Equation 11.39 on to the eigenspace of the zero eigenvalues to obtain the tune shift polynomial

$$C^{(4)} = \hat{P}_{\lambda=0}\left(H^{(4)} - \frac{1}{2}[S^{(3)}K^{(3)}, H^{(3)}]\right) . \tag{11.42}$$

Next, we invert the rest by using the inverse on the subspace spanned by the non-zero eigenvalues

$$K^{(4)} = V\text{“}\Lambda^{-1}\text{”}U^T\left(H^{(4)} - \frac{1}{2}[S^{(3)}K^{(3)}, H^{(3)}] - C^{(4)}\right) , \tag{11.43}$$

where $U, V$ and $\Lambda$ are the matrices from the SVD decomposition of $(1-S^{(4)})$. The expression "$\Lambda^{-1}$" denotes the inverse of $\Lambda$, but applying the "rule" $1/0 \to 0$ for the zero eigenvalues, as already used in Equation 8.50 in Section 8.4.2. We note that there is an ambiguity in the decomposition, because adding any Hamiltonian $K'$ that is part of the null space (or, equivalently, depending on the action variables only) to the map $K^{(4)}$ will also fulfill Equation 11.39. We resolve this ambiguity by requiring that the projection of $K$ onto the null-space of $(1 - S^{(4)})$ must be zero, or $\hat{P}_{\lambda=0}K^{(4)} = 0$. This ambiguity is sometimes referred to as gauge invariance and resolving it is called "fixing the gauge." The method can be extended to calculate higher orders by following the previous steps order by order, but this is beyond our scope.

We can easily add the functionality to find the normal forms to our one-dimensional MATLAB model from Section 11.4 up to fourth order. We therefore seek to determine $K^{(3)}, K^{(4)}$, and $C^{(4)}$ from the transfer matrix $R$ and the Hamiltonian $\hat{H}$ in Equation 11.15. From $R$ we first calculate the matrices $S^{(3)}$ and $S^{(4)}$ and then use Equation 11.35 to obtain $K^{(3)}$. Then we calculate the right-hand side of Equation 11.39 and use SVD to decompose $S^{(4)}$ according to Equation 11.40. The following function called nrnf() implements this.

```
% non-resonant normal forms
function [K,C]=nrnf(H0,R)
N=2; N1=N; N2=N*(N+1)/2; N3=N2*(N+2)/3; N4=N3*(N+3)/4; % subspace size
MM3=N1+N2+1:N1+N2+N3; MM4=N1+N2+N3+1:N1+N2+N3+N4;      % index ranges
K=0*H0; C=0*H0;                   % initialize output arrays
[S1,S2,S3,S4]=adjoint3(R);    % calculate the S matrices
K(MM3)=inv(eye(N3)-S3)*H0(MM3);    % calculate K^(3)
S3K3=0*K; S3K3(MM3)=S3*K(MM3);     % calculate S^(3)*K(3)
H3=0*H0; H3(MM3)=H0(MM3); H4=0*H0; H4(MM4)=H0(MM4); % init H^(3), H^(4)
H4tmp=0*H0; H4tmp=H4-0.5*PB(-S3K3,-H3);    % minus flips Poisson bracket
[U,LAM,V]=svd(eye(N4)-S4);    % svd of 1-S^(4)
[val,pos]=min(diag(LAM));    % find position of smallest eigenvalue
P0=zeros(N4);
if (abs(val)< 1e-10) P0=V(:,pos)*U(:,pos)'/(U(:,pos)'*V(:,pos)); end;
C(MM4)=P0*H4tmp(MM4);
H4tmp(MM4)=H4tmp(MM4)-C(MM4);    % subtract from fourth order
for j=1:N4        % invert where you can..
  if abs(LAM(j,j))>1e-10 LAM(j,j)=1/LAM(j,j); else LAM(j,j)=0; end
end
K(MM4)=V*LAM*U'*H4tmp(MM4);       % calculate K^(4)
```

The function receives the Hamiltonian H0 and transfer matrix R as input variables and returns the Hamiltonians for the phase-space distortion K and the amplitude-dependent tune shift C. The code follows the description in a straightforward fashion. One must, however, pay attention to the ordering of Hamiltonian when concatenating them.

The most important result of the normal-form procedure is the amplitude-dependent tune shift encoded in the Hamiltonian C. By construction, C only depends on the action variable $J_x^2$ in fourth order which implies that the coefficient $h_{x^4}$ of $x^4$ and $x'^4$ are equal and half the magnitude of the coefficient of $x^2 x'^2$. This in turn implies that we can write the fourth order of the tune shift Hamiltonian $C^{(4)} = 4h_{x^4}J_x^2$ and the amplitude-dependent tune shift becomes $\Delta Q = (1/2\pi)\partial C^{(4)}/\partial J_x = (4/\pi)h_{x^4}J_x$. It is a straightforward exercise to verify that this tune shift agrees with that derived from Fourier-transforming tracking data. Moreover, deriving a Taylor-map from the $K$ allows us to approximately transform the triangular-shaped phase portrait shown on the left-hand side of Figure 11.1 into circles. But this is left as an exercise.

In this chapter we considered the motion of individual particles under the influence of non-linear forces where the particles propagate independently of each other. In the following chapter we will instead look at the interaction of the same-charge particles in a beam among themselves—the so-called *collective effects*.

## QUESTIONS AND EXERCISES

1. Determine the amplitude-dependent tune shift for the model with a single sextupole from Equation 11.3 for the tunes (a) $Q = 0.31$ and (b) $Q = 0.2526$. To do so, launch particles with $\hat{x}_0' = 0$ and starting amplitudes $\hat{x}_0$ in the range $0.01 < \hat{x}_0 < 0.15$ in steps of 0.01. Track for $16 \times 1024$ turns and, for each amplitude, determine the tunes by (a) Fourier transforming the recorded positions, and (b) by using the three-point method from Section 7.5.1. Finally plot the tunes either versus the starting positions $\hat{x}_0$, or the initial action variable $J_0 = (\hat{x}_0^2 + \hat{x}_0'^2)/2$.

2. Replace the sextupolar kick in Equation 11.1 with that (a) for an octupole with $(k_3 L/6)x_n^3$, and (b) for a decapole with $(k_4 L/24)x_n^4$, transform the map to scaled variables, and prepare the Poincaré plots corresponding to those shown in Figure 11.1.

3. Plot the 1000-turn dynamic aperture in the normalized variable $\hat{x}$ as a function of the tune $Q = \mu/2\pi$ in the range $0 < Q < 1$ in steps of 0.01. (a) Do this for a sextupole, (b) octupole, (c) decapole.

4. Generalize the maps from Exercise 2 to two transverse dimensions and follow the steps outlined in Section 11.2 to prepare a survival plot, as shown Figure 11.2. Use the tunes $Q_x = 0.31$ and $Q_y = 0.28$.

5. What is the Hamiltonian $\tilde{H}$ of a thin (a) dipole corrector, (b) skew quadrupole, (c) skew sextupole, (d) skew octupole?

6. Determine the kick that each of the elements from the previous exercise produces by calculating $e^{-:\tilde{H}:}x$, $e^{-:\tilde{H}:}x'$, $e^{-:\tilde{H}:}y$, and $e^{-:\tilde{H}:}y'$.

7. A beam line consists of a dipole corrector that kicks the beam by an angle $\theta$, a drift space with length $L$, and a second corrector that kicks the beam by $-\theta$. Express the correctors by their Hamiltonians, push the upstream corrector to the end of the beam line, and use the CBH formula to determine their joint action.

8. In Exercise 7, replace the dipole correctors by (a) skew quadrupoles, (b) (one-dimensional) sextupoles, and repeat the analysis. What additional terms (monomials) appear when calculating the Poisson brackets in the CBH formula? What types of magnets would cause the same type of terms?

9. Implement the beam line with two (one-dimensional) sextupoles from Exercise 7 in the MATLAB code, available from this book's web page. Verify that you obtain the Hamiltonian after concatenation with the CBH formula you calculated in Exercise 7.

10. Discuss extending the code to handle transverse displacements of multipoles.

11. Build a beam line with three (one-dimensional) octupoles, all having the same excitation, placed at positions with equal beta functions, and mutually separated by $\Delta\mu = 60°$ phase advance. Move all magnets to the same reference point, inspect the first-order Hamiltonian, and find out, which resonances are excited.

12. Calculate the amplitude-dependent tune shift for the ring from Exercise 1 using the MATLAB function nrnf() to determine the non-resonant normal form of the Hamiltonian describing the beam line. Use the same tunes and compare with the tune shift determined earlier.

# CHAPTER 12

# Collective Effects

By collective effects, we denote the interaction among the many particles that make up a beam. For one, they repel each other by so-called *space charge* forces, because they all carry the same electrical charge. Moreover, the particles scatter from each other—in much the same way as beam particles scatter from gas particles—and this may lead to particle losses by the *Touschek effect* or to emittance growth by *intrabeam scattering*. But beyond interacting among themselves, the beam particles interact via their environment by exciting electro-magnetic fields that affect later arriving parts of the beam. These so-called *wake fields* can act within a single bunch and also on other bunches. This interaction of the beam with its environment can form a system that feeds back on itself and can become unstable. These mechanisms are called *single-bunch instabilities*, if the wake fields have a very short range, and they are called *coupled-bunch instabilities*, if the range covers consecutive bunches. In the early sections of this chapter, we will discuss the mechanisms of these different collective effects and later briefly address mitigating measures. But first, we cover space charge.

## 12.1 SPACE CHARGE

We already calculated the electro-magnetic fields from one beam on a counter-propagating beam in the first paragraph of Section 9.6 and found there that the forces from the electric and the magnetic fields add, because the field-producing beam and the deflected beam moved in opposite directions. On the other hand, for co-moving particles with the same charge, the forces of the electric and magnetic fields oppose each other and we find $\vec{F} = d\vec{p}/dt = e(1 - \vec{\beta}_0^2)\vec{E} = e\vec{E}/\gamma_0^2$, such that space-charge forces are proportional to $1/\gamma_0^2$, which makes them mostly important at low beam energies. From the change of a particle's angle $d\vec{r}'/dt = (d\vec{p}/dt)/p_0$ with $p_0 = \beta_0\gamma_0 mc$, we find $d\vec{r}' = e\vec{E}dt/\beta_0\gamma_0^3 mc = e\vec{E}ds/\beta_0^2\gamma_0^3 mc^2$, where we used that $ds = \beta_0 cdt$. Following the discussion from Section 9.6, the magnitude of the electric field for a round Gaussian beam is given by $E(r) = (Ne/(2\pi)^{3/2}\sigma_s\varepsilon_0)d(r,\sigma_r)$. For small deviations from the origin $r \ll \sigma_r$ we obtain $E(r) = (Ner/\sqrt{2\pi}\sigma_s)/4\pi\varepsilon_0\sigma_r^2$, which is linear in $r$. Hence, we obtain $dr' = Nr_p rds/\beta_0^2\gamma_0^3\sqrt{2\pi}\sigma_s\sigma_r^2 = -r\Delta k_1(s)ds$ where we introduce the focusing function $\Delta k_1(s) = -Nr_p/\beta_0^2\gamma_0^3\sqrt{2\pi}\sigma_s\sigma_r^2$. From Equation 8.33 we find that this defocusing leads to a lowering of the tune with the tune shift

$$\Delta Q = \frac{1}{4\pi}\oint \beta(s)\Delta k_1(s)ds = -\frac{Nr_p}{4\pi\beta_0^2\gamma_0^3\varepsilon_r}\left(\frac{C}{\sqrt{2\pi}\sigma_s}\right), \tag{12.1}$$

where $\sigma_r^2 = \varepsilon_r\beta(s)$ and $C$ is the circumference. $r_p = e^2/4\pi\varepsilon_0 m_m c^2$, is the classical radius of a particle with mass $m_p$. The factor $C/\sqrt{2\pi}\sigma_s$ accounts for the ratio of peak to average current in a ring, if the beams are bunched and have bunch length $\sigma_s$.

If a beam has a Gaussian distribution with an elliptic cross section, Equation 9.28 describes the electric field. Expanding it for small $x$ and $y$ leads to the following focusing functions $\Delta k_{1x}$ and $\Delta k_{1y}$ in the horizontal and vertical direction

$$\Delta k_{1x} = -\frac{2Nr_p}{\sqrt{2\pi}\sigma_s\beta_0^2\gamma_0^3}\frac{1}{\sigma_x(\sigma_x + \sigma_y)} \quad \text{and} \quad \Delta k_{1y} = -\frac{2Nr_p}{\sqrt{2\pi}\sigma_s\beta_0^2\gamma_0^3}\frac{1}{\sigma_y(\sigma_x + \sigma_y)}, \quad (12.2)$$

respectively. Here the beam sizes $\sigma_x$ and $\sigma_y$ depend on the longitudinal position $s$. The corresponding tune shifts $\Delta Q_x$ and $\Delta Q_y$ follow from Equation 8.33. For convenience, we rewrite $Nr_p/\sqrt{2\pi}\sigma_s = \hat{I}/\beta_0 I_c$ with the peak beam current $\hat{I} = Ne\beta_0 c/\sqrt{2\pi}\sigma_s$ and the Alvén current $I_c = 4\pi\varepsilon_0 m_p c^3/e = 17045$ A for electrons with mass $m_e$. This allows us to introduce the perveance $K = 2I/\beta_0^3\gamma_0^3 I_c = 2Nr_p/\sqrt{2\pi}\sigma_s\beta_0^2\gamma_0^3$, which simplifies Equation 12.2 to $\Delta k_{1x} = -K/\sigma_x(\sigma_x + \sigma_y)$ and $\Delta k_{1y} = -K/\sigma_y(\sigma_x + \sigma_y)$.

Instead, basing the discussion of space-charge effects on Gaussian distributions, the *Kapchinsky-Vladimirsky*, or KV-distribution, in the action $J_x, J_y$ and angle $\psi_x, \psi_y$ variables

$$\Psi(J_x, \psi_x, J_y, \psi_y) = \frac{1}{(2\pi)^2 J_{x0}J_{y0}}\delta\left(1 - \frac{J_x}{J_{x0}} - \frac{J_y}{J_{y0}}\right) \quad (12.3)$$

is often used in the literature [91, 92, 93]. It is normalized to unity and the projections onto the real space coordinates $x = \sqrt{2J_x\beta_x}$ and $y = \sqrt{2J_y\beta_y}$ results into a homogeneously filled ellipse, defined by its boundary $x^2/a_x^2 + y^2/a_y^2 = 1$ with half axes $a_x = \sqrt{2J_{x0}\beta_x}$ and $a_y = \sqrt{2J_{y0}\beta_y}$, and constant density $\rho_{KV} = 1/\pi a_x a_y$. The rms beam sizes are then given by $\sigma_{x,KV}^2 = \langle x^2 \rangle = a_x^2/4$ and $\sigma_{y,KV}^2 = \langle y^2 \rangle = a_y^2/4$. When comparing tune shifts derived from the KV-distribution and Gaussians we have to keep in mind that the density $\rho_{KV} = 1/4\pi\sigma_{x,KV}\sigma_{y,KV}$ in the center of the distribution is only half of that of an equivalent Gaussian, which is $\rho_G = 1/2\pi\sigma_{x,KV}\sigma_{y,KV}$. The tune shifts derived from the KV-distribution are therefore half that of an equivalent Gaussian. We need to keep this in mind when interpreting the quantities quoted in the literature.

In order to find out how space-charge forces affect the beam sizes, we follow Sacherer and derive *envelope equations* for the beam sizes starting from Equation 3.62 in Section 3.3.4 that describes the dynamics of a single particle. With the Ansatz from Equation 3.63, we derived the first of Equations 3.66, which reads $u'' + k_1(s)u - 1/u^3 = 0$ after using Equation 3.68 to replace $\psi'$. Realizing that $u = \sqrt{\beta(s)}$ and after multiplying the equation with the square root of the emittance $\sqrt{\varepsilon}$ we obtain the envelope equation $\sigma'' + k_1(s)\sigma - \varepsilon^2/\sigma^3 = 0$, where we identified $\sigma = \sqrt{\varepsilon\beta(s)}$. Including space-charge forces is now easily accomplished by adding the corresponding defocusing functions $\Delta k_{1x}$ and $\Delta k_{1y}$ to the focusing functions $k_1(s) \rightarrow k_1(s) - \Delta k_1$ with the result

$$\sigma_x'' + k_1(s)\sigma_x - \frac{\varepsilon_x^2}{\sigma_x^3} - \frac{K}{\sigma_x + \sigma_y} = 0 \quad \text{and} \quad \sigma_y'' - k_1(s)\sigma_y - \frac{\varepsilon_y^2}{\sigma_y^3} - \frac{K}{\sigma_x + \sigma_y} = 0, \quad (12.4)$$

where the opposite signs of $k_1(s)$ describe the focusing of the quadrupoles in the accelerator, which are focusing in one and defocusing in the other direction. Note also, that the interpretation of the perveance $K$ pertains to the center of a Gaussian distribution with beam sizes $\sigma_x$ and $\sigma_y$.

Integrating Equations 12.4 numerically is straightforward in MATLAB for a beamline with ten $60°$ FODO cells, each 10 m long, and with one meter long quadrupoles. The two second-order equations need to be transformed into a system of four first-order differential equations and the following lines encode the derivatives within a function called sachfun()

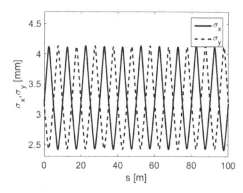 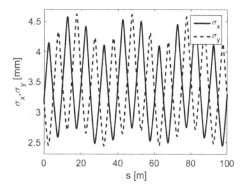

Figure 12.1 The beam sizes $\sigma_x$ and $\sigma_y$ in 10 FODO cells with $K = 0$ (left) and $K = 5 \times 10^{-8}$ (right). The initial values are equal in both cases.

```
% variables: x(1)=sigx, x(2)=sigx', x(3)=sigy, x(4)=sigy'
dxds(1)=x(2); dxds(2)=-k1(s).*x(1)+epsx^2./x(1).^3+Kperv./(x(1)+x(3));
dxds(3)=x(4); dxds(4)=k1(s).*x(3)+epsy^2./x(3).^3+Kperv./(x(1)+x(3));
```

that is passed to the Runge-Kutta integrator ode45, which integrates the system of differential equations and returns the variables along the beam line in the range from 0 to 100 m with the following call

```
[s,x]=ode45(@sachfun,[0,100],x0,odeset('MaxStep',1e-2));
```

Here x0 contains the four initial values, obtained by calculating the beta functions with the functions from Chapter 3. Moreover, we restrict the step size to 1 cm in order to achieve adequate accuracy. The function k1(s) returns the quadrupole gradients as a function of the position s along the beam line. Figure 12.1 shows the horizontal and vertical beam sizes for the beam line with for $\varepsilon_x = \varepsilon_y = 10^{-6}$ m-rad and $K = 0$ on the left-hand side and for $K = 5 \times 10^{-8}$ on the right-hand side, which corresponds to 10 mA protons with a kinetic energy of 25 MeV. We observe a distinct beating of the beam sizes, similar to a that from mismatched quadrupoles.

So far, we discussed space-charge forces that act directly between particles, but the particles in a beam also induce image charges in the beam pipe, we employed them earlier for diagnostic purposes in Chapter 7. The image charges, in return, exert so-called *indirect space-charge* forces on the beam particles. To analyze this system, we consider a beam with line density $\lambda(s) = dN/ds$ centered between two parallel horizontal plates, separated by $2h$. They represent the vacuum chamber. On their surface, the electric field lines must be perpendicular in order to satisfy the boundary conditions on perfectly conducting metallic surfaces. This is achieved by introducing a hierarchy of image charges. The boundary condition on the upper wall is satisfied by an image charge at $+2h$. But now we have two charges to compensate on the lower wall. This is accomplished by introducing an additional image charge at $-4h$ that assures that the field lines emanating from the image charge at $2h$ is perpendicular on the lower wall. The same argument holds for the image charges at $-2h$ and $4h$. Of course the charges at $\pm 4h$ require compensating charges at $\pm 6h$ and so forth. We therefore arrive at an infinite sequence of image charges with alternating polarity. The electric field $E_y$ at a vertical distance $y$ from the center of the beam, is then given by

$$E_y = \sum_{n=1}^{\infty} (-1)^n \frac{\lambda}{2\pi\varepsilon_0} \left( \frac{1}{2nh+y} - \frac{1}{2nh-y} \right) = \frac{\lambda}{4\pi\varepsilon_0 h^2} \frac{\pi^2}{12} y , \qquad (12.5)$$

where we used $\sum_{n=1}^{\infty} (-1)^n/n^2 = \pi^2/12$. We therefore find a vertical force, linear in the distance $y$ from the center of the beam pipe, that is focusing in the sense that it points back towards the center of the beam. The horizontal component of the force can be found from $\text{div } \vec{E} = 0$ and can be found to have the opposite sign but equal magnitude. It is therefore defocusing. The focusing and defocusing forces due to these indirect space-charge forces lead to tune shifts, first discussed by Laslett in [94] for a number of different beam pipe geometries and referred to as *Laslett tune shifts*. Note, that the indirect space-charge forces from a displaced beam will alter the image charges and thereby the fields. In this way the beam as a whole can affect its own centroid motion, and is therefore a coherent tune shift, that will be visible on beam position monitors.

## 12.2 INTRABEAM SCATTERING AND TOUSCHEK-EFFECT

The particles in a beam are not only affected by their average field, as covered in the previous section. They also perform longitudinal and transverse oscillations and will scatter from each other. For electrons, this process is called *Møller scattering*. Modern synchrotron light sources operate with very small bunch sizes and large number of particles per bunch, which makes the probability for scattering among particles within a bunch very large and constitutes a limiting factor for their performance. Most scattering events only change the momentum of the participating beam particles by a small amount and, since the events are random, they increase the emittances; in a process called *intrabeam scattering*. The growth rates were first calculated by Piwinski [95] and he showed that in storage rings operating below transition, mostly proton or heavy-ion storage rings, intrabeam scattering increases the emittances in all there dimensions towards an equilibrium. On the other hand, in rings operating above transition, the emittances can grow indefinitely. The reader is referred to the original literature [95, 96] and a recent comparison in [93] for the formulae to evaluate the growth rates.

Occasionally, in a scattering event, the momenta may change by a large amount and this can lead to particle losses. In most storage rings, the betatron tunes are much larger than the synchrotron tune and the transverse momenta are consequently much larger than the longitudinal momenta in the beam's reference frame. Now, if a scattering event deflects a fraction of the large transverse momenta into the longitudinal phase space, the particle is lost, because it exceeds the momentum acceptance $\delta_{max}$, the height of the separatrix from Equation 5.40 in Section 5.4. This process is called the *Touschek effect*. The detailed derivation of the Touschek lifetime $\tau_T$ is beyond the scope of this book and we only quote the result, slightly adapted from [93]

$$\frac{1}{\tau_T} = \frac{N_b \beta_0^3 r_p^2 c D(\xi)}{8\pi\gamma_0^2 \sigma_x \sigma_y \sigma_s \delta_{max}^3} \quad \text{with} \quad D(\xi) = \xi^{1/2} \int_0^1 \left[ \frac{1}{u} - 1 - \frac{1}{2} \log\left(\frac{1}{u}\right) \right] e^{-\xi/u} du \qquad (12.6)$$

and $\xi = \delta_{max}^2 \beta_x / \beta_0^2 \gamma_0^2 \varepsilon_x$. The reason for a short Touschek lifetime is a high charge density $N_b/\sigma_x\sigma_y\sigma_s$, which makes collisions, also those with large momentum transfer, very frequent. The problem can be alleviated by either increasing the bunch volume, often by increasing the bunch length $\sigma_s$. A second option is to increase the momentum acceptance $\delta_{max}$ by increasing the accelerating voltage $\hat{V}$ in Equation 5.40. Evaluating the lifetime in

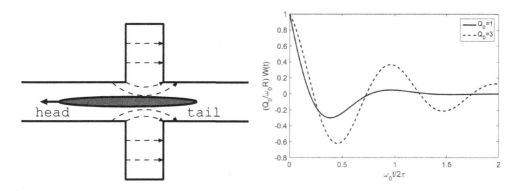

Figure 12.2 On the left-hand side a bunch (grey) passes a cavity-like structure, where it excites longitudinal fields, shown as dashed lines. On the right-hand side, the longitudinal fields, excited by a single particle, in resonating structures with $Q_r = 1$ and $Q_r = 3$ are shown.

Equation 12.6 is rather simple, after defining the parameters the integral $D(\xi)$ is evaluated numerically with the following lines of code

```
f=@(u)exp(-xi./u).*(1./u-1-0.5.*log(1./u));
D=integral(f,0,1)*sqrt(xi)
```

For a rapid estimate of the lifetime we can use average values for the beam sizes $\sigma_x$ and $\sigma_y$ and the beta function $\beta_x$, but for a more realistic estimate a weighted average $\oint(1/\tau_T(s))ds/C$ of the lifetime $\tau_T(s)$ at every longitudinal position $s$ in the ring should be evaluated.

## 12.3  WAKE FIELDS, IMPEDANCES, AND LOSS FACTORS

The indirect space charge forces, discussed at the end of Section 12.1, are caused by the beam by interacting with its environment, and then, in turn, affect beam. Beam loading, discussed in Section 6.6, is another example, where the beam excites fields in a cavity that act back on particles arriving later. This mechanism is the generic feature: the leading particles excite fields that acts back on particles arriving later. Often resonant structures in the beam pipe, such as the accelerating cavities, bellows, or steps in the cross section, are responsible for it. The fields that the beam "leaves behind," are called *wake fields*, because they resemble the wave pattern, called the "wake," behind a boat moving through water. We will base the discussion on a *resonator wake*, which is the *impulse response* of a damped harmonic oscillator, or equivalently, of the *RLC*–circuit shown in Figure 6.7. The image on the left-hand side in Figure 12.2 illustrates the concept; the bunch, shown as the grey ellipsoid, moves towards the left and the head excites fields in the cavity-like structure that subsequently act back on the tail of the same bunch. The excited fields can be either transverse or longitudinal and we will mostly discuss the latter.

We already calculated the impedance $Z(\omega)$, which is the response to a harmonic excitation with frequency $\omega$, of the *RLC*–circuit in Section 6.3 and gave the result in Equation 6.19.

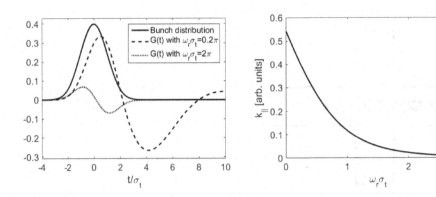

**Figure 12.3** The wake fields $W(t)$ (left) caused by a bunch with Gaussian distribution for low (dashed) and high (dotted) frequencies and the loss factor $k_\parallel$ (right) as a function of the frequency $\omega_r \sigma_t$ normalized to the bunch length $\sigma_t$.

Setting the $n^2 = 1$ and multiplying numerator and denominator with $\omega \omega_r / iQ_r$ we arrive at

$$Z_\parallel(\omega) = \frac{R_s \omega \omega_r / iQ_r}{\omega^2 - i\omega_r \omega / Q_r - \omega_r^2} = \frac{R_s}{i\sqrt{4Q_r^2 - 1}} \left[ \frac{\omega_+}{\omega - \omega_+} - \frac{\omega_-}{\omega - \omega_-} \right] \tag{12.7}$$

where $\omega_\pm = (\omega_r / 2Q_r) \left[ i \pm \sqrt{4Q_r^2 - 1} \right]$ are the roots of the denominator in the first equality. The second equality follows from a partial-fraction decomposition. Note that $\omega_- = -\omega_+^*$ and this implies that the second term in the last equation is the complex conjugate of the first one. Therefore, the wake function is *causal* and only has non-zero values for $t > 0$; only particles arriving later than the particle that excited the wake field are affected. The impulse response $W(t)$ of the resonator, often called the *wake potential*, is given by the Fourier transform of the harmonic response, the impedance, and for positive times $t$, which is behind the source particle, we obtain

$$W(t) = \frac{1}{2\pi} \int_{-\infty}^{\infty} Z_\parallel(\omega) e^{i\omega t} d\omega = \frac{R_s}{\sqrt{4Q_r^2 - 1}} \left[ \omega_+ e^{i\omega_+ t} - \omega_- e^{i\omega_- t} \right] \tag{12.8}$$

$$= \frac{2R_s}{\sqrt{4Q_r^2 - 1}} \text{Re} \left[ \omega_+ e^{i\omega_+ t} \right] = \frac{\omega_r R_s}{Q_r} e^{-t/\hat{\tau}_d} \left[ \cos \hat{\omega} t - \frac{1}{\sqrt{4Q_r^2 - 1}} \sin \hat{\omega} t \right]$$

with the abbreviations $\hat{\tau}_d = 2Q_r / \omega_r$ and $\hat{\omega} = \omega_r \sqrt{1 - 1/4Q_r^2}$. On the right-hand side in Figure 12.2, we show $W(t)$ as a function of time $t$ for values of $Q_r = 1$ and $Q_r = 3$. Discontinuities in the beam pipe are often modeled by fast decaying wake field with $Q_r = 1$ and a frequency $\omega_r \sim 1/\sigma_t = c/\sigma_s$, determined by the bunch length $\sigma_s$. Only particles arriving a time $t$ after the excitation experience a change of its energy proportional to $W(t)$.

In a bunch with finite length a particle at some position $t$ in the tail will experience the superposition of all the fields excited by the particles at $t' < t$ that are ahead of it, $G(t) = \int_0^\infty \psi(t - t') W(t') dt'$, which is commonly called the *wake function* of a bunch with longitudinal distribution $\psi(t)$. For a bunch with a Gaussian distribution

$\psi(t) = e^{-t^2/2\sigma_t^2}/\sqrt{2\pi}\sigma_t$, we obtain

$$
\begin{aligned}
G(t) &= \frac{1}{\sqrt{2\pi}\sigma_t} \int_0^\infty e^{-(t-t')^2/2\sigma_t^2} \frac{2R_s}{\sqrt{4Q_r^2-1}} \operatorname{Re}\left[w_+ e^{i\omega_+ t'}\right] dt' \\
&= \frac{R_s}{\sqrt{4Q_r^2-1}} e^{-t^2/2\sigma_t^2} \operatorname{Re}\left[w_+ w\left(\frac{\omega_+\sigma_t}{\sqrt{2}} - i\frac{t}{\sqrt{2}\sigma_t}\right)\right] , \quad (12.9)
\end{aligned}
$$

where we use the following representation of the complex error function [23] $w(z) = (2/\sqrt{\pi}) \int_0^\infty e^{-\alpha^2+2i\alpha z} d\alpha$ in the evaluation of the integral. On the left-hand side in Figure 12.3, we show the longitudinal particle distribution $\psi(t)$ (solid) and $G(t)$ for two broadband resonators. Both have $Q_r = 1$, but one is characterized by a very low frequency (dashed), such that $\omega_r \sigma_t = 0.2\pi$ and the other (dotted) by a ten times higher $\omega_r$. We see that the low-frequency wake function almost follows the bunch profile (dots) and all the trailing particles are expected to lose energy, whereas the high-frequency wake shows an oscillation within the bunch and some particles will lose energy and other will gain energy.

The average energy loss of the entire bunch is given by the wake function $G(t)$ averaged over the bunch distribution $\psi(t)$ is described by the *loss factor* $k_\|(\sigma_t) = \int_{-\infty}^\infty \psi(t)G(t)dt$. Evaluating the integral by first expressing the complex error function in $G(t)$ by its integral representation, exchanging the order of integration, evaluating the integral over $t$, and, finally, using the integral expression once again, we find

$$
k_\|(\sigma_t) = \frac{R_s}{\sqrt{4Q_r^2-1}} \operatorname{Re}\left[w_+ w(\omega_+\sigma_t)\right] \quad \text{with} \quad w_+ = \frac{\omega_r}{2Q_r}\left(i + \sqrt{4Q_r^2-1}\right) . \quad (12.10)
$$

On the right-hand side in Figure 12.3, we display $\operatorname{Re}\left[w_+ w(\omega_+\sigma_t)\right]$ as a function of $\omega_r\sigma_t$ and observe that long bunches cause much smaller losses than short ones, because the oscillations of the wake function within the bunch, mentioned at the end of the previous paragraph, will cancel and thus reduce the average energy loss.

Instead of calculating the loss factor in the time domain by first convoluting the bunch distribution $\psi$ with the wake potential $W(t)$ and then once again with $\psi$, we can write the loss factor in terms of the Fourier transforms and find

$$
k_\|(\sigma_t) = \frac{1}{2\pi} \int_{-\infty}^\infty Z_\|(\omega)\tilde\psi^2(\omega)d\omega , \quad (12.11)
$$

where $\tilde\psi(\omega)$ is the Fourier transform of the longitudinal bunch profile $\psi(t)$ and $\tilde\psi^2(\omega) = e^{-\omega^2\sigma_t^2}$ is the spectral power density of the bunch. Equation 12.11 shows that the energy loss can be written as the overlap of the spectrum of the impedance $Z(\omega)$ and power spectrum of the bunch $\tilde\psi^2(\omega)$. Conversely, if the spectra do not overlap, there is no energy loss or other interaction between bunch and impedance, a theme we will meet again in later sections. Note that $k_\|$ denotes the energy lost by a single particle. In the literature, it is often given in the units of V/pC, such that the energy lost by a bunch can be obtained by multiplying the loss factor by the bunch charge $Ne$.

Maxwell's equations intricately relate the transverse and longitudinal components of the electro-magnetic fields that transfer momentum to the beam. One consequence is the *Panofsky-Wenzel theorem*, which states that change of the transverse forces along the beamline equals the negative of the transverse changes of the accelerating forces $\frac{\partial F_\perp}{\partial s} = -\nabla_\perp F_s$. This relation links the longitudinal and transverse wake-fields, or equivalently, the impedances. For the transverse impedances this implies $Z_\perp(\omega) = (c/i\omega)Z_\|(\omega)$, and for the resonator from Equation 12.7 we obtain

$$Z_\perp(\omega) = \frac{(c/i\omega)R_s}{1 + iQ_r\left(\frac{\omega}{\omega_r} - \frac{\omega_r}{\omega}\right)} \quad \text{and} \quad W_\perp(t) = \frac{cR_s/Q_r}{\sqrt{1 - 1/4Q_r^2}}\, e^{-t/\tau_d}\sin(\hat{\omega}t)\,, \qquad (12.12)$$

where the wake function $W_\perp(t)$ is the Fourier-transform of $Z_\perp(\omega)$. For transverse impedances one usually quotes $R_\perp = (c/\omega_r)R_s$ instead of the longitudinal shunt impedance $R_s$.

The finite resistivity of the beam pipe material causes the electro-magnetic fields, especially those with high frequencies caused by very short bunches, to penetrate into the walls as a consequence of the skin-effect. The energy dissipated causes additional losses that are due to the longitudinal *resistive wall impedances*. The associated wake fields that trailing particles experience. The longitudinal impedance is given [12] by $Z_{\parallel,rw} = \mu Z_0 \delta_s/2\mu_0 b$ with the *skin depth* $\delta_s = \sqrt{2/\sigma_c\mu\omega}$ and the conductivity $\sigma_c$, the beam pipe radius $b$, the permeability $\mu$, and the impedance of free space $Z_0 = \mu_0 c$.

We saw that the discontinuities in the beam pipe and its finite resistance give rise to impedances that cause bunches to leave behind energy in the form of wake fields that acts back on later portions of the bunch. In some circumstances, this may lead to a feedback mechanism that can become unstable, even if the beam is un-bunched, as we shall see in the next section.

## 12.4 COASTING-BEAM INSTABILITY

In the absence of a radio-frequency system, the beam in a storage ring is not bunched, but evenly smeared-out around the circumference $C$. It is often referred to as a *coasting beam*. Its distribution function $\psi(\Theta, \delta)$ in the variables $\theta = 2\pi s/C$ and $\delta$ therefore only depends on $\delta$ and is denoted by $\psi_0(\delta)$. After such a beam is perturbed, we assume the perturbation to behave like a harmonic wave $e^{in\theta - i\Omega t}$ with $n$–fold periodicity and frequency $\Omega$ traveling around the ring. The distribution function $\psi(\theta, \delta, t)$ then assumes the form

$$\psi(\theta, \delta, t) = \psi_0(\delta) + \psi_n(\delta)e^{in\theta - i\Omega t} \qquad (12.13)$$

with a perturbation $\psi_n$, assumed to be small compared to $\psi_0$. The current $I(t)$ caused by the perturbed distribution is $I_n e^{in\theta - i\Omega t}$ with the Fourier transform $\tilde{I}_n = 2\pi I_n e^{in\theta}\delta(\omega - \Omega)$. In conjunction with the impedance $Z_\parallel(\omega)$ it causes the energy offset $\delta$ to change by $d\delta/dt = -(e/\beta_0^2 E_0 T_0)\int_{-\infty}^{\infty} Z_\parallel(\omega)\tilde{I}_n(\omega)e^{-i\omega t}d\omega/2\pi$. The equations of motion for a particle in the coasting beam then become

$$\frac{d\theta}{dt} = \omega_0 - \omega_0\eta\delta \quad \text{and} \quad \frac{d\delta}{dt} = -\frac{e}{\beta_0^2 E_0 T_0}I_n Z_\parallel(\Omega)e^{in\theta - i\Omega t}\,. \qquad (12.14)$$

In order to find the frequencies $\Omega$ and in particular their imaginary part, which determines the growth rate of an instability, we will determine self-consistent solutions for the distribution function with the help of the Vlasov equation

$$\frac{\partial\psi}{\partial t} + \frac{\partial\psi}{\partial\theta}\frac{d\theta}{dt} + \frac{\partial\psi}{\partial\delta}\frac{d\delta}{dt} = 0\,. \qquad (12.15)$$

It follows from the conservation of particles $d\psi/dt = 0$ in time. Inserting the equations of motion from Equation 12.14 and integrating over $\delta$ leads to a requirement for self-consistency for the perturbing current $I_n(\delta)$. After canceling $I_n$ from both sides of the equation, we find the *dispersion relation*

$$1 = -i\frac{e^2 Z_\parallel(\Omega)}{\beta^2 ET_0^2}\int_{-\infty}^{\infty}\frac{\partial\psi_0(\delta)/\partial\delta}{n\omega_0(1 - \eta\delta) - \Omega}d\delta\,, \qquad (12.16)$$

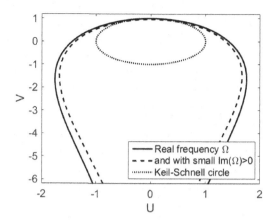

**Figure 12.4** The complex impedance plane $U + iV$ with the solid line indicating the limit of stability. The dashed line indicates stable conditions and the dotted circle shows the conservative estimate from Equation 12.20, the *Keil-Schnell stability criterion*.

where we neglected $\partial\psi_n/\partial\delta \ll \partial\psi_0/\partial\delta$. If the unperturbed momentum distribution is Gaussian with $\psi_0(\delta) = (N_b/\sqrt{2\pi}\sigma_\delta)e^{-\delta^2/2\sigma_\delta^2}$, the equation becomes after substituting $\xi = \delta/\sigma_\delta$

$$1 = -i\frac{I_0 Z_\|(\Omega)/n}{2\pi\beta^2(E/e)\eta\sigma_\delta^2}I_D(\Omega) \quad \text{with} \quad I_D(\Omega) = \frac{1}{\sqrt{2\pi}}\int_{-\infty}^{\infty}\frac{\xi e^{-\xi^2/2}d\xi}{\xi - \left(1 - \frac{\Omega}{n\omega_0}\right)/\eta\sigma_\delta}, \quad (12.17)$$

where $I_0 = N_b e/T_0$ is the macroscopic beam current. With the abbreviation $\xi_1 = \left(1 - \frac{\Omega}{n\omega_0}\right)/\eta\sigma_\delta$, we solve the dispersion integral $I_D$ by writing $1/(\xi - \xi_1) = i\int_0^\infty e^{-i\alpha(\xi-\xi_1)}d\alpha$ and add a term $e^{b\xi}$ to the integral in order to use parametric differentiation with respect to $b$ to replace the linear term in $\xi$ in the numerator under the integral. The integral over $\xi$ then has the form of a shifted Gaussian in $\xi$ that can be integrated by completing the square. The remaining integral over $\alpha$ is solved by using the following representation of the complex error function $w(z) = (2/\sqrt{\pi})\int_0^\infty e^{-\alpha^2+2i\alpha z}d\alpha$, resulting in

$$I_D(\Omega; b) = i\sqrt{\frac{\pi}{2}}e^{b^2/2}w\left(\frac{\xi_1 - b}{\sqrt{2}}\right), \quad (12.18)$$

where the auxiliary argument $b$ in $I_D$ refers to the additional parameter introduced for parametric differentiation. Thus, differentiation $I_D(\Omega, b)$ once with respect to $b$ and then setting $b = 0$ results in the sought integral

$$I_D(\Omega) = 1 + i\sqrt{\frac{\pi}{2}}\xi_1 w\left(\frac{\xi_1}{\sqrt{2}}\right) \quad \text{with} \quad \xi_1 = \frac{1}{\eta\sigma_\delta}\left(1 - \frac{\Omega}{n\omega_0}\right), \quad (12.19)$$

where we evaluate the derivative of the complex error function with the relation [23] $w'(z) = 2i/\sqrt{\pi} - 2zw(z)$.

Even with $I_D(\Omega)$ given as a function of $\Omega$, it is very difficult to solve Equation 12.17 for a given impedances $Z_\|$ for the, in general complex, frequency $\Omega$. Instead, we introduce the scaled impedance $U + iV = (I_0/2\pi\beta_0^2(E_0/e)\eta\sigma_\delta^2)(Z_\|(\Omega)/n)$, such that Equation 12.17 now reads $(U + iV) = i/I_D(\Omega)$. Displaying the real and imaginary parts of $i/I_D(\Omega)$ as a function of $\Omega$ yields the onion-shaped curves, shown in Figure 12.4. For $\mathrm{Im}(\Omega) > 0$ (dashed curve) they lie inside the solid line, and for negative values they lie outside. This indicates that the solid line, drawn for purely real $\Omega$, separates the complex plane into the inside region, where the beam is damped, and the outside region, where the instabilities grow. The shape of the curve that separates the stable from the unstable region depends on the unperturbed momentum spread $\psi_0(\delta)$, but the circle with unit radius, shown as the dotted line in Figure 12.4, serves as a conservative guess for stability. Thus, if the scaled impedance $|U + iV| < 1$ lies inside the unit circle, we expect the beam to be stable and we therefore obtain the *Keil-Schnell stability criterion*

$$\left|\frac{Z_\|(\Omega)}{n}\right| < \frac{2\pi\beta_0^2(E_0/e)|\eta|\sigma_\delta^2}{I_0}\mathcal{F} , \tag{12.20}$$

where the impedance $Z_\|(\Omega)$ needs to be evaluated at harmonics of the revolution frequency $\Omega = n\omega_0$. It is noteworthy that a larger beam current $I_0$ requires a smaller impedance for the beam to remain stable. Moreover, a larger momentum spread $\sigma_\delta$ helps to sustain a larger impedance; this stabilizing effect of a finite momentum spread is the classical application (in beam physics) for Landau damping. $\mathcal{F}$ is a form factor that varies for different momentum distributions and is unity for a Gaussian distribution.

## 12.5  SINGLE-BUNCH INSTABILITIES

In this section we will consider very short-range wake potentials that act back on the same bunch. In order to investigate the effect of the wake potential on the bunch length and momentum spread, we follow [97] and construct a simplified model of the longitudinal dynamics. As dynamic variables of the longitudinal phase space, we use the scaled arrival time $x_1 = \Omega_s t/\alpha$, and the momentum deviation $x_2 = \delta$. Here $\Omega_s$ is the synchrotron frequency and $\alpha$ is the momentum compaction factor. For two-dimensional Gaussian distributions, given by Equation 2.16, we then construct maps for the centroids $X_i$ and the sigma matrix $\sigma_{ij}$ for $i, j = 1, 2$. The motion, unperturbed by wake fields, is characterized by synchrotron oscillations, modeled by a rotation matrix $U(\nu_s)$ that only depends on the synchrotron tune $\nu_s = \Omega_s T_0$. Radiation damping and excitation affect the momentum, given by $x_2' = \xi x_2 + \sqrt{1 - \xi^2}\sigma_0\hat{P}$ with the damping decrement $\xi = e^{-T_0/\tau_d}$, where $\tau_d$ is the synchrotron radiation damping time from Equation 10.7, and $\hat{P}$ is a "random number generator," defined through its averages $\langle\hat{P}\rangle = 0$ and $\langle\hat{P}^2\rangle = 1$. The equilibrium bunch length $\sigma_0$ results from the joint action of oscillations, damping and radiation excitation. Moreover, if the bunch length is very short, such that $\omega_0\sigma_t \ll 1$, we can approximate the wake potential by a constant $W(x_1) = W_0 H(x_1)/E_0$ with the Heaviside function $H(x)$, which is unity for positive arguments and zero otherwise. $E_0$ is the beam energy. For the change in $x_2$ we thus have $x_2' = x_2 - f(x_1)$ with $f(x_1) = (f_0/\sqrt{2\pi\sigma_{11}})\int_{-\infty}^{x_1} e^{-z^2/2\sigma_{11}}dz$ and $f_0 = NeW_0/E_0$. The integral can be expressed in terms of error functions [23]. From these assumptions, Hirata [97] constructs maps for the $X_i$ and $\sigma_{ij}$ for synchrotron oscillations

$$X_i' = \sum_{j=1}^{2} U_{ij}X_j \quad \text{and} \quad \sigma' = U\sigma U^t , \tag{12.21}$$

for radiation excitation and damping

$$X_1' = X_1, \quad X_2' = \xi X_2, \quad \sigma_{11}' = \sigma_{11}, \quad \sigma_{12}' = \xi\sigma_{12}, \quad \sigma_{22}' = \xi^2\sigma_{22} + (1 - \xi^2)\sigma_0^2 , \tag{12.22}$$

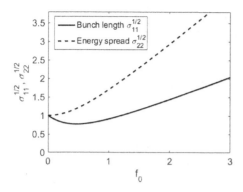 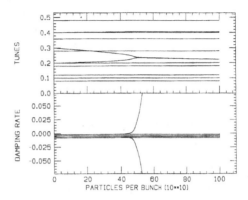

Figure 12.5  The bunch length and momentum spread of as a function of the strength $f_0$, which is proportional to the charge per bunch (left) and the mode spectrum of the transverse mode-coupling instability (right, from [98]).

and for the wake field

$$X'_1 = X_1, \quad X'_2 = X_2 - \frac{f_0}{2}, \quad \sigma'_{11} = \sigma_{11}, \quad \sigma'_{12} = \sigma_{12} - \frac{f_0\sqrt{\sigma_{11}}}{2\sqrt{\pi}},$$

$$\sigma'_{22} = \sigma_{22} - \frac{f_0\sigma_{12}}{\sqrt{\pi\sigma_{11}}} + \frac{f_0^2}{12}. \tag{12.23}$$

In [97], the equilibrium configuration, which is the period-1 fixed point of the concatenation of these three maps, is calculated analytically. Here, we use MATLAB instead to encode the three maps and to iterate for a few damping times to find the equilibrium bunch parameters numerically. We use the following function

```
function [X,sigma]=hirata_iterate(N,U,xi,sig0,f0,X,sigma)
for k=1:N
  [X2,sigma2]=hirata_synosc(U,X,sigma);
  [X3,sigma3]=hirata_radamp(xi,sig0,X2,sigma2);
  [X4,sigma4]=hirata_wake(f0,X3,sigma3);
  X=X4; sigma=sigma4;
end
```

which returns the bunch centroid X and sigma matrix sigma after N turns. Inside the functions hirata_xxx Equations 12.21 to 12.23 are implemented. In Figure 12.5, we show the equilibrium bunch length and energy spread as a function of the strength of the wake $f_0$. We observe that the bunch length initially shrinks before exceeding its equilibrium value. This shrinking may be attributed to modified longitudinal fields caused by the bunch itself and is called *potential well distortion.* The wake fields also cause the momentum spread to increase, which is eventually responsible for increasing the bunch length. This effect is called *turbulent bunch lengthening,* because the wake fields increase the momentum spread and this translates into an increased bunch length. The MATLAB functions for this simulation are discussed further in Appendix B.5.

Turning briefly to transverse wake fields, we note that they are excited by particles passing discontinuities in the beam pipe with transverse offsets $\Delta x$ or $\Delta y$. The excited fields are transverse and will transversely kick trailing particles. In an ultra-relativistic linear accelerator, where the particles maintain their longitudinal position, a transversely oscillating particle at the head will resonantly excite the particle at the tail, because both have the same betatron oscillation frequency. It is easy to show [11, 99] that the amplitude of the trailing particle will grow linearly with time and are lost in this mechanism, called *beam break-up*. By operating a few accelerating structures at the beginning of the accelerator off-crest, it can be mitigated. In the structures, the head and the tail of a bunch receive different energies through the linear energy variation—a chirp—along the bunch. The momentum dependence of the quadrupole focusing then causes the betatron frequencies to vary along the bunch and thus will reduce the resonant excitation of the trailing particles. The spread of oscillation frequencies causes the particles to decohere, another example of Landau damping. In this particular case, it is called *BNS-damping*, after the names of the inventors [100].

In the free-electron lasers based on linear accelerators the peak currents can reach several kA and very high-frequency wake fields can cause density modulations at optical wavelengths with $\mu$m periodicity. This self-modulating of the electron bunch is called *micro-bunching instability*, which can drastically increase the momentum spread and the emittance of the beam, which may prevent lasing and even cause beam loss. This instability can be usually prevented with a *laser heater*, in which the momentum spread of the beam is slightly increased in order to provide Landau damping to mitigate the growth of the instability.

In storage rings, the longitudinal positions of particles constantly change due to synchrotron oscillations. Even this system can be analyzed with a two-particle model [11, 99, 93]. Here, we will instead briefly mention a complementary approach [98], which is inspired by [97]. We consider a transverse wake potential, where a particle with transverse offset $\hat{x}$ and longitudinal position $\hat{z}$ gives a transverse kick to a particle at position $z$. For very short bunches it has the form $W_\perp(z) \propto \hat{x}(\hat{z} - z)$. From this potential we determine a map for the centroid and sigma matrix of a six-dimensional Gaussian distribution, described by Equation 2.16. The transverse kick $\Delta x' = \Delta x_2 \propto f(x_5)$ of a particle at longitudinal position $z = x_5$ is then given by integrating over all particles ahead with the result

$$f(x_5) \propto \exp\left[-\frac{(x_5 - X_5)^2}{2\sigma_{55}}\right]\left\{X_1\sqrt{\frac{2\sigma_{55}}{\pi}} + [X_1(x_5 - X_5) - \sigma_{15}]\,w\left(-i\frac{x_5 - X_5}{\sqrt{2\sigma_{55}}}\right)\right\},$$

$$(12.24)$$

where we use the canonical naming of the phase-space variables introduced in Section 2.3 and $w(z)$ is the complex error function [23]. From $f(x_5)$ we then derive the maps for centroid and sigma matrix. After adding maps for betatron and synchrotron oscillations as well as for radiation damping and excitation in order to determine the equilibrium beam sizes, we numerically deduce equilibrium beam parameters in the same way we discussed earlier in this section. From slightly perturbing the equilibrium we then derive a linearized map from which we deduce the perturbed tunes. On the right-hand side in Figure 12.5, taken from [98], we show the tunes and the growth rates as a function of the particles stored in a single bunch for a simulation of LEP at 20 GeV. At $N_b \approx 45 \times 10^{10}$ two frequencies merge, which causes this mechanism to be called *mode coupling instability*. A careful analysis shows that the tune merges with its lower synchrotron sideband. In our model the latter is related to the dynamics of the correlations between transverse and longitudinal degrees of freedom, $\sigma_{15}$ and $\sigma_{25}$, which can be visualized as the head moving opposite to the tail of the bunch and leads to coining this particular instability *fast head-tail instability*. Fast, because the growth rate, visible in the lower plot, increases dramatically.

Besides the fast head-tail instability, which has a distinct threshold for its growth rate, the "normal" *head-tail instability* additionally depends on the chromaticity $Q'$ of the storage

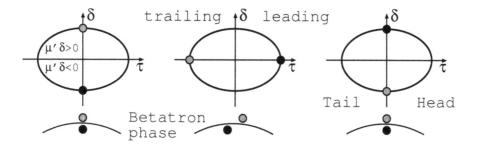

Figure 12.6 Illustration of the mechanism that causes the head-tail instability. See the text for an explanation.

ring and its growth rates are proportional to $Q'$. This instability is one of the main reasons for using sextupoles to adjust $Q' = \mu'/2\pi$ to values close to zero. The physical reason for the instability can be traced to a systematic non-zero betatron phase between leading and trailing particles in the bunch [101]. This can be explained with the help of Figure 12.6, which shows two particles (black and shaded) with opposite phases of their synchrotron oscillations in the phase space of arrival time $\tau$ and momentum offset $\delta$. We assume that they initially (left image) have the same longitudinal position $c\tau$ and additionally have the same transverse phase, shown on the lower sketch. During the first quarter of the synchrotron oscillation, the black particle is ahead of the shaded particle, but has lower energy. Therefore, it lags behind in the betatron phase proportional to $\mu'\delta$, which is indicated on the lower sketch. During the next quarter, from the middle to the right image, the black particle has higher energy and its betatron phase will increase, until it is equal to that of the shaded particle, once the situation on the right is reached. In the next two quarters (not shown) the shaded particle is ahead in $c\tau$ but lagging in betatron phase. So, the leading particle always lags in betatron phase. This is important, because only the leading particle affects the betatron motion of the trailing particle via transverse wake fields and if the relative betatron phase is always the same the energy change of the trailing particle will always be the same. Imagine pushing a child on a swing; pushing a little ahead of the turnaround point will reduce the amplitude, pushing a little after it, will increase the amplitude. We also point out that there is a second out-of-phase betatron mode which has just the inverse behavior. It is damped if the in-phase mode, described above, is excited and vice versa. In summary, the chromaticity is responsible for this instability method and needs to be compensated to small values. We refer the reader to [102] for a thorough analysis of this effect.

The short-range wake fields were responsible for the single-bunch instabilities discussed in this section. In the next, we will consider long-range wake fields due to narrow-band resonating structures that can couple the motion of multiple bunches in a storage ring.

## 12.6 MULTI-BUNCH INSTABILITIES

Most storage rings operate with a large number of stored bunches, for example, almost 3000 in LHC. Any narrow-band resonant structure with a large quality factor $Q_r$ have long decay times $\tau_r = 2Q_r/\omega_r$ and, once excited by one bunch, affect many later-arriving bunches. This configuration resembles a system of weakly coupled oscillators, in which the slowly decaying, also called *long-range wake fields*, provide the coupling mechanism. In the following, we will consider transverse wake fields, for which the equations of motion for the horizontal position $x_n$ of bunch number $n$ can be written as

$$\ddot{x}_n + \omega_\beta^2 x_n = -\frac{e^2 N_b c^2}{E_0 C} \sum_{k=0}^{\infty} \sum_{m=0}^{M-1} W_\perp \left(\frac{n-m}{M}C + kC\right) x_m \left(t - \frac{n-m}{M}T_0 - kT_0\right) \quad (12.25)$$

where $C$ is the circumference and $T_0$ the revolution time in the ring and $\omega_\beta$ is the betatron frequency. The term on the right-hand side describes the force due to the wake field that bunch $m$, having had transverse position $x_m(t - \Delta t)$ at a time $\Delta t = (n - m)T_0/M + kT_0$ earlier. Here $(n - m)T_0/M$ is the difference in travel time of bunch $n$ and $m$, and $kT_0$ takes into account very long range wakes from earlier turns $k$. If the bunch $m$ excites a wake at this time, it starts oscillating and the argument of $W_\perp(c\Delta t)$ describes the amplitude of the wake field a distance $s = c\Delta t$ after it was excited. Since Equation 12.25 is linear in the transverse position, we attempt to solve it with an exponential $x_n = A_n e^{-i\Omega t + 2\pi i p n/M}$. The phase $2\pi i p n/M$ describes the relative phase of individual bunches. For example, for $p = 0$ they all oscillate in phase and for $p \neq 0$ a snapshot in time will show an oscillation around the ring with $p$ periods. Inserting this trial for $x_n$ into Equation 12.25 and after replacing the wake function $W_\perp$ by its Fourier transform, the impedance $Z_\perp$, we obtain for the eigenfrequency $\Omega$

$$\Omega = \omega_\beta - i\frac{Me^2 N_b c}{4\pi E_0 T_0 Q_x} \sum_{q=-\infty}^{\infty} Z_\perp \left(\omega_\beta + (p - qM)\omega_r\right) . \quad (12.26)$$

Here we see that the real part of the impedance $\text{Re}(Z_\perp)$, evaluated at the betatron sidebands $\omega_q = \omega_\beta + (p - qM)\omega_0$ at multiples of the revolution harmonic $\omega_0$, causes an imaginary contribution to the eigenfrequency $\Omega$ and thus may lead to exponential growth or damping of the mode, depending on the sign. Similarly, longitudinal narrow-band impedances $Z_\parallel$ cause longitudinal coupled-bunch modes by slowly decaying wake fields that couple the synchrotron oscillations of multiple bunches, but we refer the reader to the specialized literature [102, 103].

A common source of spurious narrow-band impedances are higher-order modes in accelerating structures. For example, the pill-box cavity from Section 5.1 does not only support the "wanted" accelerating $\text{TM}_{mnp} = \text{TM}_{010}$ mode, but also all modes with larger $mnp$, and even TE–modes with many different eigenfrequencies. Any one of them may overlap with a betatron side band and potentially lead to an unstable mode. Therefore one normally tries to prevent these modes by increasing the losses of these modes by adding *higher-order mode dampers*. These are antennas that extract the power deposited by the beam into these modes, reduce the $Q_r$–value of the modes, and causes them to decay before the next bunch arrives.

In electron or positron storage rings the instabilities can only grow, if their growth rate exceeds the damping rate due to synchrotron-radiation. Furthermore, if damping higher-order modes is insufficient, an active damping system such as a *feedback systems* is used. It senses the growth of individual modes $p$ and use a pulsed magnet, a *kicker,* to provide kicks to the beam that counteract the instability. In recent years, the rise in computing speed of digital signal processors made it possible that most modern feedback systems monitor the oscillations of individual bunches and act back on them *bunch-by-bunch* [103].

Even for coupled-bunch instabilities, a spread in oscillation frequencies helps to reduce the growth rates, because the resonant coupling of the individual oscillators is reduced. We already encountered this mechanism, Landau damping, in previous sections. Sometimes octupoles are used to provide amplitude-dependent tune shifts, which, in conjunction with a finite emittance, cause a spread of the betatron frequencies.

## QUESTIONS AND EXERCISES

1. Injecting protons with a kinetic energy of 50 MeV into CELSIUS with a circumference of 82 m, we used to store a beam current of 10 mA in a single RF-bucket. The bunch length was on the order of 10 m, average beta functions of 10 m, and transverse rms beam size on the order of 5 mm. Estimate the space charge tune shift.

2. How do you have to adjust Equation 12.1 to calculate the space-charge tune shift for highly charged ion beams, such $Ar^{10+}$ or $Pb^{82+}$?

3. Compare the two types of doublet lattice shown in Figure 3.24 and investigate which one performs better in the presence of space charge. Use the same geometry as the Figures, but (a) replace the thin quadrupoles by 0.6 m long ones and re-match the phase advances in both planes to $\mu/2\pi = 0.4$. (b) Use three copies of the thus prepared beam line and prepare functions `k1(s)` to return the quadrupole gradients along the long beam line, such that you can base your code on the MATLAB script `sacherer.m`. (c) Run the simulations for both configurations and prepare plots of the beam sizes along the beam line. (d) Extract the Twiss parameters $\beta_e$ and $\alpha_e$ at the end and calculate $B_{mag}$ with respect to the zero-current Twiss parameters to quantify the mismatch. (e) Which configuration is better?

4. The MATLAB simulation to integrate Equation 12.4, discussed in Section 12.1 and used in the previous exercise, requires initial Twiss parameters to be specified. In order to be able to use the equations for a ring, on the other hand, we have to find the periodic initial conditions by minimizing a suitable `chisq()` function, which returns a value characterizing the difference of the initial Twiss parameters and those at the end of the beam line. Implement this and test it with a beam line consisting of three copies of the doublet configurations from the previous exercise, such that the tune of the periodic system is $3 \times 0.4 = 1.2$. Inspired by Equation 3.69, devise a method to extract the tune with space charge from the simulations.

5. (a) Prepare a plot of $D(\xi)$ from Equation 12.6. (b) Find the beam parameters of your favourite light source and estimate its parameter $\xi$ and Touschek lifetime $\tau_T$.

6. (a) Show that the following description of the wake function $G(t) = \int_{-\infty}^{t} \psi(s) W(t - s) ds$ is equivalent to the one given in Section 12.3. For the wake potential $W(t)$ from Equation 12.8, calculate and display the (b) longitudinal wake function $G(t)$ and (c) the loss factor $k_{\parallel}$ for a bunch with the normalized box distribution $\psi(t) = 1/2a$ for $-a < t < a$.

7. Repeat the previous exercise for a bunch with a parabolic distribution $\psi(t) = 3(a^2 - t^2)/4a^3$ for $-a < t < a$ and $\psi(t) = 0$ otherwise.

8. Consider Equation 12.16 and determine the limit of stability for the parabolic distribution $\psi_0(\delta) = 3(a^2 - \delta^2)/4a^3$ for $-a < \delta < a$ and $\psi_0(\delta) = 0$ otherwise. Display the limit in the representation used in Figure 12.4.

9. Use the Keil-Schnell criterion to find a limit for the longitudinal impedance for the ring you considered in Exercise 5.

10. Use the simulation from Section 12.5 and plot the shift of the equilibrium arrival time `X(1)` as a function of $f_0$, or equivalently, the current dependence of the synchronous phase.

11. Use the simulation from Section 12.5 and start from slightly perturbed equilibrium beam parameters X, and sigma and record X(1) for 1024 turns. Fourier-transforming the values recovers the bare synchrotron tune in the zero-current limit $f_0 \rightarrow 0$. Find out whether the synchrotron tune depends on the beam current by varying $f_0$ and extracting the synchrotron tune by Fourier-transforming the turn-by-turn data.

12. Equation 12.25 is amenable to a numerical treatment by observing that it can be cast into the form

$$\ddot{x}_n + \omega_\beta^2 x_n = -\xi \sum_{\text{earlier } m} W_{n-m} x_m \, , \qquad (12.27)$$

where $W_{n-m}$ describes the wake, seen by bunch $n$, but left behind by all bunches $m$ over multiple turns, including bunch $n$ itself. $\xi$ is proportional to the beam current and the impedance of the wakes. (a) Based on the transverse resonator wake from Equation 12.12, determine $W_{n-m}$. You may need to "wrap around" the wake over multiple turns, if $Q_r$ is large. If we consider $N$ equidistant bunches, $W_N$ is the sum of all wakes a bunch caused on previous turns and "sees" itself. (b) Use the Ansatz $x_n = A_n e^{i\omega t}$ to obtain the following set of equations

$$\omega^2 A_1 = \omega_\beta^2 A_1 + \xi \left[ W_1 A_N + W_2 A_{N-1} + \ldots W_N A_1 \right]$$

$$\vdots \qquad\qquad\qquad\qquad (12.28)$$

$$\omega^2 A_N = \omega_\beta^2 A_N + \xi \left[ W_1 A_{N-1} + W_2 A_{N-2} + \ldots W_N A_N \right] \, ,$$

which have the form of an eigenvalue equation for the eigenvalue $\omega^2$. (c) Turn these equations into matrix form, determine the eigenvalues, and plot them as a function of the current $\xi$. (d) How do you recognize an instability?

# CHAPTER 13

# Accelerator Subsystems

This chapter discusses several subsystems that are essential for operating accelerators: the control system, particle sources, the vacuum system, and, if the accelerator uses superconducting components, the cryogenic system. The accelerated beams are highly ionizing and we therefore discuss radiation protection issues and round up with mentioning the conventional facilities, such as the electricity, cooling water, and air-conditioning, as well as civil engineering.

## 13.1 CONTROL SYSTEM

A computerized *control system* makes all diagnostic devices, all magnets, and all other actuators accessible, under a common user interface, to the human operators of the accelerator.

### 13.1.1 Sensors, actuators, and interfaces

Instrumentation and diagnostic devices can be classified as *sensors,* because they intercept signals from the beam, such as BPMs, perform signal conditioning of the analog signals, and make them available as voltages or currents. Frequently, analog-to-digital converters (ADC) are used to convert the signal to a computer-readable form. In particular, this signal chain is used to monitor beam currents, positions, and sizes. Other sensors measure temperatures, radiation levels, and flow rates of liquids or gases. Also digital signals, such as the state of limit switches, or other status information, for example, fault indication, are made available to computers. Once the signals are available in digital form, they are prepared by a micro-controller, programmable-logic controller (PLC), or other front-end computer, and, once converted to a standard format, passed on to other computers. Most test and measurement equipment, such as fast digitizers to record transient wave forms, oscilloscopes, spectrum and network analyzers support computer interfaces, historically GPIB, and, nowadays, its modern Ethernet-based variants LXI or VXI. The communication to these devices is usually based on a standardized language, called *Standard Commands for Programmable Instruments,* or SCPI.

The second class of devices are *actuators.* They comprise all power supplies for magnets and for motors, but also switches to turn on or off some device or functionality. The latter are easily interfaced by simply toggling an output pin on a micro-controller and connecting the pin to suitable circuitry to adapt voltage and power levels. Most modern power supplies are already equipped with standardized interfaces, often RS-232, USB, or Ethernet, that allow interfacing to the control-system computers. Stepper motors, used for precision position control of, for example, wire-scanners, require special controllers, most of which have

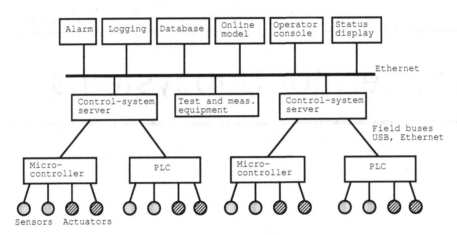

Figure 13.1   A typical control-system architecture.

standard interfaces or require only few digital control lines. Often ready-made modules for PLC systems are available to control them. Likewise, many synchronous motors, used, for example, to adjust the gap height of undulators, come with their dedicated controllers. These controllers sense the current in the motor windings and other position sensors to control the position and speed in a closed-loop feedback. Many other devices with an electronic interface are straightforward to interface with a dedicated micro-controller, but often industrial solutions, based on PLC systems, are preferred.

A particular device, sensor or actuator, needs to communicate with some other computer systems, such as micro-controllers or PLCs, and there are a number of *interfaces*, called *buses*, available. They are based on physical transmission media and on protocols of how to encode the information. Devices on the same circuit board as the micro-controller often communicate via serial bus systems, called I2C or SPI. The physical media are wires that carry electrical signals of a few volts. Data are transmitted with a synchronous serial protocol and follow a convention of how to encode the information in a transmission. The micro-controller then translates this information into a format that can be transported over larger distances using a *field-bus*. Examples are RS-232, RS485, CAN, Modbus, ProfiBus, or Ethernet. Similar to the other buses, they are based on various choices of the physical media and protocols of how to encode the information, sent from a computer to the next higher level. And that brings us to the overall architecture of control systems.

## 13.1.2  System architecture

A typical control-system architecture is shown in Figure 13.1. At the bottom we see the sensors and actuators that are connected to micro-controller or PLC systems, as discussed in the previous subsection. Power supplies or motor-controllers with built-in bus interface belong to this category as well. The controllers, in turn, communicate with control-system servers via field-buses, Ethernet or USB, as mentioned before. The servers are directly connected to a control-system network, which is usually Ethernet-based, from which the high-level clients obtain their information. In the decades before the change of the millennium, the low-level controllers were often VME computers and the servers used workstations such as VAX or Unix computers. Modern commercial systems are based on PLC systems and the WinCC software. In accelerator laboratories, a number of systems are in use, one of which is EPICS. The latter is centered around a server which is called the *input-output controller* (IOC). It provides many interfaces to the lower layers with micro-controllers and PLCs, while exposing a unified interface to the higher levels of the control system.

One of these upper-level clients is an *alarm manager*, which monitors status information of critical parameters and raises an alarm if a fault is detected, either acoustically or visually on an alarm console. Some alarms, such as those requiring rapid attention, are usually handled automatically as part of the alarm system. The second important functionality is *logging* of system parameters. This functionality is important to analyze long-term drifts of parameters that affect the performance of the accelerator, or to perform a post-mortem analysis after a fault. The set-parameters of magnets and other actuators, needed to configure the accelerator, are stored in a *database.* It is used to initialize all magnets and motors when starting up the accelerator. After tuning the performance, new configurations are saved in this database. Often, *online modeling* software for beam optics and beam physics calculations is part of the control system. It is important for the understanding of the performance of the running accelerator and to find configurations with new features. The modeling software obtains its input data either from the database, or from the hardware by directly querying the control-system servers for current values. *Operator consoles* usually run a graphical user interface with a visual representation of the accelerator supporting point-and-click access to all control-system parameters. In order to obtain a rapid overview of the status of the accelerator, *status displays* are shown in the control room and in other locations throughout the laboratory.

The layout shown in Figure 13.1 provides the basic functionality of the control system, but often additional interfaces to the control system servers from a number of programming languages, including Python and MATLAB, are available. This functionality is essential for rapid development of additional features to extend the functionality of the accelerator. Other extensions are possible through gateways to non-native communication channels, such as MQTT, a protocol commonly used to interface distributed sensors.

### 13.1.3 Timing system

A number of devices, such as pulsed magnets, must receive trigger information at very precise moments in time, synchronized with many other devices that are equally time-critical. The orchestration of this synchronized operation is handled by the *timing system.* A centrally located timing controller dispatches messages to timing-receivers that are usually placed close to, or as part of, the power supply for the time-critical device, for example, a kicker magnet. Part of the receiver is a very accurate clock that is periodically synchronized very accurately to the central timing controller. Each receiver contains tables of time delays with respect to a repetitive time-stamp, at which devices are triggered. For a pulsed accelerator this time stamp would signal the start of the pulse and each timing receiver then triggers its device with respect to it.

### 13.1.4 An example: EPICS

In the spirit of the hands-on approach, we briefly discuss the lower levels of the EPICS control system by making the analog and digital pins of an Arduino UNO micro-controller available to an IOC running on a Raspberry Pi, which, in turn, exposes an interface for the Arduino to the higher layers of the control system. A detailed discussion of such a system can be found in [104]. Here we only outline the basic functionality, but provide all code on the web site for this book.

We assume that the Arduino is connected via its USB interface to the Raspberry Pi that runs the EPICS software, see [104] for instructions of how to set this up. The Arduino is programmed to respond to a very simple protocol. Turning on or off digital output pin D05, is done by sending the command D05 1 or D05 0 to the Arduino via the serial line over the USB connection. Here we use pin 5 as an example. Reading digital input pin 4 is done by sending DI4? at which the Arduino responds with DI4 1 or DI4 0, depending on whether the voltage level on the input pin is high or low, respectively. In the same fashion, A0? requests the analog voltage on analog pin 0, and the Arduino responds with A0 1.54, where 1.54 is the voltage level on the pin in this example.

On the EPICS IOC, running on the Raspberry Pi, the communication exchange described in the previous paragraph is defined in a protocol file arduino.proto that is partially reproduced here:

```
set_bit {out "DO\$1 %u"; ExtraInput = Ignore;}
get_bit {out "DI\$1?"; in "DI\$1 %u"; ExtraInput = Ignore;}
get_analog {out "A\$1?"; in  "A\$1 %f";  ExtraInput = Ignore;}
```

Here $1 is a wildcard symbol that can assume any value. The protocol file defines the communication from the IOC to the lower layers, the Arduino. If several Arduinos are connected, we can use the same protocol file, provided the Arduinos are programmed with the same firmware. The interface towards the higher levels is defined by connecting the protocol file to a so-called database file arduino.db, which is partially reproduced here:

```
record(ai, "$(USER):A0") {
  field(SCAN, "1 second")
  field(DTYP, "stream")
  field(INP, "@arduino.proto get_analog(0) $(PORT)")
}
```

Here $(USER) and $(PORT) are placeholders for the high-level control-system name of the device that we discuss further below. The database record is of type ai or analog input, and the Arduino is of type stream and read one per second by using the input INP function get_analog(0) that is specified in the file arduino.proto. The Arduino is connected to the port specified in the variable $(PORT). The IOC process is configured and started from a command file, called st.cmd, which contains the following lines

```
drvAsynSerialPortConfigure("SERIALPORT","/dev/ttyACM0",0,0,0)
asynSetOption("SERIALPORT",-1,"baud","9600")
              :
dbLoadRecords("db/arduino.db","PORT='SERIALPORT',USER='ARDUINO'")
```

The first command defines the name SERIALPORT and connects it to the USB device /dev/ttyACM0 to which the Arduino is connected, before defining the communication parameters, such as the baud-rate in the subsequent lines. Finally, the connection between the USB port and the protocol file with the available high-level commands is established in the call to the dbLoadRecords() function, which also defines the $(PORT) and $(USER) variables referred to earlier. Once the IOC process is started, the pins of the Arduino are accessible from the control system, for example from the command line of the Raspberry Pi, by executing caget ARDUINO:A2. This call then returns the voltage on pin 2. Setting pins is done with a call to caput, again, see [104] for a more detailed discussion. Adding further Arduinos that are connected to other USB ports only requires to repeat the lines in the command file. When building or buying a new device that must be connected to EPICS, it

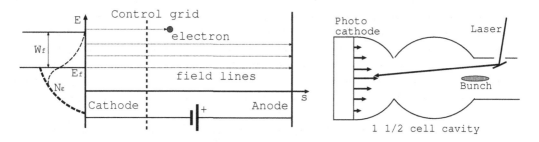

Figure 13.2 A thermionic electron gun (left) with the density of states $N_\varepsilon$ shown to the left of the cathode. At higher temperatures a few electrons in the cathode reach energies above the Fermi level $E_f$ and even exceeding the binding energy $W_f$ such that they escape the cathode and are accelerated towards the anode. In an RF photo-cathode (right) a laser provides the energy to overcome the work function and the RF-fields accelerates the escaping electrons.

often suffices to prepare protocol and database files, and hand them over to the person in charge of the IOCs.

Controlling all devices is a necessary prerequisite to operate accelerators, but we also need beams, and their generation in the different particle sources is the topic of the next section.

## 13.2 PARTICLE SOURCES

In this section we briefly discuss the basic physical processes that govern the creation and initial acceleration of electrons, protons and ions. First we consider electrons.

### 13.2.1 Electrons

Probably the simplest way to obtain electrons is to pass an electric current through a filament to heat it and extract the electrons with a positive voltage towards the anode. This configuration is often referred to as a *diode gun*. Adding a control grid, which can be used to modulate the current, turns it into a so-called *triode gun*. The geometry is illustrated on the left-hand side in Figure 13.2. At low temperatures, electrons fill all states in a conductor up to the Fermi-level $E_f$ [18] with a density of state proportional to $N_\varepsilon \propto \sqrt{E}$, which is indicated by the thick dashed line in Figure 13.2. At higher temperatures, electrons acquire higher energies, as indicated by the thin dashed line. At very high temperatures they even exceed their binding energy, the work function $W_f$. Thus they can escape the conductor and are then attracted by the positive potential of the anode. The cathode material must withstand very high temperatures and, preferably, should have a low work function $W_f$. Values for commonly used metals, such as for tungsten (W) is $W_f \approx 4.5\,\text{eV}$ and for lanthanum-hexaboroid (LaB$_6$) it is $W_f \approx 2.5\,\text{eV}$. The number of electrons within the cathode that have energies exceeding $W_f$ can escape and give rise to an extracted current density $j_c$, given by the *Richardson-Dushman* equation $j_c = AT^2 e^{-W_f/kT}$ with $A = 4\pi e m_e k^2/h^3 = 120\,\text{A/cm}^2\text{K}^2$, where $k$ is Boltzmann's constant and $h$ is Planck's constant.

Once the electrons have left the solid and are in the region between cathode and anode, they are accelerated towards the anode. If the extracted number of electrons is large, their negative charge shields the electric field and the effective voltage "seen" by the extracted electrons is reduced, which results in a *space-charge limited electron gun*. In a one-dimensional model, we describe the current density $j_c = en(s)v(s)$ through the electron density $n(s)$ and their velocity $v(s) = \sqrt{2eU(s)/m_e}$ and the electric potential $U(s)$. Solving for $n(s)$, we find $n(s) = j_c/e\sqrt{2eU(s)/m_e}$. Moreover, the potential $U(s)$ must obey Poisson's equation $d^2U/ds^2 = en(s)/\varepsilon_0$ in the region between cathode and anode. After inserting $n(s)$ we obtain $U^{1/2}d^2U/ds^2 = j_c/\varepsilon_0\sqrt{2e/m_e}$, which is solved by $U(s) = U_a(s/d)^{4/3}$ with the anode voltage $U_a$ and the distance between cathode and anode $d$. Solving for $j_c$ we obtain the *Child-Langmuir* law $j_c = (4\varepsilon_0 d^2/9)\sqrt{2e/m_e}U_a^{3/2}$. We thus find that in space-charge limited guns the current density is proportional to the anode voltage $U_a$, raised to the power of 3/2, or $U_a^{3/2}$. The constant of proportionality is called the *perveance* and depends on the geometry of the electron gun. The value we derived is only valid for the planar geometry in our one-dimensional model.

In practice the anode voltage $U_a$ is applied constantly and the electrons are therefore extracted continuously. If we later want to accelerate the beams with a radio-frequency system, we need to bunch the continuous stream of electrons on the time scale given by the frequency of the system. For this, one often uses one or several low-frequency buncher cavities to modulate the speed of the electrons while they are not yet relativistic and rely on the faster electrons to catch up with the slower ones, which is called *ballistic bunching*.

One method to create very short electron bunches already at the cathode uses lasers and employs the photo-effect to knock out electrons from the cathode. The duration, or bunch length, of the electrons then follows that of the laser pulse. The cathode is often made of cesium-based alloys and requires photon energies larger than the work function $W_F$. The latter is typically on the order of a few eV and thus requires photons in the ultra-violet range. In order to obtain bunch charges of nC, very intense UV laser pulses are needed, also because the so-called *quantum efficiency*—the number of electrons per photon—is only a few percent. A common way to produce these laser pulses is to triple the frequency of high-intensity infra-red laser pulses from Titanium-Sapphire lasers with pulse energies in the mJ range. The frequency-tripling is based on non-linear optical crystals that, simply speaking, combine three infra-red photons to form a single ultra-violet photon. The conversion efficiency is limited and yield ultra-violet pulses with energies about a third of the infrared pulse energies. Typical pulse lengths are on the order of several ps.

The electron bunches created in such a gun are often very dense and the high charge density makes them susceptible to the transverse space-charge forces, discussed in Section 12.1, and that will increase the emittance. This can, however, be avoided, provided that the electrons are rapidly accelerated to relativistic energies, where, according to Equation 12.2, the space-charge defocusing forces are suppressed by $1/\gamma_0^3$. To achieve this acceleration to relativistic energies, the cathode is embedded in an accelerating radio-frequency cavity and the timing of the laser is adjusted, such that the laser pulse impinges on the cathode when the accelerating field is maximum. Since the accelerating gradients are on the order of tens of MV/m, the electrons reach relativistic energies of 10 MeV already within the first 10 to 20 cm. A challenge for the design are the thermal loads on the accelerating structure resulting from the high fields and may require to operate the gun in a pulsed mode.

The positrons needed for colliders are typically created by impinging a high-energy electron beam onto a tungsten target, which causes a bremsstrahlung and $e^+e^-$ pair-production cascade, already mentioned in Section 9.2. A magnetic collector, based on a tapered solenoid with maximum field close to the target, concentrates the flux of positrons towards a smaller transverse cross section. Accelerating them will reduce the beam size and emittances further

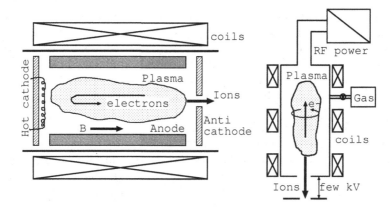

Figure 13.3  In a Penning source (left) electrons are trapped longitudinally by cathode and anti-cathode and radially by magnetic fields. At the same time, the potential difference between cathodes and anode accelerates them to energies that allow them to ionize gas. In an ECR source (right), the electrons are accelerated to higher energies by RF-fields at the cyclotron frequency.

by adiabatic damping. After injecting the positrons into a storage ring, radiation damping, discussed in Section 10.1 will reduce the beam sizes further.

### 13.2.2  Protons and other ions

Most ion sources are based on trapping electrons in a variety of magnetic and electric fields, such that they have a high probability of colliding with gas ions, thereby ionizing them. In a *hot cathode ion source* the electrons are created in a hot filament and accelerated towards the anode, without ever reaching it. The density of the gas has to be in an intermediate range of densities, as given by the *Paschen curve;* the density must be high enough to allow sufficiently many gas atoms to be ionized and not too high, to allow the electrons to gain sufficient energy, before impacting with the next gas atom or molecule. Placing this ion source on a high-voltage platform allows us to extract the ions with a well-defined energy. The ionized gas forms a plasma that is usually contained in magnetic fields. Both solenoidal and so-called multi-cusp configurations are used. The latter use permanent magnets, arranged with alternating polarity around the perimeter of the plasma chamber to create magnetic fields that deflect the electrons back into the chamber and prevents them from touching the walls. Many geometries, optimized for particular ion species are used to provide the beams for experiments with ions, but also for industrial accelerators, such as ion implanters used for doping materials in the semiconductor industry.

A variant of a hot cathode ion source is the *duoplasmatron,* where a moderately intense electron beam is guided by an intermediate electrode towards the anode. The electrons, as well as gas leaked into the chamber, pass through the same narrow orifice, such that the electrons have an increased probability of ionizing the gas and creating a plasma from which the ions are extracted and accelerated further. Duoplasmatrons are used to create protons, but also ions of other gases, such as helium or oxygen.

A *Penning source,* shown on the left-hand side in Figure 13.3, can operate either with a heated filament or as a *cold cathode ion source.* In either case, electrons are trapped by a solenoidal magnetic field between a negatively charged cathode and an anti-cathode. This configuration prevents the accelerated electrons from actually reaching the anode, and, instead, bounce between the cathodes and ionize gas that is leaked into the chamber. Positively charged protons are accelerated towards the cathode and can be extracted from the source to be accelerated further. Apart from producing protons, can Penning sources also be used to create moderately charged ions of other gases. For additional discussions of ion sources, see the comprehensive overview [105] and a recent review [106].

The anti-protons used in $p\bar{p}$–colliders, or in dedicated experiments with anti-protons, are created by impinging protons with multi-GeV energies onto a target and extracting them from the reaction products of the nuclear reactions. They are first focused by a magnetic horn and subsequently passed to a storage ring in which their momentum spread and their emittances are reduced by either stochastic cooling or by electron cooling.

### 13.2.3  Highly charged ions

In order to obtain highly charged ions, the electrons that need to ionize the low-lying orbitals of the gas must have sufficiently high energies. In an *electron cyclotron resonance*, or ECR ion source, the energy for the electrons, trapped in a longitudinal solenoidal $B_s$ and transverse sextupolar fields, is provided by an external radio-frequency source, tuned to cyclotron frequency $\omega_c = eB_s/m_e$. This causes the electrons to move very rapidly and ionize the gas, which causes even more electrons to be present that can ionize more gas. Whereas the electrons are confined by the magnetic fields, the ions can escape the plasma volume by diffusion. They are sub-sequentially accelerated by a moderate voltage of about 20 kV and then guided to experiments or accelerated further, for example, in a cyclotron.

Ions with very high positive charge states, and even fully stripped states, are available from *electron beam ion sources*, or EBISs. Here a highly intense electron beam, with currents up to ampere and energies up to 100 keV, serves the dual purpose of first ionizing the gas atoms and then transversely trapping the positive ions in its negative electrostatic potential well. Longitudinally, the ions are initially trapped by electrodes on positive potential and, once a sufficient number of highly ionized ions is available, the electrode potential is lowered to allow the ions to escape. From the escaping ions, which have a mixture of charge states, a particular one can be selected by passing the ions through a *Wien filter,* consisting of crossed electric and magnetic fields that deflect most charge states away from a narrow aperture. Only for a particular combination of velocity and charge state the forces cancel, and thus permit the ions to pass the aperture.

In the ion sources discussed so far, only gases were ionized. Solid materials can, however, also be converted to ions in a sputtering ion source, where an ion beam with energies of a few keV impinges on a solid surface and knocks out ions from the surface by a process called *sputtering*. Positively charged ions can be extracted and further accelerated.

### 13.2.4  Negatively charged ions

A *source of negative ions by cesium sputtering* (SNICS) is based on first creating neutral cesium atoms in gaseous form that are subsequently positively ionized on a hot cathode at lower negative potential than a second cold cathode. The cesium ions, impinging on the cold cathode, sputter positively charged beam ions from the cathode material. The beam ions subsequently have to pass through a cloud of the neutral gaseous cesium, where they pick up electrons from the cesium, which is a generous electron donor. This process leaves positive cesium ions behind. They sputter further beam ions. The beam ions become negatively charged and can be accelerated further.

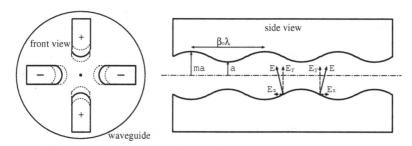

**Figure 13.4** Left: the view onto the front of a radio-frequency quadrupole with the dotted lines indicating the extreme pole-tip radii due to their longitudinal modulation, shown on the right for facing electrodes. Note how the modulation is responsible for a non-zero longitudinal component $E_z$ of the electric field, which bunches and accelerates the beam. The modulation of adjacent poles (not shown) is shifted by $90^o$.

If the primary beam ions are already available in a positively ionized charge state, passing them through a *charge exchange cell*, filled with neutral cesium or potassium gas transfers electrons from the gas to the beam ions, thus creating negatively charged beam ions, that can be used in tandem Van-de-Graaff accelerators or used in charge-exchange injection schemes, covered in Section 13.3.

### 13.2.5  Radio-frequency quadrupole

In order to mitigate the effect of space charge, discussed in Section 12.1, the beams created in the ion sources must be accelerated as quickly as possible. Simultaneously, they need to be strongly focused transversely. Finally, they must be longitudinally bunched in order to efficiently accelerate them further in later stages of the accelerator. In the late 1960s Kapchinskii and Tepliakov first discussed the *radio-frequency quadrupoles* (RFQ) to satisfy these requirements. RFQs are based on exciting a TE$_{210}$ mode in a circular waveguide, which is loaded by quadrupolar electrodes, as shown on the left-hand side in Figure 13.4. The electric field vector of this mode has an azimuthal $\cos 2\phi$–dependence and a linear radial dependence, such that it is zero in the center, which is the behavior one expects for a quadrupole. Moreover, this mode excites facing electrodes to the same and adjacent electrodes to opposite polarities. Every half period of the RF the polarity of the quadrupolar force reverses, such that the ions experience an alternating sequence of focusing and defocusing fields, resembling that of FODO cells. High RF power thus provides strong transverse focusing. By longitudinally modulating the transverse distance between the electrodes, the electric field $E$ acquires a longitudinal component $E_z$, as shown on the right-hand side in Figure 13.4. The period of the modulation is chosen to equal $\beta_0\lambda$, where $\lambda$ is the wavelength of the RF, such that the direction of $E_z$ reverses after a distance $\beta_0\lambda/2$. The magnitude of $E_z$ is adjusted by varying the modulation $m$ of the pole-tip radius, which varies between $a$ and $ma$. In early parts of the RFQ, the modulation is small and the RFQ mostly focuses transversely while the longitudinal field only slightly modulates the energy of the ions, such that they will slowly bunch. In later stages the modulation depth is increased, which also increases the accelerating field $E_z$. With increasing energy of the ions, also their speed $\beta_0$ increases and the period $\beta_0\lambda$ must be adjusted accordingly.

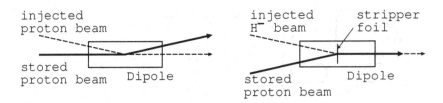

Figure 13.5  Using a static dipole field to deflect an incoming beam onto the closed orbit does not work, because on the next turn the field would deflect the returning beam into the wall (left). If on the other hand the polarity of the incoming beam is changed, for example by reversing the polarity of negatively charged ions in a stripper foil, the returning beam is deflected in the opposite direction and stays on its closed orbit (right).

RFQs are used as the first acceleration stage in many proton and ion accelerators. They are specifically designed for one ion species, but then they can capture around 90 % of the ions in bunches. The beam energies at the exit from the RFQ are in the range of a few MeV.

## 13.3  INJECTION AND EXTRACTION

Injecting a beam from a transfer line into a storage ring and later extracting it from the ring to direct it to an external experiment or yet another ring for further acceleration is a common exercise in many accelerator complexes. Unfortunately, it is not possible to inject with static magnetic fields only. See, for example, the left-hand graphics in Figure 13.5, which shows the beam coming from a transfer line as a thin dashed line that is deflected towards its left by the dipole magnet onto the closed orbit in the ring. After one turn, the stored beam returns to the injection position, and the same dipole used for injection, again deflects the beam towards its left and thereby guides it into the beam pipe. In order to overcome this problem we can either change the type of particle from, for example $H^-$ to a proton, use pulsed magnets, or inject off-orbit and rely on dissipative forces to move the beam onto its closed orbit.

Let us first consider *charge-exchange injection* based on passing an $H^-$ beam from the ion source through a very thin carbon stripper-foil, embedded in a dipole magnet, as shown on the right-hand graphics in Figure 13.5. The foil removes both electrons from the $H^-$ ions with very high efficiency and turns them into protons. Since the dipole magnet deflects the protons and $H^-$ in opposite directions, the protons can continue to circulate in the ring. In order to avoid heating the foil by excessive traversal of the protons, the closed orbit in the ring is only temporarily placed onto the foil during the injection and retracted once the injection has ended. The stored beam then circulates unimpeded by the foil, which prevents the beam from continuously losing energy in the foil. Apart from placing the injected beam onto the closed orbit, its phase space ellipse must match that of the stored beam. In short, the beta functions of the injected and stored beam must match in order to prevent the emittance from increasing due to filamentation, as discussed in Section 8.2.4.

Another option to inject into a ring is based on using very fast pulsed magnets, so-called *kicker magnets,* with the ability to increase and decrease their maximum field in times much shorter than the revolution time in the ring. In this way, the magnetic field that deflects the

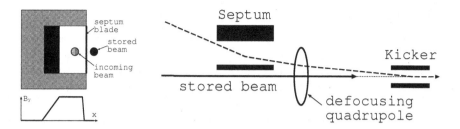

**Figure 13.6** A septum magnet (left) provides a strong magnetic field on one side of the "septum blade," while maintaining a field close to zero on the other side, where the stored beam circulates. It is used in conjunction with pulsed kicker magnets (right) to deflect the incoming beam onto the closed orbit.

injected beams onto the closed orbit is gone, once the stored beam returns after one turn. This method is often referred to as *single-turn injection.* Since the kickers must have very rapid rise and fall times of the fields, their amplitude is usually small and *septum magnets* with static magnetic fields are used to provide most of the required deflection angle and make the injected beam almost parallel to the stored beam, often aided by a defocusing quadrupole of the ring lattice. Septum magnets, a sketch is shown on the left-hand side in Figure 13.6, provide a large magnetic field in the region where the injected beam passes, but close to zero field for the stored beam, because a so-called septum blade carries a large current that shields the region with the stored beam. Matching the transverse beta functions of injected and stored beams is also necessary in this mode of operation.

If the injected beam has a much smaller emittance than the desired emittance of the stored beam, instead of a single and very fast kicker magnet, one can use a moderately fast closed bump, as discussed in Section 8.4.1. The injected beam then fills different parts of the transverse phase space of the stored beam on consecutive turns and therefore the final emittance of the stored beam, in this *multi-turn injection* mode, is larger compared to the single-turn mode from the previous paragraph. Nevertheless, multi-turn injection is often used if dissipative forces, for example due to the emission of synchrotron radiation, damp particles with large betatron amplitudes back towards their equilibrium positions around the closed orbit, as discussed in Section 10.1.

*Extracting the stored beam* from a ring is normally based on operating the single-turn injection mode in the reverse order; a fast kicker magnet, aided by a defocusing quadrupole, deflects the beam into a septum magnet that increases the separation of stored and extracted beam further until normal-sized dipole and quadrupole magnets can be used to guide the beam to its new destination, either an experiment or the next accelerator. This mode extracts the entire beam during a very short time.

If, on the other hand, an experiment, for example, to slowly irradiate some sample, requires a *slowly extracted beam,* it is possible to operate the storage ring close to a third-order resonance, such that the transverse phase-space exhibits the triangular shape shown in Figure 11.1. Operating even closer to the third-order resonance makes the escaping tails more pronounced. Properly adjusting the sextupoles that drive this resonance then causes one of the tails to pass the blade of a septum such that the extracted beam particles can subsequently be guided to the experiment.

## 13.4 BEAM COOLING

In Chapter 10, we found that the emission of synchrotron radiation provides a friction force that damps oscillations of the beam and thus "cools" the electrons. This process, does not, however, work for protons or other ions. Therefore, other methods of cooling were invented to improve the beam quality, for example, to reduce the momentum spread in ion storage rings used to study narrow nuclear resonances. Another application, stochastic cooling, is the reduction of the emittance of anti-proton beams. This reduces the transverse beam sizes and makes reasonable luminosities in a collider feasible.

*Stochastic cooling* [107] of the transverse motion is based on detecting the position of an ensemble of oscillating particles in an un-bunched beam with a transverse Schottky monitor. After amplifying and filtering this signal, it is applied to a kicker, placed at a location 90 degrees different in betatron phase, which gives a kick to the same ensemble in order to reduce the oscillations. Since the particles within such an ensemble should stay close together, the kick must be applied as soon as possible. Therefore, the correction signal takes a short-cut across a ring to the diametrically placed kicker. Since the ensembles are longitudinal slices of an un-bunched beam, both the position monitor and the kicker must be sufficiently fast—having a large bandwidth—in order to distinguish the different slices. The cooling times are rather large, starting from seconds for low-intensity beams but can also exceed several minutes for higher intensities. The discovery and implementation of stochastic cooling made the accumulation of anti-protons possible, which subsequently led to the discovery of the W–bosons; and, soon after, both discoveries were jointly honored with a Nobel Prize. Initially developed for un-bunched beams, today it is also used for bunched beams.

A second method to improve the beam quality in ion storage rings is *electron cooling* [108]. It is based on superimposing a high-intensity electron beam with the ion beam, which transfers kinetic energy from the random motion of the protons to the electrons, provided the electron velocity spread is much smaller than that of the ions. Calculating the friction force closely resembles the derivation of the energy loss, described by the Bethe-Bloch equation in Section 9.2, but with a different Coulomb-log. In the derivation of the Bethe-Bloch equation, the electrons are at rest, whereas in a cooler they move with a velocity, given by the acceleration voltage of the cooler. If the electrons are slower than the ions, they reduce the velocity of the ions, if they are faster, the ions appear to come from the opposite direction and are again forced to match the velocity of the electrons. The range of relative velocities over which cooling is efficient, is rather narrow, and is determined by the velocity spread of the electrons. This spread is a consequence of the finite temperature of the cathode in which the electrons are generated. The typical magnitude of energy that an ion exchanges with the electrons is on the order of eV for protons and increases quadratically with the charge state of the ions. In small rings typical cooling times are on the order of ms to seconds.

*Ionization cooling* is based on alternating ionizing material with accelerating cavities in a moderately long, say tens of meters long, section. Beam particles with large transverse momenta $p_x$ or angles $x' \approx p_x/p_z$ lose momentum in their forward direction, which reduces both $p_x$ and $p_z$, but the cavity only restores the longitudinal momentum $p_z$, such that the particle's $x'$ is reduced. The mechanism is similar to damping by synchrotron radiation, discussed in Section 10.1.2. Ionization cooling is the only known method to cool muon beams.

## 13.5 VACUUM

The charged particles in accelerators always travel in an evacuated beam pipe in order to avoid collision with gas atoms, because the residual gas behaves just like a spread-out target. In Chapter 9 we found that the beam loses energy by *ionizing* the target material. This increases the beam's momentum spread and creates positive ions that may accumulate in an electron beam and cause a much higher ion density than the original gas density of the neutrals. The second effect is transverse *Rutherford scattering* of the beam particles on gas atoms or ions. If the minimum distance between beam particle and gas atom, the impact parameter, is small, this leads to large deflection angles that may lead to an immediate loss of the beam particle. Much more often the impact parameter is large and the deflection angle is small and randomly distributed. This leads to an increase of the oscillation amplitude of the individual particle and increases the emittances of the beam.

In order to minimize these detrimental consequences of gas atoms, a large number of pumps are connected to the beam pipe to provide high or ultra-high vacuum conditions. In the following subsections we will discuss the basics of gas dynamics, pumps, gauges, and a method to simulate vacuum systems.

### 13.5.1 Vacuum basics

We assume that the atoms or molecules, having mass $M$, that are left in the beam pipe, are in thermal equilibrium with the environment at temperature $T$. Since each degree of freedom in equilibrium on average carries energy $k_B T/2$, with the Boltzmann constant $k_B$, the rms velocity of the gas molecules in three dimensions is given by $M v_{rms}^2/2 = 3 k_B T/2$, or $v_{rms}^2 = 3 k_B T/M$. The gas particles follow a Maxwell velocity distribution and conventionally the mean velocity $\bar{v}$, rather than the rms velocity is used. They are related by $\bar{v} = (8/3\pi) v_{rms}$. For $H_2$ molecules at $T = 300\,\mathrm{K}$ the mean velocity is $\bar{v} = 1850\,\mathrm{m/s}$ and for $N_2$ molecules we find $\bar{v} = 475\,\mathrm{m/s}$, which is a value we use for estimating vacuum-related quantities.

The *mean free path* $\bar{l}$ between collisions among gas particles is a characteristic number for the density $N/V$ of the gas, where $N$ is the number of gas molecules in a volume $V$. During the traversal of the distance $\bar{l}$ a gas molecule with diameter $d$ sweeps through a volume $\delta V = \pi d^2 \bar{l}$, and, on average, takes part in one collision. Thus, we expect $N/V = 1/\delta V = 1/\pi d^2 \bar{l}$, and, after solving for $\bar{l}$ we find for the mean free path $\bar{l} = 1/\pi d^2 (N/V)$. For rarefied gases we can use the *equation of state* $PV = N k_B T$ of an ideal gas to relate the density $N/V$ to the pressure $P$. Diatomic nitrogen molecules $N_2$ have a diameter of $d \approx 3.7 \times 10^{-10}\,\mathrm{m}$ and the mean free path at atmospheric pressure becomes $\bar{l} \approx 60\,\mathrm{nm}$, while, at a pressure of $10^{-6}\,\mathrm{mbar}$, it is $\bar{l} \approx 60\,\mathrm{m}$. In accelerators the pressure is almost always in the latter range, or even lower, such that the mean free path $\bar{l}$ of the gas molecules is larger than the size of the vacuum enclosure, such as the beam pipe. The dynamics is therefore entirely determined by collisions with the walls of the container and its geometry. Note that using the equation of an ideal gas is vindicated retroactively for this regime of operation.

In the framework of the international system of units the fundamental unit of pressure is the Pascal, defined by $1\,\mathrm{Pa} = 1\,\mathrm{N/m^2}$. Historically, also bar and torr were used as a unit of pressure, where one bar equals $10^5\,\mathrm{Pa}$ and $1\,\mathrm{torr} = 133\,\mathrm{Pa}$. A third commonly used unit is based on the ambient pressure at sea level, which is approximately 1 bar or 1000 mbar, which equals $10^5\,\mathrm{Pa}$. The mbar is commonly used when discussing technical vacuum systems, such as those found in particle accelerators. Pressure ranges that occur in a technical context are *rough vacuum* in the range of 1000 to 1 mbar, *medium vacuum* in the range of 1 to $10^{-3}\,\mathrm{mbar}$, *high vacuum* in the range of $10^{-3}$ to $10^{-7}\,\mathrm{mbar}$, and *ultra-high vacuum* (UHV) in the range below $10^{-7}$ mbar, the pressure range almost always found in beam pipes of accelerators.

From the equation of state, $PV = Nk_BT$, we see that the product of pressure $P$ and volume $V$ is proportional to the number of particles in the volume $N$, provided the temperature $T$ is constant, which we always assume here. The rate of moving gas molecules around is thus characterized by the *gas flow* $Q = d(pV)/dt$ which is proportional to $dN/dt$. The gas flow through a pipe with different pressure levels on the left $P_L$ and the right-hand side $P_R$ is proportional to the pressure difference, such that we have $Q = C(p_R - p_L)$ with the proportionality constant $C$, called the *conductance* of the pipe, conventionally expressed in units of l/s. Microscopically, the surface roughness of the pipe will cause molecules to scatter with random angles. Therefore, their motion can be described as a diffusion process, with the conductance $C$ taking the role of the diffusion constant. For practical calculations with $N_2$ molecule at room temperature, the conductance $C_p$ of a pipe with radius $r$, given in cm, and length $L$, also given in cm, is approximately given [109] by $C_p[\text{l/s}] \approx 100\,r^3[\text{cm}]/L[\text{cm}]$, an equation often used for estimates. Note that the conductance is large for short pipes with large radius. Likewise, the conductance $C_A$ of an aperture with area $A$, given in cm$^2$, is given by [109] $C_A[\text{l/s}] \approx 11.7\,A[\text{cm}^2]$.

The gas molecules are originally adsorbed to the pipe walls, but are slowly released into the volume of the pipe. This is called *outgassing* and depends on the preparation history and storage of the pipe material, as well as the temperature or exposure to radiation, either from the beam or from synchrotron radiation. The desorbed molecules constitute a gas load $\Delta Q$ which quantifies the number of molecules released from the walls per unit time. It is given in the same units as the gas flow, in mbar l/s. Despite the strong dependence on the preparation and cleaning of the pipe, we quote measured outgassing rates [110] for aluminum and stainless steel to be able to estimate the performance of vacuum systems. There, the outgassing rate of aluminum after one hour of pumping is given by $55 \times 10^{-10}$ mbar l/s cm$^2$. After ten hours of pumping the rate is quoted to be reduced to one tenth. Baking the pipe—pumping, while heating it up to around 200 °C for a few hours—will reduce the outgassing by a factor $10^5$. The corresponding outgassing rates for stainless steel are reported to be worse, but have corresponding orders of magnitude.

The gas molecules are removed from the vacuum system by *pumps*, which act as traps for gas molecules. The latter diffuse towards the pump, but once captured in the pump, do not return to the gas volume. The number of gas molecules captured constitutes a gas flow $Q_p$ that is removed from the vacuum system and moved to a gas reservoir at higher pressure. This could be the entrance of another pump or ambient pressure. The ratio of the pressure at the outlet-port to that at the inlet-port is called the *compression ratio* of the pump. Furthermore, the number of removed molecules is proportional to the pressure $P$ at the inlet-port of the pump, such that we have $Q_p = -SP$, with the pump speed $S$ as proportionality constant. It is commonly given in units of l/s. The minus sign indicates that the gas molecules are removed from the system. In the next subsection we will have a closer look at the pumps used in accelerators.

## 13.5.2 Pumps and gauges

In order to reach rough-vacuum conditions above 1 mbar, *roots* or *screw pumps* are used. They move gas volumes mechanically and are limited by limited sealing capabilities; gas returns from the high- to the low-pressure side, which limits their compression ratio to below 100. These pumps are used as fore-pumps for other pumps that reach lower pressure levels and are sometimes referred to as *roughing pumps*.

In *oil diffusion pumps,* vaporized oil drags gas molecules from the inlet port towards the high-pressure side, where it is removed with, for example, root pumps. The operating range of these pumps covers $10^{-3}$ to $10^{-7}$ mbar. Their disadvantage is that the oil can contaminate the low-pressure side.

This contamination is avoided by entirely mechanical *turbo-molecular pumps.* They are based on fast-rotating inclined turbines that knock the gas molecules towards the high-pressure side. The speed of the blades must be comparable to the mean velocity of the gas molecules, despite operating with rotation speeds of up the $10^5$ revolutions per minute. These pumps operate in the range of $10^{-4}$ to $10^{-10}$ mbar. They work better for heavier gases. Conversely, the small mass of $H_2$–molecules and their therefore large mean velocities makes pumping of hydrogen difficult.

*Sputter-ion pumps* are based on Penning-traps with two titanium cathodes on either end of a cylindrical anode at high positive potential. The configuration resembles the one, shown on the left-hand side in Figure 13.3. Trapped electrons ionize the gas molecules that enter the pump volume and the positively charged ions subsequently impact on the cathodes, where they are either buried or sputter titanium. The titanium is deposited as a chemically highly reactive thin film on the anode, where it adsorbs gas molecules, which is one pumping mechanism. The other is directly burying the gas ions deep in the cathode. The pump speed of ion pumps depends on the type of gas, where CO, $N_2$, and $O_2$ are among the "getter-able" gases. The operating range of ion pumps starts around $10^{-4}$ mbar and extends to $10^{-11}$ mbar. The pump speed depends, however, on the pressure range.

*Non-evaporable getter* (NEG) pumps are based on sputtering a highly porous Zirconium-Aluminum alloy on the inner walls of a pipe which creates a labyrinth in which gas molecules are trapped and thus removed from the beam pipe. Since sputtering then happens at ambient pressure, NEG pumps need to be activated by heating the material while pumping the released gas. This process must be repeated once the NEG material is "full," either after long use or accidentally venting the vacuum system. In many accelerators with extremely small vacuum chambers, the entire inner beam pipe walls are covered with NEG material and provide pumping to reach low pressure levels. This would be impossible with localized pumps, because the gas molecules would only very slowly diffuse to the pumps through the small pipes with their limited conductance.

Cooling surfaces to very low temperatures reduces their probability to desorb gases and therefore traps gas molecules. This is the operating principle of *cryogenic pumps.* As a matter of fact, all super-conducting elements in an accelerator, either magnets or accelerating structures, act as pumps and adsorb gas molecules, which is especially undesirable in the latter, because it may create field emitting sites that may cause quenches. After this brief discussion of commonly used pumps in accelerators we turn to the gauges. They tell us about how well the pumps work by reporting the measured pressure level.

*Pirani gauges* operate in rough and medium vacuum, down to the $10^{-4}$ mbar range. They are based on measuring the resistance of a heated wire, which is cooled by transferring energy to the gas molecules. In equilibrium the resistance is related to the density of molecules, or, equivalently, the pressure.

*Ionization gauges* are the standard pressure measurement device at pressure levels in the high and ultra-high regime below $10^{-3}$ mbar and down to $10^{-12}$ mbar. Their principle of operation is based on ionizing the gas and measuring the current of the ionization products. On the one hand, the ionization is done with an external electron source, in which case the gauge is called a *hot-cathode gauge.* On the other hand, the gas is ionized in a Penning-trap with electric and magnetic fields in which a plasma burns and results in measurable currents that are proportional to the pressure. The latter are called *Penning* or *cold-cathode gauges.*

The composition of the gas in a vacuum system—or the partial pressures—is measured with a *residual gas analyzer*. One operating principle is based on accelerating ionized gas molecules or atoms and passing them through a long electro-static quadrupole that is excited by a sinusoidal voltage on top of a constant voltage. The motion of the gas ions with a specific mass is only stable for certain excitation frequencies and measuring the current exiting from the quadrupole as a function of the frequency contains information about the gas composition.

After having discussed the basic concepts and the hardware we now turn to calculating the pressure profile for a given set of pipes, pumps, and outgassing sources.

### 13.5.3 Vacuum calculations

In order to simulate a linear array of pipes, pumps and gas sources, we split the system into small longitudinal slices and consider how the number of gas particles in each slice changes with time. If the temperature is constant, the number of particles in each slice is related to the pressure and volume by the ideal-gas law $PV = Nk_BT$ and for each slice we find

$$v\frac{\partial P}{\partial t} = \frac{\partial}{\partial z}c\frac{\partial P}{\partial z} - sP + q , \qquad (13.1)$$

where we introduce the per-unit-length quantities for the conductance $c = CL$, the pump speed $s = S/L$, outgassing $q = \Delta Q/L$, and the volume $v = V/L$, which is also the cross-section of the pipe. The interpretation of the terms is straightforward. On the left-hand side $v\partial P/\partial t$ is the change of the number of particles in the slice, $q$ is the rate per unit length by which gas molecules are injected, and $sP$ is the number of particles removed by a pump. The first term on the right-hand side describes the diffusion into adjacent slices, because $c\,\partial P/\partial z$ is the gas-flow in the slice, as discussed in Section 13.5.1, and its derivative denotes the difference of the flow towards the right and towards the left, which is just the change of the number of particles due to diffusion. If we assume that the coefficients are piecewise constant, we can integrate this partial differential with standard methods [111] and obtain the temporal evolution of the system. We leave the general treatment to the specialized literature and consider a very simple system only.

That system consists of a pump with pump speed $S = sL$ connected to a volume $V = vL$. We assume that all gas is already in the volume $V$ and we can neglect further outgassing ($q = 0$). Furthermore, the volume is directly connected to the pump, such that we can ignore diffusion ($c = 0$). We thus find that the pressure $P$ is determined by $VdP/dt = SP$ or $P(t) = P_0e^{-t/\tau}$ with $\tau = V/S$ and the *pump-down time scale* is given by the ratio of the volume, given in liters, and the pump speed $S$, given in l/s.

Since the time scale $\tau$ is often very small, equilibrium conditions are reached within a very short time, such that the equilibrium pressure is the important quantity. It is given by Equation 13.1 where the temporal derivative on the left-hand side equals zero. Rewriting the equation for a pipe of finite length $L$ we find

$$0 = c\frac{d^2P}{dz^2} - sP + q . \qquad (13.2)$$

We assume piece-wise constant values for the conductance $c$, the pump speed $s$ and outgassing $q$. Then Equation 13.2 becomes an ordinary differential equation with piece-wise constant coefficients, which can be solved by exponential functions. From the solutions, we derive transfer matrices that relate the pressure $P_R$ and its derivative $dP_R/dz$ on the right-hand side of the pipe to the values $P_L$ and $dP_l/dz$ on the left-hand side. Here we use the gas flow $Q = CdP/dz$ instead of the derivative and, furthermore, add a third column to account for the inhomogeneous term, the outgassing rate $q$. A little algebra results in the general transfer matrix in the following equation

$$
\begin{pmatrix} P_L \\ Q_L \\ 1 \end{pmatrix} = \begin{pmatrix} \cosh(\sqrt{s/cL}) & -\frac{L}{c}\frac{\sinh(\sqrt{s/cL})}{\sqrt{s/cL}} & -\frac{qL^2}{c}\frac{\cosh(\sqrt{s/cL})-1}{(s/c)L^2} \\ -c\sqrt{s/c}\sinh(\sqrt{s/cL}) & \cosh(\sqrt{s/cL}) & qL\frac{\sinh(\sqrt{s/cL})}{\sqrt{s/cL}} \\ 0 & 0 & 1 \end{pmatrix} \begin{pmatrix} P_R \\ Q_R \\ 1 \end{pmatrix}.
$$

(13.3)

A vacuum system containing many elements is then described by the multiplication of the matrices for the respective elements. Note that for short elements, the conductance becomes very large, which allows us to simplify the matrices for such elements. Once we have calculated the transfer matrix for the vacuum system we find that we have two equations relating the four unknown quantities $P_R, Q_R, P_L$, and $Q_L$. In order to determine these quantities uniquely, we need to specify boundary conditions. The simplest, and most frequently used, is to flange-off the ends of the vacuum system, which implies $Q_R = Q_L = 0$. No gas flows into or out of the vacuum system at the ends. The computer program *vaktrak*, which implements this method, is described in [112]. In the software accompanying this book we provide a simplified version using MATLAB, which follows the same strategy as the beam optics code from Chapter 3. Figure 13.7 shows the pressure profile in a system with leaks and pumps connected by 10 m long pipes. The geometry and component values can be found in the MATLAB file. An important observation from Figure 13.7 is that the pressure varies considerably and is lowest near the pumps and that is where the gauges are frequently located. This may lead to an underestimate of the true pressure in the system.

A useful relation immediately follows from analyzing the matrix for pure conductance, which is given by the matrix in Equation 13.3 in the limit $q \to 0$ and $s \to 0$, which has the same form as the matrix of a "drift space" with a "length" of $-1/C$. Analogous to the way drift spaces are concatenated by adding the lengths, the conductance $C_t$ of two pipes with conductances $C_1$ and $C_2$ is given by adding the inverse conductances $1/C_t = 1/C_1 + 1/C_2$. A second useful relation follows from inspecting the matrix for a short pump and concatenating it with a conductance

$$
\begin{pmatrix} P_L \\ Q_L \end{pmatrix} = \begin{pmatrix} 1 & -1/C \\ 0 & 1 \end{pmatrix} \begin{pmatrix} 1 & 0 \\ -S & 1 \end{pmatrix} \begin{pmatrix} P_R \\ 0 \end{pmatrix} = \begin{pmatrix} (1+S/C)P_R \\ -SP_r \end{pmatrix},
$$

(13.4)

where we used the reduced $2 \times 2$ transfer matrices, and assume that the right-hand side is flanged off ($Q_R = 0$). If we furthermore assume that the pressure on the left-hand side $P_L$ is known we can solve for the effective pump speed $1/S_{eff,L} = -P_L/Q_L = 1/S + 1/C$. This implies that connecting a pump through a pipe with conductance $C$ to the point where we need to pump, reduces the effective pump speed, and if $C \ll S$ the effective pump speed will be determined by the conductance of the pipe $C$ and not by the pump speed $S$.

In passing, we point out that the equations that govern vacuum systems closely resemble those that describe electric circuits, where the pressure $P$ takes the role of the voltage and the gas flow $Q$ that of the electric current. The analogy is apparent, when comparing Equation 13.4 with Equation 6.15, and can be used to employ electric circuit codes to simulate vacuum systems [113]. The analogy even extends to Kirchhoff's laws. In multiple-connected vacuum systems with a number of pipes connecting multiple nodes, the sum of the gas flow into a node must add up to zero and the pressure difference around a circular loop of pipes must add up to zero. These laws allow us to uniquely calculate the pressure in complicated vacuum systems [112].

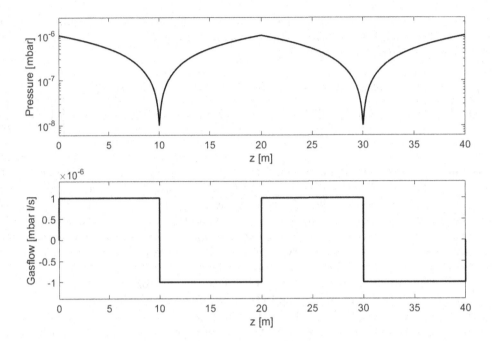

**Figure 13.7** The pressure profile (top) and the gas flow (bottom) as a function of the longitudinal position $z$ for a vacuum system with gas sources at $z = 0, 20$, and $40\,\text{m}$ and $200\,\text{l/s}$ pumps at $z = 10$ and $30\,\text{m}$. The $10\,\text{m}$ long pipes have a specific conductance of $c = 10\,\text{m\,l/s}$.

Whereas the vacuum system is present in every accelerator, is a cryogenic system only needed if super-conducting components, such as magnets or accelerating structures, are used and require cooling by liquid Helium. Such refrigeration systems are the topic of the next section.

## 13.6  CRYOGENICS

The super-conducting magnets and accelerating structures, used in some accelerators, are cooled with liquid Helium to temperatures of $4\,\text{K}$ or even below $2\,\text{K}$. If liquid Helium is removed from the system, the system operates as a *liquefier;* if the Helium circuit is closed, it operates as a *refrigerator*. Here we consider the latter and, inspired by Chapter 6 in [114], we sketch a refrigeration plant in Figure 13.8 with the components and their connections shown on the left-hand side and the temperature-entropy phase diagram on the right-hand side.

The mode of operation of this plant is based on the compressor receiving gas at its low-pressure side, labeled by (1), and compressing it at constant temperature to a higher pressure (2). Since gas normally warms up when compressed, the compressor is connected to a water-cooled heat bath that keeps the temperature of the gas constant. The gas then passes through a heat exchanger and enters a turbine (3) or other expansion engine, where energy is extracted and the gas thereby cooled. The cooled gas is guided to the low-pressure branch (7) and on its path back to the compressor (1), it is passed through the uppermost

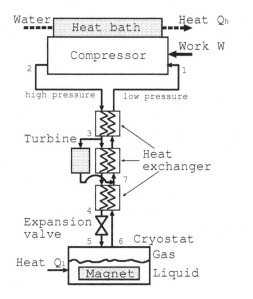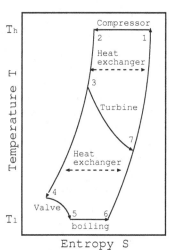

Figure 13.8  The refrigeration process.

heat exchanger in order to cool the high-pressure gas. This shortened cycle (1-2-3-7-1) repeats until the gas is sufficiently cold and a fraction is passed through the second and third heat exchanger to an expansion valve, where the temperature is lowered further and liquid Helium is produced to cool magnets or accelerating structures in a cryostat. Any heat entering the system at low temperatures boils off some of the Helium (6), which, still very cold, is used in the middle and lower heat exchangers to cool down the gas before the expansion valve.

The plot on the right-hand side in Figure 13.8 shows the corresponding operation principle in the $T - S$–phase diagram. On the top, at high temperature, the horizontal arrow from (1) to (2) illustrates the isothermal compression of the gas. The turbine that connects (3) and (7) ideally operates at constant entropy, but this would require very fast expansion of the gas. Since this happens at finite speed the line connecting (3) and (7) describes a so-called poly-tropic mode that changes the entropy. The heat exchangers connect the left and the right branch of the diagram and cause the cold gas returning to the compressor in the right branch to pre-cool the gas coming from the compressor in the left branch. The operation of the expansion valve keeps the enthalpy constant but increases the entropy. In this process of cooling down, part of the gas becomes liquid at (5). Heat entering the system boils off the liquid at constant temperature, resulting in a volume in the cryostat where liquid and gaseous phase coexist.

We can estimate the *efficiency* of of the refrigeration process by making the rough approximation that the heat bath in the compressor extracts the heat $Q_h$ from the process and that operating the compressor requires the work $W$, which is normally needed to drive motors in the compressor. Furthermore, we assume that the only place where heat $Q_l$ enters the system is at low temperature. These approximations treat the process as a Carnot-cycle with heat $Q_h$ and $Q_l$ added and removed at constant temperature and the cooling process approximated adiabatically—at constant entropy—such that the curved line segments (2-4) and (6-1) in the phase diagram on the right-hand side in Figure 13.8 are replaced by vertical lines that leave the entropy $S$ unchanged. Since energy is conserved, the heat and work entering and leaving the system must balance and we have $W + Q_h + Q_l = 0$, where heat entering the process is positive and heat leaving the system is negative. Moreover, the heat $Q_h$ exits the system at temperature $T_h$ and $Q_l$ enters at $T_l$. Since the heat $Q_h$ enters at temperature $T_h$, it changes the entropy by $S_h = Q_h/T_h$ and likewise at low temperature $S_l = Q_l/T_l$. In our approximation, where all processes are assumed to be reversible, the entropy must also balance, which leads to $Q_h/T_h + Q_l/T_l = 0$. Inserting this equation into the energy balance to eliminate $Q_h$, and solving for the ratio of work $W$ needed to remove heat $Q_l$ from the low temperature, we find $W/Q_l = (T_h - T_l)/T_l$, the well-known efficiency ratio for a Carnot-cycle. Thus, for a process that cools from ambient temperature $T_h = 293\,\mathrm{K}$ to liquid Helium temperature $T_l = 4\,\mathrm{K}$ we obtain $W/Q_l \approx 73$ and almost twice that value for $T_l = 2\,\mathrm{K}$. Even under the idealizing approximations any heat entering the system at low temperature $T_l = 4\,\mathrm{K}$ requires more than 70 times that energy to drive the compressor $W$. In practice, our idealizing approximations are violated by a significant margin and a rule of thumb is, that a kW at high temperature is needed to remove one watt at liquid Helium temperature. Refrigeration plants are normally specified by their ability to cool the heat entering per unit time, the power, at low temperatures, such as "100 W at 4.2 K." Considering the latent heat of vaporization $Q = 20.8\,\mathrm{J/g}$ of Helium, we find that 100 W evaporate 4.8 g/s of Helium and with the density $\rho_{LHe} = 0.125\,\mathrm{g/cm}^3$ we see that this corresponds to evaporating 138 liters of liquid Helium per hour.

Since the heat $Q_l$ entering the system requires a large expenditure in energy $W$, it is mandatory to carefully insulate all components at low temperature. Cryogenic containers are often dewars, which consist of an inner container with the cryogenic gas or liquid separated by an evacuated volume from an outer shell, in order to minimize thermal contact between inner and outer containers. At very low temperatures, even infra-red radiation from the high-temperature outer will heat the inner, cold container. This can be prevented from happening by adding one or several layers with multi-layer insulation in the space between the containers. Finally, we need to consider the large difference in volume between the liquid and gas phase of Helium at room temperature. The latter requires a 770 times larger volume and appropriate safety measures need to be taken into account when designing the containers.

## 13.7   RADIATION PROTECTION AND SAFETY

Accelerated beams are highly ionizing and therefore pose a significant threat to both personnel and the components of the accelerator. One reason is the creation of free radicals. For example, in materials with a high content of water, $OH^-$ and $H^+$ are created, both of which are chemically very reactive and potentially damage cells in living organisms and causes them to malfunction to the point of causing cancer. Moreover, if cells containing hereditary information, DNA, are exposed, the reproduction of the organism is compromised or results in mis-formed offspring. Materials used in accelerators, which are based on hydro-carbons, may change their properties as a result of being exposed to radiation. Epoxy, used to insulate coils in magnets, becomes brittle and must be replaced, if exposed to too high doses.

We discuss the units to quantify the radiation and its effect on biological material in the next section. To complement our brief discussion, we refer to Section 36 on "Radioactivity and Radiation Protection" in [60].

## 13.7.1   Units

The energy loss from ionizing the target material $dE/dx$, as described in Section 9.2, is the fundamental process. The *absorbed dose D*, is the deposited energy density, measured in J/kg, assigned the special unit *Gray* (Gy) and defined by 1 Gy=1 J/kg. Historically, the unit *rad* was used with the conversion 100 rad = 1 Gy. Photons and electrons lose their energy rather slowly compared to alpha particles (Helium nuclei) or heavier ions. The latter are characterized by a large *linear energy transfer* (LET) from radiation to target material. The LET is typically measured in J/m or keV/$\mu$m and, for example, for protons is given by Equation 9.6. In biological material, the high local energy deposition of high-LET radiation is more damaging.

This variation in damaging effect of the absorbed dose on biological material is characterized by a *biological weighting factor w* that depends on the specific type of organ and describes the risks to develop cancer from long-time exposure to radiation. The product of weighting factor $w$ and absorbed dose $D$ is denoted by *equivalent dose $H = wD$* and is measured in *Sievert* (Sv). Historically, the unit *rem* was used with the conversion 100 rem=1 Sv. Currently used values [60] for the weighting factors are $w = 1$ for photons and electrons, $w = 2$ for protons, and $w = 20$ for high-LET particles, such as alpha particles or heavy ions.

The natural background radiation level of the equivalent dose is on the order of 5-10 $\mu$Sv/day and largely depends on the altitude, with minimum values below 2 $\mu$Sv/day at sea level and several $\mu$Sv/hour during airplane travel. On the ground, the main source of radiation comes from natural sources, such as Radon. The annual average dose for the general population is 2.4 mSv/year [60]. The annual permitted dose for *radiation workers* is 20 mSv/year in Europe and 50 mSv/year in the US. The *lethal whole body dose* is on the order of a few Gray causing a 50 % probability to die within a month.

## 13.7.2   Range of radiation in matter

Keeping the ionizing radiation away from humans or sensitive equipment requires shielding and the necessary thickness of the shielding depends on the depth of penetration, or the *range* of the radiation in matter.

The high nuclear charge $Z$ of heavy ions limits their range to below a mm and normally the beam pipe shields adequately and screens to view beam profiles, as discussed in Section 7.3, stop the beams in most cases. Even alpha particles emitted by radioactive material are mostly absorbed in the outer layers of the skin and pose only a moderate risk, because the skin is replaced regularly anyway. On the other hand, alpha emitters that enter the human body, either by breathing or by imbibing, are an extremely serious threat, because this high-LET radiation is absorbed in small volumes deep inside the body and causes severe local damage to the nearby tissue with a high probability to damage the cells and cause cancer.

Protons lose their energy at a rate given by Equation 9.6 and, for example, with a kinetic energy of 200 MeV, they produce the Bragg-peak, such as the one shown in Figure 9.2, at a depth of about 28 cm in water or in human tissue, which is largely water-based. This feature is used in accelerators for proton-therapy, as already discussed in Section 9.2. From Equation 9.6 we see that the energy loss per unit length $dE/dx$ only grows logarithmically for high energies and that causes protons with very high energies to lose their energy only slowly and they require very thick shielding. In LHC, even several meters of material in the collimators do not stop the beam.

High-energy electrons and photons cause bremsstrahlung cascades and showers when they impinge on an obstacle. Their attenuation in many materials can be characterized by an exponential decay length, called the *radiation length,* and often denoted by $X_0$. Values of the radiation length are tabulated, for example, in Chapter 6 of [60], and for water we have $X_0 = 36\,\text{cm}$, for Iron (Fe) $X_0 = 1.76\,\text{cm}$, and for shielding concrete $X_0 \approx 11\,\text{cm}$.

When losing beam particles, even the shielding material can be activated and becomes radioactive itself. Materials with a high nuclear charge $Z$ can break up when hit by a beam particle and form a multitude of fission products, some of which may be radioactive. Materials with lower $Z$ are therefore preferentially used in regions exposed to a high beam losses. Using an Aluminum beam pipe instead of one made of stainless steel limits the activation in exposed regions. Moreover, lost beam particles can excite giant photo-nuclear resonances that produce neutrons, which are difficult to shield because they are electrically neutral and difficult to stop. In concrete, their attenuation length for low energies is on the order of 10 cm, which increases to about 50 cm for energies above 100 MeV.

### 13.7.3  Dose measurements

The primary source of ionizing radiation close to the accelerator beam pipe are lost beam particles and they are detected by the *beam loss monitor system* [115]. The main monitors are *ion chambers* consisting of two electrodes at high voltage of several kV, separated by a volume containing an inert gas, such as Argon. Charged particles or photons, passing the volume, ionize the gas. The electrons then move to the anode, where their arrival is detected as a current pulse. At intermediate voltages the pulse is proportional to the energy loss of the passing radiation and saturates at high voltages, which defines the *Geiger-Müller* regime of operating the ion chamber.

Another option to directly detect ionizing radiation is based on reverse-biased *pin–diodes* that produce a current pulse if hit by either electrons or photons, which causes a small current to flow. Operational amplifiers, operated with high amplification factors, provide voltages that can can be detected and digitized.

Outside the accelerator enclosure, or with the accelerator turned off, only neutrons and photons are present, where the latter are often detected with calibrated Geiger-Müller detectors or ion chambers. Small hand-held devices are available for personal use. A further detector for photons is based on a combination of scintillating crystals, such as NaI or CsI, and photo-multiplier tubes.

Neutrons must be moderated with a hydrogen-rich material, such as polyethylene, to low energies where their cross section for nuclear reactions is large and causes secondary and charged particles to be created. These reaction products are sub-sequentially detected with an ion chamber or scintillating crystals embedded within the moderator.

### 13.7.4  Personnel and machine protection

The exposure of personnel working in accelerator environments is typically monitored by *film badges* with sensitive areas that are blackened, if exposed to radiation. They need to be analyzed periodically, typically once every few months. A second type of monitor are *track etched detectors.* Etching the detector material, after ionizing radiation and neutrons have penetrated it, makes tracks left by the radiation visible. Later analysis under a microscope allows us to correlate the track density with the exposure. Small hand-held ion chambers are used to monitor the radiation levels when working in exposed areas. Apart from periodically monitoring the accumulated dose levels of radiation workers, such as the accelerator maintenance personnel, their health level is regularly monitored by medical personnel.

The accelerator enclosure is a *controlled area,* whose access is meticulously overseen by interlock systems. Before the beam is allowed to enter a section of the accelerator, the enclosure is searched to ensure that no colleague, who may have fallen ill and fainted, is left behind next to the accelerator. Usually a number of switches, placed throughout the enclosure, must be pushed in a particular order during the *search procedure* to ensure that all hidden areas are checked. Once the accelerator is searched and the access interlock is enabled, temporary access during periods, while beam is absent, is often controlled by key banks. Colleagues who need temporary access, remove a key from the key bank while they work next to the accelerator and operation of the accelerator is impossible until all keys are returned to the key bank.

While these systems ensure that no personnel are harmed during the operation of the accelerator, the *machine protection system* provides interlocks that prevent the accelerator from harming itself by excessive beam loss or by sending beams to areas unsuited to transport it. In order to monitor the losses a number of beam loss monitors are installed throughout the accelerator and any excessive loss level will either dump the stored beam or prevent the source from producing more beam. Other systems are based on measuring the beam current at different positions along the accelerator and if there is a significant discrepancy, operation is stopped. Finally, special diagnostics such as diagnostic screens may be damaged if the beam current is too high and their insertion in the beam path enables an interlock to limit the beam current to safe levels.

## 13.8   CONVENTIONAL FACILITIES

A sizeable fraction of the construction cost of accelerators is spent on the buildings and other civil engineering, while operation costs are to a large extent due to the price of electricity. We therefore very briefly touch upon these points.

### 13.8.1   Electricity

Magnets and the radio-frequency systems of accelerators are the main consumers of wall-plug power that is provided by the power grid. The power levels range from below 1 MW for small accelerators and large facilities with multiple accelerators, such as CERN, exceed 100 MW. Often special agreements with utility companies exist in order to limit the cost, but in return the accelerators may have to stop operation at times of high demand for electricity from other customers.

### 13.8.2   Water and cooling

Many consumers of power, such as normal-conducting magnets and radio-frequency structures, generate a substantial amount of heat and must be water-cooled. Often de-ionized water is required to prevent sparking due to the high voltages or creepage currents by large currents, especially when cooling magnets by passing water through holes within the conductors, shown in Figure 4.13, that make up the coils. Moreover, the temperature and the pressure of the cooling water must be stabilized to prevent thermal drifts of components. Especially radio-frequency components are prone to detune with varying cooling water temperatures. Devices that cannot be water-cooled will deposit the heat they generate into the ambient air and make an elaborate air-conditioning system necessary, which often has to deal with a wide range of conditions.

Since only a fraction of the wall-plug power from the grid ends up in the beam, most of the power will have to be removed by water-cooling or air-conditioning. In modern accelerators this has triggered investigations to find ways to recuperate some of the generated heat, for example, to heat nearby buildings.

### 13.8.3 Buildings and shielding

Finally, the accelerator and personnel operating it, must be housed. Large accelerators, such as the LHC or the European XFEL are placed in deep underground tunnels, that are dug by special tunnel-boring machines, known as "moles"; a rather expensive method, but unavoidable, if accelerators are placed in densely populated areas. On the other hand, if placed above ground, significant shielding is required to prevent the radiation from reaching personnel or users of the accelerator, who occupy buildings adjacent to the accelerator.

## QUESTIONS AND EXERCISES

1. Consider the thermionic electron gun from Section 13.2.1 and prepare plots of the charge density $n(s)$, the potential $U(s)$, and the velocity $v(s)$ of the electrons for $0 < s < d$. Discuss the divergence of $n(s)$ near the cathode.

2. You have a 100-liter pump connected through a 10 m long round pipe with a diameter of 10 cm connected to a large volume of $1 \, m^3$ that contains gas that you want to remove. (a) What is the conductance of the long pipe? (b) What is the effective pump speed with which you actually pump the large volume? (c) What is the exponential pump-down time scale? (d) How much more efficient would the pump be, if you connect it directly to the large volume?

3. You need to expose a delicate sample directly to synchrotron radiation, such that you cannot use a vacuum window in the 10 m long photon beam line. Unfortunately, the sample also outgasses significantly at a rate of $10^{-3}$ mbar l/s. Luckily, you have two large 200 l/s pumps at your disposal to design a *differential pumping* section and ensure that the pressure at the shutter, close to storage ring, is below $10^{-8}$ mbar. Only then are you allowed to open it and receive photons to expose the sample. Your plan is to insert 1 m long narrow pipes with a diameter of 1 cm into the 10 m long line, which otherwise has a diameter of 10 cm. (a) Test different sequences of wide and narrow pipes and placement of pumps and find out whether you can achieve your goal. (b) Can you reach the required pressure with smaller and therefore and less expensive pumps.

4. Find out (a) the average received dose for the population in your country of residence; (b) the dose received during a transatlantic 10-hour flight.

# CHAPTER 14

# Examples of Accelerators

After discussing many aspects of accelerators, let us look at examples of accelerators that are presently in operation and are based on the principles and methods discussed in the previous chapters. We refer to the relevant chapters by pointers in square brackets.

## 14.1   CERN AND THE LARGE HADRON COLLIDER

The European Organization for Nuclear Physics (CERN) was founded in June 1953 and charged with the pursuit of the high-energy frontier of elementary particle physics. Over the decades the proton synchrotron (PS) and the Super Proton Synchrotron (SPS) were built. The latter discovered the force carriers of the electro-weak interaction, the $Z$ and $W$–bosons. In the 1980s the $e^+e^-$–collider LEP with a circumference of 27 km was constructed to carry out precision studies of the previously found bosons until 2002. Almost simultaneously with the start of LEP's construction, the idea was born to replace the $e^+e^-$–collider with a proton storage ring to collide multi-TeV protons, the *Large Hadron Collider* LHC. By the time its construction began to replace LEP in the same tunnel, the only missing particle within the so-called standard model of particle physics, was the Higgs boson, which indeed was found in the LHC by 2012.

The protons that eventually collide at high energy are created in a Duoplasmatron [Section 13.2.2], placed on a high-voltage platform to provide protons with an energy of 90 keV energy. The protons are further accelerated to 750 keV in an radio-frequency quadrupole [Section 13.2.5], which simultaneously accelerates, bunches, and transversely focuses the protons before injecting them into Linac 2, which is a drift-tube linac [Figure 1.2]. In Linac 2, the protons are accelerated to 50 MeV and then beam currents of up to 150 mA are injected into the four rings of the PS booster. They operate with harmonic number $h = 1$ [Section 5.3] with a single, long bunch in each ring. Splitting the beam into four rings is necessary to mitigate the large space-charge tune shift $\Delta Q$ [Equation 12.1] due to the very high charge per bunch and low energy. After the four rings are filled, their energy is increased to 1.4 GeV and the bunches are transferred to the PS in two batches of four. Two bunches each fill six of seven available RF-buckets in the PS. In the PS, an "RF-gymnastics" scheme [Section 5.5] is used to split the six long bunches into 18 shorter ones. Then, the energy is raised to 25 GeV and the bunch structure is adapted once again to create 72 shorter bunches with 25 ns spacing. Before transferring the bunches from the PS to the SPS, the bunches are rotated by 90 degrees in longitudinal phase space [Section 5.5] in order to match the acceptance of the RF system [Figure 5.5] of the SPS and to avoid emittance growth. Three or four consecutive fills with 72 bunches each from the PS are then transferred to the SPS. There, their energy is increased to 450 GeV, before they are extracted, pass the 3 km long transfer lines from SPS to LHC, and, after matching the transverse beta functions [Section 3.6], are finally injected into the LHC. This procedure is repeated 13 times to fill 2808 bunches into the LHC, each containing more than $10^{11}$ protons. The total stored current is on the order of 0.5 A. All the extractions from one accelerator and injections into the next one use single-turn injections [Section 13.3] onto the closed orbit in order to maintain small emittances.

Once all protons are assembled in the two rings of the LHC, the fields in the magnets are increased and the beams are accelerated [Section 5.6] in about 20 minutes to the collision energy, presently up to 6.5 TeV per beam. At this energy, the beams carry up to 350 MJ each and the protons actually emit a few kW of synchrotron radiation that must be prevented from being deposited in the magnet cold mass at 1.9 K, because that requires a large cooling effort [Section 13.6]. Instead, the radiation is intercepted by so-called beam screens at temperatures between 5 K and 20 K, which requires less power to cool. During the ramp to the final energy, the beam optics is detuned to reduce the maximum beta functions in the final focus quadrupoles and prevent excessive losses. But once the collision energy is reached with beams circulating stably, the minimum beta functions at the collision points are "squeezed" to small values. This also causes the maximum beta functions in the adjacent quadrupoles to increase. With small beta functions at the interaction points the two counter-propagating beams are moved transversely and brought into collision. Now they start producing luminosity [Section 9.5] at a level of $\mathcal{L} = 10^{34}/\text{cm}^2\text{s}$ per experiment and the detectors start collecting data. Unless something unforeseen occurs, such as magnet quenches, the beams stay in collision for about 12 hours before they are dumped, the magnets are ramped down to injection energy, and the entire procedure to fill LHC starts again at the proton source.

The LHC was conceived to reach the highest possible proton energy in the available 27 km-long tunnel. Therefore, the crucial components are the dipole magnets that must reliably reach the highest achievable fields. Moreover, the large number of super-conducting magnets [Section 4.4] requires them to be produced on an industrial scale. The 1232 dipoles, installed in LHC today are 15 m long, reach fields of 8.3 T, and are cooled by super-fluid Helium at 1.9 K. They have two apertures to guide the two counter-propagating proton beams in the same magnet. In the LHC arcs, groups of three dipoles are interspersed between alternating focusing and defocusing magnets to form FODO cells [Section 3.3.3] with a cell length of 106.9 m with 23 cells forming one of the eight arcs of LHC. In between two adjacent arcs are straight sections, four of them are occupied by the large experiments ATLAS, CMS, LHCb, and ALICE. These straight sections consist of matching sections with dispersion suppressors [Figure 3.30] to adapt the optics in the arcs to the telescopes [Section 3.7.1] that demagnify the beams before the respective collision points. The close temporal spacing of the bunches by 25 ns makes it necessary to collide with small crossing angles [Section 9.5] in order to avoid excessive perturbations from the long-range collisions between successive bunches in the two beams before and after their collisions. The crossing angle reduces the luminosity, but is necessary to ensure the stability of the beams. One of the other four straight sections house the super-conducting acceleration structures [Section 6.5.2] needed to accelerate the beam and to provide longitudinal phase stability [Section 5.3]. The large number of stored protons in many bunches makes the beam susceptible to multi-bunch instabilities [Section 12.6] that are counteracted by feedback systems. The high per-bunch current is responsible for a number of single-bunch instabilities [Section 12.5]. One is based on ionizing the residual gas in the beam pipe. The electrons, accelerated in the electrostatic potential of the beam, hit the beam pipe and desorb even more gas. This and other perturbations cause the protons in the beam to stray away from their reference orbit. Therefore, two straight sections are used to collimate stray particles and prevent them from hitting the cold mass of the magnets and causing quenches of super-conducting components. The last straight section contains the beam dump, which is needed to abort the beams, either before ramping the magnets down for a new cycle, or if a quench or some other fault happens. In order to dump the beams, a sequence of 15 kicker magnets deflect a beam into a sequence of 15 septum magnets, before dilution kickers spread out the train of bunches transversely. The beams then pass a composite-carbon window and are stopped in a carbon-core beam dump shielded by concrete-filled iron yokes.

Apart from protons, the LHC also stores lead (Pb) ions that are brought into collision and produce a plasma consisting of the constituents of atomic nuclei: quarks and gluons. In this state, which resembles that of the universe immediately after the Big Bang, the quarks and gluons are no longer bound inside the nuclei and therefore allow us to investigate their coalescing into hadrons. The lead ions are produced by heating ultra-pure lead and subsequentially ionize it in an ECR source [Figure 13.3] to produce $Pb^{29+}$. These ions are accelerated in an RFQ [Section 13.2.5] to 250 keV and injected into Linac 3 that accelerates the lead ions to 4.2 MeV per nucleon. The ions then pass a stripper foil to remove the remaining electrons from the lead ions and produce the charge state $Pb^{54+}$. These ions are subsequently injected [Section 13.3] into the low-energy ion ring LEIR, where they are accelerated to 72.2 MeV per nucleon. After further acceleration and bunch manipulations in the PS and the SPS, they are finally injected into the LHC, where they are brought into collisions at beam energies of 2.75 TeV per nucleon, mostly to provide luminosity for the ALICE detector, which is dedicated to heavy-ion physics.

In the coming years the LHC will be upgraded to increase the luminosity [Section 9.5] ten-fold. This stimulated an ambitious program to reduce the beta function at the interaction point [Section 3.7.8] with stronger quadrupoles, to increase the stored current [Section 12.5 and 12.6], the beam safety system, and to install crab-cavities in order to compensate a necessary increase of the crossing angle at the interaction points [Section 9.5]. Moreover, the entire injector chain of linacs, PS booster, PS and SPS will be upgraded to cope with higher beam currents. For the more distant future, the 100 km *future circular collider*, FCC, is under discussion in order to reach proton energies of 50 TeV. Furthermore, the CLIC collaboration targets an $e^+e^-$ linear collider with beam energies up to 1.5 TeV.

## 14.2 EUROPEAN SPALLATION SOURCE

Because neutrons are electrically neutral they can penetrate deep into samples and because they have magnetic moments they can probe magnetic properties of materials. Moreover, with momentum $p$ corresponding to ambient temperatures, their de Broglie wavelength $\lambda = h/p$ is on the order of one Ångström, the typical spacing of atoms in matter. Thus, neutrons probe properties of matter that are complementary to those probed by x-rays, having the same ranges of wavelengths. Historically, the constant flux of neutrons from nuclear reactors was used to analyze samples from material and life sciences. Later accelerator-based sources provided neutrons delivered in short pulses, which allows us to select the neutron energy with time-of-flight methods. Presently, the European Spallation Source ESS is under construction in Sweden. It will deliver proton beams with a beam power of 5 MW to the target where nuclear spallation reactions will produce a large flux of neutrons. Once operational, it will be the most powerful source of neutrons, world-wide.

The protons are created in an ECR source [Section 13.2.2] placed on high-voltage platform that causes the protons to be accelerated to a kinetic energy of 75 keV, before they are injected into an RFQ that increases their energy to 3.6 MeV, followed by a normal-conducting drift-tune linac [Figure 1.2], that brings their energy to 90 MeV. In the next section, 28 super-conducting spoke-cavities, optimized to accelerate protons at low velocities, increase their energy to 216 MeV. From this point onwards they are accelerated with super-conducting elliptic cavities to their maximum energy of 2000 MeV and are guided onto the neutron-production target, a fast-rotating tungsten wheel. The neutrons, created in nuclear spallation reactions, initially have the high energy of the incident proton beam, but are slowed down, or moderated, to thermal energies, comparable to $k_B T$ with T=300 K, in an enclosure made of hydrogen-rich material. A number of holes in the moderator allow neutrons to escape and travel towards the experiments. Time-of-flight selection of the neutrons is possible, because the accelerator operates in a pulsed mode, where 14 times per second a 2.86 ms long pulse train of proton bunches is accelerated and directed to the target.

Since high-intensity protons can also be used to produce neutrinos and to study their properties—especially their oscillations—investigations are underway to double the repetition rate of the linear accelerator and use every other pulse to produce neutrinos. For this purpose, the "neutrino pulses" are directed into a small storage ring, where the 2.86 ms long pulses from the linac are compressed to about 1 $\mu$s and sub-sequentially extracted and used to create first pions. The pions then decay into muons, whose charge state can be selected in a so-called magnetic horn, which, in turn, produces neutrinos or anti-neutrinos. There is one detector for the neutrinos close to the accelerator and a second one several 100s of km away. Comparing the number of detected neutrinos will then allow precision measurements of neutrino oscillations.

## 14.3   SLAC AND THE LINAC COHERENT LIGHT SOURCE

In the first half of the 1960s, a 3 km-long linear accelerator for electrons was constructed at the Stanford Linear Accelerator Center (SLAC). During its first decades of operation, electrons with maximum energy of 20 GeV were used to probe the sub-structure of nucleons. In the 1980s a novel scheme to multiply the power level from the klystrons increased the maximum energy to 50 GeV, which made it possible to create $Z$–bosons, one of the carriers of the electro-weak force, by accelerating and colliding electrons and positrons in the first linear collider, the SLC. After completion of the program, part of the linear accelerator was turned into the first X-ray free-electron laser, which started operation in 2009. The short wavelength and the unprecedented intensity of the x-ray pulses makes it possible to obtain diffraction images from individual samples, rather than from crystallized assemblies of multiple samples. And the ultra-short duration of the radiation pulses in the fs-regime allows us to collect the diffraction patterns before the high-intensity pulse destroys the sample. By synchronizing the radiation from the FEL with an external conventional laser, snapshots of chemical reactions can be collected that are later sorted and assembled into movies.

At a rate of 120 times per second, electrons are created by a ultra-violet laser pulse impinging on a photo-cathode that is embedded in a radio-frequency accelerating structure [Section 13.2.1]. In order to stabilize the transport of the bunch, having very small momentum spread, through the linear accelerator, the spread is increased in a controlled way with a laser heater [Section 12.5]. Two bunch compressors [Section 3.7.9] at locations with beam energies of 250 MeV and 3.4 GeV, compress the bunch to sub-ps lengths. RF structures that operate at a frequency of 2856 MHz [Sections 6.4.3 and 6.5.1] accelerate the bunches up to their maximum energy of 16 GeV. Diagnostic equipment [Section 7.3] is installed in multiple locations along the linac to ensure the beam quality: wire scanners to measure the emittances [Section 7.4] and transversely deflecting structures [Section 7.3] to measure the bunch lengths. After accelerating bunches with intensities of several 100 pC to 16 GeV, the bunch enters the undulator magnets [Section 10.2.2] that consists of 33 modules with a fixed gap and a length of 3.4 m each. While passing the undulator, the electrons incite the SASE process [Section 10.4] and produce highly intense and ultra-short radiation pulses with wavelengths down to 0.1 nm or 1 Å, the typical distance between atoms in matter. The radiation is guided by mirrors to a number of different experimental stations. Several novel schemes for arrival-time diagnostics on the fs time scale and for self-seeding [Section 10.4] were successfully implemented in LCLS.

Presently, construction of a second free-electron laser, LCLS-II, is underway at SLAC. It uses super-conducting acceleration structures [Section 6.5.2] in order to continuously accelerate electron bunches with a charge of 100 pC at a rate of up to a million times per second to an energy of 4 GeV. A particular challenge will be to achieve high beam quality at high repetition rate, as well as high gradients in the accelerating structure when operating in a continuous mode. Moreover, three bunch compressors cause very high peak currents that make the beam susceptible to micro-bunching instabilities. A beam switchyard will send beams from both the already operating normal-conducting accelerator and the new super-conducting accelerator to two variable-gap undulators, one for the soft x-ray and the other for the hard x-ray regime, expanding the spectral range.

## 14.4  MAX-IV

Max-IV is a third-generation synchrotron light source that is equipped with a number of synchrotron-radiation beam lines, dedicated either to spectroscopy in wide spectral range from UV to X-rays or to diffraction experiments to a determine the three-dimensional structure of proteins. In Max-IV, a normal-conducting linac provides electrons for two rings, one operating at 1.5 GeV to produce UV and soft x-ray photons, and a second one, operating at 3 GeV to produce harder x-rays.

A thermionic cathode [Section 13.2.1], embedded in a radio-frequency accelerating structure, generates the electrons for the rings. Up to one nC with a normalized emittance of $\varepsilon_n = 10\,\mu$m-rad is accelerated in the normal-conducting linear accelerator that operates at a frequency of 3 GHz [Section 6.4.3]. After reaching the energy of 260 MeV the first bunch compressor [Section 3.7.9] reduces the bunch length. After the bunch compressor, the electrons are accelerated to an energy of 1.5 GeV, where they are either injected into the 1.5 GeV ring, or accelerated further to 3 GeV, to be injected into the 3 GeV ring with so-called top-off injection, where losses in the ring are continuously replenished by a new beam from the linac. This larger ring has a circumference of 528 m and the magnet lattice consists of multi-bend achromats [Section 3.7.5] with seven short dipoles and a number of quadrupoles. Using a large number of short dipoles is beneficial to reach small emittances [Section 3.7.5] and the horizontal emittance is as low as $\varepsilon_x = 330$ pm-rad. In order to keep the circumference limited, strong combined-function dipole magnets are used in the achromats. The strong fields are made possible by keeping the magnet gaps small, which, in turn, requires distributed pumping with NEG pumps [Section 13.5.2]. Moreover, the iron yoke for large sections of a multi-bend cell are prepared in numerically controlled milling machines (CNC) and are internally aligned, which simplifies assembling the accelerator. The equilibrium beam parameters, such as emittance, damping times, and momentum spread are then entirely determined by the magnet lattice [Section 10.1]. The RF system in the 3 GeV ring uses a 100 MHz system, which causes the bunches to be fairly long in order to prevent intra-beam scattering [Section 12.2]. This permits us to maintain small emittances, despite large single-bunch beam currents. Undulators and wigglers [Section 10.2.2] generate the synchrotron radiation to serve the users. The small vertical height of the beam pipe permits using narrow gap undulators with high fields to generate photons into the hard X-ray regime. Since the small emittances are only achievable with a well-corrected lattice, methods from Section 8.5.6 are used to find deviations of the real accelerator from its model. Furthermore, the synchrotron radiation is often focused to micro-meter spot sizes on to samples and this requires stabilization of the beam orbit with feedback systems [Section 8.5.7]. Moreover, coupled-bunch instabilities [Section 12.6], which can arise due to the high current stored in many bunches, are stabilized with the coupled-bunch feedback systems [Section 12.6].

The smaller 1.5 GeV ring is designed following the same strategy as the 3 GeV ring, but has a 5 times smaller circumference and produces photons in the soft X-ray regime. Since the linear accelerator is only used occasionally to top-off the stored currents in the two rings, the rest of the time, a so-called short-pulse facility passes the 3 GeV beams at the end of the linac through an undulator, where it produces very short light pulses. In this mode the electron bunches can have a smaller charge per bunch, but must have smaller emittances than the beams that fill rings. Therefore, a second RF-gun with a laser-driven photo-cathode [Section 13.2.1] is installed next to the thermionic gun. The laser cathode produces bunches with charges of about 0.15 nC and emittances of 1.5 $\mu$m-rad. Single bunches are produced at a rate of 100 Hz, accelerated and compressed in the first bunch compressor. Instead of directing them to the 3 GeV ring, they pass a second bunch compressor and are compressed further to sub-ps lengths, whence they enter an undulator magnet [Section 10.2.2] to produce radiation in the soft X-ray regime with the duration mimicking that of the compressed electron bunches. Discussions to augment this short-pulse facility with a soft x-ray free-electron laser are ongoing.

## 14.5 TANDEM ACCELERATOR IN UPPSALA

Ions accelerated to a few MeV are wonderful probes of material composition, especially of the near-surface region of materials. *Accelerator mass spectroscopy* (AMS) allows us to detect the fraction of, for example, $^{12}C$ in samples with a sensitivities better than $10^{-16}$. Depth profiling of materials containing heavy atoms on the nm-scale is possible using projectiles of light ions with *Rutherford Backscattering* (RBS). Conversely, heavy ions are used to probe materials containing light atoms using *Elastic Recoil Detection Analysis* (ERDA). Furthermore, ions excite the emission of specific x-rays (PIXE) or nuclear reactions (NRA) that are specific to the target material and beam energy. The ions used for these methods are often provided by a tandem accelerator, already mentioned in Chapter 1.

In the tandem laboratory in Uppsala, four ion sources provide the beams that are used as projectiles. Two duoplasmatrons, one of which is equipped with a potassium-filled charge-exchange cell [Section 13.2], generate beams from gaseous materials. Two sputtering sources (SNICS) [Section 13.2.4] are available to provide beams from solids. The sources are placed on a high negative potential, which causes the ions to be accelerated to an energy, typically between 20 and 40 keV. Dipole magnets, and for some low-energy beams, electrostatic fields, then select the mass to charge ratio of the ions injected into the pelletron, a tandem accelerator based on a Van-de-Graaff accelerating column [Figure 1.3]. In the center of the tandem accelerator, located at high potential of up to 4.5 MV, either a thin conversion foil or a gas stripper convert the negative ions to positive ions. This allows using the same acceleration voltage once again in order to reach energies of up to several 10s of MeV, provided several electrons are removed to create higher positive charge states. At the exit of the pelletron, a triplet [Section 3.7.2] of electro-static quadrupoles is used to transversely focus the beam. A carefully calibrated spectrometer-dipole magnet selects the particle type and charge state that is directed to a switchyard used to distribute the beams to the experimental stations, each devoted to one of the analysis methods mentioned above. One of the beam lines is equipped with a micro-beam system that uses collimators and a triplet [Section 3.7.2] to focus the ion beam down to spot sizes on the order of a $\mu$m, which is used to analyze sensitive microstructures, often organic materials.

The second accelerator is a *mini carbon dating system* (MICADAS) [116], a compact AMS system, dedicated to satisfying the high demand for radiocarbon dating. The beam is produced as a negatively charged ion in a sputtering source [Section 13.2], located at a few tens of kV and the charge state is selected in a low-energy mass spectrometer with a 90-degree bending magnet. It uses permanent magnets with super-imposed electro-static fields, which can be adjusted on a time scale of ms and allows rapid switching between $^{12}C, ^{13}C$, and $^{14}C$ in order to minimize systematic errors. Then the carbon ions are accelerated in a tandem configuration, where 170 kV are provided by a commercial high-voltage power supply. In this configuration, the negatively charged ions are accelerated towards a gas stripper, located at high potential, where they are converted to positively charged ions and accelerated once again with the same acceleration voltage. The ions are then analyzed in a 90-degree magnet and a second 90-degree electrostatic deflector, before they are counted with a semiconductor detector.

A third accelerator is a 350 kV implanter that delivers beam currents of mA. It is used for *Medium Energy Ion Scattering,* a technique similar to RBS, but provides better depth resolution on the order of 0.5 nm, which allows us to analyze, for example, gate stacks for transistors and other samples, such as thin films. Moreover, it is used to dope a substrate or modify its conductivity, where the depth of the modification is energy dependent and therefore adjustable. A feature that is particularly attractive to develop *pn*–junctions in semiconductors. The ion-implanter [117] is equipped with a commercial high-current ion source, that can produce beams of most elements by operating either in gas, oven, or in sputter modes. The ion source is placed on a 20 kV high-voltage platform followed by an dipole magnet to select the ion species and a Cockroft-Walton [Figure 1.4] accelerator reaching voltages of up to 330 kV. The accelerated ions can be chopped with an electrostatic deflector and bunched with a drift-tube buncher [Section 3.7.9] in order to produce short, down to 0.3 ns long, bunches, which allow us to make time-of-flight measurements of the back-scattered ions from a sample. The samples are mounted on a rotating stage inside a scattering chamber with a two-dimensional position-sensitive detector that allows us to record the arrival time of the scattered ions.

The fourth accelerator is also used for time-of-flight measurements with ions, accelerated by even lower voltages of 0.5 to 10 kV, which improves the sensitivity to sample properties near the surface. The beams are produced in a commercial ion source, followed by a Wien-filter. It uses crossed electric and magnetic fields that deflect the particles, except for one velocity, at which the electric and magnetic forces cancel. This device selects the velocity of the beam particles that are chopped to produce short bunches that, after being focused in so-called einzel-lenses, directed onto a sample. The backscattered ions are post-accelerated and guided to a detector.

## 14.6   ACCELERATORS FOR MEDICAL APPLICATIONS

A large number of accelerators, many more than those used in fundamental research, are used in industrial and medical applications. Nowadays, many large hospitals operate facilities to irradiate cancer patients with x-rays generated by electrons impinging on a target.

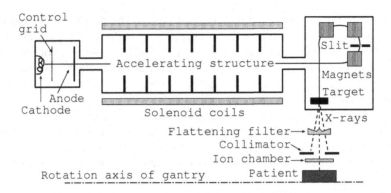

Figure 14.1 Medical electron linac for the irradiation of tumors with x-rays.

Figure 14.1 shows a sketch of such a system. The electrons are created in a thermionic electron gun [Section 13.2.1] that is equipped with a control grid in order to modulate and control the beam current. A small linear accelerator, often based on accelerating structures operating at 3 GHz [Section 6.4.3], accelerate the electrons to energies of about 5-20 MeV. At these energies they can still be focused with solenoids [Section 3.1.7]. After the linac, quadrupole and dipole magnets focus and deflect the electrons by 270 degrees. A slit is used to define the momentum of the particles, before they are directed onto a target, where they create x-rays [Section 9.2]. Following the target, a flattening filter scatters the x-rays in such as way as to make the transverse distribution more homogeneous, apertures and collimators define the lateral extent of the irradiation, and ion chambers measure the dose [Section 13.7.3] before the radiation impinges on the patient. The whole linac assembly is sufficiently small to be built into a so-called *gantry* that can be rotated around the patient who remains fixed in order to prevent the patient's organs and the tumor from moving. A second use of electron beams with energies of several MeV, albeit at much higher beam intensities, is the sterilization of single-use material in hospitals, such a disposable syringes [118].

Instead of irradiating tumors with x-rays, protons are used as well. They deliver the highest dose at a certain depth in the Bragg-peak [Section 9.2]. For protons with a kinetic energy of about 200 MeV the maximum dose [Section 13.7.1] is deposited at a depth of 28 cm in water-dominated material, such as human tissue. The protons are typically generated in a Penning ion-source [13.2.2] in the center of a cyclotron [Chapter 1] that accelerates them to their maximum energy in the range of 200-250 MeV. After extracting the protons from the cyclotron with an electrostatic deflector, they traverse a beam line to the patient treatment room. The depth of the Bragg-peak can be modulated by passing the protons through a stack of Plexiglass or other water-rich material, where it loses part of its energy. Collimators, specifically manufactured for each patient, define the transverse field of the irradiation. The tumor adaptation stage with the range modulator and the collimator can be either placed in a horizontally oriented beam line, or built into a gantry that rotates the beam around the patient. Since the rigidity $B\rho \approx 2.4$ Tm of 200 MeV protons is much larger than that of the electron with 20 MeV or less, the gantries for protons require much larger magnets and are very large, often weighing several 100s of tons.

The width of the Bragg-peak, caused by ions with higher mass number, for example, carbon, is narrower than that of protons and this stimulated the construction of centers for ion therapy. The energy of carbon ions with a range of 30 cm in water is about 430 MeV per nucleon, which corresponds to a rigidity of $B\rho = 6$ Tm. At these energies, synchrotrons are more suitable to accelerate the ions and, additionally have the ability to vary the extraction energy and thereby adjust the depth of the Bragg peak, albeit at the expense of higher complexity of the accelerator. After the ion source [Section 13.2.2], an RFQ [Section 13.2.5] accelerates the ions before injecting [Section 13.3] into the synchrotron that accelerates them to the desired extraction energy. Carefully controlled resonance extraction [Section 13.3] close to the half- or third-order resonance creates a constant flow of accelerated ions towards the patient treatment rooms. Due to the higher rigidity of the beams the gantries for carbon ions are even a few times larger and heavier than those for protons.

Whereas Synchrotrons and larger cyclotrons are used to treat patients, smaller cyclotrons are used to create radionuclides that are used both for diagnostics and for patient treatment. The latter is based on implanting radioactive material in the tumor. This treatment method, called *brachytherapy,* uses iodine $^{125}$I, $^{103}$Pd,$^{106}$Ru, and a number of other radio-isotopes. For diagnostic purposes, radioactive isotopes, such as $^{123}$I, can replace the stable isotope $^{127}$I and are used as tracers for metabolic pathways. $^{123}$I decays with the emission of a very hard photon with energy 159 keV that is detected in *Single Photon Emission Computed Tomography* (SPECT) cameras. They allow us to reconstruct the location where the isotope decays. Using isotopes that decay by emitting a positron, allows us to detect photons created in the annihilation of the positrons with nearby electrons. Two photons with an energy of $E = 511$ keV, equal to the electron's rest mass, are emitted back-to-back and from recording them in coincidence, the locus of the decay can be determined. This method is called *Positron emission tomography* (PET). The needed isotopes are created in small cyclotrons, typically producing protons with energies between 7 and 70 MeV. Nowadays the Penning ion sources [13.2.2] in the center of the cyclotrons are optimized to produce $H^-$ ions and the cyclotron accelerates them. Once they reach their maximum energy, a stripper foil [13.3] converts the $H^-$ ions to protons that are deflected in the opposite direction and easily extracted and guided to the target. The target material and beam energy determine the type of radioactive isotope generated in the process. The high demand for radionuclides caused more than 1000 of these small cyclotrons to be built and operate world-wide [118].

## 14.7 INDUSTRIAL ACCELERATORS

The vast majority of all accelerators are used in industry to manufacture, modify, and characterize materials. The largest group with more than 10000 systems built are ion implanters, similar to the one already mentioned in Section 14.5. They are the workhorse of the semiconductor industry, where they are used to implant dopants into the wafers from which integrated circuits are manufactured. Not only the total number of systems, but also their variety is large with many different ion species, energies, and currents. Many different ion sources [Section 13.2.2], usually based on a plasma column, where electrons are trapped transversely by a magnetic field and longitudinally by negatively biased electrodes [Figure 13.3] create the ions that are implanted. The acceleration voltage then determines the depth at which they are implanted. The current, jointly with the exposure time, determines the density of the implants.

Electrons with energies up to a few 100 keV are used to locally deposit heat in a material and thereby modify its properties. *Hardening* of some materials is accomplished by quickly heating the surface locally which forces the materials to undergo a phase transition to a structure that is harder. Other materials are hardened by rapidly melting and re-solidifying the surface, a process called *glazing.* The majority systems using low-energy electrons, however, are *electron beam welding* machines. They are able to locally melt two metals, even different metals, and permanently join them upon solidification of the weld. Moreover, the

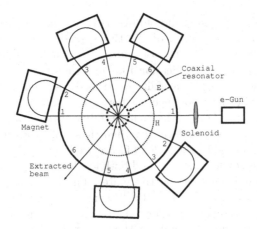

**Figure 14.2** Top view on to a Rhodotron$^R$. See the text for an explanation.

weld can be as deep as several tens of cm, which allows us to join rather thick metal sheets, such as those used in the manufacture of large steam turbines [118].

Apart from heating, as mentioned in the previous paragraph, electrons also ionize the material and thereby change its physical, chemical, and biological properties. One class of examples comes from cross-linking polymer chains. This increases the heat resistance, for example, of cable insulation, it is used to cure the polymers in tires for automobiles, and it is used to manufacture heat-shrinkable plastic tubing. We already mentioned its use in sterilization of medical waste prior to its disposal, which requires very high doses. Similarly, irradiating waste water decomposes toxic compounds and pathogenic microorganisms. Lower doses are used to irradiate food products in order to kill dangerous bacteria. The range of the charged electrons in matter is limited and converting the electrons to x-rays in a target increases their ability to penetrate thicker samples, which allows us, for example, to make x-ray photographs of entire trucks at customs stations in search for illicit material. The accelerators to produce electron beams with energies up to about 300 keV are based on an acceleration gap connected to a high-voltage generator. For mid-range energies between 300 keV and 5 MeV, for example, Cockroft-Walton generators [Chapter 1] and Dynamitrons$^{(R)}$ are used. The latter are based on transformers with primary windings at ground potential, which are inductively coupled to the secondary windings with rectifiers, connected in series. Above energies of 5 MeV electron linacs are used at moderate currents. If, on the other hand, high continuous currents of up to 100 mA at energies of up to 10 MeV are required, Rhodotrons$^{(R)}$ are used. They are based on a coaxial $\lambda/2$–resonator operated at frequencies in the range 100-200 MHz. Figure 14.2 illustrates their operation. The electrons are created in a triode gun [Section 13.2.1] and focused into the coaxial [Figure 6.2] resonator. The inner conductor is shown as the dashed small circle in the center, through which the electrons pass, just as the radial electric field changes polarity. During their first traversal of the resonator, indicated by the label "1," the electrons are accelerated by the radial electric field $E$ and gain up to 1 MeV of energy. Their direction is reversed in the first magnet, which deflects and focuses the electrons. On its second traversal through the resonator on the trajectory labeled "2," the electrons gain 1 MeV once again, as they do on the subsequent traversals. Since the magnet on the bottom left is missing, the electrons are extracted and guided to the site where they irradiate samples.

Neutron generators often use deuterium, ionized in a source [Section 13.2.2], and accelerated to moderate energies. Impinging the neutrons onto targets that contain deuterium or tritium, produces neutrons that are versatile probes to analyze materials. Low-energy neutrons, captured and inelastically scattered, produce a characteristic $\gamma$–radiation that carries information about the elemental and isotopic composition of the irradiated sample. Moreover, when using pulsed neutron sources, the time delay between the secondary $\gamma$–rays and re-emitted neutrons carries information about fissionable materials. It is created as a by-product when treating radioactive waste or manufacturing nuclear weapons. Moreover, probing for the ratios of C, N, H, and O in samples allows us to detect explosives, gives indications about chemical weapons, and possibly narcotics. A large number of small neutron sources are used in the exploration of oil fields. They are enclosed in the drilling assembly deep down in the borehole, where the secondary $\gamma$–rays provide information about the surrounding rock formation at great depths and the prospect to find oil or gas. These smaller neutron generators use accelerating voltages in the $100\,\text{kV}$ range and produce about $10^8$ neutrons per second in pulsed mode. Larger systems, often several meters long, use RFQs [Section 13.2.5] or Dynamitrons to accelerate much larger numbers of deuterium ions, and produce rates above $10^{13}$ neutrons per second.

## QUESTIONS AND EXERCISES

1. Let's assume that you will visit an accelerator laboratory, either for research or to see a friend, and you want to make yourself knowledgeable about the lab. Therefore, you write a short 2-page essay, similar in spirit and organization to the sections in this chapter. You structure the essay by first discussing the purpose of the lab and the research done there, then follow the beam from its generation to its destination—an experiment. Exciting places you might visit, are

   (a) the *Relativistic Heavy Ion Collider* RHIC at Brookhaven National Laboratory;

   (b) the *Continuous Electron Beam Facility* CEBAF at Jefferson Lab;

   (c) the *Facility for Anti-proton and Ion Research* FAIR at GSI in Darmstadt;

   (d) the nuclear physics facility *Spiral-2* in Caen;

   (e) the microtron MAMI in Mainz;

   (f) the *Accelerator Test facility* ATF-2 in Tsukuba;

   (g) any of the third-generation synchrotron light sources, for example ALS, Bessy-2, SLS, PLS, or their brethren operating at higher beam energies ESRF in Grenoble or APS in Argonne;

   (h) the SASE FEL facilities SACLA in Harima and the SwissFEL near Zürich;

   (i) the $\Phi$–factory Daphne in Frascati;

   (j) the B-factory KEKB in Tsukuba;

   (k) the *Heidelberg Ion Therapy* center HIT;

   (l) the *International Fusion Materials Irradiation Facility* IFMIF;

   (m) the large cyclotrons at TRIUMF in Vancouver or at PSI near Zürich.

   Good places to search are the web site www.jacow.org with proceedings of practically all accelerator conferences that have ever taken place and inspirehep.net with a vast data base of research articles.

2. There are a few accelerators in the planning stages and you might wonder what they are about. Briefly write about them as well.

(a) The *International Linear Collider* ILC project.

(b) The *Compact Linear Collider* CLIC.

(c) *Accelerator Driven Systems* ADS to process nuclear waste.

# The Student Labs

In these student labs we describe a few experiments that illustrate activities, relevant to building or operating accelerators, that can be done without access to a real machine. Most are based on inexpensive and readily available components. Note, however, that the purpose of the labs is to obtain hands-on experience with these topics and not so much to perform a high-precision measurement. Keeping this in mind, we start with beam size measurements and use a laser pointer as a substitute for a particle beam.

## A.1  BEAM PROFILE OF LASER POINTER

In this lab we measure the transverse profile of the beam emitted from a laser pointer and determine the beam profiles by recording the image on a screen with a camera, as discussed in Section 7.3. We direct the laser pointer onto a sheet of paper and record it with a web camera. The intensity of the laser is controlled with a pulse-width modulated supply voltage. In order to reduce interference from stray light, we encapsulate this setup in a box. Before recording the laser spot, we place a paper with markers with known distances between them on the screen in order to determine the conversion from pixels on the camera to mm on the screen. In our case, we found 6.6 pixel/mm horizontally and 7.7 pixel/mm vertically.

After recording the image, we can process it further with MATLAB, such that we can cut out the region of interest with the spot, display the image as a contour plot and show the projections onto the horizontal and vertical axis. Figure A.1 shows the result of running the following MATLAB script:

```
% analyze_image.m, V. Ziemann, 181128
im=rgb2gray(imread('set3/power1.jpg')); roi=im(260:400,320:500);
subplot(2,2,1); imshow(imcomplement(roi+4));
subplot(2,2,4); contour(flipud(roi));
xlabel('x [pixel]'); ylabel('y [pixel]')
subplot(2,2,2); plot(sum(roi,2),'k'); xlim([0,size(roi,1)]);
camroll(270); set(gca,'YTick',[]); xlabel('y [pixel]');
title(['FWHM = ',num2str(fwhm(sum(roi,2))),' pixel'])
subplot(2,2,3); plot(sum(roi,1),'k'); xlim([0,size(roi,2)]);
set(gca,'YTick',[]); xlabel('x [pixel]')
title(['FWHM = ',num2str(fwhm(sum(roi,1))),' pixel'])
```

After reading the image with the `imread()` function, the `rgb2gray()` function converts it to grayscale. We then cut out the *region of interest,* store the result in `roi`, and display it with inverted grayscale and a 4 pixel baseline added in the top-left subplot. The bottom

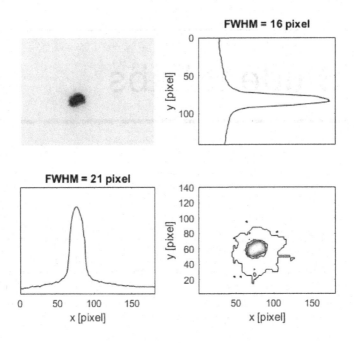

Figure A.1 The image (top-left) and contour plot (bottom-right) of a laser spot recorded with a webcam and the projections (top-right, bottom-left).

right subplot contains the same image, but displayed as a contour plot. The `flipud()` function reverses the vertical axes, because images, by convention, have their origin in the top left corner. The other two axes show the projections within the region of interest onto the respective axes. Note that the `camroll()` function rotates the axes in the top-right plot. The `fwhm()` function determines the *full-width at half-maximum* of the profiles. It is calculated by first finding the baseline and the maximum and the curve and then searching to the left and to the right from the maximum until the half-height points are found. Their difference serves as an estimate for the FWHM. An implementation in MATLAB is given in [104] and also available from this book's web page. Using the FWHM is more robust than calculating the central second moment, because the latter is easily biased by noise in the tails of the curve, whereas the FWHM focuses on the central region. Using the FWHM in pixels and the conversion factors from above, we find that the widths are 3.2 mm and 2.1 mm, respectively. Note, that the FWHM of a Gaussian distribution is 2.35 times its standard deviation $\sigma$.

The image on which Figure A.1 is based already had the laser intensity at the lowest possible value and still the contour plot on the lower right shows a very rapid transition between the lowest and the largest intensity levels in the center of the spot. This indicates that the image is saturated and is not a faithful representation of the beam spot. This is a common problem when recording images with cameras. Either attenuating the laser with polarizers and neutral density filters or carefully adjusting the camera gain and shutter can alleviate the problems with saturation. Adjusting the camera is unfortunately unavailable on the inexpensive webcam used in this experiment, but inserting a polarizer and rotating it to an angle that significantly reduces the laser intensity allows us to record profiles without significant saturation. We do not dwell on this topic further, but point out that

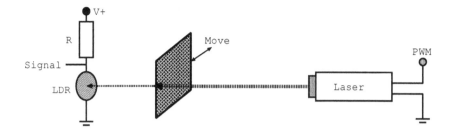

Figure A.2 Measuring the width of a laser pointer by moving an opaque obstacle across the laser. The obstacle obscures the light-sensitive resistor (LDR) that is part of a voltage-divider resistor and changes the signal, which is recorded by an Arduino micro-controller (from [104], used with permission).

they illustrate the practical issues one often encounters when using cameras to record beam profiles.

Instead of improving this method, we consider a second method, which resembles the operation of wire scanners, also mentioned in Section 7.3. But instead of a wire, we move an obstacle across the beam and record the changing signal from a light-sensitive resistor (LDR). The geometry is illustrated in Figure A.2. The laser module is powered by a pulse-width modulated signal in order to adjust its intensity. The shaded obstacle is mounted on a movable wagon, such that its position can be correlated with the signal from the voltage divider on the left-hand side, consisting of a $10\,\mathrm{k\Omega}$ resistor and the LDR. Building such a setup with a salvaged chassis from an old CD-drive, which contains a wagon driven by a stepper motor and controlled with an Arduino, is described in [104]. We can, however, use any linear actuator with a step size of 0.1 mm or better to position the obstacle.

On the upper plot in Figure A.3 we show the raw signal from the LDR as a function of the horizontal position of the obstacle. The transition around 6 mm from illumination on the left-hand side to occultation on the right is clearly visible. After smoothing data with the MATLAB function `smooth(rawdata,3)`, which averages three consecutive points, we calculate its derivative by calculating the difference between two adjacent points. On the lower plot in Figure A.3, also taken from [104], we show the derivative, which reveals a somewhat asymmetric profile of the laser beam. The width of the profile, here given as the FWHM, is reported above the lower graph. Here the horizontal FWHM size is found to be 0.8 mm.

This rather general setup allows us to address further questions: How does varying the laser intensity by adjusting the PWM signal to power the laser change the beam size? Or, does diffraction off the edge of the obstacle play a role? Do we find the same profile, when crossing the laser from the opposite direction? Experiment with different methods to extract the width. Compare different laser pointers. Compare with different methods to determine the beam size.

Despite these open questions, which are left as suggestions for exercises, we will use the scanner to determine the beam sizes while changing the focusing in order to determine the "beam matrix" and the "emittance" of the laser in the next lab.

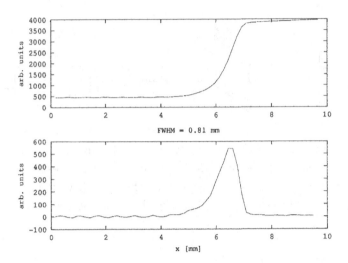

**Figure A.3** Top: The illumination of the LDR as a function of the position of the obstacle. Bottom: the derivative of the upper curve, giving the beam profile (from [104], used with permission).

## A.2   EMITTANCE MEASUREMENT WITH A LASER POINTER

In the lab, we describe a method to measure the "beam matrix" of the laser. The method resembles a quadrupole scan, described in Section 7.4, but instead of changing the excitation of a quadrupole, we change the longitudinal position of a focusing lens, as shown in Figure A.4. In this way, we vary the transfer matrices between the laser and the beam size scanner and obtain independent constraints that allow us to determine $\hat{\sigma}_{xx}, \hat{\sigma}_{xx'}$, and $\hat{\sigma}_{x'x'}$ at a reference plane from the horizontal beam size measurements. Of course, we need to carefully align the lens on the optical axis, in order to avoid displacing the laser laterally, when moving the lens. This was discussed in Sections 8.1.1 and 8.2.1.

In order to correlate the measured beam sizes $\sigma_x(s)$ to the position $s$ of the lens, we need to determine the transfer matrix $R(s)$ from the reference plane to the beam size measurement as a function of $s$. Here we refer to Figure A.4 for the definition of the used symbols. For $R(s)$ we then find

$$R(s) = \begin{pmatrix} 1 & L-s \\ 0 & 1 \end{pmatrix} \begin{pmatrix} 1 & 0 \\ -1/f & 1 \end{pmatrix} \begin{pmatrix} 1 & s \\ 0 & 1 \end{pmatrix} \tag{A.1}$$

and, following the same strategy laid out in Section 7.4, we pretend to know $\hat{\sigma}_{xx}, \hat{\sigma}_{xx'}$, and $\hat{\sigma}_{x'x'}$ at the reference plane, and derive the measured beam sizes from it. We obtain

$$\sigma_x^2(s_i) = R_{11}^2(s_i)\hat{\sigma}_{xx} + 2R_{11}(s_i)R_{12}(s_i)\hat{\sigma}_{xx'} + R_{12}^2(s_i)\hat{\sigma}_{x'x'} \tag{A.2}$$

with the transfer matrix elements, found from evaluating Equation A.1

$$R_{11}(s) = 1 - \frac{L}{f} + \frac{s}{f} \quad \text{and} \quad R_{12}(s) = L - \frac{Ls}{f} + \frac{s^2}{f} . \tag{A.3}$$

Measuring the beam sizes $\sigma_x(s_i)$, while moving the lens to position $s_i$ leads to the following

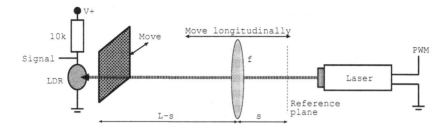

Figure A.4 Emittance measurement of a laser pointer.

linear set of equations

$$
\begin{pmatrix} \vdots \\ \sigma_x^2(s_i) \\ \vdots \end{pmatrix} = \begin{pmatrix} \vdots & \vdots & \vdots \\ R_{11}^2(s_i) & 2R_{11}(s_i)R_{12}(s_i) & R_{12}^2(s_i) \\ \vdots & \vdots & \vdots \end{pmatrix} \begin{pmatrix} \hat{\sigma}_{xx} \\ \hat{\sigma}_{xx'} \\ \hat{\sigma}_{x'x'} \end{pmatrix} , \tag{A.4}
$$

which are just multiple copies of Equation A.2, arranged in a convenient way. Since the measurements $\sigma_x^2(s_i)$ and the corresponding transfer matrix elements are known, we can determine the beam matrix elements at the reference plane $\hat{\sigma}_{xx}, \hat{\sigma}_{xx'}$, and $\hat{\sigma}_{x'x'}$ by matrix inversion or by using the pseudo-inverse discussed in Section 8.4.2 and given by Equation 8.48. Measurement uncertainties $\Delta\sigma_x(s_i)$ of the beam size measurements can be taken into account in a straightforward fashion by dividing each one of the Equations A.2 by the corresponding uncertainty, given by $\Delta\sigma_x^2(s_i) = 2\sigma_x(s_i)\Delta\sigma_x(s_i)$, as previously discussed in Section 8.4.2 as well. Moreover, the measured values $\sigma_x(s_i)$ should comprise the minimum beam size and significantly larger values on either side of the minimum in order to accurately determine both the beam size and angular divergence. Finally, take notice that the measure values $\sigma_x$ are standard deviations, not FWHM.

Once the beam parameters at the reference point are known we can calculate the "emittance" of the laser beam $\hat{\varepsilon}$ from Equation 7.8 and find $\hat{\varepsilon}^2 = \hat{\sigma}_{xx}\hat{\sigma}_{x'x'} - \hat{\sigma}_{x'x'}^2$, which can be expressed in units of the "emittance" of a diffraction limited laser beam $\varepsilon_\lambda$ as $\hat{\varepsilon} = M^2\varepsilon_\lambda$ with $\varepsilon_\lambda = \lambda/4\pi$. The Twiss parameters $\hat{\beta}$ and $\hat{\alpha}$, which can also be found from Equation 7.8, are related to the Rayleigh length and the wave front curvature of the laser beam [119] at the reference plane. We discuss the relation between light optics and charged-particle optics further in Appendix B.4, available online from this book's web page.

## A.3  HALBACH MULTIPOLES AND UNDULATORS

In this lab we build small magnets using readily available, and inexpensive, cubes of permanent magnet material. Despite their small size the magnets have remarkably high magnetic fields. We base the design on the discussion of the Halbach multipoles from Section 4.5, but realize that we have to adapt the calculations for the trapezoidal magnet shapes to the cubic shapes.

We follow the calculation from Section 4.5.2, but have to change the trapezoidal geometry shown in Figure 4.20 to that shown in Figure A.5 with two cubes that rotate by a suitable "tumbling angle" $k\phi$ that depends on their azimuthal position $\phi$. We first determine the contribution of a single cube to the multipolar order, characterized by the exponent $m$, of the complex potential $\hat{z}^m$, by calculating the field $\underline{\tilde{B}}^*$ from Equation 4.46, but using the square area $\Omega$ from Figure A.5 instead. The integral then becomes

$$\int_\Omega \frac{dxdy}{(x+iy)^{m+2}} = \int_{r-h/2}^{r+h/2} dx \int_{-h/2}^{h/2} \frac{dy}{(x+iy)^{m+2}} = \begin{cases} 2\arctan\left(\frac{h^2}{2r^2}\right) & \text{for } m=0 \\ \frac{h^2 r}{r^4+h^4/4} & \text{for } m=1. \end{cases} \tag{A.5}$$

The integrals are straightforward to evaluate and lead to the results shown for the dipolar ($m = 0$) and quadrupolar ($m = 1$) component. Higher orders are straightforward to evaluate but the expressions become rather lengthy and we confine ourselves to dipoles and quadrupoles.

Adding the contributions from $M$ cubes, located at angles $2\pi j/M$ with $j = 0, \ldots, M-1$, where each cube is rotated by its tumbling angle $k\phi$, gives us the contribution from all the cubes to the field $\hat{\underline{B}}^*$

$$\hat{\underline{B}}^* = \tilde{\underline{B}}^* \sum_{j=0}^{M-1} e^{2\pi i(k-m+2)j/M} . \tag{A.6}$$

Following the reasoning from the end of Section 4.5.2, the sum of the exponential factors equals $M$, if $(k-m-2)/M$ is an integer, and zero otherwise. For the dipole and quadrupole fields we therefore obtain

$$\hat{\underline{B}}^* = B_r \frac{M}{\pi} \arctan\left(\frac{h^2}{2r^2}\right) \quad \text{for a dipole,} \tag{A.7}$$

$$\hat{\underline{B}}^* = B_r \frac{M}{\pi} \frac{h^2 r}{r^4 + h^4/4} \quad \text{for a quadrupole.}$$

The achievable fields are rather impressive. For a dipole with $M = 8$ permanent magnet blocks, similar to the one shown on the right-hand side in Figure 4.19, but made of $h = 5$ mm cubes and with $r = 9.5$ mm. We find that the field is given by $0.35 \times B_r$ and for NdFeB magnets with $B_r \approx 1.2$ T the dipole fields almost reach $0.42$ T. For quadrupoles with similar geometry, but different tumbling factor, gradients above $70 \times B_r$ T/m can be reached, albeit in small apertures, only.

We now use these parameters and design a frame to hold the 5 mm permanent magnet cubes with OpenSCAD. We base the design on a large cylinder and punch out the central hole as well as small square holes to hold the magnets. The following MATLAB script implements this.

```
% halbach_M8_simple.m, V. Ziemann, 180124
clear all
M=8;        % number of permanent magnet cubes
k=2;        % tumble factor: k=2 -> dipole, k=3 -> quadrupole
h=5;        % size of cube
h2=h+0.4;   % little extra space for tolerances
fp=fopen('test.scad','w');
fprintf(fp,'difference(){\n');
fprintf(fp,'  cylinder(h=%6.2f,r=14);\n',h+1.5);
fprintf(fp,'  translate([0,0,-0.5]) {cylinder(h=%6.2f,r=4.5);}\n',10*h);
for j=0:M-1      % loop over segments
  phi=j*360/M;
  psi=k*phi-phi;  % subtract one because rotations add up
  fprintf(fp,'  rotate([0,0,%8.3f]) {translate([9.5,0,1.5]) {\n',phi);
  fprintf(fp,'    rotate([0,0,%8.3f]) {translate([-2.7,-2.7,0]) {
```

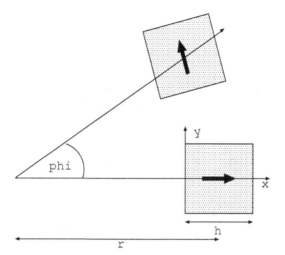

Figure A.5 The geometry with the magnetic cubes.

```
                    cube([5.4,5.4,5.4]);}\n',psi);
     fprintf(fp,'      translate([-0.25,%6.2f,%6.2f]) {cube([0.5,0.5,10]);}
                    }}}\n',h2/2-0.02,h2-1);
   end
   fprintf(fp,'}\n'); fclose(fp);
   system('openscad test.scad &');
```

After clearing the workspace we define the number $M$ of cubes to use, the tumble factor $k$ to determine the multipolarity of the magnet, and the size of the cube. We define h2 to be 0.4 mm larger to make the holes a little bigger to allow for finite tolerances. Then we open the output file test.scad to which the OpenSCAD commands are written and immediately write the command difference() and the definition of the large cylinder. We give it the radius 14 mm and make it 1.5 mm higher than the size of the cube. From the first cylinder we subtract a displaced second cylinder with radius 4.5 mm and large height to ensure that it punches a hole into the larger cylinder. In the following loop over the $M$ cubes, each rotated by angle phi, we define the 5.4 mm cubes and a smaller 0.5 mm cube to indicate the direction of the easy axis. Note that each cube is rotated by an additional angle psi with respect to the rotation with phi. Finally, after the loop, we write the closing brace, close the file, and automatically open the file in OpenSCAD with the system() function call.

Running the above MATLAB script with $k = 2$ and $k = 3$ to obtain frames for a dipole and quadrupole magnet respectively produces the images shown in Figure A.6. We clearly observe the different tumbling factors for the two cases. In particular, on the right-hand side, the notch that indicates the easy axis points downwards in the upper and lower cube, while the easy axis on the left- and right-most cube point upwards. This is the same pattern we already observed in Figure 4.19. The frame for the quadrupole is shown on the right-hand side in Figure A.6. The easy axes of the cubes in the top left and the bottom right are facing each other, just as expected for quadrupoles.

Running the script, pressing F6 in OpenSCAD to create the mesh and export the geometry as a .stl file from the *File→Export* menu point allows us to subsequently load the .stl file in a slicer program and create a 3D print of the frame. Examples are shown in

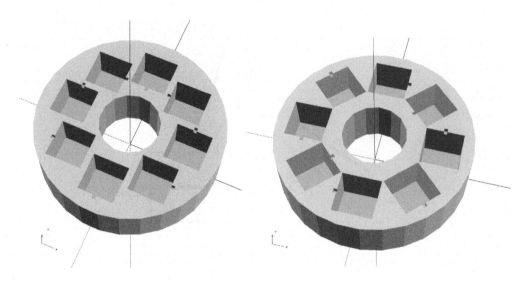

Figure A.6 The frames for a small dipole (left) and quadrupole (right) modeled with OpenSCAD. The square holes have sides with a length of 5.4 mm and the inner diameter is 9 mm and the outer diameter is 28 mm.

Figure A.7, for the dipole with permanent magnets installed on the left-hand side and a quadrupole without magnets on the right. These magnets are very short, the length of the magnetically active material—the permanent magnets—is only 5 mm, but making the large cylinder higher and subtracting higher cubes will give us a frame into which we can stack several permanent magnets on top of each other. We must take care, however, that the easy axes in each stack point in the same direction. In this way we increase the magnetic length in steps of 5 mm.

Our next task is to build a small undulator magnet with two full periods and one tapered entrance and one exit period to adjust the field integrals as discussed in Section 4.6. We will use 5 mm cubes of permanent magnet material and use four cubes per period $\lambda$ and allow for 1 mm space between the magnets. This results in a period of $\lambda = 24$ mm and the total length of the undulator will be a little under 100 mm. Equation 4.57 describes the magnetic flux density for such an undulator, but only for permanent magnets that touch each other. It turns out that additional space between permanent magnets can be accounted for by a packing factor $\varepsilon$. It is given by the length of the permanent magnet material to the total length of a period. Since there is one extra mm of space between adjacent magnets, the packing factor is $\varepsilon = 5/6 \approx 0.83$. Taking it into account turns Equation 4.57 to

$$\underline{\hat{B}}^*(\hat{z}) = 2i\underline{B}_r \frac{\sin(\varepsilon\pi/4)}{\pi/4}\left(1 - e^{-2\pi h/\lambda}\right)e^{-2\pi g/\lambda}\cos(2\pi z/\lambda) . \tag{A.8}$$

We recall that $h$ is the height of the permanent magnet material, here the size of a cube, and $g$ is the half-gap of the undulator. We will assume it to be $g = 5$ mm in the following design. For the peak flux density $B_p$ on the undulator axis, we then obtain $B_p = 0.305 \times B_r$ and for a half gap of $g = 10$ mm we find $B_p = 0.083 \times B_r$.

We use the following OpenSCAD program to design the undulator with half-gap $g = 5$ mm

```
// Halbach undulator with 10mm gap, uncompensated field integrals
```

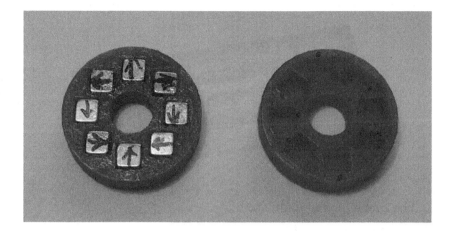

Figure A.7 The printed frame for a dipole with permanent magnet cubes inserted (left) and the empty frame for a quadrupole (right). The side of a cube has a length of 5 mm.

```
difference() {
union() {
  cube([94,40,4]);
  color("blue") {
    translate([0,23,4]){cube([94,10,6]);}
    translate([0,7,4]){cube([94,10,6]);}
  }
}
for (i=[0:14]) {translate([4.95+i*6,27.5,8]) {
  rotate([0,0,-90*i-90]) {
    cube([5.2,5.2,6],center=true);
    translate([2.65,0,6]) {cube([0.5,0.5,10],center=true);}
  }}}
for (i=[0:14]) {translate([4.95+i*6,12.5,8]) {
  rotate([0,0,90*i-90]) {
    cube([5.2,5.2,6],center=true);
    translate([2.75,0,6]) {cube([0.5,0.5,10],center=true);}
  }}}
}
```

The undulator frame is constructed as the difference of the union of the base plate with the blue blocks and the holes for the 5 mm permanent magnet cubes. The hole is slightly larger to ease insertion of the magnets. We also add the small cubical 0.5 mm holes to indicate the direction of the easy axis for each magnet. All we now have to do is to order the magnets, wait for their arrival, and insert them in their respective frames.

## A.4 MAGNET MEASUREMENTS

Once the magnets from the previous lab are assembled, we need to characterize them. In this lab, we determine the longitudinal variation of the magnetic flux density $B$ along the

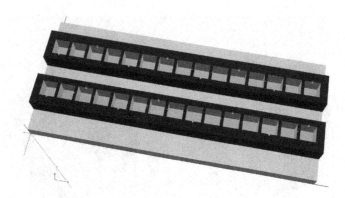

Figure A.8  The frame for the undulator.

"beam axis" of the Halbach dipole, which is shown on the left-hand side in Figure A.7. We therefore mount a A1302 Hall sensor onto the chassis of a salvaged CD-drive and mount the magnet onto the movable wagon. The left-hand side of Figure A.9 illustrates the setup. This configuration resembles the profile measurement of the laser pointer in Appendix A.1 and we, again, use an Arduino to control the stepper motor on the CD-drive. A limit switch on the chassis allows us to always start a measurement from a well-defined reference position. A measurement is initiated by sending the command SCAN? to the Arduino, which first moves the wagon to the reference, or home, position, echos SCAN to the serial line, and then slowly steps in the opposite direction, while simultaneously reading out the Hall sensor via an analog input and sending the values, converted to Tesla, back across the serial line. The code on the Arduino closely follows the discussion in [104] and is available from this book's web page.

The protocol of sending SCAN? and receiving a known number of values is easily implemented in MATLAB. We first have to define a serial device with s=serial('/dev/ttyUSB0', 'BaudRate',9600), open that device with fopen(s), before initiating the scan, reading the magnetic field values from the serial line, and producing a plot of the longitudinal field profile. The following code implements this

```
fprintf(s,'SCAN?'); fscanf(s);      % read the "SCAN" echoed
Bfield=zeros(1,150);
for i=1:150                         % loop over the data points
  Bfield(i)=str2double(fscanf(s)); % string to double
end
xscale=0.18*(0:149);                % 0.18 mm/motor step
plot(xscale,Bfield)
```

and produces the plot shown on the right-hand side in Figure A.9. We see that the peak field, which occurs when the Hall sensor is inside the magnet, reaches 0.16 T. With a commercial Gaussmeter we measured about 0.14 T. The discrepancy is likely due to the tolerance in the sensitivity of the Hall sensor, which can vary between 1.0 and 1.6 mV/Gauss, where we used 1.3 mV/Gauss in the Arduino sketch. The measured values, however, are consistent with remanent fields in the range of $\underline{B}_r \approx 0.4$ to 0.5 T, according to Equation A.7. This appears to be a reasonable value for the low-grade magnets used.

This lab illustrates the methodology of measuring magnets, but in order to make it use-

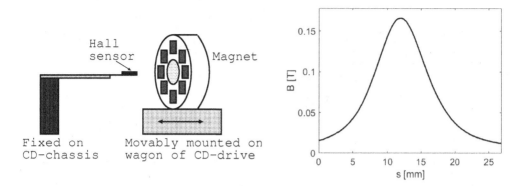

Figure A.9 The magnetic measurement system and the vertical field in the dipole shown on the left-hand side in Figure A.7.

ful, the sensor must be calibrated properly, a sturdier translation stage and micro-stepping controller should be used, but this is left as a project for the interested reader.

## A.5   COOKIE-JAR CAVITY ON A NETWORK ANALYZER

In this lab we calculate and measure the frequency and $Q$–value of the fundamental $TM_{010}$–mode of a tin can that, once upon a time, contained butter cookies. We use this cookie jar as a substitute for the pill-box cavity we theoretically analyzed in Section 5.1.

The height of our jar is approximately $l = 65\,mm$ and the radius is $R = 95\,mm$. See the left-hand side in Figure A.10 for an illustration and compare to Figure 5.1. The eigen-frequencies of the resonator are given by Equation 5.22. For the fundamental mode with $n = 0, m = 1$, and $p = 0$ we find $f_{010} \approx 1.2\,GHz$. Before measuring the resonance frequencies, we need to couple power into the cavity. To this end, we drill holes into the lid of the jar and install BNC connectors with their outer contact, the shield, electrically connected to the metal of the jar. To its inner contact, we solder short wires with a length of a few mm, shown as the dotted lines below the BNC connectors on the top of the jar. One connector (BNC1) is placed in the center of the lid. The longitudinal electric field in the cavity can excite signals in the antenna, which can be detected on BNC1. The second BNC connector (BNC2) serves as the "power coupler," discussed in Section 6.3, to excite the fields in the cavity through the small antenna, shown as the dotted line below BNC2. The antennas, which couple power into the cavity and simulate the beam, interact with the electric field in the cavity, whereas the loop couples to the magnetic field. The loop is electrically connected to the inner wall of the jar and to the center connector of BNC3. The coupling factor $\beta$, introduced in Section 6.3, can be adjusted by changing the length of the small antennas soldered to BNC1 and BNC2, for example, by cutting off some of the wire of the antenna. The $\beta$ of the loop-coupler can be changed by increasing or decreasing the size of the loop, for example by bending the wire of the loop to a different shape.

We use a *network analyzer* (NA) to experimentally determine the resonance frequency and other parameters. NAs have two or more connectors, referred to as *ports,* equipped with directional couplers, mentioned near the end of Section 6.2. Sensors, connected to the couplers, measure the signals flowing out of, and into the ports. Connecting the cookie-jar and exciting one port, while simultaneously recording the signal reflected from the jar,

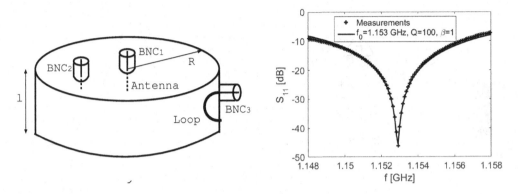

Figure A.10  The cookie jar used as a pill-box cavity.

allows us to determine the reflection coefficient $\Gamma$ defined in Equation 6.20. It is equal to one of the S–parameters, $S_{11}$, that is returned from the NA. Plotting $S_{11}$, while varying the excitation frequency $f$, the resonance frequencies of the jar will appear as dips, because on resonance, the cavity absorbs the power and dissipates it in the cavity walls, as described in Section 6.4.1. This determines the $Q$–value of the cavity, discussed both in Sections 6.3 and 6.4.1. On the right-hand side of Figure A.10 we show the measured values of the fundamental mode of our cookie-jar as the crosses. The raw data was saved from the NA to a USB-stick as a text file and imported into MATLAB, where it is displayed and processed further. From Figure A.10, we find the measured resonance frequency to be 1.153 GHz, which is in the vicinity of the value we earlier estimated from the geometry. The solid line denotes a fit of the real part of Equation 6.20 to the raw data. It is calculated in the following MATLAB script

```
d=importdata('f1b.csv');
f=d.data(:,1)/1e9; amp=d.data(:,2)
g=@(f0,Q,beta,x)20*log10( ...
    abs((beta-1-i*Q*(x./f0-f0./x))./(beta+1+i*Q*(x./f0-f0./x))))
res=fit(f,amp,fittype(g),'Start',[1.153,200,1.1])
plot(f,amp,'k+',f,g(res.f0,res.Q,res.beta,f),'k');
```

which determines the fit parameters $f_0, Q$, and $\beta$, including their confidence intervals. After importing the file f1b.csv with the raw data, we define the function g() to encode Equation 6.20 and assign it as the function to fit in the call to fittype(). After providing meaningful starting values [f0,Q,beta] to the MATLAB function fit(), it returns the fitted parameters in the structure res. Fit parameters are accessible as, for example, res.Q for the $Q$–value, which turns out to be around 100. The coupling $\beta = 1$ equals the critical value, at which reflections are minimized.

The quality of our make-shift cavity, the cookie-jar, is rather modest, but this is expected, since we did not pay special attention to its design and we only removed the dielectric inside, the cookies. Furthermore, the accuracy of the measurement can be improved. For high-precision measurements, one must *calibrate* the NA by measuring the S–parameters close to the cavity. This is done by terminating the connecting cable with an open connection, with a short, and with a 50 Ω resistor. This allows the NA to automatically remove the influence of imperfect connectors and cables on the measurements.

This lab can be extended in multiple ways, for example, by varying the length of the antenna to change the coupling $\beta$. Furthermore, the transmission coefficient from one BNC connector on the cookie-jar to another one is described by the $S_{21}$ and can be measured by connecting the BNC connectors to two ports on the NA. This simulates exciting the cavity through one connector and extracting it from the second, thus mimicking the power extracted by the beam.

## ROUGH IDEAS FOR FURTHER LABS

1. Mount mirrors on stepper motors and use them for steering the beam from a laser pointer.

2. Observe the laser spot on a screen and design a laser pointer orbit correction system to move it in a controlled way on the screen.

3. Experiment with model servos. Investigate pointing stability. The servo mimics a noisy power supply.

4. Design and build a Halbach dipole with two concentric dipoles, where a second "ring" of magnet blocks is mounted outside the first eight magnet blocks and will increase the field.

5. Make the outer ring such that it can be rotated with respect to the inner ring. This allows you to continuously adjust the dipole strength. Verify this and calibrate the magnet of the magnet test bench.

6. Design and build a variable permanent-magnet quadrupole with two concentric, but movable, rings of magnet blocks, where the outer magnets can be rotated with respect to the inner blocks.

7. Use a model-servo or stepper motor to press a lever against the bottom of the cookie-jar in order to tune its frequency.

# Appendices Available Online

All appendices are available from this book's web page at `https://www.crcpress.com/9781138589940`.

B.1   LINEAR ALGEBRA

B.2   MATLAB PRIMER

B.3   OPENSCAD PRIMER

B.4   LIGHT OPTICS, RAYS, AND GAUSSIAN

B.5   MATLAB FUNCTIONS

# Bibliography

[1] Mathworks Inc. Project web site: https://www.mathworks.com/.

[2] Scientific Programming Language GNU Octave. Project web site: https://www.gnu.org/software/octave/.

[3] MADX. Project web site: http://mad.web.cern.ch/mad/.

[4] K. Brown. The general first and second order theory of beam transport optics. Technical Report SLAC-TN-64-027, SLAC, 1964.

[5] K. Brown et al. Transport: A computer program for designing charged particle beam transport systems. Technical Report CERN-73-16, CERN, 1973.

[6] OpenSCAD User Manual. Project web site: https://en.wikibooks.org/wiki/OpenSCAD_User_Manual.

[7] L. Susskind. *Special Relativity and Classical Field Theory*. Penguin Books, London, 2018.

[8] H. Goldstein, J. Safko, and C. Poole. *Classical Mechanics*. Pearson, London, UK, 3rd edition, 2014.

[9] E. Courant and H. Snyder. Theory of the alternating gradient synchrotron. *Annals of Physics*, 3:1–48, 1958.

[10] I. Gradstein and I. Rishik. *Tables of Series, Products and Integrals*. Verlag MIR, Moskau and Harri Deutsch, Thun, first edition, 1981.

[11] H. Wiedemann. *Particle Accelerator Physics I + II*. Springer, Berlin, 2nd edition, 2003.

[12] F. Zimmermann, K. Mess, and M. Tigner. *Handbook of Accelerator Physics and Engineering*. World Scientific, 2nd edition, 2013.

[13] D. Edwards and L. Teng. Parametrization of linear coupled motion in periodic systems. *IEEE Transactions on Nuclear Science*, 20:885, 1973.

[14] R. Helm. Note on reversible dispersion matching with arbitrary phase advance. Technical Report SLAC-PUB-3278, SLAC, 1984.

[15] P. Raimondi and A. Seryi. Novel final focus design for future linear colliders. *Physical Review Letters*, 86:3779, 2001.

[16] J. D. Jackson. *Classical Electrodynamics*. John Wiley & Sons, New York, 2nd edition, 1975.

[17] G. Fischer. Iron dominated magnets. Technical Report SLAC-PUB-3726, SLAC, 1985.

[18] C. Kittel. *Introduction to Solid State Physics.* Wiley, 8th edition, 2005.

[19] S. Russenschuck. *Field Computation for Accelerator Magnets: Analytical and Numerical Methods for Electromagnetic Design and Optimization.* Wiley-VCH, Weinheim, 1st edition, 2010.

[20] G. H. Hoffstaetter et al. CBETA Design Report, Cornell-BNL ERL Test Accelerator. Technical report, Cornell University and BNL, 2017. arXiv:1706.04245.

[21] K. Halbach. Design of permanent multipole magnets with oriented rare earth cobalt magnets. *Nuclear Instruments and Methods,* 169:1, 1980.

[22] D. Zangrando and R.P. Walker. A stretched wire system for accurate integrated magnetic field measurements in insertion devices. *Nuclear Instruments and Methods,* A 376:275, 1996.

[23] M. Abramowitz and I. Stegun. *Handbook of Mathematical Functions.* Dover Publicatons Inc., New York, 1972.

[24] A. Bambini, A. Renieri, and S. Stenholm. Classical theory of the free-electron laser in a moving frame. *Physical Review A,* 19:2013, 1979.

[25] S. Hancock, M. Lindroos, E. McIntosh, and M. Metcalf. Tomographic Measurements of Longitudinal Phase Space Density. *Comput. Phys. Commun.,* 118:61–70, 1999.

[26] R. Garoby. RF Gymnastics in Synchrotrons. *CERN Accelerator School, RF for accelerators, Ebeltoft, 2010,* page 431, 2011. CERN Yellow-Report CERN 2011-07.

[27] D. Dancila et al. A compact 10 kW solid-state RF power amplifier at 352 MHz. *Proceedings of the Particle Accelerator Conference IPAC 2017 in Copenhagen,* page 4292, 2017.

[28] P. Marchand et al. High power 352 MHz solid state amplifiers developed at the Synchrotron SOLEIL. *Physical Review Special Topics - Accelerators and Beams,* 10:112001, 2007.

[29] S. Peggs (ed). ESS Technical Design Report. Technical Report ESS-0016915, ESS, 2013.

[30] R. Yogi et al. Tetrode Based Technology Demonstrator at 352 MHz, 400 kWp for ESS Spoke Linac. *Proceedings of the 15th IEEE International Vacuum Electronics Conference IVEC14 in Monterey, USA,* 2014.

[31] D. Pozar. *Microwave Engineering.* Wiley & Sons, New York, 2nd edition, 1998.

[32] D. Alesini. Power coupling. *CERN Accelerator School, RF for Accelerators, Ebeltoft, 2010,* page 125, 2011. CERN Yellow-Report CERN 2011-07.

[33] T. Wangler. *RF Linear Accelerators.* Wiley-VCH, Weinheim, 2nd edition, 2008.

[34] P. Wilson. High energy linacs: Applications to storage rings RF systems and linear colliders. *AIP Conference Proceedings 87,* page 450, 1981. Also available as SLAC-PUB-2884, revised 1991.

[35] G. Franklin, J. Powell, and A. Emami-Naeini. *Feedback Control of Dynamic Systems.* Pearson, Boston, 7th edition, 2015.

[36] A. Bosco et al. Laser wire. *Nuclear Instruments and Methods*, A 592:162, 2008.

[37] A. Andersson et al. Determination of a small vertical electron beam profile at the Swiss Light Source. *Nuclear Instruments and Methods*, A 591:437, 2008.

[38] T. Shintake. Proposal of a nanometer beam size monitor for e+e- linear colliders. *Nuclear Instruments and Methods*, A 311:453, 1992.

[39] R. Akre et al. A Transverse RF deflecting structure for bunch length and phase space diagnostics. *Particle Accelerator Conf., Washington*, page 2353, 2001.

[40] A. Chao et al. Experimental Investigation of Nonlinear Dynamics in The Fermilab Tevatron. *Phys. Rev. Lett.*, 61:2752, 1988.

[41] P. Castro et al. Betatron function measurement at LEP using the BOM 1000 turns facility. *Particle Accelerator Conf., Washington*, page 2103, 1993.

[42] A. Langner and R. Tomás. Optics measurement algorithms and error analysis for the proton energy frontier. *Physical Review Special Topics - Accelerators and Beams*, 18:031002, 2015.

[43] P. Bambade et al. Observation of beam-beam deflections at the interaction point of the SLAC linear collider. *Phys. Rev. Lett.*, 62:2949, 1989.

[44] V. Ziemann. Beyond Bassetti and Erskine: Beam-Beam Deflections for Non-Gaussian Beams. *Proceedings of the workshop on beam-beam and beam-radiation interactions, Los Angeles 1991*, page 36, 1991. Also available as SLAC-PUB-5582.

[45] Richard Blankenbecler and S. D. Drell. A quantum treatment of beamstrahlung. *Phys. Rev.*, D36:277, 1987.

[46] R. C. Field. Beamstrahlung monitors at SLC. *Nucl. Instrum. Meth.*, A 265:167, 1988.

[47] M. Furman et al. Beam-beam diagnostics from closed-orbit distortion. *15th International Conference on High Energy Accelerators, Hamburg*, page 36, 1992. also SLAC-PUB 5742.

[48] D. Boussard. Schottky noise and beam transfer function diagnostics. *CERN Accelerator School, Rhodes, 1993*, page 749, 1995. CERN Yellow-Report CERN 95-06, Vol.2.

[49] R. Pasquinelli. Review of microwave Schottky beam diagnostics. *Proceedings, 3rd International Conference on Particle Accelerator (IPAC 2012): New Orleans, USA, May 2-25, 2012*, page 4175, 2012.

[50] M. Betz, O. Jones, T. Lefevre, and M. Wendt. Bunched-beam Schottky monitoring in the LHC. *Nuclear Instruments and Methods*, A 874:113, 2017.

[51] J. Billan, J. P. Gourber, J. P. Koutchouk, and V. Remondino. Suppression of the main LEP coupling source. *Proceedings of the 1993 Particle Accelerator Conference (PAC 93): May 17-20, 1993 Washington D.C.*, pages 68–70, 1993.

[52] T. Raubenheimer, F-J. Decker, and J. Seeman. Beam distribution after filamentation. *Proceedings of the Particle Accelerator Conference*, 1995.

[53] V. Ziemann. Response matrices in strongly coupled storage rings with a radio-frequency system constraining the revolution time. *Physical Review Special Topics - Accelerators and Beams*, 18:054001, 2015.

[54] W. Press et al. *Numerical Recipes*. Cambridge University Press, Cambridge, second edition, 1992.

[55] Y. Marti and B. Autin. Closed orbit correction of a.g. machines using a small number of magnets. Technical Report CERN-ISR-MA-73-17, CERN, 1973.

[56] J. Ögren and V. Ziemann. Optimum resonance control knobs for sextupoles. *Nucl. Inst. and Methods*, A 894:111, 2018.

[57] W. Corbett, M. Lee, and V. Ziemann. A fast model calibration procedure for storage rings. *Particle Accelerator Conf., Washington*, 1993.

[58] J. Safranek. Experimental determination of storage ring optics using orbit response measurements. *Nuclear Instruments and Methods*, A 388:27, 1997.

[59] E. Segré. *Nuclei and Particles*. The Benjamin/Cummings Publishing Company, Reading, 2nd edition, 1982.

[60] M. Tanabashi et al. (Particle Data Group). The review of particle physics (2018). *Phys. Rev. D*, 98:030001, 2018.

[61] R. Brun et al. GEANT - Detector Description and Simulation Tool. Technical Report CERN Program Library Long Writeup W5013, CERN, 1993.

[62] M. Furman. The hourglass reduction factor for asymmetric colliders. Technical Report SLAC-ABC-41-Rev, SLAC, 1991.

[63] E. Keil. Beam-beam dynamics. *CERN Accelerator School, Fifth Advanced Accelerator Physics Course, Rhodes, 1993*, page 539, 1995. CERN Yellow-Report CERN 1995-06.

[64] J. T. Seeman. Observations of the beam-beam interaction. *Proceedings of the Joint US-CERN School on Particle Accelerators, Sardinia, 1985, Springer Lecture Notes 247*, page 121, 1985. also available as SLAC-PUB-3825.

[65] M. Bassetti and G. Erskine. Closed Expressions for the electric field of a two-dimensional Gaussian charge. Technical Report CERN-ISR-TH/80-06, CERN, 1980.

[66] K. Yokoya and H. Koiso. Tune shift of coherent beam-beam oscillations. *Particle Accelerators*, 27:181, 1990.

[67] M. Furman, Y. Chin, J. Eden, W. Kozanecki, J. Tennyson, and V. Ziemann. Beam-beam diagnostic from closed-orbit distortion. *Presented at the 15th International Conference on High Energy Accelerators, Hamburg*, 1992. Also available as SLAC-PUB-5742.

[68] K. Hirata and E. Keil. Barycenter motion of beams due to beam-beam interaction in asymmetric ring colliders. *Nuclear Instruments and Methods*, A 292:156, 1990.

[69] C. Adolphsen (ed.). The international linear collider technical design report 3.ii: Accelerator baseline design. Technical Report SLAC-R-1003, SLAC, 2013.

[70] M. Aicheler, P. Burrows, M. Draper, T. Garvey, P. Lebrun, K. Peach, N. Phinney, H. Schmickler, D. Schulte, and N. Toge. A Multi-TeV linear collider based on CLIC technology: CLIC Conceptual Design Report. Technical Report CERN-2012-007, CERN, 2012.

[71] J. Schwinger. On the classical radiation of accelerated electrons. *Phys. Rev.*, 75:1912, 1949.

[72] K. W. Robinson. Radiation effects in circular electron accelerators. *Phys. Rev.*, 111:373–380, 1958. [,205(1958)].

[73] A. Hofmann. *The Physics of Synchrotron Radiation.* Cambridge University Press, Cambridge, UK, 1st edition, 2004.

[74] W. Rindler. *Essential Relativity.* Springer, Heidelberg, 2nd edition, 1977.

[75] D. Vaughan. *X-Ray Data Booklet.* Lawrence Berkeley Laboratory, 1st edition, 1986. updated version available on-line at http://xdb.lbl.gov.

[76] J. Madey. Relationship between mean radiated energy, mean squared radiated energy and spontaneous power spectrum in a power series expansion of the equations of motion in a free-electron laser. *Nuovo Cimento*, 50B:64, 1979.

[77] A. Kondratenko and E. Saldin. Generating of coherent radiation by a relativistic electron beam in an ondulator. *Part. Accel.*, 10:207–216, 1980.

[78] R. Bonifacio, C. Pellegrini, and L. Narducci. Collective instabilities and high gain regime in a free electron laser. *Opt. Commun.*, 50:373–378, 1985.

[79] E. Saldin, E. Schneidmiller, and M. Yurkov. *The Physics of Free Electron Lasers.* Springer, Heidelberg, 1999.

[80] P. Schmüser, M. Dohlus, and J Rossbach. *Ultraviolet and Soft X-Ray Free-Electron Lasers.* Springer Verlag, Berlin, 2008.

[81] W. B. Colson and A. M. Sessler. Free electron lasers. *Ann. Rev. Nucl. Part. Sci.*, 35:25–54, 1985.

[82] G. Stupakov. Using the beam-echo effect for generation of short-wavelength radiation. *Physical Review Letters*, 102:074801, 2009.

[83] G. Geloni, V. Kocharyan, and E. Saldin. Experimental demonstration of a soft x-ray self-seeded free-electron laser. *Journal of Modern Optics*, 58:1391, 2011.

[84] D. Ratner et al. Experimental demonstration of a soft x-ray self-seeded free-electron laser. *Physical Review Letters*, 114:054801, 2015.

[85] S. Milton et al. Exponential gain and saturation of a self-amplified spontaneous emission free-electron laser. *Science*, 292:2037, 2001.

[86] J. Andruszkow et al. First observation of self-amplified spontaneous emission in a free-electron laser at 109 nm wavelength. *Physical Review Letters*, 85:3825, 2000.

[87] P. Emma al. First lasing and operation of an ångstrom-wavelength free-electron laser. *Nature Photonics*, 4:641, 2010.

[88] T. Ishikawa et al. A compact x-ray free-electron laser emitting in the sub-angstrom region. *Nature Photonics*, 6:540, 2012.

[89] M. Berz, K. Makino, and W. Wan. *An Introduction to Beam Physics*. CRC Press, Boca Raton, 2015.

[90] Alex J. Dragt. Lectures on nonlinear orbit dynamics. *AIP Conf. Proc.*, 87:147–313, 1982.

[91] S. Y. Lee. *Accelerator Physics*. World Scientific, 2nd edition, 2004.

[92] M. Reiser. *Theory and Design of Charged Particle Beams*. Wiley-VCH, Weinheim, 2004.

[93] A. Wolski. *Beam Dynamics in High-Energy Accelerators*. Imperial College Press, London, 1st edition, 2014.

[94] L. J. Laslett. On intensity limitations imposed by transverse space-charge effects in circular particle accelerators. Technical Report BNL-Report 7534, Brookhaven National Laboratory, 1963.

[95] A. Piwinski. Intra-beam-Scattering. In *Proceedings, 9th International Conference on the High-Energy Accelerators (HEACC 1974): Stanford, California, May 2-7, 1974*, pages 405–409, 1974.

[96] James D. Bjorken and Sekazi K. Mtingwa. Intrabeam Scattering. *Part. Accel.*, 13:115–143, 1983.

[97] K. Hirata. A theory of bunch lengthening in electron storage rings with localized wake force sources. *Particle Accelerators*, 22:57, 1987.

[98] V. Ziemann. A theory for the transverse mode coupling instability using moment maps. Technical Report CERN-SL-94-21-AP, CERN, 1994.

[99] D. Edwards and M. Sypers. *An Introduction to the Physics of High Energy Accelerators*. John Wiley & Sons, Inc., New York, 1993.

[100] V. E. Balakin, A. V. Novokhatsky, and V. P. Smirnov. VLEPP: Transverse Beam Dynamics. *Conf. Proc.*, C830811:119–120, 1983.

[101] M. Sands. The head-tail effect: an instability mechanism in storage rings. Technical Report SLAC-TN-69-008, SLAC, 1969.

[102] A. Chao. *Physics of Collective Beam Instabilities in High Energy Accelerators*. Wiley & Sons, Inc., New York, 1993.

[103] J. Fox, T. Mastorides, C. Rivetta, D. Van Winkle, and D. Teytelman. Lessons learned from positron-electron project low level RF and longitudinal feedback. *Phys. Rev. ST Accel. Beams*, 13:052802, 2010.

[104] V. Ziemann. *A Hands-On Course in Sensors Using the Arduino and Raspberry Pi*. CRC Press, Boca Raton, 2018.

[105] R. Bailey (ed.). *CERN Accelerator School on Ion Sources in Senec, Slovakia*. CERN, European Organization for Nuclear Research, 2012. CERN-2013-007.

[106] G. Machicoane and P. Ostroumov. *Beam Dynamics Newsletter 73*. International Committee for Future Accelerators, 2018.

[107] D. Möhl. Stochastic cooling for beginners. *CERN Accelerator School, Antiprotons for Colliding Beam Facilities, 1983*, page 97, 1983. CERN Yellow-Report CERN 1984-15.

[108] H. Poth. Electron cooling: Theory, experiment, application. *Physics Reports*, 196:135, 1990.

[109] E. Trendelenburg. *Ultrahochvakuum*. Verlag G. Braun, Karlsruhe, 1963.

[110] R. Elsey. Outgassing of vacuum materials-II. *Vacuum*, 25:347, 1975.

[111] V. Ziemann. Vakdyn, a program to calculate time dependent pressure profiles. *Vacuum*, 81:866, 2007.

[112] V. Ziemann. Vacuum tracking. Technical Report SLAC-PUB-5962, SLAC, 1992.

[113] K. Kakihara G. Horikoshi, Y. Saito. An analysis of a coplex network of vacuum components and its application. *Vacuum*, 41:2132, 1990.

[114] T. Flynn. *Cryogenic Engineering, second edition*. CRC Press, Boca Raton, 2005.

[115] K. Wittenburg. Beam loss monitors. *CERN Accelerator School on Beam Diagnostics, Dourdan, 2008*, page 249, 2009. CERN Yellow-Report CERN 2009-05.

[116] H. Synal, M. Stocker, and M. Suter. Micadas: A new compact radiocarbon dating system. *Nucl. Inst. and Methods*, B 259:7–13, 2007.

[117] M. Linnarsson et al. New beam line for time-of-flight medium energy ion scattering with large area position sensitive detector. *Review of Scientific Instruments*, 83:095107, 2012.

[118] R. Hamm and M. Hamm. *Industrial Accelerators and Their Applications*. World Scientific, Singapore, 2012.

[119] V. Ziemann. Charged particle transport, Gaussian optics, error propagation: It's all the same. *Proceedings of the Seventh Particle Accelerator Conference IPAC 2016 in Busan, South Korea*, page 3324, 2016.

# Index

$B_{mag}$, 199–203
$Q$ of cavity, 155
    External, 155
    Loaded, 155
    Relation to coupling, 155

ABCD matrices, 150–152, 154–155, 162
Acceleration, 134–137
Achromat, 76, 77
ADA, 7
Alvarez, L., 3
ANSYS, xi
Anti-proton production, 302
Arduino, 297, 298, 340
ATF-2, 329

Beam loading
    Fundamental theorem, 165–166
    Steady-state, 166–167
    Transient, 167–170
Beam position monitor, 39, 175–180, 209
Beam profile monitor, 180–185
Beam-beam
    Beamstrahlung, 238–239
    Coherent pairs, 239
    Crossing angle, 232
    Diagnostics, 188–189
    Disruption, 238
    Hourglass effect, 230
    Linear collider, 237
    Luminosity, 229–232
    Tune shift, coherent, 235–237
    Tune shift, incoherent, 233–235
Beta beating, 50, 198–200
Betatron, 7
Bethe-Bloch formula, 222–226
Bucket height, 128
Bunch compressor, 81–82, 84, 254, 322–324
Bunch length, 132, 134, 182, 230, 238,
    242–244, 257, 279–289

Carnot efficiency, 313
CEBAF, 9, 329
Central limit theorem, 21, 26

CERN, xiii, 6–8, 319
Chromaticity, 52–54, 80, 187, 189, 197, 201,
    220, 259
    Calculating, 54
    Correction, 80
CLIC, 8, 80, 330
Cockroft-Walton, 4, 325
Complex potential, 87–89, 103, 264–265,
    272, 336
Control system, 295–299
Cookie-jar cavity, 341–343
Correction
    Beta function, 212–213
    Chromaticity, 213, 216–217
    Closed orbit, 214–215
    Coupling, 217
    Dispersion-free steering, 215
    Orbit in beam lines, 209
    Trajectory knobs, 208–209
    Tune, 215–216
Courant-Snyder invariant, 44, 47, 59, 227,
    244
Cryogenics, 312–314
Cyclotron, 4, 5, 9, 76, 302, 326, 327

Daphne, 9, 329
de'Broglie, L., 2
DESY, 7, 8, 258
Dipole
    Combined function, 35
    Rectangular, 35–36
    Sector, 32–34
Disk-loaded waveguides, 160–163
Dispersion, 54–58
    Suppressor, 79–80, 320
Disruption, 238
Doublet, 74, 75, 78, 79, 137, 138, 258, 293

Electron cooling, 174, 193, 228, 302, 306
Emittance, 49–52, 246
    Growth, 58–60, 226–228, 244–245, 282
    Measurement, 183–185
EPICS, 297–299

Database file, 298
Protocol file, 298
ESS, 167, 321

FAIR, 7, 329
FCC, 8, 321
Feed down, 195–196
Feedback, 218–219
Fermilab, 6
FFT, 44, 187
FODO cell, 14, 25, 40–68, 70, 79, 82, 280,
    320
Free-electron laser
    Phase space, 252
    SASE, 254–257, 323
    Small gain, 251–254
Frenet-Serret tripod, 14, 16

Gaussian distribution
    Moments, 22
    Multi-variate, 22
    One-dimensional, 21
Gray, 315
Greinacher circuit, 4
GSI, 7, 329

Hamiltonian, 17, 18, 133, 263–266, 268–278
    Concatenating, 267
    In MATLAB, 267
    Knobs, 272
    Moving, 265
    Normal forms, 275
    Resonances, 272
Higgs-boson, 6, 7, 232, 319
Histogram, 19–21, 25, 133
HIT, 329

IFMIF, 329
ILC, 8, 74, 80, 330
Impedance
    Matching, 155
    Of wake field, 285
Instability
    Coasting beam, 286–288
    Multi bunch, 291–292
    Single bunch, 288–291
Interaction region, 80–81, 221
Intrabeam scattering, 282
Ion chamber, 316
Ionization cooling, 306

ISR, 7, 236

Jefferson laboratory, 9, 329

Keil-Schnell stability criterion, 287
KEKB, 8
Kerst, D., 7
Knobs, xii, 209, 212

Larmor, J., 241
Lawrence, E., 4
LCLS, 258
LEP, 8, 194, 290, 319
LHC, 8, 11, 70, 215, 229, 232, 241, 318–321
Lifetime
    Gas scattering, 228
    Luminosity, 222
    Quantum, 246
Liquefier, 312
Liquid helium, 101, 156, 311–314
Loss factor, 285
Louvre, 10
Low-level RF system, 170–172
Luminosity, 188, 197, 227, 229–232, 238, 320
    Beam-beam, 229–232
    Fixed target, 221–222
    Lifetime, 222

Machine protection, 316
MADX, xi
Magnet
    Measurements, 339–341
    Normal-conducting, 89–101
    Permanent magnets, 106–114
    Super-conducting, 101–105
MAMI, 329
Matching
    Beta functions, 66
    Impedance, 155
    Phase advance, 65–66, 68
MATLAB simulation
    Beam loading, 169
    Beam position monitor, 177
    Beam-beam, 238
    Cavity, 159–160
    Chromaticity, 54
    Fitting functions, 26, 342
    Hamiltonians, 263–277
    Magnet saturation, 98
    Multi-bunch instability, 294

Normal forms, 277
Normal-cond. dipole, 91–98
Normal-cond. quadrupole, 98–99
Single-bunch instability, 289
Space charge, 280
Super-cond. dipole, 103–105
Super-cond. quadrupole, 105
Surface and contour plots, 23, 331
Tracking, 259–263
Transverse 2D, 40–52
Transverse 3D, 55–58
Transverse 4D, 60–65
Waveguide coaxial, 147–149
Waveguide round, 146–147
MAX-IV, 323
Maxwell's equations, 30, 85–87, 107,
120–122, 156–158, 175, 285
Maxwell, J., 6, 241
McMillan, E., 5
Measuring
Beam current, 173–174, 183
Beam position, 175–180
Beam size, 180
Beam size of laser pointer, 333
Beta function, 183–185
Betatron phase advance, 187
Emittance, 183–185
Magnets, 114–116
Tune, 44, 179, 186
Turn-by-turn positions, 186–188
MICADO, 215
Misaligned components, 194–197
Momentum compaction factor, 60, 119, 126,
127, 220, 288
Momentum spread, 242–244
Multi-bend achromat, 77–78

Network analyzer, 11, 152, 295, 341–342
Normal forms, 275–277
Normal-conducting
Cavities, 163–164
Magnets, 89–101

Octupole, 37, 88, 267
Oliphant, M., 5
OpenSCAD, 15–16, 25, 336–339
Orbit errors
Beam lines, 198
Rings, 203
Orbit response matrix, 217–218

Particle sources, 299–303
PEP-II, 9
Pepper pot, 183
Permanent magnets, 106–114
Building small magnets, 335–339
Manufacturing, 106
Multipoles, 108–112
Undulator and wiggler, 112–114
Personnel protection, 316
PETRA, 8
Phase stability, 5, 124–127
Phase-slip factor, 126–127, 135, 189, 244
Phase-space
coordinates, 24, 27, 31, 32, 38, 42, 131,
151, 187, 194, 196, 263
longitudinal, 126, 130
normalized, 44
transverse, 201, 203
Photo cathode, 300
Pill-box cavity
$Q_0$–value, 158, 342
Coupling, 342
Losses, 156–159
Measuring, 341–343
Modes, 120–124
Shunt impedance, 158
Pin diode, 316
Poincaré plot, 42–44, 48, 187, 260, 275, 278
Positron production, 300
Propagating Moments, 38–39
PSI, 329

Quadrupole, 5, 13
Thick, 31–32
Thin, 29–30
Quadrupole errors
Beam lines, 198
Rings, 204
Quadrupole scan, 184

Radiation protection, 314–317
Refrigerator, 312
RF gymnastics, 133–134, 319
RHIC, 329
Rutherford scattering, 226

Sacherer, F., 280
SACLA, 329
SASE, 255, 257, 258, 323, 329
Schottky signals, 189–192

Sextupole, 37, 88, 267
Shims, 97, 117, 194
Shunt impedance, 158
Sievert, 315
Skew quadrupole errors
    Beam lines, 200
    Rings, 205
SLAC, xi, 8, 9, 80, 236, 322
SLC, 8, 237
Solenoid, 37, 193, 197, 300
Space charge
    Direct, 279
    Indirect, 282
SPEAR, 8, 9
Spiral-2, 329
SPS, 319–321
Stochastic cooling, 7, 228, 302, 306
Super-conducting
    Cavities, 164–165
    Magnets, 101–105
SwissFEL, 329
Synchrotron oscillations
    Large amplitude, 127–133
    Matched, 131
    Small amplitude, 124–127
Synchrotron radiation, 112, 241–251, 323
    Damping, 242–246
    Dipole magnets, 248–249
    Excitation, 58–60, 244–246
    Power, 247–248
    Spectrum, 247
    Undulator and wiggler, 249–251

Tandem accelerator, 3, 303, 324, 325
Telescope, 70–73, 80
Tevatron, 7
Thermionic cathode, 299–300
Timing system, 297
TME cell, 78–79
Tomography, 133
Touschek effect, 282
Transfer matrix
    Thin quadrupole, 29–30
    Combined function dipole, 34–35
    Coordinate rotation, 36
    Rectangular dipole, 35–36
    Sector dipole, 32–34
    Solenoid, 37
    Thick quadrupole, 31–32
    Vacuum systems, 311

Transit-time factor, 124
Transversely deflecting structure, 182
Triplet, 72–74, 324, 325
TRIUMF, 329
Twiss parameters, 44, 45, 49–52, 59,
    183–185, 199–200, 202, 204, 212,
    335

Vaktrak, xi, 310–311
Van-de-Graaff accelerator, 119, 303
Van-de-Graaff, R., 4
Vecksler, V., 5
VEPP, 7

Wake field
    longitudinal, 283
    resonator, 283
    transverse, 286, 292
Wake function, 284
Wake potential, 284
Widerøe, R., 2, 3, 7

Printed in the United States
by Baker & Taylor Publisher Services